18th World Textile Conference (AUTEX 2018)

AA006528

IOP Conference Series: Materials Science and Engineering Volume 460

Istanbul, Turkey
20 – 22 June 2018

ISBN: 978-1-5108-8006-1
ISSN: 1757-8981

Printed from e-media with permission by:

Curran Associates, Inc.
57 Morehouse Lane
Red Hook, NY 12571

Some format issues inherent in the e-media version may also appear in this print version.

Printed by Curran Associates, Inc. (2019)

For permission requests, please contact the Institute of Physics at the address below.

Institute of Physics
Dirac House, Temple Back
Bristol BS1 6BE UK

Phone: 44 1 17 929 7481
Fax: 44 1 17 920 0979

techtracking@iop.org

Additional copies of this publication are available from:

Curran Associates, Inc.
57 Morehouse Lane
Red Hook, NY 12571 USA
Phone: 845-758-0400
Fax: 845-758-2633
Email: curran@proceedings.com
Web: www.proceedings.com

TABLE OF CONTENTS

Dear Authors and Readers,

It is a great pleasure for me to write this foreword for Autex 2018, 18th World Textile Conference, held in Istanbul, Turkey between 20-22 June 2018. Autex Conference brings together academicians, researchers, industry professionals, experts in textile field from all over the world and makes a significant contribution to realization of Autex objectives. It is a great honour for Istanbul Technical University, Textile Technologies and Design Faculty, to host this organization, which is one of the most important events in the field of textiles.

The papers presented in the conference contributed to the knowledge and novelty in textile science in the sessions of Protective Textiles, Innovative Functional Textiles, Ecological and Environmental Textiles, Recycling and Life Cycle Analysis, Surface Functionalization and Coating, Smart, Interactive and Multifunctional Textiles, Technical Textiles, Composites and Membranes, Nanotechnology, Nanotextiles, Electrospinning, Textile and Clothing Machinery, Comfort Science, Modelling and Simulation, Fashion - Garment Industry and Marketing, Textile Testing, Measuring Technology, Fashion - Design and Garment Industry, Textile Economy, Textile Supply Chain Management, Advanced Fibers and Materials. Medical Textiles. Tissue Engineering, Implants, Textile Processing, Advanced Fibres and Materials, Modern Textile Education and Training, Innovation & Entrepreneurship, Fashion, Garment Industry, Simulation & Modeling. Seven plenary speakers made presentations in addition to the contributed papers.

I deeply hope the proceedings included in this book will conribute as a reference book for novel studies and research.

Prof. Dr. Fatma Kalaoğlu
Conference Chair

IOP Conf. Series: Materials Science and Engineering **460** (2018) 011002 doi:10.1088/1757-899X/460/1/011002

Peer review statement

All papers published in this volume of *IOP Conference Series: Materials Science and Engineering* have been peer reviewed through processes administered by the proceedings Editors. Reviews were conducted by expert referees to the professional and scientific standards expected of a proceedings journal published by IOP Publishing.

Screen printing of UV curable polyurethane acrylate binder prepared with different pigment concentrations on synthetic leather and gloss and hardness properties of printed films

B Gülçin[1], K Berdan[1] and K K Burçak[1]

[1]Istanbul Technical University, Faculty of Textile Technologies and Design, Department of Textile Engineering, Gumussuyu Campus, Gumussuyu, Beyoglu, Istanbul, Turkey

bkayaoglu@itu.edu.tr

Abstract. In this paper, UV curing technology and screen printing method were firstly together applied on synthetic leather, in order to determine the gloss and hardness values of screen printed UV curable polyurethane acrylate binder films prepared with different pigment concentrations. UV curable water-borne pigmented formulations were prepared with different pigment concentration (3%, 5% and 7%) and two types of photoinitiators (Omnirad® 500 and Omnirad® 819 DW). The curing process were conducted under different combinations of lamps (Ga, Hg, GaHg and GaGaHg) at three power levels. Pendulum hardness and gloss values of printed films were investigated. In gloss measurements, higher gloss values were obtained at the pigment concentration of 3% at all lamp combinations. Moreover, the highest gloss value (20.96) of the sample printed with 3% pigment concentration was obtained at GaHg lamp combination at 679 mJ/cm^2 energy density. The higher hardness values were obtained at a pigment concentration of 3% at all lamp combinations. The highest hardness value (9.8) of the pigmented polymeric films with 3% pigment concentration was obtained at GaGaHg lamp combination at 966 and 1663 mJ/cm^2 energy densities. The results show that high gloss and hardness values can be achieved at lower pigment concentration of 3%. Considering energy efficiency, curing under the GaHg lamp combination at an energy intensity of 679 mJ/cm^2 can be recommended for synthetic leather samples, which are printed with low pigment concentration (3%).

Keywords— **Pigment concentration, polyurethane acrylate binder, screen printing, synthetic leather, UV curing.**

1. Introduction

Pigment printing has attracted interest and the materials in the printing paste, such as pigments, binders, thickeners and etc., have been studied since mid-1940's. According to a research conducted around mid-1980s, it has been presumed that more than half of the textile printed products in the world are produced by pigment printing. Pigment printing offers some advantages such as simplicity, removal of inexpensive wet post-processing, application to all types of fibers, printing styles and printing methods, and a wide choice of pigments in terms of color and light fastness preferences [1]. Conventional thermal curing techniques are generally used to bond the pigments to the textile surface. After printing, the product is dried and then subjected to a short-time fixing process with dry air at high temperature. At the end of fixing, the binder used polymerizes to form a film layer and allows the pigments to be held on the fiber [2,3]. UV curing technology is used to cure inks, coatings, adhesives and other UV-sensitive materials by polymerization. The major advantages of UV cured coatings can be listed as, low cost and energy requirements, very fast cross-linking, application at room temperature, and the absence of volatile organic compounds which is harmful to the environment or health. UV curable formulations consist of oligomers, diluent monomers and photoinitiators. The main function of the photoinitiators is to initiate polymerization by forming a reactive group by absorbing the energy of the photons coming from the radiation during curing. All photochemical reactions are based on the activation of certain energy of excited molecules in the structure. Many types of photoinitiators contain at least one aromatic ring in which the carbonyl (C=O) group is present. For this reason, they strongly absorb UV rays. Molecules which are called as free radicals are stimulated by absorption of light energy by the photoinitiator and polymerization reaction occurs [3]. Almost all UV lamps used in UV curing units are based on mercury tubes. The standard mercury lamp is used in approximately 80% of the existing applications on the market. In a coating system, pigments or light stabilizers may absorb UV radiation and thus they can reduce the amount of free radicals produced by a photoinitiator. The main spectral output of the mercury lamp is below 300 nm and at 365 nm. In clearcoat applications, this spectrum can be used well with photoinitiator absorption. However longer wavelength absorption can be obtained with a gallium addition. Then intense output is added in the wavelength range of 400–450 nm. These lamp systems are suitable for pigmented coating applications which requires high UV intensity sources during polymerization reaction. In UV curing technology, one of the important factors for UV curing is the applied energy density at the coating surface, and the unit for the energy density is mJ/cm^2 [4].

Augusto et al. investigated changes in hardness, gloss and color values of water-borne clear and pigmented printing inks cured at different UV radiation levels. Pigmented formulations were prepared by using carbon black and organic pigments of low opacity and applied to the glass plates as a film of about 7.0 μm ± 1.3 μm by means of a film applicator. Coated films were cured in a device with a conveyor belt, under medium pressure mercury lamp and the effect of each pigment under UV light was examined. According to the results of the experiments, the average König hardness values of the clear polymeric films were much higher than those of the cured pigmented films. The pigments in the formulations led to the less absorption of UV light by the photoinitiators and this resulted in reduction of the curing level. Hardness values of the films obtained with formulations prepared with red and yellow pigments were lower than those obtained from blue and black pigmented formulations. Consequently, it was stated that the chemical structure of the pigments affected UV light absorption and the degree of curing could be changed. The color and gloss values in all formulations varied according to the applied energy density during curing. It was observed that films exposed to higher energy density during curing lost their gloss and exhibited yellowing due to deterioration on the surface [5].

Synthetic leathers have wide range of area of use – e.g., shoes, upholstery, automotive, leathercraft, and handbags, which features the patterns and the coloration methods. Although UV curable inks are applied with digital printing technology onto the synthetic leathers, the production speed is lower than that of conventional technologies, such as roller printing or screen printing. Particularly in coloration of shoes and upholstery products, conventional technologies are still in use. However, the printing pastes include solvent based binder and they are thermal curable [6,7]. Hence, in the current study, UV curing technology and screen printing method were firstly together applied on the synthetic leather, in order to determine gloss and hardness values of screen printed UV curable polyurethane acrylate binder films prepared with different pigment concentrations.

2. Experimental

2.1. Materials

Polyurethane (PU) based synthetic leather samples (Flokser Tekstil, Turkey) with a fabric weight of 545 g/m^2, were used as they are supplied. In the preparation of water-borne UV curable clear and pigmented formulations, a flexible, water-borne, aliphatic PU acrylate binder (Laromer® UA 9059, BASF, Germany) was used. Bisacyl phosphine (Omnirad® 819 DW, BASF, Germany) effective for deep curing and α-hydroxyketone (Omnirad® 500, BASF, Germany) effective for surface curing [4] were used as the photoinitiators. Wetting agent (Exosel 54, Acar Kimya, Turkey), defoamer (Foamaster® 8034, BASF, Germany), anionic acrylic copolymer thickener (Orgaclear P 460, Organik Kimya, Turkey), deionized water and ammonia solution (NH_4OH, Kimetsan, Turkey) were also used in the formulation. Pigmented formulations were prepared using a red pigment (Irgazin® Red K 3840, CI Pigment Red 254, BASF, Germany).

3. Method

3.1. Preparation of printing pastes and printing

In this study, formulations were prepared with different pigment concentrations (%3, %5 and %7) separately and printed on synthetic leather samples by a flat screen printing machine. In the printing paste formulation, a flexible, water-borne, aliphatic PU acrylate resin (Laromer® UA 9059) as a binder and photoinitiators of bisacyl phosphine (Irgacure® 819 DW) and α-hydroxyketone (Irgacure® 500), pigment, wetting agent, defoamer, deionized water and ammonia solution (NH4OH) were used. The pH and the viscosity of the formulations were arranged as 8.0-9.0 (with WTW Inolab pH 7110 pH meter) and 20.000- 25.000 cP (at 20 rpm, with spindle 6 with a Brookfield DV-E viscometer), respectively. Water-borne UV curable printing paste formulation is given in Table 1.

Table 1. Water-borne UV curable printing paste formulations.

Materials	Quantity (%)
Deionized water	26
Binder (Laromer® UA 9059)	66
(Irgacure® 500)	1.8
(Irgacure® 819 DW)	1.8
(Irgazin® Red K 3840)	3, 5 and 7
Thickener	2.63
Wetting agent	0.49
Defoamer	0.1
Ammonia	0.33
Deionized water	26
Binder (Laromer® UA 9059)	66

3.2. UV Curing

After printing, samples were cured at room temperature using a UV curing machine equipped with gallium and mercury (Ga/Hg) lamps and a conveyor belt with an adjustable speed. In this study, total applied energy amount was measured with a UV-Integrator Type D radiometer (Table 2). UV curing of printed samples was conducted at 3 power levels (60, 90 and 120 W/cm) at a belt speed of 10 m/min under different UV lamp combinations (Hg, Ga, GaHg, GaGaHg).

3.3. Hardness measurement

For hardness measurements, pigmented films were prepared with a film applicator (Byk Gardner) on glass plates with a thickness of 120 μm and cured in UV curing machine. After UV curing, König hardness values of pigmented polymeric films were evaluated with a Byk Pendulum Hardness Tester according to ASTM D4366 [8] and the hardness values were reported in seconds. The procedure of the hardness tester is based on the oscillation of a pendulum which is released on the surface. The oscillation depends on the surface hardness. The period of oscillation is 1.4 seconds. In the measurements, the total period of oscillation is calculated as multiplying one oscillation period by the number of oscillations [9, 10].

Table 2. Water-borne UV curable printing paste formulations.

Power levels (W/cm)					
60		90		120	
Irradiation sources and Applied energy (mJ/cm^2)					
Ga	186	Ga	287	Ga	497
Hg	288	Hg	439	Hg	731
GaHg	444	GaHg	679	GaHg	1166
GaGaHg	630	GaGaHg	966	GaGaHg	1663

3.4. Gloss measurements

In principle, gloss is defined as the ratio of light intensity reflected from the surface of an object to light intensity coming onto the surface. Hence, gloss is expressed by the amount of light reflected from the surface. Smooth and very glossy surfaces reflect more light. Gloss properties of surfaces are measured by a glossmeter which measures the specular reflection from the surfaces. There are three different geometries of the glossmeters. The widespread application of these geometries are 20°, 60°, 85° degrees which are used for the measurements of high, semi and low gloss values respectively. In addition, depending on the industrial applications, there are two more specific measurement geometries of 45° and 75° [11]. The gloss measurements of printed synthetic leather samples were made according to ASTM D523 [12], "Standard Test Method for Specular Gloss" standard using a Rhopoint, Novo-GlossTM glossmeter at 60° geometry.

4. Results and Discussions

4.1. UV Curing process and pendulum hardness measurement

In the preparation of the water-borne printing paste formulations, three pigment concentration were selected as 3%, 5% and 7% in order to determine the effect of pigment amounts. The pigmented pastes were prepared with equal amounts of photoinitiator (Omnirad® 819 DW and Omnirad® 500). Pigmented films were applied with the film applicator on glass plates at a thickness of 120 μm and cured in UV curing machine. The hardness of the pigmented films was measured by pendulum hardness tester and the hardness values were reported in seconds. The hardness values of UV cured pigmented films are shown in Table III. Distribution of the hardness values of pigmented films was shown in Fig. 1.

The hardness of polymeric films is directly related to the degree of polymerization reaction during UV curing. As a result of polymerization reaction, higher crosslinking density generally increases film hardness and it provides better abrasion resistance properties [13].

According to König hardness values of pigmented films (Table III), the increase in pigment concentration from 3% to 7% results in a decrease of hardness values. This can be explained by a reduction of the UV light absorption of the photoinitiators on the film surface due to pigment presence in the pigmented formulations. Decrease in hardness is related to decrease in curing degree due to the insufficient amount of free radicals which produced by photoinitiators required for polymerization. The higher hardness values (minimum 7 - maximum 9.8) were obtained at a pigment concentration of 3% at all lamp combinations. The highest hardness value (9.8) of the pigmented polymeric films with 3% pigment concentration was obtained at GaGaHg lamp combination at 966 and 1663 mJ/cm^2 energy densities. According to the results, although GaGaHg combination yielded the highest hardness value at 3% pigment concentration, the use of GaHg combination is suggested for curing considering the energy efficiency. The results suggested that higher gloss and hardness values can be obtained at lower pigment concentration of 3% and curing under GaHg lamp combination at 679 mJ/cm^2 energy density.

Table 3. Hardness values of UV cured pigmented films (B: UV cured pigmented film, P: Pigment).

Lamp sources	Applied energy (mJ/cm²)	B %3P	B %5P	B %7P
	186	8.4	7	7
Hg	287	8.4	8.4	8.4
	497	8.4	8.4	8.4
	288	8.4	7	7
Ga	439	8.4	7	7
	731	8.4	7	7
	444	8.4	7	8.4
GaHg	679	8.4	8.4	8.4
	1166	8.4	8.4	8.4
	630	8.4	7	7.
GaGaHg	966	9.8	9.8	9.8
	1663	9.8	9.8	9.8

Figure 1. Distribution of pendulum hardness values of pigmented films prepared with different pigmented (3%, 5% and 7%) formulations (B: UV cured pigmented film, P: Pigment).

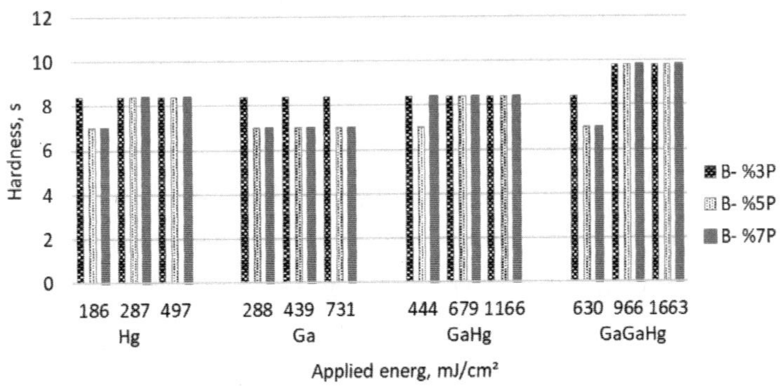

4.2. Gloss measurements

Gloss values of printed and UV cured synthetic leather samples were shown in Table IV. Distribution of the gloss values of printed and UV cured synthetic leather samples was shown in Fig. 2. According to gloss measurement results (Table IV), increasing the pigment concentration affected the gloss values adversely and lower values were obtained in every lamp combination at different power levels. The pigment volume concentration (PVC) is the ratio of the total volume of pigment to the total dried material (film) present in printing. PVC affects various properties such as gloss, washability, strength, reflection and rheological properties. If there is enough amount of binder available in formulation, it completely encloses the pigment at a given pigment volume concentration and fills the gaps between the particles [14]. In this case the gloss of the surface is high because there is no light scattering in this concentration. However, as the pigment concentration increases, the pigment particles begin to protrude out of the binder surface, distributing the light falling on the surface and preventing the true reflection. For this reason, as the pigment concentration increases, the gloss decreases [15]. The higher gloss values (minimum 14.68 - maximum 20.96) were obtained at the pigment concentration of 3% at all lamp combinations. Moreover, the highest gloss value of the sample printed with 3% pigment concentration was obtained at GaHg lamp combination at 679 mJ/cm^2 energy density.

Table 4. Gloss test results of printed and UV cured synthetic leather samples (A: printed synthetic leather sample, P: Pigment).

Lamp sources	Applied energy (mJ/cm^2)	B %3P	B %5P	B %7P
	186	8.4	7	7
Hg	287	8.4	8.4	8.4
	497	8.4	8.4	8.4
	288	8.4	7	7
Ga	439	8.4	7	7
	731	8.4	7	7
	444	8.4	7	8.4
GaHg	679	8.4	8.4	8.4
	1166	8.4	8.4	8.4
	630	8.4	7	7.
GaGaHg	966	9.8	9.8	9.8
	1663	9.8	9.8	9.8

Figure 2. Distribution of gloss values of printed and UV cured synthetic leather samples (A: printed synthetic leather sample, P: Pigment).

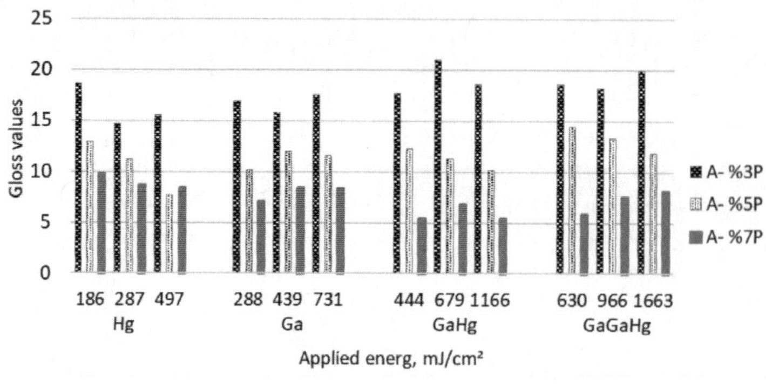

5. Conclusion

In the current study, water-borne UV curable formulations prepared with different pigment concentrations in order to determine the effects of pigment concentrations on gloss and hardness values of screen printed UV curable polyurethane acrylate binder films. UV curing technology and screen printing method were successfully applied together on the synthetic leather. According to the gloss measurement results, the increase in pigment concentration negatively affected the gloss values and low values were obtained in each lamp combination at different power levels. As the pigment concentration increases, the pigment particles begin to protrude from the surface and cause the light falling on the surface to disperse. For this reason, the actual reflection from the surface was inhibited and therefore low gloss values were obtained. High gloss values (minimum 14.68 - maximum 20.96) were obtained at a pigment concentration of 3% in all lamp combinations. Furthermore, the highest the gloss value of the sample printed with a pigment concentration of 3% was obtained at an energy density of 679 mJ/cm^2 in a GaHg lamp combination.

When the results of the König hardness measurement of the pigmented films were evaluated, it was observed that the hardness values decreased by increasing the pigment concentration from 3% to 7%. These results are explained by the fact that in pigmented formulations the absorption of UV radiation in the film surface via photoinitiators is reduced by the pigments. Due to the reduced UV absorption, the hardness values decreased because of the insufficient amount of free radicals generated by the photoinitiators required for the polymerization. The better hardness values (minimum 7 - maximum 9.8) were obtained at a pigment concentration of 3% in all lamp combinations. The highest hardness value (9.8) of pigmented polymeric films with a pigment concentration of 3% was obtained at 966 and 1663 C energy densities in the GaGaHg lamp combination. According to the results, although the GaGaHg combination gives the highest hardness value at the 3% pigment concentration, the use of the GaHg combination for energy efficiency improvement is recommended.

Acknowledgments
The authors acknowledge The Scientific and Technological Research Council of Turkey (TUBITAK) for the financial support (Project No. 215M995).

References

[1] L. V Tyrone, "Textile processing and properties: Preparation, dyeing, finishing and performance," Text. Sci. and Tech., pp. 175.
[2] Pigment Printing Handbook, 1995, Research Triangle Park, NC: American Association of Textile Chemists and Colorists, pp. 32.
[3] B. Neral, S. Sostar-Turk and B.Voncina, "Properties of UV-cured pigment prints on textile fabric," Dyes and Pigm., vol. 68, pp.143-150, 2006.
[4] R Schwalm, UV Coatings: Basics, Recent Developments and New Applications (Amsterdam/London: Elsevier, 2007), 39.
[5] M. A. G. Bardi and L. D. B. Machado, "Accompanying of parameters of color, gloss and hardness on polymeric films coated with pigmented inks cured by different radiation doses of ultraviolet light," Radiat. Phys. Chem., vol. 81, pp. 1332-1335, 2012.
[6] Gujral, R.J. 2009. Artificial Leather, Real Comfort, Science Tech Entrpreneur
[7] Dinçerler Tekstil, < http://www.fezasunideri.com.tr/>, accessed 07.12.2017
[8] ASTM D4366-95: Standard Test Methods for Hardness of Organic Coatings by Pendulum Damping Tests, Test Method A-König Pendulum Hardness Test, ASTM International, West Conshohocken, PA, 1995, (Withdrawn 2003).
[9] https://www.gardco.com/pages/hardness/pendulumhardness.cfm, accessed 14.01.18
[10] http://www.worldoftest.com/pendulum-hardness-tester, accessed 14.01.18
[11] http://www.byk.com/fileadmin/byk/support/instruments/theory/appearance/en/Intro_Gloss.pdf, accessed 14.02.18
[12] ASTM D523-14, Standard Test Method for Specular Gloss, ASTM International, (West Conshohocken, PA, 2014).
[13] W. Schaeffer, "Radiation Curable Components and Their Use in Hard, Scratch Resistant Coating Applications," UV & EB Technical Conference Proceedings, Sartomer USA, LLC, 2014.
[14] J. S. Preston, C. Nutbeeın, and R. Chapman, "Impact of pigment blend and binder level on the structure and printability of coated papers," PaperCon, pp. 418-437, 2011.
[15] E. Farmand, "Examination of the factors determining the degree of gloss in paint films," Diploma Work in Chemical Engineering, Göteborg, Sweden, 2012.

Shape-variable adaptive fiber-reinforced plastics based on shape memory alloy hybrid yarns

M Ashir, M Vorhof, A Nocke and C Cherif

Institute of Textile Machinery and High Performance Material Technology, Faculty of Mechanical Science and Engineering, Technische Universität Dresden, 01069 Dresden, Germany

Corresponding author: moniruddoza.ashir@tu-dresden.de

Abstract In the transport and industrial sectors, moving parts and kinematics play an important role. In this context, weight savings and a reduced number of joints of moving parts are essential factors, enabling lowered fuel consumption and precise dynamic movements of machine parts. Hence, single-axis and intrinsic adaptive fiber-reinforced plastic (FRP) components with structurally integrated actuators can replace conventional moving parts. Thus, this paper aims at presenting shape-variable, adaptive FRP based on shape memory alloy (SMA) hybrid yarns. In order to protect the matrix from direct contact with SMA, to ensure free and even mobility of SMA in FRP, and to fully exploit the deformation capability of SMA and therefore the adaption potential of the entire FRP structure, wire-shaped SMA was converted into textile-based actuators in a core-sheath structure. The functionalized preforms were developed by means of weaving technology. The SMA hybrid yarn was introduced into the woven fabric structure in the warp direction. Subsequently, the functionalized preforms were infused. The adaptive FRP were characterized in terms of their deformation behavior with fully and partially sheathed SMA in lengths of 240 mm and 120 mm, respectively. Results revealed that adaptive FRP with fully sheathed SMA deform stronger than those with partially sheathed SMA.

***Keywords*—Adaptive fiber-reinforced plastics, shape memory alloys, weaving, friction spinning technology.**

I. Introduction

THE importance of fiber-reinforced plastics (FRP) for different novel products in automobile, aerospace or wind energy has been growing steadily in recent years [1]. The integration of functional materials into the reinforcement fabrics of FRP poses a breakthrough for the development of different innovative products for lightweight applications. Electroactive materials, magnetostrictive materials, piezo electric materials, bimetals, shape memory polymers and shape memory alloys (SMA) are examples of functional materials being promising candidates for actuators in FRP. However, SMA are more suitable than other functional materials due to their higher energy density (10^7 J/m^3) [2]. Moreover, SMA are commercially available in wire form, which is one of the most important criteria for the textile technical integration of SMA into reinforced fabrics.

SMA is a material that remembers its original shape after being heated over its phase transformation temperature [3]. The thermally induced shape-memory effect (SME) is activated by Joule heating. Below the transformation temperature, the twinned martensite alloy is deformed plastically without damaging its crystalline structure. By the thermal induced activation of the detwinned SMA above its transformation temperature, a phase change into an austenite state is included, accompanied by the regression of deformation. As a result of cooling the SMA to temperatures below the transformation temperature, a phase transition into martensite and thus into the initial state is triggered. This phenomenon is illustrated in Fig. 1.

Different alloys are commercially available, which can be used as SMA, such as copper-zinc-aluminum (CuZnAl), copper-aluminum-nickel (CuAlNi) or nickel-titanium (NiTi). However, NiTi-based SMA is preferable for most FRP applications due to its higher elongation at break of 12.5% and maximum strain recovery of 8% [4].

The development of adaptive FRP with textile-technically integrated SMA into reinforced fabrics has been expanded steadily in recent years [5-10]. However, this research project presents the development of adaptive FRP with SMA hybrid yarn (SMA-HY), which was structurally integrated into textile-reinforced fabrics by means of weaving technology. The developed adaptive FRP were electro-mechanically characterized based on fully and partially sheathed SMA.

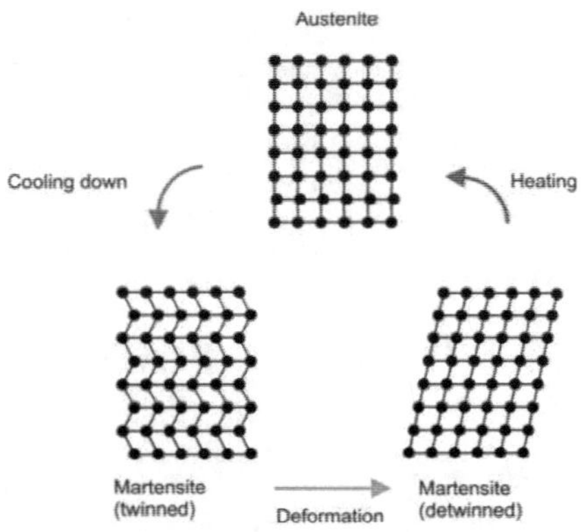

Fig. 1. Schematic representation of shape memory effect (SME) [5].

II. MATERIALS AND METHODS

A. *Materials*

In this research project, E-glass rovings type EC17-1200-350 (PD Glasseiden GmbH, Germany) were used for the formation of a base structure. E-glass was selected as this type of glass fibers (GF) is most frequently used for FRP applications due to its favorable mechanical properties, good processability and beneficial price-performance ratio compared to other GF.

A commercially available Alloy H ox. sa. (Memry GmbH, Germany) type of SMA was used for the functionalization of the adaptive preform. The basic properties of the SMA are listed in Table 1.

TABLE 1
PROPERTIES OF THE SMA USED

Properties	Value
Diameter in mm	0.305
Mass fraction of Ni in %	54.8
Mass fraction of Ti in %	45.2
Austenite start temperature in °C	82
Austenite finish temperature in °C	90
Martensite start temperature in °C	65
Martensite finish temperature in °C	55
Heat treatment	Straight annealed

The sheath materials for the development of SMA-HY are E-glass and polypropylene (PP) staple fiber rovings of 2000 tex and 4000 tex, respectively. The PP staple fiber rovings were used as sheath material to realize a more compact and dense fiber sheath. The diameters of glass and PP fibers were 17.12 mm and 47.62 mm, respectively.

As resin and hardener for the infusion of the functionalized preform, MGS® RIMR 135 and MGS® RIMH 137 (Hexion a. s., Sokolov, Czech Republic) were employed. They were mixed at a ratio of 10:3 by weight prior to infusion.

The most important properties of glass fiber rovings used for the production of functionalized preforms are listed in Table 2, which were determined by authors. The stress-strain behavior of the glass rovings was measured following the norm ISO 3341 on the tensile testing device Zwick Z 100 (Zwick GmbH & Co. KG, Germany). The test speed was set to 200 mm/min and the initial load was kept at 0.5 cN/tex. The test length was set to 500 mm. 15 measurements were executed in order to determine the average value and standard deviation.

TABLE 2
PROPERTIES OF THE GLASS FIBER ROVINGS USED FOR THE PRODUCTION OF FUNCTIONALIZED PREFORMS

Properties	Value	Standard deviation
Fineness in tex	1201	2.46
Breaking strength in N	540	34.1
Elongation at break in %	1.96	0.08

B. Production of SMA-hybrid yarn (SMA-HY)

In order to protect the matrix from direct contact with SMA, to ensure free and even mobility of SMA in FRP, and to fully exploit the deformation capability of SMA and therefore the adaption potential of the entire FRP structure, the wire-shaped SMA was converted into textile-based actuators in a core-sheath structure based on the friction spinning technology - DREF 2000. The spinning head of the DREF 2000 friction spinning machine for the production of SMA-HY is shown in Fig. 2. During this process, SMA is wrapped with glass and PP fibers to form SMA-HY. However, SMA were form-fitted to the reinforced fabrics to ensure force transmission from SMA to FRP. Production details of SMA-HY can be found in ref. [11].

Fig. 2. Spinning head of the DREF 2000 friction spinning machine for the production of friction-spun SMA-HY (1: core feeding, 2: sheath feeding, 3: opening roller, 4: spinning drum) [11].

The microscopic view of produced SMA-HY is shown in Fig. 3.

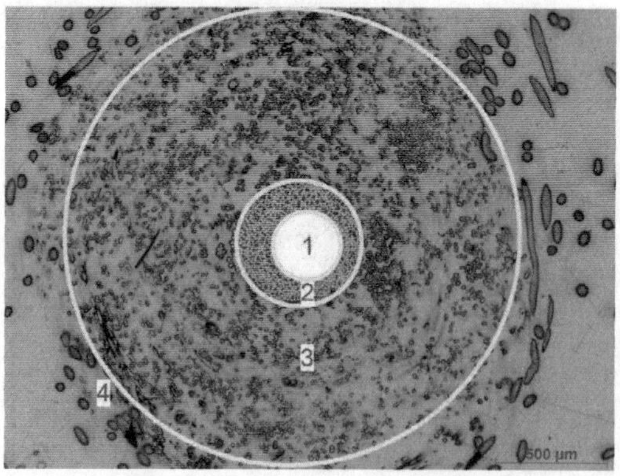

Fig. 3. Microscopic view of SMA-HY (1: SMA, 2: glass filament rovings, 3: glass staple fiber rovings, 4: PP staple fiber rovings).

C. Production Functionalized preform

The functionalized preform was generated on a rapier weaving machine type P1 (Lindauer Dornier GmbH, Germany). A hinged structure was developed during weaving for the realization of structure-integrated joint in adaptive FRP. The warp and weft yarn densities of the hinged and non-hinged areas are shown in Table 3.

TABLE 3
YARN DENSITIES FOR THE DEVELOPMENT OF A FUNCTIONALIZED PREFORM

Yarn density in yarn/cm	Hinged area	Non-hinged area
Warp	4	8
Weft	2.4	4.8

The hinged area was realized by plain weaving with floating warp yarns, whereas the non-hinged area was achieved with a multi-layered fabric including 3 sets of weft yarns. The floating warp yarns were cut after the weaving process. The functionalized preform during the weaving process is shown in Fig. 4.

Fig. 4. Functionalized preform.

Prior to the infusion process, two types of functionalized preforms were prepared based on fully and partially sheathed SMA in lengths of 240 mm and 120 mm, respectively. By stripping the both sided sheath of SMA in non-hinged area, the partially sheathed SMA was realized. The reason behind this variation was to investigate the deformation behavior of adaptive FRP with variable sheath length of SMA. The electrical contating of two adjacent SMA was executed by means of welding in a meandering pattern.

D. Infusion

The adaptive preforms were infiltrated by the Seemann Corporation Resin Infusion Molding Process (SCRIMP), being a variation of the VARI process (Vacuum Assisted Resin Infusion). The decisive advantage of this method is the additional flow aid, which ensures the proper distribution of resin within the reinforcement structure. After infusion, the system was cured at room temperature for 24 h. Subsequently, the system was annealed at 50 °C for 15 h in a laboratory oven, so that the maximum degree of cross-linking of epoxy resin was achieved. Annealing at higher temperatures generally leads to reduced tempering times. However, in this particular case, this option was inapplicable since the SME of the SMA wire starts at approx. 80 °C. Adaptive FRP with fully and partially sheathed SMA after the infusion process are shown in Fig. 5. The hinged width and surface area of infused sample were 100 mm and 240*40 mm^2, respectively.

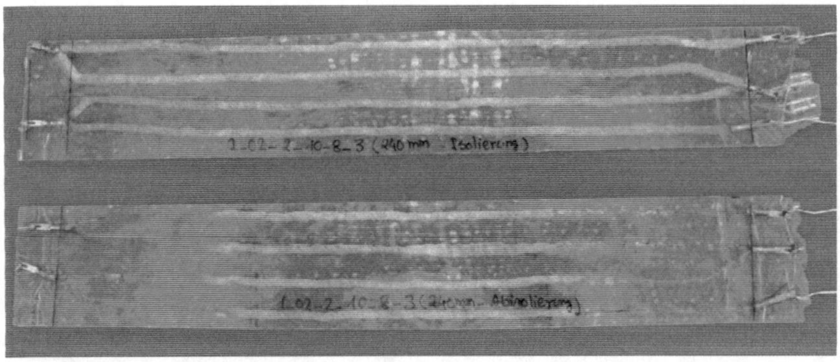

Fig. 5. Adaptive FRP with fully (top) and partially (bottom) sheathed SMA after the infusion process.

E. Electromechanical characterization

The SMA in the adaptive FRP was activated electrothermally by a laboratory power supply unit in addition to a current flow controlling unit. On- and off-times of the power supply were set to 60 s and 90 s, respectively. During on-time, the applied current flow was 1.9 A. No current flows were applied during off-time, causing passive cooling of the adaptive FRP. During the periodic heating and cooling cycles, the resulting deformation of the adaptive FRP was measured simultaneously by means of a laser triangulator. The test setup for the deformation behavior characterization of adaptive FRP is shown in Fig. 6.

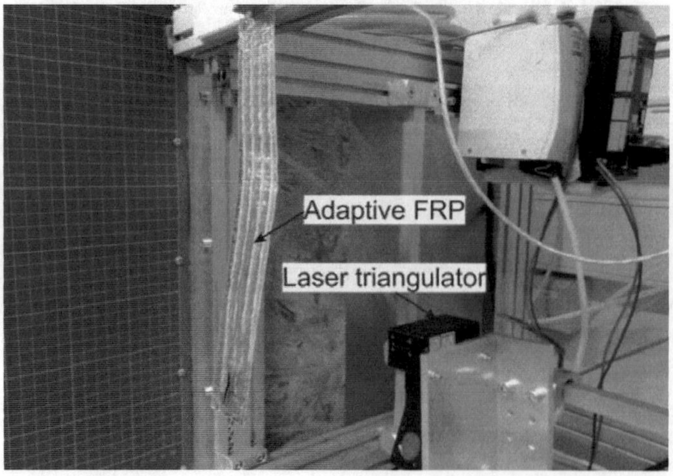

Fig. 6. Test setup for the deformation characterization of the adaptive FRP.

III. RESULTS AND DISCUSSION

The deformation behavior of adaptive FRP varies depending on the sheath length of SMA. Fig. 7 reveals that, by increasing the sheath length of SMA in adaptive FRP, the deformation of adaptive FRP is increased. This phenomenon can be attributed to the distance of the force transmission point from the hinged area. By increasing the sheath length of SMA, the force during the thermal induced activation of SMA is primarily applied on the area of the electrical contact.

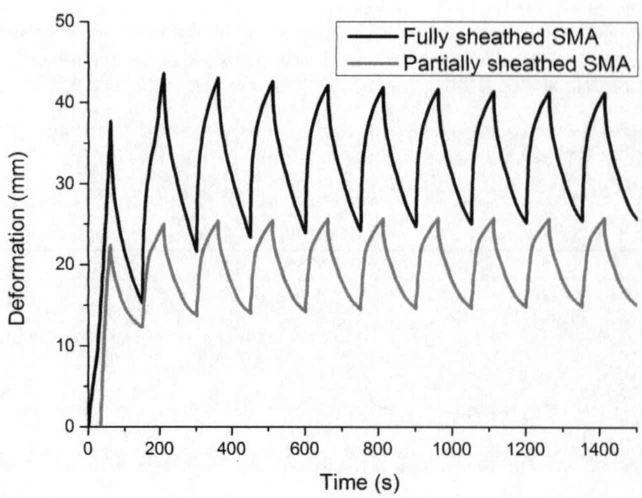

Fig. 7. Deformation curves of both types of adaptive FRP.

The deformation of both types of adaptive FRP is illustrated by a bar diagram (see Fig. 8). This bar diagram is based on the deformation curves of Fig. 7. The deformation of adaptive FRP is taken from the fifth heating and cooling cycle because from this cycle the deformation curves are relatively homogeneous over the remaining testing periods. It can be concluded from Fig. 8 that, by reducing the sheath length of SMA from 240 mm to 120 mm, the maximum deformation of adaptive FRP is reduced from 17 mm to 11 mm.

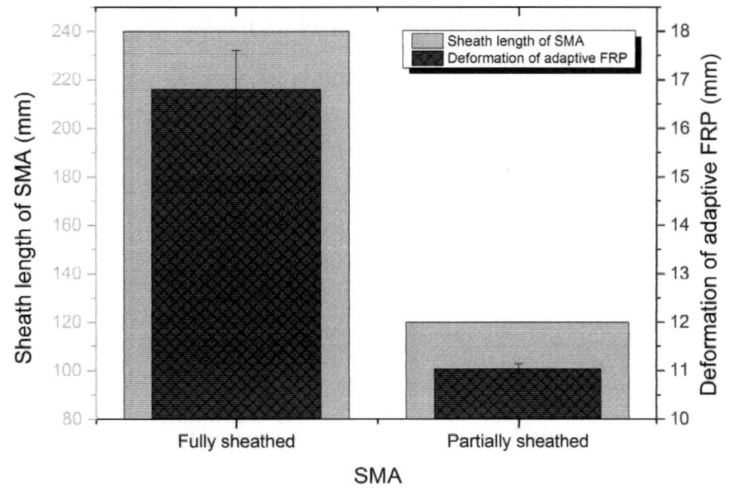

Fig. 8. Bar diagram for the deformation of adaptive FRP depending on sheath length of SMA.

IV. Conclusion

The development, realization and characterization of adaptive FRP were carried out within the research project presented in this paper. For optimum use of SMA in adaptive FRP, it was converted into core-sheath hybrid yarn. Subsequently, adaptive FRP were characterized by fully and partially sheathed SMA in lengths of 240 mm and 120 mm, respectively. After the thermally induced activation of SMA in adaptive FRP, it can be concluded that more deformation of adaptive FRP is achieved by increasing the sheath length of SMA.

Acknowledgment

The IGF research projects 18808 BR and 19832 BR of the Forschungsvereinigung "Forschungskuratorium Textil e. V." are funded through the AiF within the program for supporting the „Industrielle Gemeinschaftsforschung (IGF)" from funds of the Federal Ministry of Economic Affairs and Energy (BMWi) by a resolution of the German Bundestag. Financial support is gratefully acknowledged.

Supported by:
Federal Ministry for Economic Affairs and Energy
on the basis of a decision by the German Bundestag

References

[1] C. Cherif. Textile Materials for Lightweight Constructions. Berlin, Heidelberg: Springer-Verlag GmbH, 2016. pp.1-35.
[2] M. Ashir, L. Hahn , A. Kluge, A. Nocke and C. Cherif, "Development of innovative adaptive 3D Fiber Reinforced Plastics based on Shape Memory Alloys", Composites Science and Technology 126 pp 43-51, 2016.
[3] M. Ashir, A. Nocke, C. Theiss and C. Cherif, "Development of adaptive hinged fiber reinforced plastics based on shape memory alloys", Composite Structures 170 pp 243-24, 2017.
[4] M. Ashir, J. Hindahl, A. Nocke and C. Cherif, "Development of adaptive pleated fiber reinforced plastic composites", Composites Science and Technology 148 pp 27-34, 2017.
[5] A. Kluge, A. Nocke, C. Paul, C. Cherif, T. Linse, V. Ulbricht, "Development and characterization of textile-processable actuators based on shape-memory alloys for adaptive fiber-reinforced plastics", Textile Research Journal 83(18), pp 1936–1948, 2013.
[6] M. Ashir, A. Nocke and C. Cherif, "Development and mechanical properties of adaptive fiber-reinforced plastics", Journal of Industrial Textiles, Epub ahead of print, DOI: 10.1177/1528083718757523.
[7] M. Ashir, J. Hindahl, A. Nocke, C. Sennewald and C. Cherif, "Development of adaptive pleated woven fabrics with shape memory alloys", Textile Research Journal, Epub ahead of print, DOI: 10.1177/0040517518792736.
[8] M. Ashir, D.M.P. Vo, A. Nocke and C. Cherif, "Adaptive fiber-reinforced plastics based on open reed weaving functionalization", Materials Science and Engineering 406 012063, doi:10.1088/1757-899X/406/1/012063, 2018.
[9] M. Ashir, J. Hindahl, A. Nocke and C. Cherif, "A statistical approach for the fabrication of adaptive pleated fiber reinforced plastics", Composite Structures 207, pp 537-545, 2019.
[10] M. Ashir, L. Hahn , A. Kluge, A. Nocke and C. Cherif, "Electro-bending characterization of adaptive 3D fiber reinforced plastics based on shape memory alloys", Smart Materials and Structures, 25 pp 035-041, 2016.
[11] M. Ashir, A. Nocke and C. Cherif, "Development of shape memory alloys hybrid yarns for adaptive fiber-reinforced plastics", Textile Research Journal, Epub ahead of print, DOI: 10.1177/0040517518770678.

IOP Conf. Series: Materials Science and Engineering **460** (2018) 012003 doi:10.1088/1757-899X/460/1/012003

Development of methods for the digital representation of the thermal wearing comfort of outdoor clothing

E. Wendt, M. Awais and S. Krzywinski

TU Dresden, Institute of Textile Machinery and High Performance Material Technology (ITM), 01062 Dresden, Germany

Corresponding author: ellen.wendt@tu-dresden.de

Abstract A multi-model pore system, which includes the micro, meso, macro and mega pores, develops between fibers, yarns, fabrics or multilayer fabrics and human body respectively during wearing clothes. This multi-model pore system significantly affects thermodynamic processes of human and consequently the thermal wearing comfort of outdoor and protective clothing. The objectives of the research are to analyze this effect and to develop application-oriented principles and processes for the holistic simulation of outdoor clothing according to geometrical, textile-physical, and thermal parameters. To achieve the required aim clothing samples for five selected test persons are developed and sewed and the textile materials are comprehensively characterized. The size distribution of the mega pores are identified by the fit simulation and the wear trials are conducted in climate chamber to analyze the heat transfer between body and clothing system. In this paper the thermal simulation of the jacket and test person1 is presented. This innovative approach will give the opportunity to analyze the effect of body shape, design of clothing, environmental conditions and level of activity of human.

***Keywords*— microclimate, thermal simulation, thermo physiology, wear trial.**

I. INTRODUCTION

Numerous researchers have focused on wear comfort for humans in the recent past. The researchers have been working intensively to make complex development processes more transparent and efficient. However, there is still need of a comprehensive research, especially in functional textiles. During the design process of functional textiles for outdoor and protective clothing, both constructive (fit, ergonomic wear comfort) as well as functional aspects (weather protection, thermal wear comfort) are very important. Therefore the realization of holistic product design, predominantly based on empirical knowledge is very complex. During the process of pattern making, the body shapes and working postures have been considered by using computer-assisted 3D design/fit simulations. Therefore, today there are a lot of solutions available regarding the assessment of ergonomic comfort by using virtual modeling as well as analytical tools. But currently no computer-assisted solution exists for the realistic prediction of the thermal comfort. Hence, the cross-linking of both approaches requires extensive research.

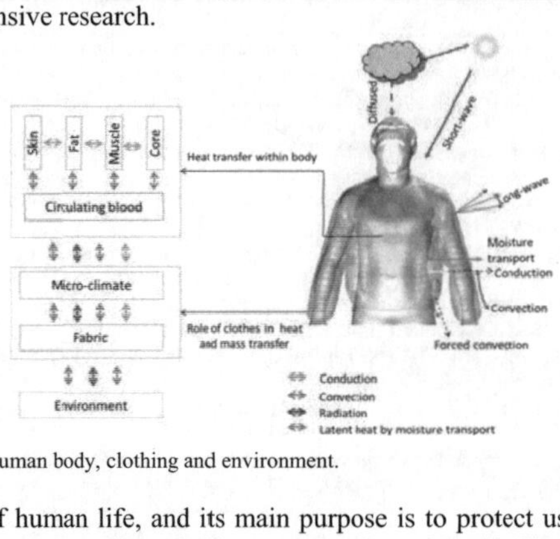

Fig. 1. Thermal regulation between human body, clothing and environment.

Clothes are an integral part of human life, and its main purpose is to protect us against environmental hazards [1]. It is easy to achieve this purpose when clothing comfort is neglected. Thermal comfort is one of the most important and experimentally investigated aspects of clothing comfort, which is affected by many factors, e.g.

environment, characteristics of clothes, microclimate, activity level etc. In addition to other factors, the thermal comfort of clothes considerably depends on the microclimate. A microclimate is the small layer of air between human body and clothing. This air layer is an important factor for thermo-physiological comfort as it is extremely close to the body. The properties of the microclimate (e.g. as insulator in winter and heat exchanger in summer) are affected by several aspects, for example thermal and draping properties of the fabric, garment design, environmental conditions, etc.

The human body constantly interacts with the environment due to convection, conduction, radiation, and sweat evaporation. It is therefore continuously subjected to heat stresses, such as solar radiation, air temperature, humidity, and metabolic heat. A thermal regulation system helps maintain the heat balance between human body and environment in order to achieve a body core temperature of 37°C [2]. The heat exchange between human body and environment during the human thermal regulation can be expressed with the following heat balance equation[3].

$$S = (M - W) - (E + C + R + B) \quad (1)$$

Here, heat storage (S) is the difference between heat production (metabolism (M) corrected for work (W)) and heat loss (evaporation (E), convection and conduction (C), radiation (R) and respiration (B)).

Many experiments have been conducted to understand the heat regulation process of the human body. Moreover, problems associated with clothing thermal comfort have been investigated, yet there is still a great comprehensive need for knowledge and research in order to resolve these problems. The use of thermal manikins or humans is normal practice in the investigation of thermal insulation through clothing and evaporative resistance [4–6]. The only advantage of using thermal manikins is the reproducibility with identical parameters and conditions; however, these tests are very expensive and time-consuming. The manikins have an average shape of a human body and are divided into different sections, which are equipped with heating sources as well as temperature and heat flux sensors. Modern thermal manikins can also sweat, walk, and even exhibit physiological responses similar to humans [7–9].

For the purpose of highly realistic investigations, subjective studies/tests involving humans can be conducted, which are however expensive and include ethical restrictions. In addition, suchlike tests can exhibit widely varying test results due to intra- and inter-subjective variations [10].

Some efforts have also been directed towards simulating the thermoregulation during the interaction of the human body or a cylinder (which has the same temperature as the surface of the human body) with the environment. The researchers have used computational fluid dynamics and investigated the thermal interaction of the nude human body in a naturally ventilated building [11], the heat transfer phenomena within the microclimate [12], and the effect of slits in garment on air flow and temperature distributions [13]. Within these research projects, the influence of different body shapes (body surface, volume), clothing design and fit have not been considered. 3D scanning techniques and an infrared thermal camera have also been used for investigating thermal insulation processes [14]; it has been reported that, if the microclimate is thicker than about 8 mm, internal convection will take place.

To avoid errors by simulation of thermal regulation of the human body, researchers have employed human thermo-physiological models with varying accuracies. There are a large number of human thermoregulation models that were developed based on the first model of 1948 [15]. A multi-node model by Fiala et al [16] is the best performing model among others [17]. This model comprises 15 idealized spherical or cylindrical body elements: head, face, neck, shoulders, arms, hands, thorax, abdomen, legs, and feet. In accordance with former studies, a division was therefore made whenever a significant change of body tissue properties occurred. As a result, the present multilayer model consists of annular concentric tissue layers and seven different tissue materials: brain, lung, bone, muscle, viscera, fat, and skin.

The main focus of this research project is to introduce a method for the application of simulation techniques, which can not only simulate the thermal regulation of the human body in terms of microclimate, clothing, and environment, but also consider the fit of the clothing influencing the ergonomic comfort of outdoor clothing. This simulation process offers an enormous potential to sectors involving outdoor, working, and protective clothing as well as sportswear.

II. Methods and Procedures

A. Selection of Test Persons and 3D Scanning

Five test persons were selected for investigations and wear trials. These test persons were scanned by using the 3D-bodyscanner VITUS in two standing positions (standard and relaxed) as shown in fig. 2. The data were further used for determination of the body measurements according to the ISO 8559 and DIN EN ISO 7250. With the help of the software *Geomagic*, the data of all test persons were prepared according to the requirements for further simulations.

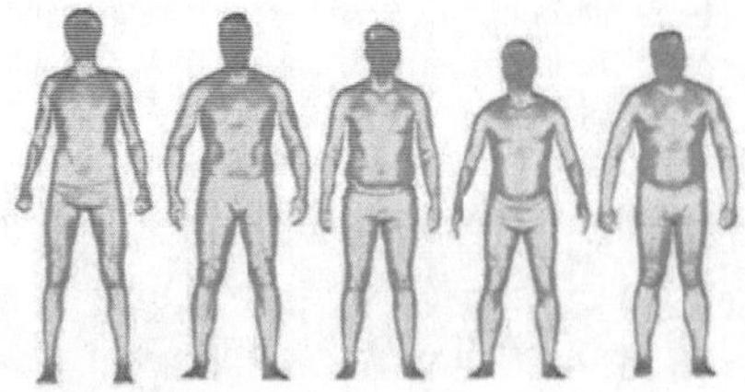

Fig. 2. 3D CAD data of all test persons (standard position).

B. Selection of Material and Clothing

Four different types of fabrics T1 (material: PES, Fabric type: R/L Knit), T1H (material: PES, Fabric type R/R Knit), T2 (material: PA6, fabric type: plain weave) and T3 (material: PES, Fabric type: fleece) were selected to construct a long sleeve shirt, long underpants, a trekking trousers and a jacket respectively. In the further process the physical, mechanical, and thermal characteristics of all the materials were tested (Table I).

Furthermore two different clothing systems for two environment conditions were defined. Clothing system-I for warm climate consists on long sleeve shirt and long trekking trousers. Clothing system-II for cool climate consists on long sleeve shirt, long underpants, trekking trousers, and jacket.

TABLE I
PHYSICAL, THERMAL, AND MECHANICAL PROPERTIES OF FABRICS

	Properties		Fabric			
			T1	T1H	T2	T3
Physical	Fabric thickness [mm]		0,46	0,56	0,29	1,55
	Average mass per unit area of the fabrics [g m^{-2}]		161	184	125	168
	Absolute water vapor permeability [Pam^2 W^{-1}]		2,82	3,14	4,42	6,41
	Air permeability [mm sec^{-1}]		1077	221	75	1057
Thermal	Thermal conductivity [10^{-3} W m^{-1} K^{-1}]		46,24	55,10	36,58	42,18
	Thermal diffusivity coefficient [10^{-6} m^2 s^{-1}]		0,23	0,092	0,25	0,25
	Thermal absorptivity [W m^{-2} s$^{1/2}$ K^{-1}]		161,2	186,8	230,6	84,48
	Specific heat [Jkg^{-1}K^{-1}]		493	1643	295	1640
	Thermal resistivity [10^{-3} K m^2.W^{-1}]		8,50	8,62	7,00	39,38
Mechanical	Bending stiffness [mN cm]	warp	0,12	0,21	0,32	0,29
		weft	0,07	0,10	1,03	0,29
	Tensile stiffness (Ɛ=1%) [kPa]	warp	761	577	6445	646
		weft	448	513	38766	289

C. 2D Pattern Development

According to the defined clothing systems and the body measurements of five test persons, 2D patterns developed by using the 2D CAD software *Grafis*, as shown in fig. 3. *Grafis* is an user-friendly software that offer an easy opportunity to consider the individual body measurements during pattern making. Then, 2D pattern cuts were sewn together and made the clothing`s ready for the wear trial. It was taken into account that the pattern pieces for the different test persons were produced in the same sequence and with the same sewing techniques to avoid design variations (fig. 4).

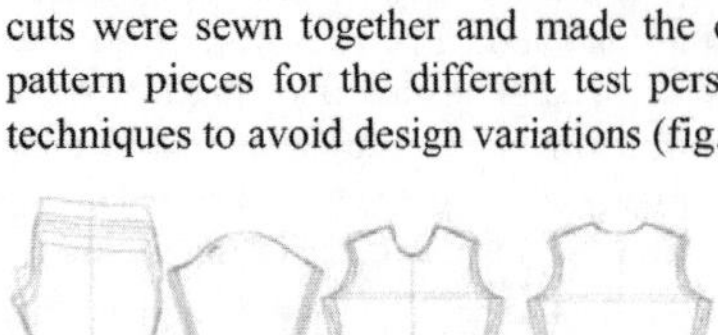

Fig. 3. 2D patterns for long underpants, long sleeve shirt, trekking trousers and jacket for test person 1

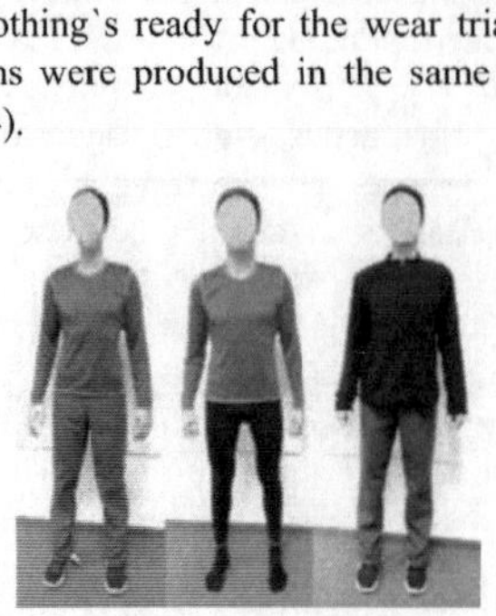

Fig. 4. Fit test of the clothing for the test person 1; Clothing system-I (left); Clothing system-II (middle + right)

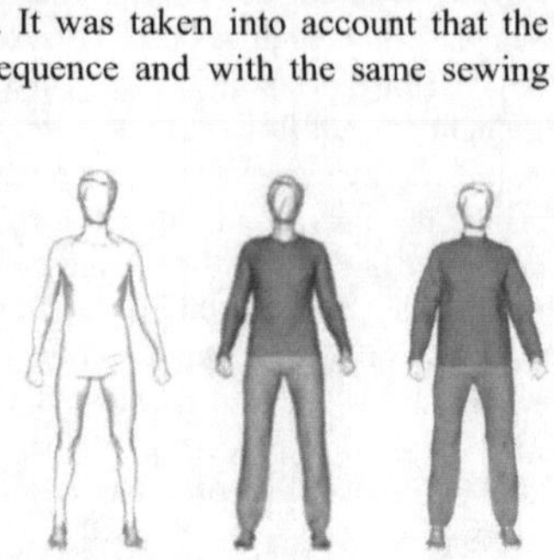

Fig. 5. 3D fit modeling of clothing`s on test person 1. a) CAD model, b) Clothing system-I, c) Clothing system-II.

D. 3D Fit Simulation of Clothings

The 3D fit simulation of clothings was completed by using the software *Modaris V8*, which gives the opportunity of a digital 3D visualization of garments by virtual sewing and also considering the mechanical properties of the textile materials. The developed 2D patterns (*Grafis*) were exported as an AAMA/DXF file and then imported into *Modaris V8* for 3D fit simulation. After assigning the sewing information and the fabric properties (Table I), the pattern cuts were simulated on the virtual test person 1 (fig. 5). As a result an uneven space is observed between the body and the clothing`s. The shape and the thickness of the space depend on the draping properties of the fabric. This space is quite important for the development of the microclimate.

E. Wear Trial

Wear trials were conducted to investigate the thermo-physiological response of the five selected test persons by the Hohenstein Institut für Textilinnovation GmbH. For this purpose each of the test persons has performed two trials in the selected climate; warm (23°C, 50% RH) and cool (15°C, 40% RH), in a climatic chamber. The test procedure, lasted for a total 125 minutes, was carried out by each test person on a motor driven treadmill (Kettler Track Experience, Ense-Parsit). The total duration of wear trial consisted on five phases of sitting and walking. In phase 1, 3 & 5, the test person must be seated for 15 min. Whereas in phase 2 and 4, test person walked on treadmill for 40 min with 4km/h and 6km/h respectively.

The skin temperature, the humidity of microclimate, and the wetness of the skin were recorded with help of sensors MSR 345 Datalogger (MSR Electronics GmbH, Seuzach, Swizerland). These sensors were attached on the body of the subjects at 10 different points according to ISO 9886 [18]. To calculate the temperature and humidity of the microclimate, the sensor was pointed towards microclimate and with a distance from the body. The calculation of the average skin temperature, the microclimate temperature and the microclimate humidity was performed according to the 8-point method from ISO 9886. A SpotOn system (3M Medica, Neuss, Germany) was used to measure the body core temperature. Furthermore the total amount of sweat that was produced by the subject, was calculated by weighing the nude subject before and after the wear trial. The amount of sweat that was entrapped by the clothing was calculated as the weight difference of the clothing before and after the trial. The clothing were always washed and conditioned for 24 hours before the next trial.

F. Thermal Simulation of Test Person and the Clothings

In first step of thermal simulation, there is a need of thermo-physiological virtual model of a human that. For this purpose, already developed CAD models (*Geomagic /Design Concept 3D*) of test persons were used (fig. 2).

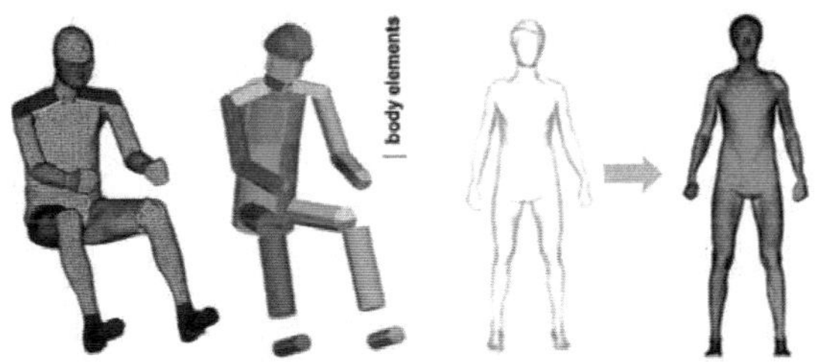

a b

Fig. 6. a) Thermo-physiological model in Theseus-FE, left: surface model, right: body elements [19], b) Thermo-physiological model based on individual data of test person 1.

Subsequently, a thermo-physiological model was developed by means of the Fiala Model (developed at HFT Stuttgart and De Monfort University, UK [20]) and the framework for the software *Theseus FE*. For this purpose, a NASTRAN format of the body model was imported into Theseus FE, and each body part (body division was executed according to the Fiala Model) was assigned to the Fiala thermo-physiological properties. This generated human model (figure 6b), that is based on individual scan data, can now perform active and passive thermoregulation.

The 3D model of the jacket that already developed by 3D fit simulation, was imported into Theseus FE software also via NASTRAN interface. The thermal regulation of body-clothing-environment system is strongly affected by the thermal characteristics of the clothing and the microclimate. As, the microclimate is not uniform across the entire body, so different air zones had to be defined within the microclimate in terms of thickness and their location on the body.

Every air zone has different thickness and also associated with different part of body and jacket surface. Therefore heat transfer because of conduction or convection did not occur constantly across all the air zones. The heat transfer coefficient of each air zone was calculated using the following correlation [21].

$$h = \frac{(Nu_L K)}{L} \qquad (2)$$

Here, h is heat transfer co-efficient, L is characteristic length (thickness of air zone), K is conductivity, and Nu_L is the Nusselt number (which is the function of Prandtl and Rayleigh number). The boundary conditions defined for the thermal simulations follow.

1) Thermal and physical properties, specific heat, conductivity, thickness, and mass per unit area were defined to the jacket surface and the air zone.
2) An environment was defined with air velocity of 0.3m/sec, which has a temperature of 15˚C and 40% relative humidity.
3) Metabolic rates (Met) were assigned to the thermal manikin according to the three different levels of activity in the wear trials (Table III), which were calculated against the weight of test person 1.
4) The boundary conditions to realize heat transfer between outer surface of jacket & external environment, inner surface of jacket & relevant air zone, and air zone & relevant body segment/surface were defined as linear convection.
5) Heat flux due to long wave radiation was considered by defining the view factor cavity. For this purpose, radiation properties were defined to the jacket shell and human model.

III. Results and Discussion

A. *Thermal Simulation*

The simulation was run for 7500 seconds (125 min) and following (fig. 7) results were obtained.

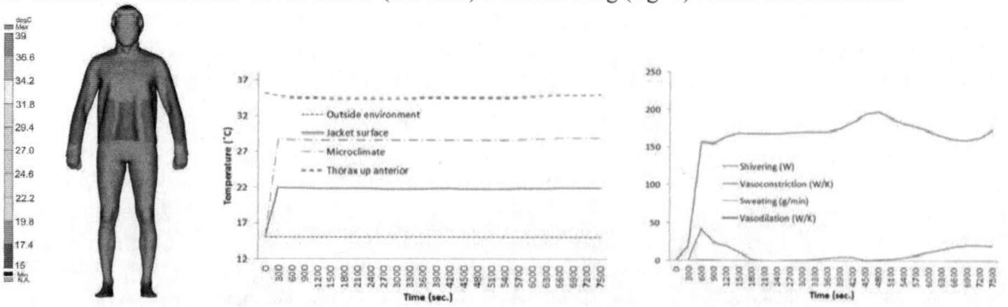

a b c

Fig. 7. a) Surface temperature of body and jacket during thermal simulation at time steps 900s, b) Temperature behavior of surface at thorax up anterior during the simulation, c) Response of active system of human model during simulation.

It can be noticed (fig. 7b) that with the start of simulation, the temperatures of jacket surface and microclimate increase to 22˚C and 28˚C respectively, then maintain almost constant temperature line throughout the simulation. This is because of the influence of two constant temperatures; the body surface temperature (35-36 ˚C) at thorax up anterior and the outside temperature (15˚C). As, microclimate is closer to the body as compare to the jacket surface, therefore it has higher temperature than the jacket surface temperature. Moreover, the trend of the graph shows that jacket surface and microclimate work like an insulator to preserve the body heat. Although, the metabolic rate of test person increases during the wear trial (Table III), the body skin temperature (thorax up anterior) is remain in rang of 34.5-35.5 ˚C, because of the active system of human body that maintain the core body temperature at 37˚C. The active system consists of four processes; shivering, vasoconstriction, sweating, and vasodilation. Shivering and vasoconstriction is activated in the cool environment, but sweating and vasodilation are activated when body feels too warm. The fig. 7c indicates the function of active system during simulation. In phase 1 (sitting, 0 – 900 sec) of wear trial, the metabolic rate of test person is 1 Met. Therefore the body heat is transferred to the microclimate (heat transfer from high to low temperature). Hence, the vasoconstriction and the shivering are activated with the start of simulation to preserve or to produce the body heat. The shivering starts to decrease with the start of phase 2 (900 – 3300 sec; walking with 4km/h) and meets to horizontal line. In this phase, the vasoconstriction stops to increase and maintains constant throughout the phase. This is because of the increasing of the metabolic rate of the test person during walking (3.2 Met). It can also be observed from the fig. 7c, that at the end of phase 2 (3300 sec), the vasodilation activates but soon ended with the end of phase 3 (4200 sec). As, in this phase, metabolic rate is 1 Met, therefore vasoconstriction is further increase so that body heat can be preserved. With the start of phase 4 (4200 – 6600 sec), vasoconstriction decreases and vasodilation starts to release the heat from the body, which is produced due metabolic rate of 4 Met (walking at 6km/h). In phase 5 of wear trial, the metabolic rate is 1 Met, therefore once again the vasoconstriction starts to increase and vasodilation starts to decrease so that body heat can be preserved.

B. Validation

To validate the simulation results, the wear trials were conducted. At present, the results are still on evaluating stage, so that they cannot yet be presented. A comprehensive report we will published at the end of this project.

IV. Conclusion

This paper presented the simulation of heat regulation between human body, microclimate, clothing and environment. The transportation of heat in the microclimate is a complex phenomenon. The mass flow within the microclimate mainly depends on the surrounding environment temperature, the specific body parts, and the properties of the clothing a person wears. In the past, all thermal simulations involving the human body, its microclimate, clothing, and the environment have been performed based on the assumption that the microclimate is a uniform layer of air underneath the clothing. In reality, the thickness of the microclimate is not consistent all over the body, but it varies and depends on numerous factors, especially the draping properties of the fabric and the shape of the body. To simulate the actual scenario, a fit simulation of the jacket on an individual scanned model was performed by giving the deformation properties of fabric, and the microclimate was subsequently divided into many small air zones according to thickness and location. Moreover, to realize the scanned model as actual human model, it was coupled with the Fiala model.

In order to validate the simulation results, the wear trials of five test persons already conducted by the Hohenstein Institut für Textilinnovation GmbH. The validation will be presented soon in separate research paper.

Acknowledgment

The IGF research project 19472 BG of the Forschungsvereinigung Forschungskuratorium Textil e. V. is funded through the AiF within the program for supporting the „Industriellen Gemeinschaftsforschung (IGF)“ from funds of the Federal Ministry for Economic Affairs and Energy by a resolution of the German Bundestag.

References

[1] Adams, P.S., Slocum, A.C., Keyserling, W.M.: A Model for Protective Clothing Effects on Performance. Int. J. Cloth. Sci. Technol. 6, 6–16 (1994). doi:10.1108/09556229410054495
[2] Chen, M..: The tissue energy balance equation. Heat Transf. inMedicine Biol. Plenum Press. NewYork. 156–157 (1985)
[3] Parsons, K.: Human thermal environments. Taylor and Francis Group, Lodon and New York (2003)
[4] ASTM: Standard Test Method for Measuring the Thermal Insulation of Clothing Using a Heated Manikin. 1–7 (2015). doi:10.1520/F1291-15.1
[5] Holmer, I., Elnas, S.: Physiological evaluation of the resistance to evaporative heat transfer by clothing. Ergonomics. 24, 63–74 (1981). doi:10.1080/00140138108924831
[6] McCullough, E.A., Jones, B.W.: A Comprehensive Data Base for Estimating Clothing Insulation. ASHRAE Trans. 2888, 29 (1984)
[7] Holmér, I.: Thermal manikin history and applications. Eur. J. Appl. Physiol. 92, 614–618 (2004). doi:10.1007/s00421-004-1135-0
[8] M, V.: A new generation sweating thermal manikin for the evaluation of the thermoregulation properties of protective clothing. In: (ed.) 19 Proceedings of the 10th International Meeting on Thermal Manikin and Modeling (10i3m). Tampere, Finland. In: 19Proceedings of the 10th International Meeting on Thermal Manikin and Modeling (10i3m). pp. 1–5. , Tampere, Finland (2014)
[9] MGM, R., N, M.: A Sweating Agile Thermal Manikin (SAM) Developed to Test Complete Clothing Systems Under Normal and Extreme Conditions. RTO HFM Symp. Blowing Hot Cold Prot. against Clim. Extrem. 4, 1–7. doi:ADP010487
[10] Psikuta, A., Allegrini, J., Koelblen, B., Bogdan, A., Annaheim, S., Martínez, N., Derome, D., Carmeliet, J., Rossi, R.M.: Thermal manikins controlled by human thermoregulation models for energy efficiency and thermal comfort research – A review. Renew. Sustain. Energy Rev. 78, 1315–1330 (2017). doi:10.1016/j.rser.2017.04.115
[11] Yang, T., Cropper, P.C., Cook, M.J., Yousaf, R., Fiala, D.: A new simulation system to predict human-environment thermal interactions in naturally ventilated buildings. Proc. 10th Int. Conf. Build. Simul. 751–756 (2007)
[12] Mayor, T.S., Couto, S., Psikuta, A., Rossi, R.M.: Advanced modelling of the transport phenomena across horizontal clothing microclimates with natural convection. Int. J. Biometeorol. 59, 1875–1889 (2015). doi:10.1007/s00484-015-0994-x
[13] Lim, J., Choi, H., Roh, E.K., Yoo, H., Kim, E.: Assessment of airflow and microclimate for the running wear jacket with slits using CFD simulation. Fash. Text. 2, 1 (2015). doi:10.1186/s40691-014-0025-2
[14] Zhang, Z., Li, J., Wang, Y.: Improving garment thermal insulation property by combining two non-contact measuring tools. Indian J. Fibre Text. Res. 40, 392–398 (2015)
[15] Pennes, H.H.: Analysis of Tissue and Arterial Blood Temperatures in the Resting Human Forearm. J. Appl. Physiol. 1, 5–34 (1948)
[16] Fiala, D.: Dynamic Simulation of Human Heat Transfer and Thermal Comfort (Thesis). Sustain. Dev. 45, 1 (1998)
[17] Psikuta, A., Fiala, D., Laschewski, G., Jendritzky, G., Richards, M., Blazejczyk, K., Mekjavic, I., Rintamaki, H., de Dear, R., Havenith, G.: Validation of the Fiala multi-node thermophysiological model for UTCI application. Int. J. Biometeorol. 56, 443–460 (2012). doi:10.1007/s00484-011-0450-5
[18] Standard, I.: ISO 9886 Ergonomics-Evaluation of thermal strain by physiological measurements. (2004)
[19] Paulke, S.: Finite element based implementation of Fiala's Thermal Manikin in Theseus-FE, http://www.theseus-fe.com/simulation-software/human-thermal-model
[20] Fiala, D., Lomas, K.J., Stohrer, M.: Computer prediction of human themoregulatory and temperature responses to a wide range of environmental conditions. Int J Biometeorol. 45, 143–159 (2001). doi:10.1007/s004840100099
[21] Incropera, F.P., DeWitt, D.P., Bergman, T.L., Lavine, A.S.: Fundamentals of Heat and Mass Transfer. John Willey & Sones (2007)

The Effects of Quat-Silane Antimicrobials on the Physical and Mechanical Properties of Cotton and Cotton/Elastane Fabrics Used for Clothing

N. Yıldız Varan[1]and S.H. Eryuruk[2]

[1]*Pamukkale University, Department of Textile Engineering, Denizli, Turkey*

[2]*İstanbul Technical University, Department of Textile Engineering, İstanbul, Turkey*

Corresponding author: nvaran@pau.edu.tr

Abstract Samples were treated with quat-silane to achieve antimicrobial property for further designs to help bacteria resistant during use. In this study, the physical and mechanical properties of quat-silane treated cotton and cotton/elastane fabrics in comparison with untreated control samples were investigated. Results showed that a small significant decrease was observed for tensile strength (strip and grab methods), tear strength and seam strength. A small significant decrease was observed with the increase in quat-silane concentration for all samples. Panama weaves showed the lowest tensile strength and the highest tear strength and a small significant decrease was observed for all treated samples. Antimicrobial tests showed that all treated samples have a very good antimicrobial activity.

***Keywords*— Quat-silane, tensile strength, tear strength, seam strength, cotton, elastane.**

I. Introduction

Antimicrobial textiles are classified as those textile and fibrous materials subjected to various finishing techniques to afford protection for both the user of textile materials (against bacteria, yeast, dermatophytic fungi and other related microorganisms for aesthetic, hygienic or medical purposes) [1]. The textile itself (biodeterioration caused by mould, mildew and rot producing fungi) without negatively affecting the other important characteristics of the textiles [2-5].

With a view to develop antimicrobial textile materials, considerable research has been carried out by making use of organic and inorganic compounds, antibiotics, heterocyclics, quaternary ammonium compounds [6]. Several studies have been carried out ranging from fundamental aspects to development of antimicrobial fabrics. Antibacterial polyester fabrics have been developed by imbuing antibacterial agents into the structure of fibers rather than depositing on their surface for longer durability and effect. It is stated that the efficacy of the finished fabric to arrest the growth of Staphylococcus aureus and Escherichia coli is about 5 times higher than the conventional materials. A synergistic system of formulation comprising of inorganic chemicals involving a metal salt of a monocarboxylic acid, a carbonic acid derivative, a chelating agent, a boron compound, a dimethylene siloxane derivative and an alkane polymer has been proved to serve as an effective antimicrobial agent in arresting the growth of several bacteria, fungi and mildew. Hospital trials showed a dramatic decrease in bacteria, fungi and mildew growth in treated fabrics. The treatment also prevents the deterioration of fabrics by microorganisms [7]. Chitosan treatment on cotton renders antimicrobial activity. Chitosan treated cotton fabric showed a high reduction rate in the number of colonies [8]. Fabrics made from viscose fibers containing polysilicic acid (Visil) and aluminum silicate (Visil AP) have been given urea peroxide treatment to make them antibacterial as well as deodorizing. Instead of treating the surface of the fabrics with polymer coating, antibacterial additives have been imbedded into the fabric's polymer fibers for the production of antibacterial gowns [9].

The ideal biocidal textile materials for medical use should possess rapid inactivation of a broad spectrum of microorganisms [10,11], non-selective and non-immutable to pathogens, non-toxic and environmentally friendly, durable to repeated washes and easy to recharge in laundering or disinfection processes [12]. In addition, the recharging agents should be non-toxic, available at home, and compatible with our laundering chemicals such as detergents or bleaching agents [13]. The number of healthcare equipment manufacturers incorporating antimicrobial properties in their products has increased dramatically in recent years. Various types of antimicrobial chemical agents are used to impart antimicrobial properties to textiles. They have different properties and mechanisms [14,15]. Quaternary ammonium cations, also known as quats, are positively charged polyatomic ions of the structure NR4+, R being an alkyl group or an aryl group (Figure 2.6) [16,17]. Quats have antimicrobial effect against; Vegetative bacteria, yeast, molds, algae, viruses [18-22]. The chemical structure is presented in Figure 1.

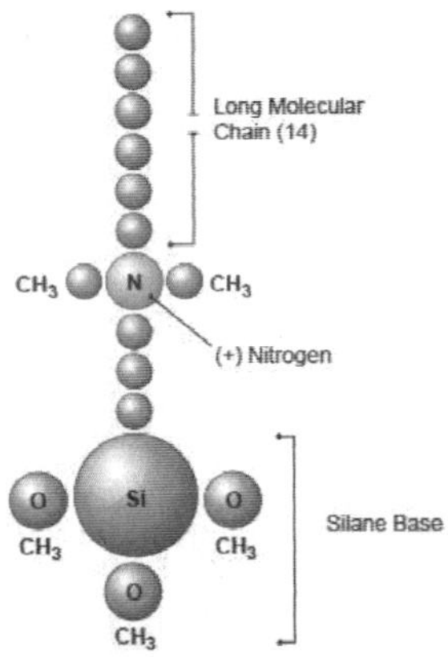

Fig.1. Chemical Structure of QAC [23].

QAC ANTIMICROBIAL MECHANISM

QACs have antimicrobial effect against a broad range of microorganisms including vegetative bacteria, yeast, molds, algae, and viruses [24-26]. The growth inhibitory activity of QACs is higher for gram-positive bacteria and algae compared with gram-negative bacteria and molds. Quats are positively charged polyatomic ions of the structure. It disrupts the virus, bacteria etc, by use of chemicals [27-29]. Silanes are extremely efficient bonding agents that can be coupled to other molecules and then used to permanently bond those molecules to a target surface [30-33].

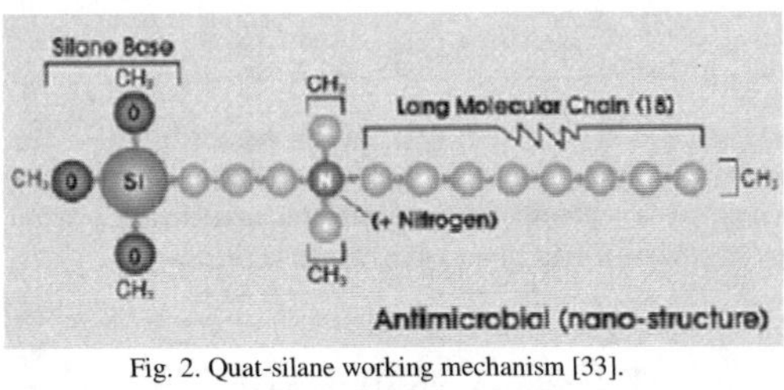

Fig. 2. Quat-silane working mechanism [33].

In this study, the physical, and mechanical properties of quat-silane treated samples in comparison with untreated control samples was investigated in order to point out the change in performances of cotton and cotton elastic fabrics that will be used for clothing for further designs.

II. EXPERIMENTAL PART

A. Materials

Quat-silane was obtained from Aegis Microbeshield. Fabrics were obtained from Wallmart. They are desized, scoured and bleached 100% cotton (Co) and 97% Co / 3% cotton/elastane fabrics. Initial properties of the fabrics are presented in Table 1.

TABLE 1.
INITIAL PROPERTIES OF FABRICS

Fabric Code	Fabric Structure	Fiber Content	Weight (g/m^2)
A	1/1 plain	97% Co / 3% elastane	132.5
B	1/1 plain	100% Co	112.8
C	2/2 panama	100% Co	120.3
D	2/1 twill	100% Co	262.1

B. Treatment of fabric samples with quat-silane

First quat-silane hydrolyzed to 2% active quat-silane for 24h at room temperature and stirred continuously for 10 min at magnetic mixer. Pad-dry-cure method was used to apply quat-silane antimicrobial using a laboratory type padding machine. The fabrics were squeezed to a wet pickup of %85. The samples were treated with 2,0 wt. %, 2,5 wt.%, 3,0 wt. % quat-silane antimicrobial agents for 30min. for 1h, dried at 85oC for 5 min and cured at 160oC for 3 min.

In the referencing system for fabric samples, the cotton/elastane plain wovens were prefixed with a 'A', 'B' for the cotton plain wovens, 'C' for the cotton 2/2 panama wovens, 'D' for the cotton 2/1 twill wovens, for fabric samples treated with different concentrations, the samples treated with 2,0 wt. % quat-silanes were prefixed with a 'X', 'Y' for the fabric samples treated with 2,5 wt.% quat-silanes, 'Z' for the fabric samples treated with 3,0 wt. % quat-silanes and 'U' for the untreated samples.

C. Methods

Antimicrobial Activity

The experimental method used to determine the antimicrobial effects was AATCC Test Method 100: 2004 "Assessment of Antibacterial Finishes on Textiles" Standards, using Staphylococcus Aureus ATCC 6538 (2.00 x 105 CFU/ml) test inoculum. The variables were calculated from Equation (1) respectively.

$$R = (100\,(C-A))/C \qquad (1)$$

R is the percentage reduction of bacteria,
A is the number of bacteria recovered from the inoculated treated sample,
C is the number of bacteria recovered from the inoculated untreated control sample.

Physical and Mechanical Properties

Physical and mechanical properties were tested in order to evaluate the fabric properties in terms of tensile strength (strip and grab method), stitch strength and tear strength before and after antimicrobial treatments. The samples were conditioned for 24 hours at 20°C, 65% relative humidity in the physical testing lab before testing. Tensile strength (strip method), tensile strength (grab mehod), tear strength, and seam strength were measured according to relevant standards (ISO 13934-1, 2013; ISO 13934-2, 2014; ISO 13937-1, 2000; ISO 13936-1, 2004).

Statistical Analysis

The statistical analysis of the experimental data was performed using JMP version 8.0.2 software package (SAS Institute, Inc., Cary, NC). The statistical analysis includes the analysis of variance (ANOVA). For the one-way ANOVA, p-values less than 0.05 were considered statistically significant.

III. RESULTS AND DISCUSSION

A. Antimicrobial activity

The antimicrobial properties of the control and quat-silane samples bound with cotton and cotton/elastane samples are presented in Table 2. The total population of Staphylococcus aureus (S. aureus) ATCC 6538 on each sample was determined. After the antimicrobial tests were performed, the live vibrio concentration of the standard blank sample at zero contact time, as well as that of a standard blank sample oscillated for 24h and that of the antimicrobial fabric sample oscillated for 24h, were compared. It has found that quat-silanes were strongly fastened to cotton and cotton/elastane fabrics and the modified fabrics have a very good antimicrobial activity. No significant change was observed with the increase in quat-silane concentration from X to Y and when the quat-silane concentration increased from Y to Z, the percentage of bacteria showed a small significant decrease.

TABLE 2.
ANTIMICROBIAL TEST RESULTS OF THE CONTROL AND QUAT-SILANE BOUND WITH COTTON AND COTTON/ELASTANE SAMPLES.

Samples	Test Organism: ATCC 6538 Staphylococcus Aureus (CFU/sample)		Percentage Reduction of Bacteria (R)
	Inoculated Sample at 0 contact time (cfu/mL)	Inoculated Sample at 24 hours oscillation (cfu/mL)	
U	0	0	0
X	$1.94x10^5$	$1.40x10^3$	99
Y	$1.94x10^5$	$1.40x10^3$	99
Z	$1.93x10^5$	$1.03x10^4$	95

B. Tensile strength (strip method)

Results are presented in Figure (3-4). A small significant decrease in tensile strength was observed after quat-silane treatments and a small significant decrease was observed with the increase in quat-silane concentration. It was observed the quat-silane treatment caused a restriction in the movement of yarns in the structure resulting a significant decrease in the elasticity. Cotton plain fabrics showed the highest tensile strength in warp direction. Panama fabrics showed the lowest tensile strength in warp direction while they showed the highest tensile strength in weft direction.

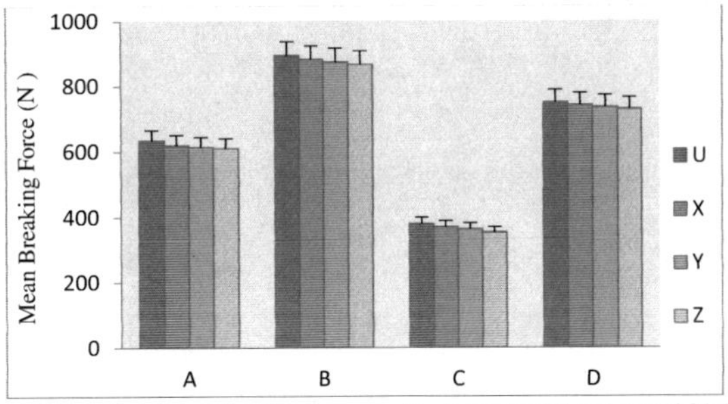

Fig. 3. Tensile strength (strip method) in warp direction for untreated and treated fabric samples

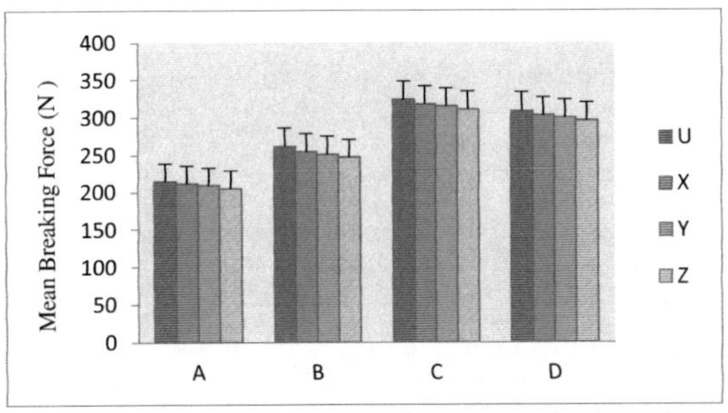

Fig.4. Tensile strength (strip method) in weft direction for untreated and treated fabric samples.

C. Tensile strength (grab method)

Results are presented in Figure (5-6). A small significant decrease in tensile strength was observed after quat-silane treatments and a small significant decrease was observed with the increase in quat-silane concentration. This is attributed to quat-silane treatment caused a swelling of fibers in the fabric structure which decreased the number of yarns per area resulting a restriction in the movement of the yarns. Cotton plain fabrics showed the highest tensile strength and panama fabrics showed the lowest tensile strength in warp direction. Twill fabrics showed the highest tensile strength and cotton/elastane fabrics showed the lowest tensile strength in weft direction.

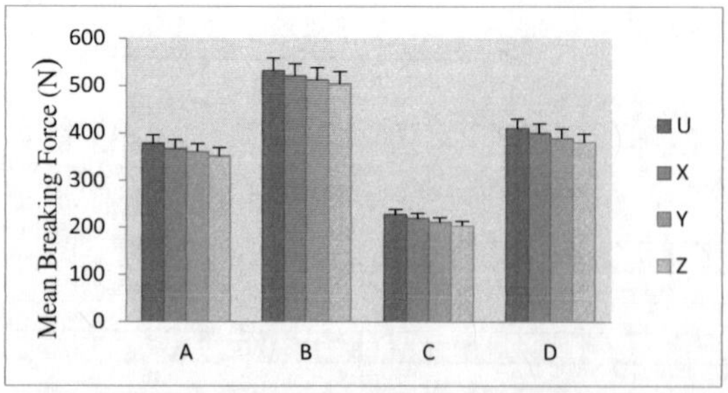

Fig.5. Tensile strength (grab method) in warp direction for untreated and treated fabric samples.

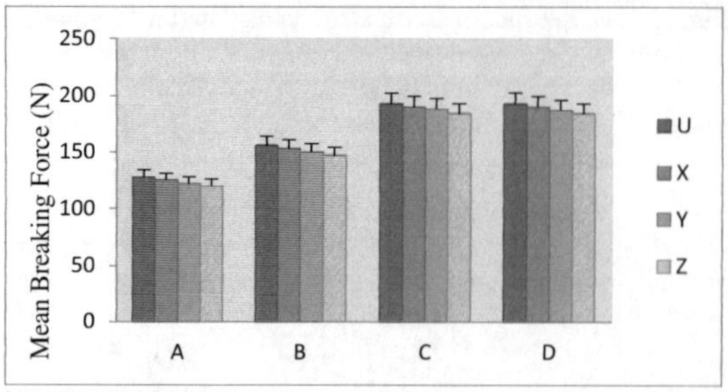

Fig.6. Tensile strength (grab method) in weft direction for untreated and treated fabric samples

D. Effect of quat-silane treatment on tensile strength

ANOVA one-way was used to analyze the effect of quat-silane treatment on tensile strength. Using one way analysis of variance, p-value was found as 0.01. The result of the analysis is shown in Table 3. As p-value is smaller than 0.05, it can be estimated that quat-silane treatment has a significant effect on tensile strength.

TABLE 3.
ANOVA AND ESTIMATION OF PARAMETERS FROM TENSILE STRENGTH.

ANOVA one way

Source of Variance	SS	df	MS	F	P-value	F crit
Between Groups	80.0833	1	80.0833	8.1028	0.0173	4.9646
Within Groups	98.8333	10	9.8833			
Total	178.9166	11				

E. Tear strength

Results are presented in Figure (7-8). A small significant decrease in tear strength was observed after quat-silane treatments and a small significant decrease was observed with the increase in quat-silane concentration. It was observed the quat-silane treatment caused a significant decrease in the number of yarns in the del shape in the region of tear. Panama fabrics showed the highest tear strength and cotton/elastane plain fabrics showed the lowest tear strength in warp direction and panama fabrics showed the highest tear strength and cotton plain fabrics showed the lowest tear strength in weft direction.

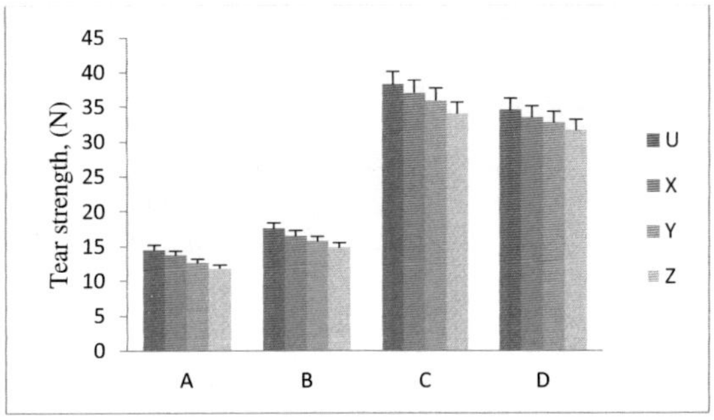

Fig.7. Tear strength in warp direction for untreated and treated fabric samples

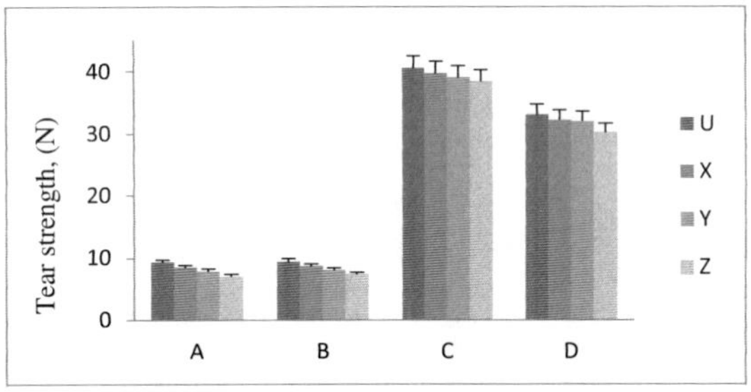

Fig.8. Tear strength in weft direction for untreated and treated fabric samples

F. *Effect of quat-silane treatment on tear strength*

ANOVA one-way was used to analyze the effect of quat-silane treatment on tear strength. Using one way analysis of variance, p-value was found as 0.008. The result of the analysis is shown in Table 4. As p-value is smaller than 0.05, it can be estimated that quat-silane treatment has a significant effect on tear strength.

TABLE 4.
ANOVA AND ESTIMATION OF PARAMETERS FROM TEAR STRENGTH

ANOVA one way

Source of Variance	SS	df	MS	F	P-value	F crit
Between Groups	560.3333	1	560.3333	21.8595	0.0008	4.9646
Within Groups	256.3333	10	25.63333			
Total	816.6666	11				

G. *Seam strength*

Results are presented in Figure (9-10). A small significant decrease in seam strength was observed after quat-silane treatments and a small significant decrease was observed with the increase in quat-silane concentration. This is attributed to quat-silane treatment caused a restriction in the movement of yarns in the fabric structure resulting a significant decrease in the elasticity. Cotton/elastane plain fabrics showed the highest seam strength and panama fabrics showed the lowest seam strength in warp direction. Twill fabrics showed the highest seam strength and panama fabrics showed the lowest seam strength in weft direction.

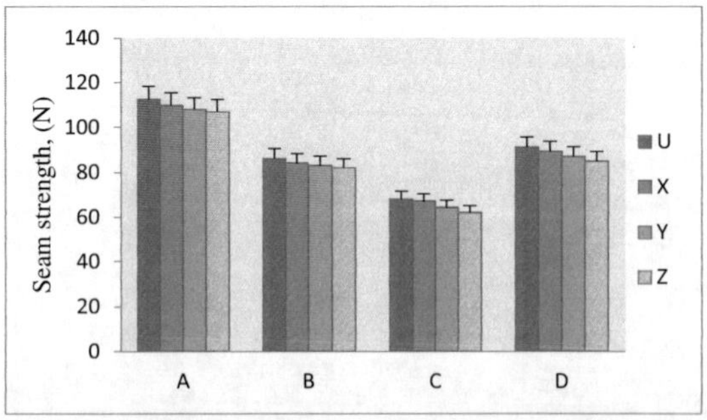

Fig. 9. Seam strength in warp direction for untreated and treated fabric samples

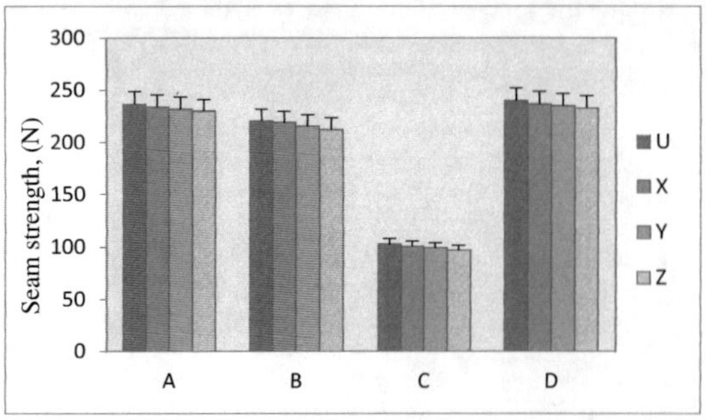

Fig.10. Seam strength in weft direction for untreated and treated fabric samples

H. Effect of quat-silane treatment on seam strength

ANOVA one-way was used to analyze the effect of quat-silane treatment on seam strength. Using one way analysis of variance, p-value was found as 0.0005. The result of the analysis is shown in Table 5. As p-value is smaller than 0.05, it can be estimated that quat-silane treatment has a significant effect on seam strength.

TABLE 5.
ANOVA AND ESTIMATION OF PARAMETERS FROM SEAM STRENGTH.

ANOVA one way

Source of Variance	SS	df	MS	F	P-value	F crit
Between Groups	425.0417	1	425.0417	16.4363	0.0005	4.3009
Within Groups	568.9167	22	25.85985			
Total	993.9583	23				

IV. CONCLUSION

Physical and mechanical properties of cotton and cotton/elastane fabrics used for clothing in further designs and treated with quat-silane agent procedure, were tested in terms of tensile strength (strip method), tensile strength (grab method), tear strength and seam strength. The tensile strength (strip and grab method), tear strength and seam strength showed a small significant decrease after quat-silane treatments. This is attributed to a small significant decrease in elasticity after processes which are related with the restriction of the slippage of yarns in the fabric structure which decreased the elasticity of the fabric samples resulting a limitation in the movement of yarns. Cotton plain fabrics showed the highest tensile strength (strip and grab methods) in warp direction before and after quat-silane treatments. This is attributed to higher fabric density caused a less fabric shrinkage after quat-silane treatments thus a higher breaking load per yarn. Panama fabrics showed the highest tear strength both in warp and weft directions before and after treatments which is attributed to higher number of yarns in the del shape in the region of tear. Twill fabrics showed the highest tensile strength (grab method) and seam strength in weft direction before and after treatments which is attributed to higher elasticity of the fabric structure in weft direction. Cotton/elastane plain fabrics showed the highest seam strength in warp direction before and after treatments which is attributed to elastane in the fabric structure provided more elasticity resulting a significant increase in seam strength. A small significant decrease was observed with the increase in quat-silane concentration. The results showed that all samples showed a very good antimicrobial activity since quaternary ammonium compounds are very durable due to covalent bonding with textiles. Finally, it's found that the quat-silane treatments altered the mechanical and physical properties of cotton and cotton/elastane fabrics but the treated samples still protected their physical and mechanical properties which will still provide comfort during use as clothings in further designs.

References

[1] Specified Requirements of Antifungal Textiles, http://www.ftts.org.tw/images/fa006E.pdf (2004, accessed 10 January 2017).
[2] W. Schnidler and P. Hauser, "Chemical Finishing of Textiles", Woodhead Publishing Ltd, Cambridge, England, 2004.
[3] L. Teufel, K.C. Schuster, P. Merschak, "Development of a fast and reliable method for the assessment of microbial colonization and growth on textiles by DNA quantitication", *Journal of Molecular Microbiology and Biotechnology*, 14, 2008, pp. 193-200.
[4] Sterilization and Sterility Assurance of Compendial Articles Pharmacopeial Forum, www.pharmacopeia.cn (2008, accessed 24 October 2016).
[5] J.D. Payne, 1997, "Antimicrobial treatment of textile materials", Patent 5700742, USA, 1997.
[6] E. Elliott and C. Woodryff, Onlines, Antimicrobials replace dyes at AATCC international conference, Textile World, pp. 88-90.
[7] M.P.C. Pliego, "Effect of natural antimicrobials against Salmonella", Escherichia coli 0157: H7 and Listeria monocytogenes. MSc diss, Texas A & M University, USA, 2007.
[8] A. Demir, T. Oktem, N. Seventekin, "Kitosanın tekstil sanayiinde antimikrobiyal madde olarak kullanımının araştırılması", *Tekstil ve Konfeksiyon*, 2, 2008, pp. 94-102.
[9] T. Abel, J.I. Cohen, R. Engel, "Preparation and investigation of antibacterial protein-based surfaces", *Journal of Textile and Apparel, Technology and Management*, 3, 2003, pp. 1-8.
[10] Y.Gao and R. Cranston, "Recent advances in antimicrobial treatments of textiles", *Textile Research Journal*, 78, 2008, pp. 60-72.
[11] N.A. Ibrahim, M. Hashem,W.A. El-Sayed, "Enhancing antimicrobial properties of dyed and finished cotton/polyester fabrics", *AATCC Review*, 10, 2010, pp. 55-63.
[12] R. Shamey and T. Hussein, "Critical solutions in the dyeing of cotton textile mateials", *Textile Progress*, 37, 2010, pp. 1-84.
[13] L. A. Davidson, "Microban Team on towels", *Home Textiles Today*, 2000, 1,p. 44.
[14] Chemical Name Index. Textile finishing chemicals, pp. 656-669.
[15] B. Simoncic and B. Tomsic, "Structures of novel antimicrobial agents for textiles – a review", *Textile Research Journal*, 80, 2010, pp.1721- 1737.
[16] C. Cortesia, G.J. Lopez, J.H. Waard, "The use of quaternary ammonium disinfectants selects for persisters at high frequency from some species of non-tuberclous mycobacteria and may be associated with outbreaks of soft tissue infections", *Journal of Antimicrobial Chemotherapy*, Vol. 65, 2010, pp. 2574-2581.
[17] K. Hegstad, S. Langsrud, B.T. Lunestad, "Does the wide of quaternary ammonium compounds enhance the selection and spread of antimicrobial resistance and thus threaten our health?", *Microbial Drug Resistance*, Vol. 16, No. 2, 2010, pp. 91-104.
[18] S. Yadav, "A study of antimicrobial properties of fabric treated with silane and N-halamine complex", PhD diss.,2009, Eastern Michigan University.
[19] B.G. Roe, "Durable non-fluorine water-repellent fabric finishing: surface treatment using silica nanoparticulates and mixed silanes", PhD diss., 2008, North Carolina State University.
[20] News, "Antimicrobial polymer emulsions", *Paint and Coatings Industry*, 2007, pp. 66-68.
[21] N. P. Prorokova, S. Y. Vavilova, V. N. Prorokov, "Effect of quaternary ammonium compound preparations on poly(ethylene terephthalate) fibre", *Fibre Chemistry, Fiber Chemistry*, Vol. 39, No. 6, 2007, pp. 445-449.
[22] R. G. Mansfield, "Keeping it fresh, Nonwovens Technical Textiles", *Textile World*, 2002, pp. 42-45.
[23] Generation of Antimicrobial Surfaces Using Dendrimer Biocides, www.surechem.org (2006).
[24] K. K. Leonas, "Using laser scanning confocal microscopy to evaluate microorganism transmission through surgical gown fabrics", *Medical Textiles International Conference*, Jul. 17-18, 1996, Bolton, UK, pp. 60-68.
[25] C.W.White and R. L. Gettings, "Evaluating the antimicrobial properties of silane modified surfaces", Industrial Report.
[26] Medical Microbiology / Bacterial Morphology, micro.digitalproteus.com
[27] Antimicrobial Treated Uniforms / Nursing Vision, www.nursingvision.org
[28] Y. H. Kim, and G. Sun, "Durable antimicrobial finishing of Nylon fabrics with acid dyes and a quaternary ammonium salt", *Textile Research Journal*, Vol. 71, No. 4, 2001, pp. 318-323.
[29] Y. H., Kim, H. Choi, J. H Yoon, "Synthesis of quaternary ammonium derivative of chitosan and its application to a cotton antimicrobial finish", *Textile Research Journal*, Vol. 68, No. 6, 1998, pp. 428-434.
[30] Monticello, R. A. (2011). Teleconference interview.
[31] Weber, M. (2011). Teleconference interview.
[32] Lamba, N. (2010). Teleconference interview.
[33] S. A. Bolkan, "Stable Aqueous Solutions of Silane Quat Ammonium Compounds", US Patent, No: 0028462 A1, 2010, accessed: 02.04.2010.
[34] AATCC Test Method 100: 2004. Assessment of antibacterial finishes on textiles.
[35] ISO 13934-1: 2013. Tensile properties of fabrics – Part 1: Determination of maximum force and elongation at maximum force using the strip method.
[36] ISO 13934-2: 2014. Tensile properties of fabrics – Part 1: Determination of maximum force using the grab method.
[37] ISO 13937-1: 2000. Tear properties of fabrics – Part 1: Determination of tear force using ballistic pendulum method (Elmendorf).
[38] ISO 13936-1: 2004. Determination of the slippage resistance of yarns at a seam in woven fabrics – Part 1: Fixed seam openning method.

IOP Conf. Series: Materials Science and Engineering **460** (2018) 012005 doi:10.1088/1757-899X/460/1/012005

Evaluating the effect of fabric type on thermal insulation properties of sports clothing

M J Abreu[1], A Catarino[1] and D Tama[1,2]

[1] University of Minho, Center for Textile Science and Technology, Campus de Azurem, Guimaraes, Portugal
[2] Ege University, Department of Textile Engineering, 35100 Bornova / Izmir, Turkey

Corresponding author: derya.tama@ege.edu.tr

Abstract. In the present study, evaluating the fabric characteristics on thermal insulation properties of garments was aimed. For that purpose, two fiber combinations and seven different knitting structures were chosen and seven long sleeve basic shirts were manufactured from these fabrics. As a water-repellent finishing treatment, all shirts were treated with 5% of a fluorocarbon-based product. A thermal manikin of PT-Teknik made in Denmark was used to measure the heat loss of the shirts and the surface of manikin was kept at a constant temperature. Afterwards, the total and effective clothing insulation values were calculated for each shirt. The comparisons of heat loss values according to the body parts, and also for the effective clothing insulations were done. Moreover, the heat flux according to the body parts were determined. In the light of these obtained data, a new shirt design was designed and proposed by combining the evaluated fabrics in order to enhance the comfort and performance of the user.

Keywords. Thermal comfort, thermal manikin, heat loss, total thermal insulation, effective clothing insulation.

1. Introduction

In active sports, the clothing is one of the key elements that affects the performance of the athlete. Slater (1985) defined comfort as "a pleasant state of physiological, psychological, neurophysiological and physical harmony between a human being and the environment" [1-3].

Depending on the exercise intensity, the human body produces energy and heat. In order to provide thermoregulation, the heat should be released to the environment. Core temperature can be smoothly raised in the range of 36–39°C with motor activities and chemical reactions, and greater than 39°C becomes critical for human body [4]. Therefore, maintenance of core temperature within the referred range and provide heat flux is crucial for thermo-physiological comfort.

As Bhatia and Malhotra (2016) concluded in their overview research about thermo-physiological wear comfort of clothing, heat and moisture transmission mechanism through clothing, need to be analyzed together with material properties and other influencing parameters [5]. There are also several methods for measuring heat and moisture transmission through clothing. The mostly used is measuring heat loss by using thermal manikins.

Thermal manikins provide a good estimate of the total dry heat loss from the body and the distribution of heat flow over the body surface. In a standard environment, these measures can be used to describe the thermal characteristics of clothing [6]. The operation of the thermal manikins can be set with the constant skin temperature, the constant heat flux or the thermal comfort regulation modes in the different body parts [7,8].

2. Methods

2.1. Specimens

In order to evaluate the influence of fabric characteristics on thermal insulation properties of active sportswear, two fiber combinations and seven different knitting structures were chosen. The details of the compositions and knitting structures were given in Table 1. Seven rounded-neck raglan sleeve tight-fitting shirts were manufactured from these fabrics. Since the active sport garments evaluated in this research were to be used for outdoor sports, a water-repellent finishing treatment was applied to all specimens. All shirts were treated with 5% of a fluorocarbon-based product.

Table 1. The compositions, knitting structures and the codes of shirts

Code	Composition	Knitting Structures
F1	100% Cotton	Single Jersey
F2	60% Polyamide	False Rib 1
F3	35% Polyester	False Rib 2
F4	5% Elastane	Single Jersey jacquard 1
F5	60% Polyamide	Single Jersey jacquard 2
F6	35% Polypropylene	Single Jersey jacquard 3
F7	5% Elastane	False Rib 3

2.2. Thermal Manikin

A female model thermal manikin (PT-Teknik made in Denmark) was used in this study in order to develop heat in a homogeneous distribution. The thermal manikin was installed in the research laboratory of the 2C2T- Center for Textile Science and Technology of the University of Minho in Portugal. The body shell of the thermal manikin is fiber glass armed polystyrene which is responsible for a light and strong mechanical construction and simultaneously a low thermal conductivity and thermal capacity. There are 20 independent sections, which can be controlled separately and independently. The thermal manikin was placed inside the climate chamber, placed around 0.1m above the floor with hanging arms and legs.

2.3. Experimental Procedure

The tests were carried out according to ISO 15831 standard. The trials were performed in constant skin temperature mode and the skin temperature of thermal manikin was set at 33 ±0,2°C. The specimens were dressed up to the thermal manikin and the manikin was kept in a stationary standing position. The tests were performed in a climatic chamber at constant ambient temperature of 24,5±1°C and relative humidity of 60±5%. The mean skin temperature, the heat flux of the manikin, and the room physical parameters were continuously monitored on computers. The obtained heat loss data was used to calculate the total thermal clothing insulation (IT) and the effective thermal clothing insulation (Icle).

2.4. Data analysis
In the present research, statistical analysis was performed by using PASW Statistics 18 statistical analysis package software. A repeated measures ANOVA (Analysis of variance) was conducted to compare the difference of effective clothing insulation values on structures and body parts. Post hoc least significant difference tests were also carried out by using repeated measures ANOVA and Bonferroni correction.

2.5. Garment Design
In the present study, a new shirt was proposed for active sportswear by combining the structures evaluated in this research. In order to design the shirt, the present researches about thermoregulatory responses of human body were analyzed. In the research conducted by Ouzzahra et al. (2012), the thermal sensitivity of regional body parts was analyzed by stimulating body with a 20˚C thermal probe applied onto the skin. The tests were performed in cold environment (Temperature of laboratory: 21,5±0,8˚C, Relative humidity: 44,2±4,8%) and the stimulation were applied during rest and during light exercise (30% VO2 max). The 14 male athletes participated the study and they were asked to score an 11-point cold sensation scale according to the thermal sensations [9]. The thermal sensation was lower and more homogenous during exercise (Figure 1).

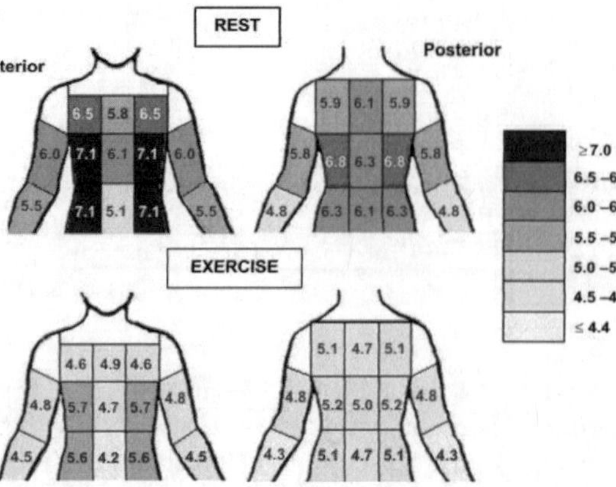

Figure 1. Body map of mean thermal sensations at rest and during exercise [9].

In another outstanding research, Fournet et al. (2013) evaluated the thermoregulatory response of males and females during 40 minutes running in the cold environment (Temp.: 10˚C, R.H.: 50%, Air Vel.: 2.8 m s−1) at 70% VO2 maximum work rate. The skin temperature was recorded by an infrared camera (Figure 2) and population-averaged body maps were developed with the obtained data [10].

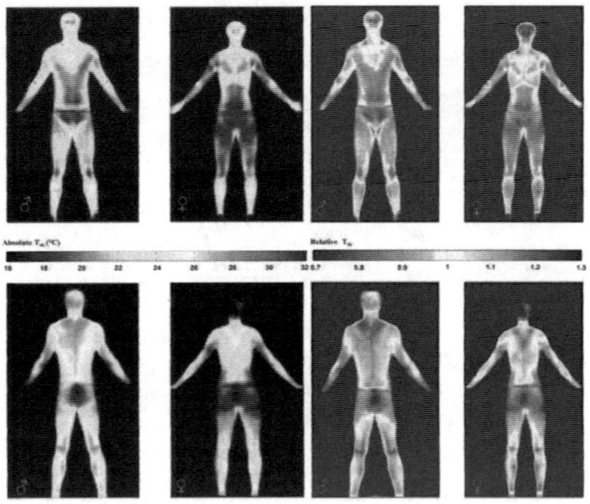

Figure 2. Group averaged body maps of absolute (left panel) and relative (right panel) skin temperature [10].

Smith and Havenith (2011), performed the research in order to develop a sweating and a thermal map of human body. All subjects completed 60 minutes treadmill run involving two exercise intensities. The laboratory were set at 25°C and 50% RH. The regional sweating rates were measured by absorbent pads and infrared images were captured by a thermal camera. With respect to obtained data, they have developed regional temperature and regional sweat guidelines (Figure 3) in order to use garment-designing process [11].

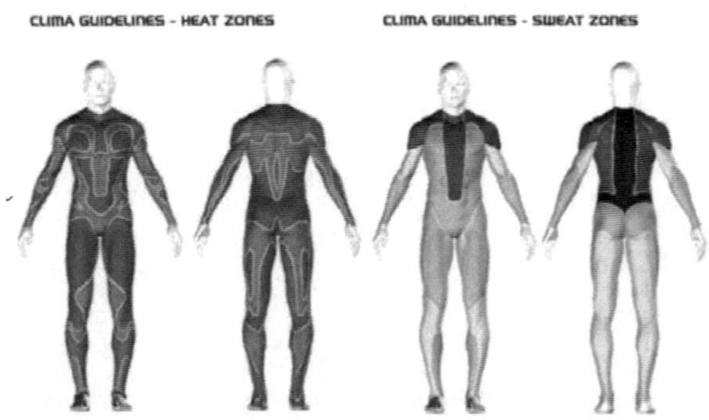

Figure 3. Regional temperature and regional sweat guidelines for garment design [11].

3. Results

The heat loss data of each specimen were analyzed according to eight body parts namely pelvis, back, back side, chest, right upper arm, left upper arm, right forearm, left forearm (Figure 4). As heat flux decreases, heat transfer from body to the environment also decreases, therefore the microclimate temperature between the body and the garment rises. For this reason, higher heat loss values are desired for active sports' clothing. According to obtained data, F1 had the least heat loss value than other structures. This result can derive from the composition (100% cotton) and the structure (single jersey) of F1. It can be said that, at the body parts right and left forearms, the heat flux had the greatest values for F5, F6 and F7.

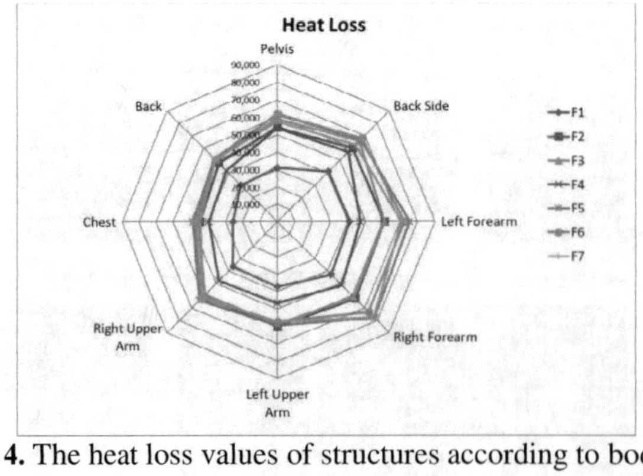

Figure 4. The heat loss values of structures according to body parts.

Figure 5 demonstrates the effective clothing values of the seven structures for eight different upper body parts. The effective clothing insulation values were higher at the back and the chest, right and left forearm had the lowest effective clothing insulation values. According to the statistical treatment (Table 2), the structure had significant difference on the effective clothing insulation properties (p=0,000). The body parts also had significant difference on the effective clothing insulation properties (p=0,000). Moreover, both the structure and body parts had a significant difference on the effective clothing insulation properties (p=0,000).

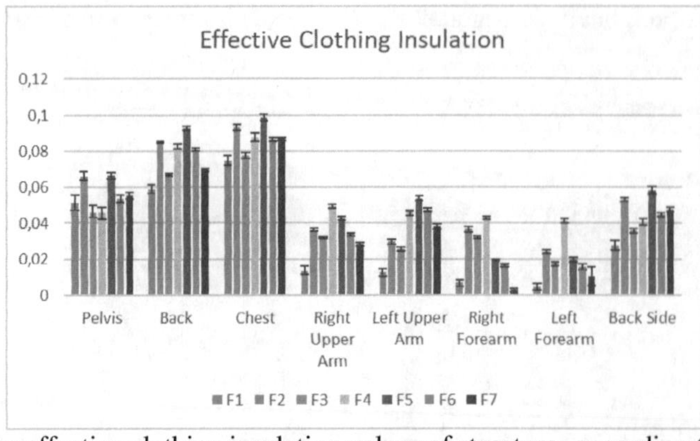

Figure 5. The effective clothing insulation values of structures according to body parts.

With respect to the pairwise test results, it can be said that the structure influences the effective clothing insulation values according with each body part. Specifically, at pelvis between F2 and F5 (p=0,699) and between F3 and F4 (p=0,135) there were no significant differences. At chest, there was no significant difference between F3 and F6 and at left upper arm, there was also no significant difference between F6 and F7.

Table 2. The repeated measures ANOVA results for effective clothing insulation values.

	Sum of squares	Degrees of freedom	Mean square	F value	Sig.
Structure	0,158	2,900	0,055	3387,729	0,000
Body Parts	1,097	2,223	0,493	16941,467	0,000
Structure * Body Parts	0,102	7,864	0,013	630,874	0,000

Considering the tests conducted in this study and the related literature review, a shirt was designed (Figure 6), two fabric constructions were suggested to use in different areas of the body with dissimilar characteristics, in order to enhance the comfort and performance of the user.

4. Conclusion

In active sports, the clothes need to cover and assist the active body and also support extreme physical and environmental performance. In order to provide this, the researches about clothing comfort of active sportswear has become an important research field for textile industry. Moreover, the necessity of considering the fabric characteristics and also human body thermoregulation in the garment design process shows up. Therefore, in the present research, thermal comfort properties of chosen fabrics were evaluated by a thermal manikin and also a literature review the thermal sensations of human body were analyzed.

As literature review indicates, during the exercise the most heated body parts for upper body are shoulders, middle chest, middle back, front sides and lateral arms. The sweating rates of upper body show that the lower and middle back are the parts that sweated most, followed by the middle chest, shoulder, neck and backsides.

Regarding to thermal behaviors of the fabrics tested in the present research and the related literature review, a rounded neck, raglan sleeve shirt for outdoor sports was designed and proposed.

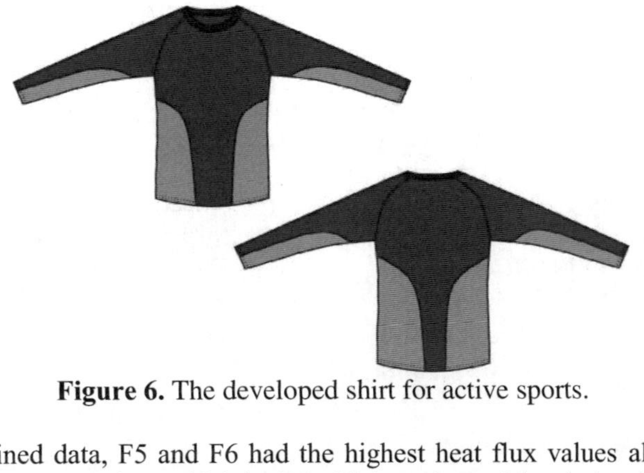

Figure 6. The developed shirt for active sports.

With respect to obtained data, F5 and F6 had the highest heat flux values almost in each body part. Additionally, F5 was the construction, which had the highest thermal insulation in pelvis, back, backside, chest and left upper arm. Moreover, F3 mostly had smaller thermal insulating. Thus, regarding to all data, it was suggested to use F3 on shoulders, front sides, armpits, middle back and lateral arms. In the remaining parts, F5 was used to manufacture the sports garment.

IOP Conf. Series: Materials Science and Engineering **460** (2018) 012005 doi:10.1088/1757-899X/460/1/012005

5. References

[1] Das, A. and Alagirusamy, R., 2010, Science in Clothing Comfort, The Textile Institute, Woodhead Publishing Limited, England, 173p.

[2] Choudhury, A.K.R., Majumdar, P. K. and Datta, C., 2011, Improving Comfort in Clothing, Chapter 1: Factors affecting comfort: human physiology and the role of clothing, The Textile Institute, Woodhead Publishing Limited, England, 459p.

[3] Li, Y. and Wong, A. S. W., 2006, Clothing Biosensory Engineering, Chapter 1: Introduction to Clothing Biosensory Engineering, The Textile Institute, Woodhead Publishing Limited, England, 381p.

[4] Balcı, G. A., Başaran, T. and Colakoglu, M., 2016, Analysing visual pattern of skin temperature during submaximal and maximal exercises, Infrared Physics & Technology 74, pp.57–62, http://dx.doi.org/10.1016/j.infrared.2015.12.002

[5] Bhatia, D. and Malhotra, U., 2016, Thermo-physiological wear comfort of clothing: an overview, J Textile Sci Eng 2016, 6:2, http://dx.doi.org/10.4172/2165-8064.1000250

[6] Abreu, M. J., 2008, Validation of the thermal comfort behavior of active wear through thermal manikin measurements, 2nd International Scientific Conference Textiles of the Future, 13-15 November 2008, Kortrijk, Belgium.

[7] Oliveira, A. V. M, Gaspar, A. R. and Quintela, D. A. (2008b) Measurements of clothing insulation with a thermal manikin operating under the thermal comfort regulation mode: comparative analysis of the calculation methods, European Journal of Applied Physiology, Vol. 104, pp. 679–688, DOI 10.1007/s00421-008-0824-5.

[8] Melikov A, 2004, Breathing thermal manikins for indoor environment assessment: Important characteristics and requirements, European Journal of Applied Physiology 92:710–713. doi:10.1007/s00421-004-1142-1

[9] Ouzzahra Y., Havenith G. and Redortier B., 2012, Regional distribution of thermal sensitivity to cold at rest and during mild exercise in males, Journal of Thermal Biology, Volume 37, Issue 7, pp. 517-523, https://doi.org/10.1016/j.jtherbio.2012.06.003.

[10] Fournet D., Ross L., Voelcker T., Redortier B. and Havenith G., 2013, Body mapping of thermoregulatory and perceptual responses of males and females running in the cold, Journal of Thermal Biology, Volume 38, Issue 6, pp. 339-344, https://doi.org/10.1016/j.jtherbio.2013.04.005.

[11] Smith C. and Havenith G., 2011, Sweat mapping in humans and applications for clothing design, The Fourth International Conference on Human-Environment System, 3-6 October, Japan.

Acknowledgment

This work was supported by FEDER funds through the Competitivity Factors Operational Program - COMPETE and by national funds through FCT – Foundation for Science and Technology within the scope of the project POCI-01-0145-FEDER-007136.

Numerical micro-scale modelling of the mechanical loading of woven fabrics equipped with particles

T. Gereke[1], O. Döbrich[1], S. A. Malik[2], R. T. Kocaman[1], D. Aibibu[1], K. Schmidt[3], S. Antonyuk[3], S. Ripperger[3], C. Cherif[1]

[1]*Technische Universität Dresden, Institute of Textile Machinery and High Performance Material Technology, 01062 Dresden, Germany*
[2]*Mehran University of Engineering & Technology, Department of Textile Engineering 76062 Jamshoro, Pakistan*
[3]*Technische Universität Kaiserslautern, Institute of Particle Process Engineering, Gottlieb-Daimler-Straße, 67663 Kaiserslautern, Germany*

Corresponding author: thomas.gereke@tu-dresden.de

Abstract The equipping of polyester multifilament fabrics with functionalized micro particles for partial and targeted reduction of the size of continuous inter-yarn pores in the fabric, utilizing the filtering effect of textile surfaces, has proven to be a very effective method for increasing the barrier properties of fabrics for use as surgical reusable textiles, cleanroom clothing or cleanable filter media. The objectives of this study were to understand the interrelationships of weaving parameters and fabric properties and to model the porous fabrics under loading conditions. Artificial Neural Networks were trained and tested in order to link the weaving and processing parameters with the fabric properties using the experimental data. Mean flow pore sizes and permeability values of woven fabrics under biaxial loading conditions were measured according to a method previously developed using a special sample holder connected to a conventional porosimetry. Finite element models of unit cells were developed and loaded virtually. Those models were transferred into voxel models and imported into software DNSlab, where they were virtually equipped with particles. For the simulation of the fluid flow, the Navier-Stokes equations were solved numerically by the Lattice-Boltzmann method. The experimental and numerical studies show that pore size of barrier fabrics can be tailored using the developed methods and models. Important parameters are the fabric construction parameters and the particle size.

Keywords— **barrier fabric, finite element method, fluid flow, micro-scale model, particle, woven fabric**

I. INTRODUCTION

The development of liquid- and particle-tight fabrics with excellent textile-physiological properties and an adjustable porosity for use as surgical reusable textiles, cleanroom clothing or cleanable filter media is of particular economic relevance. The equipping of polyester multifilament fabrics with functionalized micro particles for partial and targeted reduction of the size of continuous meso pores (i. e. inter-yarn pores) in the fabric, utilizing the filtering effect of textile surfaces, has proven to be a very effective method for increasing the barrier properties. For the design of the fabrics, realistic modelling with its pore morphology is necessary. The adjustable process parameters of the weaving machine and the yarn parameters as well as the fabric construction determine significantly the shape, size and distribution of the meso pores and micro pores (i. e. intra-yarn pores) and thus the flow behaviour of the fabrics [1,2].

Throughout the entire service life, membrane forces act on the barrier textiles. Combined loads (tension, bending and shear) lead to a deformation of the pore structure in the fabric, which can result in a change in the fabric permeability and thus an increased passage potential for contaminated particles and liquids [3]. First orienting studies on the changes of the fabric morphology induced by mechanical loading were conducted [3-5]. Complex mechanical loadings were determined on surgical textiles during operations, which are e. g. between 0.9 N/cm² and 1.2 N/cm² for elbows [6-8]. The experimental investigation and virtual simulation of these complex loading conditions based on real service loads and their effect on pore morphology is not known. Their influence on the barrier fabrics treated with particles and their durability are also still unexplored.

A numerical macro- or meso-scale approach is inadequate for an exact replication of the microporous morphology of fabrics. A micro-scale approach, in which the individual filaments are taken into account, considers not only the meso pores but also the micro pores. Methods that model the multifilament character of the yarns have proven notable accuracy in predicting the geometry of 2D and 3D woven fabrics [9-17]. Each yarn is discretized into multiple 1D element chains. Each of these element chains approximates the behaviour of a bundle of fibres and contact models are used to capture the interactions between the element chains. This allows for simulating the interactions and low cohesion between the individual fibres.

The objectives of this study are, firstly, to understand the interrelationships of weaving parameters and fabric properties and, secondly, to model the porous fabrics under loading conditions.

II. MATERIALS AND METHODS

A. Materials

Woven fabrics made of polyester multifilament yarns were manufactured using different yarns, weave parameters and machine processing parameters (machine speed, shed closing time) on a Dornier PTS4/S EasyLeno® double rapier weaving machine (Lindauer Dornier GmbH, Germany). Fabrics were loaded with biaxial tensile stresses and porosity was measured in unloaded and loaded conditions. Artificial Neural Networks were trained and tested in order to link the weaving and processing parameters with the fabric properties using the experimental data [18].

B. Experimental

Cross-sectional images of the fabrics were taken and provided for the virtual modelling of the fabrics. For this, fabrics were cut out and embedded into epoxy resin in a sample carrier. After 8 hours of hardening time, samples were smoothened and polished with a grinding device (Struers, Germany). Cross-section images were taken with a light microscope Axio Imager M1m (Zeiss, Germany) with a 500 fold magnification. An image can be seen in Fig. 1.

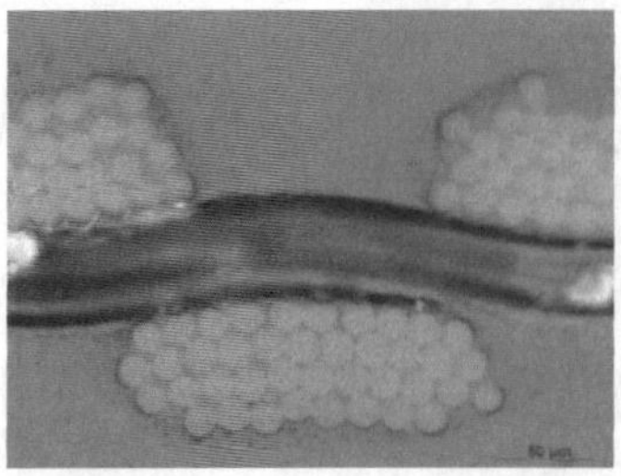

Fig. 1. Microscopic picture of barrier fabric

Two plain woven fabrics with fabric index 0.64 ('sample 1') and 0.78 ('sample 2') were chosen for experimental tests. Breaking loads of the fabrics were measured with a uniaxial tensile test machine UPM 1445 (Zwick, Germany). Samples were then loaded with biaxial tensile stresses up to 30% and 50% of the breaking forces of the fabrics in a biaxial tensile test machine (Zwick, Germany). Experimental results were then considered as a reference for the virtual loading of developed models.

Mean flow pore sizes and permeability values under biaxial loading were measured according to the method developed in [19] using a special sample holder connected to the conventional porosimetry PSM 165 (Topas, Germany). Software PSMWin enables the recording of permeability values of the porosity tests which were performed under biaxial loading.

C. Numerical

For the second part of the study, numerical models on a unit cell level were developed. The models were generated according to the method provided in [9] using digital elements and the weave parameters determined from real fabrics (Fig. 1 and Fig. 2). Fabrics were then virtually loaded in LS-Dyna (Livermore Software Technology Corporation, USA) using finite element method. Finite element models were transferred into voxel models and imported into software DNSlab (IT for Engineering (it4e) GmbH, Germany), where they were virtually equipped with particles applying Lattice-Boltzmann method.

Fig. 2. Micro-scale model of plain woven fabric

III. Results and Discussion

A. Artificial Neural Networks

Neural networks have learned well the under lying interaction between training inputs and output variables. The mean pore size of fabrics was decreased by increasing the weave density (Fig. 3). An opposite trend was observed with weft filament fineness (Fig. 4). Weave process parameters only have a minor influence on pore size. The model findings were in agreement with the experimental results, which suggest that the ANN has simulated well the input-output relationship.

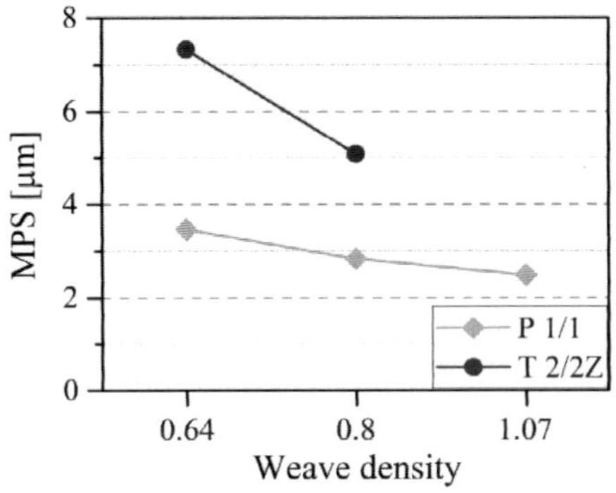

Fig. 3. Trend analysis using developed ANN model: effect of weave type on mean pore size (MPS) of barrier woven fabrics (P 1/1 – plain weave, T 2/2 – twill weave)

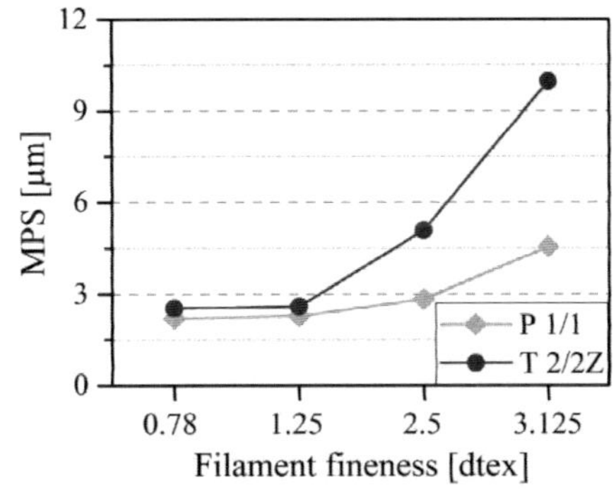

Fig. 4. Trend analysis using developed ANN model: effect of weft filament fineness on mean pore size (MPS) of barrier woven fabrics (P 1/1 – plain weave, T 2/2 – twill weave)

B. Experimental

Fig. 5 shows the force-elongation diagrams of each axis for the woven sample 1 as an example after biaxial tensile loading. It was observed that sample 1 elongates up to ca. 1.1% in the warp direction and up to 1.3% in the weft direction at 50% biaxial loading. This elongation value is up to 0.7% in the warp direction and up to 0.9% in the weft direction at 30% biaxial loading. Sample 2 showed an elongation value up to 0.9% in the warp direction and up to 0.8% in the weft direction at 30% biaxial loading. At 50% of biaxial loading, these values became 1.4% in the warp direction and 1.6% in the weft direction.

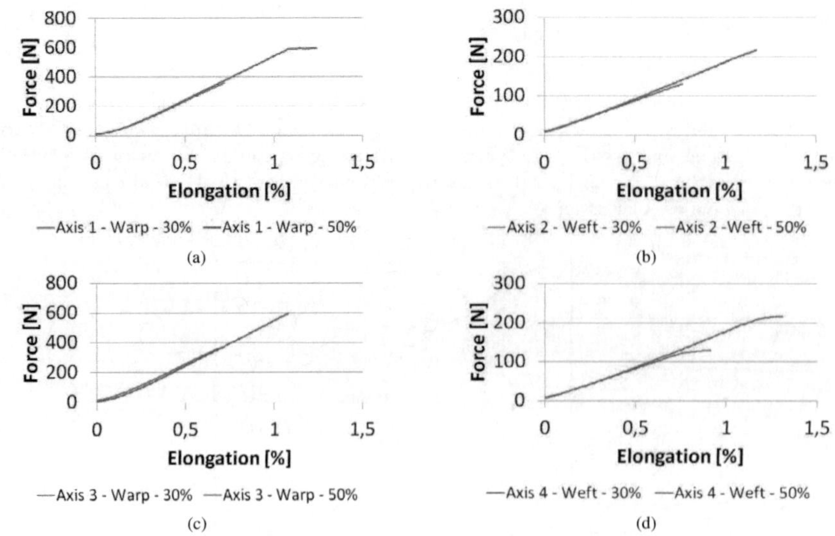

Fig. 5. Force-elongation diagrams of woven sample 1 for each axis: a) axis 1 – warp direction, b) axis 2 – weft direction, c) axis 3 – warp direction, d) axis 4 – weft direction

Permeability values were also recorded during porosity measurements under biaxial loading. Fig. 6 illustrates the permeability of woven samples without loading and under 30% and 50% biaxial loading depending on the pressure difference.

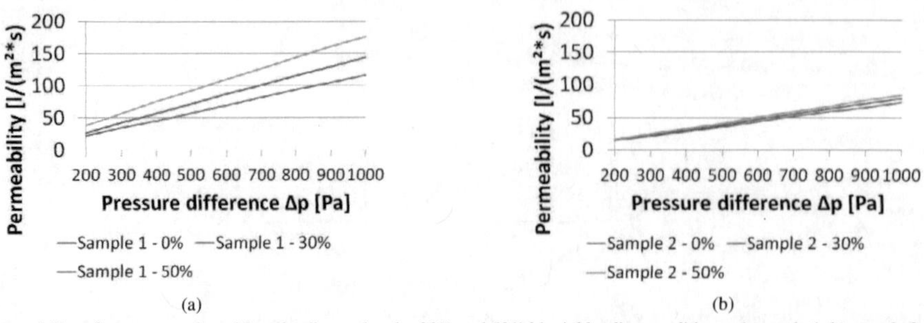

Fig. 6. Permeability of woven samples without loading and under 30% and 50% biaxial loading conditions: a) sample 1, b) sample 2

It was observed that permeability tendencies of sample 2 showed reduced permeability values comparing to sample 1. The reason for this is the fabric index; increased fabric indices reduced permeability values [18]. Woven samples under biaxial loading showed increased permeability values compared to samples without loading. Permeability value of sample 1 changes from ca. 22 to 38 $l/(m^2s)$ at a pressure difference of 200 Pa depending on the loading. This interval becomes smaller in sample 2 as from ca. 14 to 16 $l/(m^2s)$ at a pressure difference of 200 Pa.

C. Numerical

To observe the behaviour of the barrier fabrics under loading conditions on a virtual level, the microscale models were used in a numerical loading simulation according to the introduced uniaxial and biaxial tensile experiments. The detailed fabric models (Fig. 2) are symmetric and periodic boundary conditions were applied to account for the unit-cell deformation character. The loading conditions for the uniaxial and biaxial load cases are in accordance to the conducted experiments. The stiffness behaviour of the single filaments, which are regarded as tension-only threads according to the digital-element approach, was obtained by single filament tensile tests. The results of the periodic deformation under uniaxial and biaxial load can be seen in Fig. 7a and 7b. The result of a virtual shear deformation is shown in Fig. 7c.

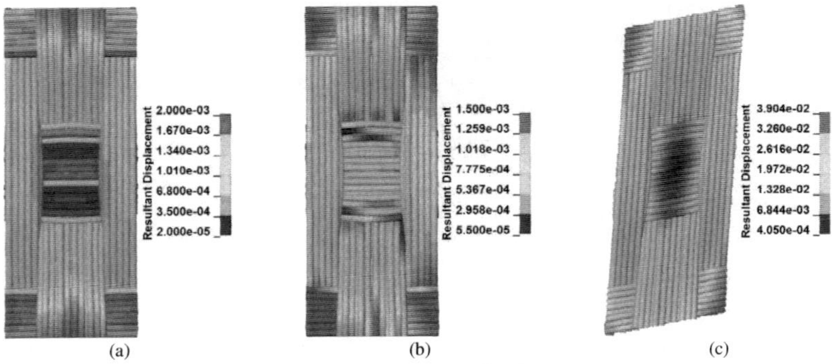

Fig. 7. Simulated deformations under a) uniaxial tension, b) biaxial tension, and c) shear

The simulation of the particle suspension flow across a woven fabric was conducted with DNSlab, where the unloaded or loaded model was imported from LS-Dyna. For the simulation of the fluid flow, the Navier-Stokes equations are solved numerically by the Lattice-Boltzmann method. The particle transport and deposition can be modeled either by a simplified Euler-Lagrange approach, where the effect of the particles on the fluid flow is neglected, or by highly resolved 4-way CFD-DEM coupling, which includes all fluid-particle and particle-structure interactions but demands more computational power. Figs. 8 - 11 show a particle deposition simulation with the Euler-Lagrange approach, where a cross-flow velocity of 0.1 $m \cdot s^{-1}$, a pressure difference of 300 Pa and a particle size of 15 μm were applied.

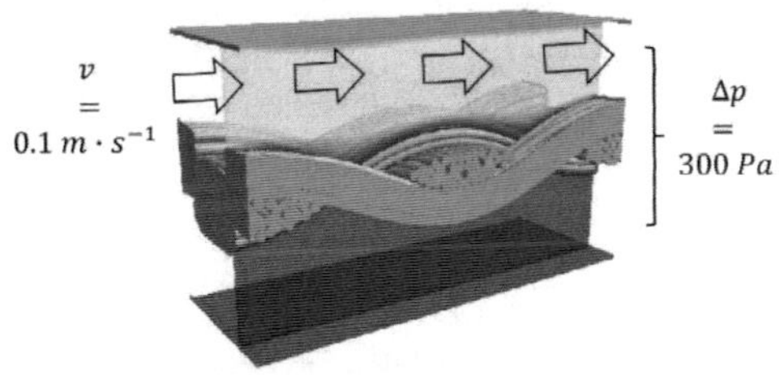

Fig. 8. 3D model of a woven fabric in a cross-flow channel with flow field computed by the Lattice-Boltzmann method.

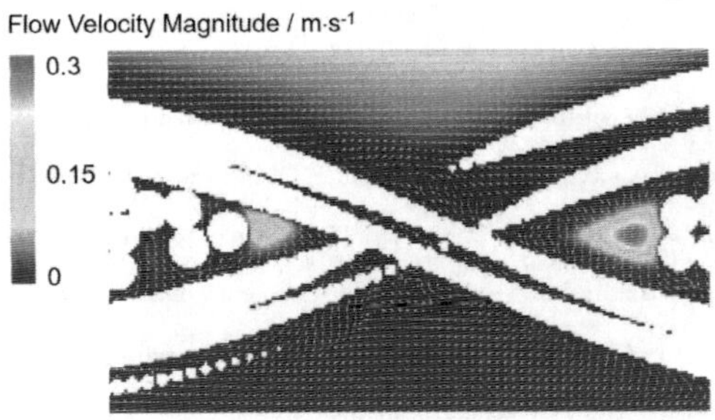

Fig. 9. Detail of the flow field shown in Fig. 8 at a crossing point of the fiber bundles. Most of the permeating fluid passes through the crossing points.

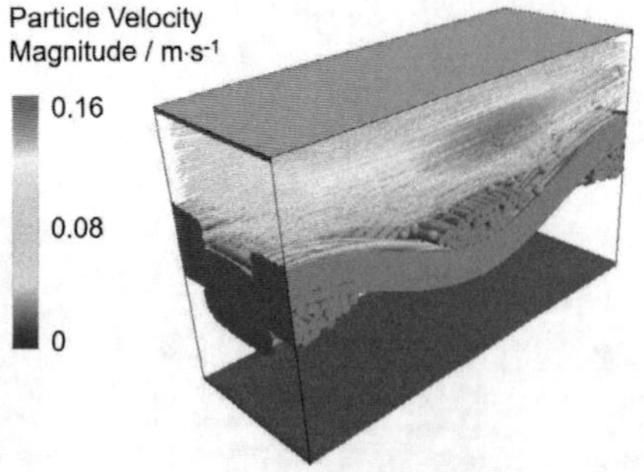

Fig. 10. Computed trajectories of 15 µm particles.

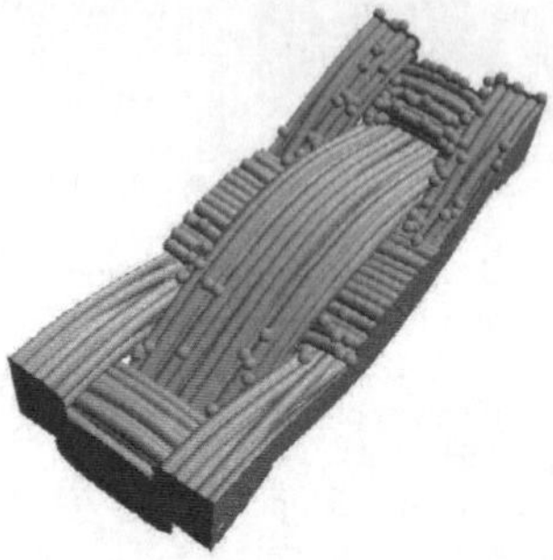

Fig.11. Particles deposited in particular at the crossing points of the fiber bundles.

IV. Conclusions

The findings suggest that it is plausible to predict fabric properties of PES multifilament fabrics using yarn, fabric construction and processing parameters with excellent accuracy by means of ANN. Numerical studies show that pore size of barrier fabrics can be tailored using the developed models. Important parameters are the fabric construction parameters and the particle size.

Acknowledgment

The authors gratefully acknowledge the financial support of the German Research Foundation (Deutsche Forschungsgemeinschaft, DFG) in scope of the projects CH 174/26-2 and CH 174/38-1.

References

[1] H. Gooijer, M.M.C.G. Warmoeskerken, J. Groot Wassink, "Flow resistance of textile materials", *Textile Research Journal*, vol. 73 no. 6, pp. 480–484, 2003.
[2] R. B. Turan, A. Okur, "Investigation of pore parameters of woven fabrics by theoretical and image analysis methods", *The Journal of The Textile Institute*, vol. 103, no. 8, pp. 875–884, 2012.
[3] K. Pietsch: Untersuchungen zur Reproduzierbarkeit der Struktur-Eigenschaftsrelation von Operationsschutztextilien unter Gebrauchsbedingungen. PhD Thesis, Technische Universität Dresden, 2010.
[4] P. D. Dubrovski, M. Brezočnik, "The usage of genetic methods for prediction of fabric porosity", In: Genetic programming - New approaches and successful applications, S.Ventura, Ed., Rijeka: InTech, 2012, pp. 171–198.
[5] M. Havlová , "Air permeability and constructional parameters of woven fabrics", *FIBRES & TEXTILES in Eastern Europe*, vol. 21, no. 2, pp. 84–89, 2013.
[6] K. W. Altman, J. H. McElhaney, J. A. Moylan, K. T Fitzpatrick, "Transmural surgical gown pressure measurements in the operating theater", *American Journal of Infection Control*, vol. 19, no. 3, pp. 147–155, 1991.
[7] J. W. Smith, R. L. Nichols, "Barrier efficiency of surgical gowns: Are we really protected from our patients' pathogens?", *Archives of Surgery*, vol. 126, no. 6, pp. 756–763, 1991.
[8] J. W. Smith, W. A. Tate, S. Yazdani, R. Y. Garcia, A. C. Muzik, R. L. Nichols, "Determination of surgeon-generated gown pressures during various surgical procedures in the operating room", *American Journal of Infection Control*, vol. 23, no. 4, pp. 237–246, 1995.
[9] O. Döbrich, T. Gereke, C. Cherif, "Modelling of textile composite reinforcements on the micro-scale", *Autex Research Journal*, vol. 14, no. 1, pp. 28–33, 2014.
[10] O. Döbrich, T. Gereke, C. Cherif, "Modeling the mechanical properties of textile-reinforced composites with a near micro-scale approach", *Composite Structures*, vol. 135, pp. 1–7, 2016.
[11] G. Zhou, X. Sun, Y. Wang, "Multi-chain digital element analysis in textile mechanics", *Composites Science and Technology*, vol. 64, no. 2, pp. 239–244, 2004.

[12] D. Durville, "Simulation of the mechanical behaviour of woven fabrics at the scale of fibers", *International Journal of Material Forming*, vol. 3 (SUPPL. 2), pp. 1241–1251 , 2010.
[13] Y. Mahadik, S. R. Hallett, "Finite element modelling of tow geometry in 3D woven fabrics", *Composites Part A: Applied Science and Manufacturing*, vol. 41, no. 9, pp. 1192–1200, 2010.
[14] B. El Said, S. Green, S. R. Hallett, "Kinematic modelling of 3D woven fabric deformation for structural scale features", *Composites Part A: Applied Science and Manufacturing*, vol. 57, pp. 95–107, 2014.
[15] N. Naouar, E. Vidal-Sallé, J. Schneider, E. Maire, P. Boisse, "Meso-scale FE analyses of textile composite reinforcement deformation based on X-ray computed tomography", *Composite Structures*, vol. 116, no. 1, pp. 165–176 , 2014.
[16] L. Daelemans, J. Faes, S. Allaoui, G. Hivet, M. Dierick, "Finite element simulation of the woven geometry and mechanical behaviour of a 3D woven dry fabric under tensile and shear loading using the digital element method", *Composites Science and Technology*, vol. 137, pp. 177–187 , 2016.
[17] S. Joglekar, M. Pankow, "Modeling of 3D woven composites using the digital element approach for accurate prediction of kinking under compressive loads", *Composite Structures*, vol. 160, pp. 547–559, 2017.
[18] S. A. Malik, R. T. Kocaman, H. K. Kaynak, T. Gereke, D. Aibibu, O. Babaarslan, C. Cherif, "Analysis and prediction of air permeability of woven barrier fabrics with respect to loom dynamics, fabric construction and material parameters", *Fibers and Polymers*, vol. 18, pp. 2005-2017, 2017.
[19] R. T. Kocaman, S. A. Malik, D. Aibibu, C. Cherif, "In situ determination of pore sizes of high density polyester woven fabrics under biaxial loading", IOP Conference series: Materials Science and Engineering, vol. 254, pp. 142011, 2017.

Efficient silver nanoparticles deposition method on DBD plasma-treated polyamide 6,6 for antimicrobial textiles

A I Ribeiro[1], D Senturk[1], K S Silva[2], M Modic[3], U Cvelbar[3], G Dinescu[4], B Mitu[4], A Nikiforov[5], C Leys[5], I Kuchakova[5], M Vanneste[6], P Heyse[6], M De Vrieze[6], A P Souto[1] and A Zille[1*]

[1]2C2T-Centro de Ciência e Tecnologia Têxtil, University of Minho, Guimarães, Portugal

[2]Federal University of Rio Grande do Norte, Textile Engineering Department, Natal/RN, Brazil.

[3]Jožef Stefan Institute, Ljubljana, Slovenia

[4]National Institute for Lasers, Plasma and Radiation Physics, Măgurele, Romania

[5]Department of Applied Physics, Ghent University, Ghent, Belgium

[6]Centexbel Ghent, Technologiepark 7, 9052, Ghent, Belgium

azille@2c2t.uminho.pt

Abstract. Several methods for silver nanoparticles (AgNPs) deposition on dielectric barrier discharge (DBD) plasma treated and not treated polyamide 6,6 fabric were tested for the production of durable antibacterial textiles. Ultrasound-assisted, dip-coating, exhaustion at 70°C and spray deposition methods were applied. The treated fabrics were analysed in terms of coating uniformity, plasma effect, nanoparticles concentration onto the fabrics and nanoparticles release. Scanning electron microscope (SEM), energy dispersive X-ray (EDX), atomic absorption spectroscopy (AAS), inductively coupled plasma optical emission spectrometry (ICP-OES) and reflectance spectroscopy(%R) were performed. In the fabrics impregnated with AgNPs by dip coating or spray methods there are no significant differences in the plasma treated and untreated samples. Ultrasound tip and exhaustion at 70°C presented higher AgNPs loading. Better AgNPs distribution, higher concentration and less agglomeration were observed in the plasma treated samples and an improved controlled release as well. The plasma treated fabrics with AgNPs deposited with the ultrasound method showed the highest concentration of nanoparticles onto the fabrics as well as the lowest NPs agglomeration. However, this deposition is limited to certain areas of the fabric, the samples were not uniform. The exhaustion method showed the best equilibrium between the NPs distribution and state of agglomeration.

1. Introduction

The World Health Organization and the European Centre for Disease prevention are really concerned about the growing healthcare-associated infections. Over 4 million people developed an infection associated to this environment and their treatment contributed for the intensification of microbial resistance worldwide [1,2]. Therefore, the increasing concern in the widespread use of biocides leads to the development of new strategies [3,4].

The study of antimicrobial fabrics with silver nanoparticles (AgNPs) incorporation has shown excellent properties in medical, pharmaceutical, cosmetics and electronics applications due to their good action against pathogens, preventing and treating infections [5]. The high surface-to-volume ratio from

AgNPs promotes an easy release of silver ion, responsible for the antimicrobial effect [6]. Three different mechanisms are proposed. The first hypothesis is linked to contact killing surfaces by the generations of reactive oxygen species (ROS) and Ag^+ ions release that degrade bacteria cell membranes. The second theory is related with oxidative intracellular reactions. The last hypothesis is associated with the bonds between Ag^+ ions and sulphur, oxygen or nitrogen atoms in biological molecules in bacteria [7]. Unfortunately, NPs could penetrate cells and display cytotoxicity and genotoxicity in human body limiting practical applications. Properties including size, agglomeration and chemical structure are responsible for NPs toxicological effect [8]. Thus, it is important to develop a new class of antibacterial surfaces with firmly nanoparticles loading and with precise local release of antibacterial effect.

The most traditional method for nanoparticles deposition onto fabrics is the pad-dry-cure technique [9]. Other methods were developed such as dip coating, electrochemical methods and layer-by-layer depositions. However, these methods showed several limitations for nanoparticles depositions [10-12].

Dielectric barrier discharge (DBD) plasma-assisted deposition of AgNPs has shown to enhance the NPs adhesion prepared in water dispersions dispersions. DBD plasma is a dry environmental friendly technology that evenly modifies textile surfaces without conventional chemical procedures [13]. DBD plasma treatment activates the fabric surface by the introduction of new polar functional groups such as carbonyl, carboxyl, ether, amine and hydroxyl according to the used gases (e.g. air, helium, nitrogen or oxygen) [14]. These polar groups are able to increase the surface energy thus altering wettability. It can also promote etching and cross-linking, without deteriorating the bulk properties of the material [15]. In addition, the attractiveness of plasma treatment for AgNPs deposition is also due to the formation of micro-roughness that associated to the chemical changes on the surface improves the AgNPs fixation. AgNPs can be deposited on different parts of the textile substrate being mechanically locked by attractive forces such as Van der Walls or electrostatic ones. Moreover, it can also be assisted by chemical interactions such as ionic, covalent, metallic and hydrogen bonds [16].

In this work, several methods for the AgNPs deposition on DBD plasma pre-treated polyamide 6,6 (PA66) were tested for the production of durable antibacterial textiles. Ultrasound-assisted deposition, dip-coating, spray and exhaustion at 70°C methods were tested in terms of coating uniformity, plasma effect, nanoparticles concentration onto the fabrics and nanoparticles release. Scanning electron microscope (SEM), energy dispersive X-ray (EDX), atomic absorption spectroscopy (AAS), inductively coupled plasma optical emission spectrometry (ICP-OES) and reflectance (%R) were performed.

2. Methods and Procedures

2.1. Materials

Commercial PA66 (warp density of 50 threads cm^{-1}, a weft density of 32 threads cm^{-1} and an areal density of 110 g m^{-2}) fabric was used. The samples were pre-washed with a 1 g L^{-1} of non-ionic detergent solutions at 60°C for 60 min and then rinsed with water and dried at 40°C before DBD plasma treatment, in order to minimize contaminations of manufacture. All the other reagents were analytical grade purchased from Sigma–Aldrich, St. Louis, MO, USA and used without further purification.

2.2. DBD plasma treatment

The DBD plasma treatment was performed in a semi industrial prototype machine (Softal GmbH/University of Minho) working at room temperature and atmospheric pressure in air, using a system of metal electrode coated with ceramic and counter electrodes coated with silicon with 50 cm effective width, gap distance fixed at 3 mm and producing the discharge at high voltage 10 kV and low frequency 40 kHz. The discharge power supplied by the electrodes and the speed may change, with maximum discharge of 1.5 kW and speed of 60 m min^{-1}. The dosage applied onto PA66 fabric was 5 kW min m^{-2}. The machine was operated at optimized parameters: 1 kW of power and velocity of 4 m min^{-1}. The plasmatic dosage (D) was calculated according to the following equation 1 [17]:

$$D = \frac{P \cdot N}{V \cdot W} \quad (1)$$

Where,
P = power (kW); N = number of passages; V = velocity (m min-1); W = width of treatment (m).

2.3. Preparation of AgNPs dispersions
Water dispersions of AgNPs with 20 nm as particle size (10 mg L^{-1}) were prepared in an ultrasonic bath for 30 min followed by an ultrasonic tip for more 30 min.

2.4. AgNPs deposition
AgNPs dispersions were deposited onto untreated and plasma treated PA66 (0.3 g per sample) by i) ultrasound tip assisted deposition, ii) dip coating iii) spray and iv) exhaustion at 70°C: i) The fabrics were immerged in 250 mL of nanoparticle dispersion for 5 minutes with an intensity of ultrasound tip at 50%, 60%, 70% or 80% with different geometries. The samples were removed and dried at 30°C; ii) The fabrics were immerged in 50 mL of nanoparticle dispersion for 10 minutes. The samples were removed and dried at 30°C; iii) AgNPs dispersion was applied by spray directly. The system was pressurized at 1.5 bar and maintained at the distance of 5 cm to the substrate. AgNPs were applied in both sides. The samples were dried at 30°C; iv) AgNPs dispersion was applied by exhaustion for 60 min, 100 rpm, a 1:100 ratio bath, at 70°C. Then, the samples were removed and dried at 30°C.

2.5. Reflectance
The samples with AgNPs were analysed by Datacolor Spectraflash SF 600 Plus CT spectrophotometer with D65 light, over the range of 390–700 nm and expressed in reflectance (%R). Reflectance measurements were made three times in different positions of the fabric and the average was calculated.

2.6. SEM and EDX
Morphological analyses of fabrics were carried out with an Ultrahigh Resolution Field Emission Gun Scanning Electron Microscopy (FEG-SEM), NOVA 200 Nano SEM, FEI Company. Secondary electron images were performed with an acceleration voltage at 5 kV. Backscattering Electron Images were realized with an acceleration voltage of 15 kV. Samples were covered with a film of Au-Pd (80-20 weight %) in a high-resolution sputter coater, 208HR Cressington Company, coupled to a MTM-20 Cressington High Resolution Thickness Controller. Atomic compositions of the membrane were examined with the energy dispersive spectroscopy (EDX) capability of the SEM equipment using an EDAX Si(Li) detector and an acceleration voltage of 5 kV.

2.7. AAS and ICP-OES analysis
Prior to analysis for AAS and ICP the AgNPs dispersions were digested in 10.00 mL of concentrated nitric acid at 95°C for 30 minutes with magnetic stirring. The samples were cooled, transferred to a 30.00 mL falcon and diluted to 30 mL of distilled water. The procedure was made in triplicate.

A flame atomic absorption spectrophotometer Analytic Jena model nova 350 was used to perform AAS analyses. Silver nanoparticles concentration in the dispersions was measured before and after the deposition in samples with and without DBD plasma treatment by exhaustion at 70°C.

ICP-OES was performed for study of AgNPs release with time. Seven samples, 5x5 cm, without and with 5 kW min m^{-2} of DBD plasma dosage with AgNPs deposited by exhaustion at 70°C were immerged separately in a flask with 20.00 mL of distilled water. After 1, 3, 5, 7, 10, 15 and 30 days, the respective sample was removed and AgNPs in the liquid were digested. A single-element standard solution was diluted to prepare the calibration curve of silver with 1, 0.750, 0.500, 0.25, 0.100, 0.075 and 0.050 mg L^{-1}. The analyses were performed in a ICP-OES spectrometer PERKIN ELMER OPTIMA 8000 equipped with an auto-sampler S10. WinLab software was used to process and for data acquisition.

3. Results

AgNPs agglomeration and its uncontrollable release represent an important limitation in their application. In this work, it was intended evaluate different methods for AgNPs deposition in order to obtain a higher and strong attached concentration of NPs onto the fabric surface and a controlled release of the NPs. The impregnation of PA66 with AgNPs was performed by i) ultrasound tip assisted deposition, ii) dip coating iii) spray and iv) exhaustion at 70°C according to figure 1 and 2.

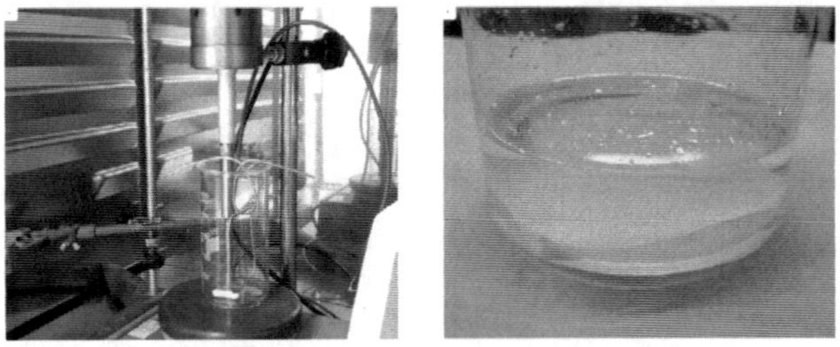

ULT **DIP**

Figure 1. Methods for AgNPs deposition. ULT: ultrasound; DIP: dip coating.

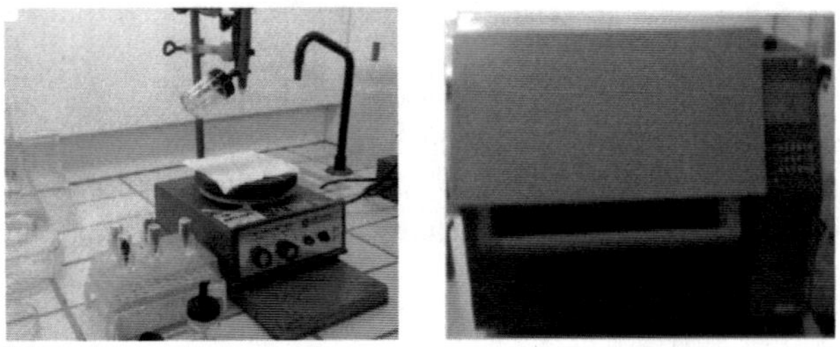

SPR **EXH**

Figure 2. Methods for AgNPs deposition. SPR: Spray; EXH: Exhaustion at 70°C.

The effect of ultrasonic tip was tested with 50%, 60%, 70% and 80% power intensity in an ultrasound bath with different geometries of the tip/sample positions as schematized in the figure 3.

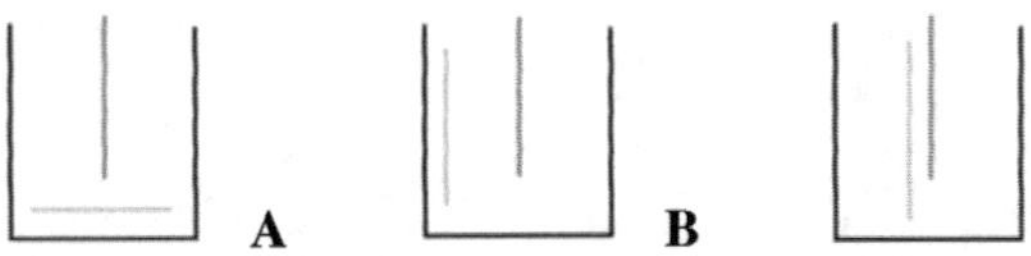

A **B** **C**

Figure 3. Main geometries of ultrasound tip tested.

All tested configurations did not show uniform AgNps distribution on the fabric surface. However, significant differences between plasma untreated and plasma treated samples were observed. Plasma sample showed higher local AgNPs dispersion but only in some specific spots of the PA66 fabric (Fig. 4). The ultrasound tip method pointed out the importance of the mass transfer in the AgNPs adhesion on the plasma treated fabric surfaces. The best sample obtained at 70% of ultrasound intensity and geometry B was used to the characterization measurements.

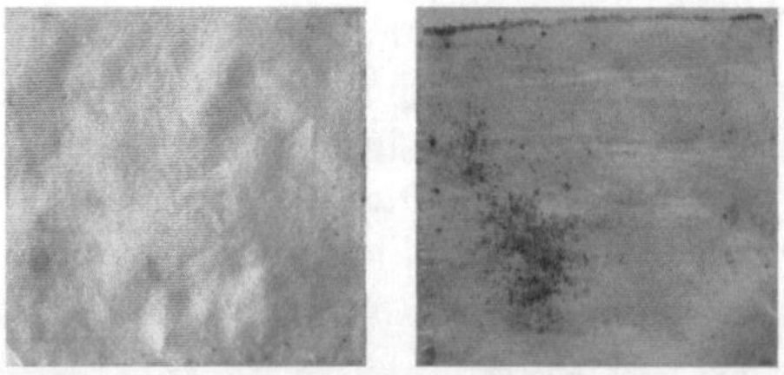

Figure 4. Tip ultrasound-assisted AgNPs deposition at 70% of ultrasound intensity using geometry B. Left sample: without plasma treatment. Right sample: with plasma treatment.

All the other tested methods showed uniform distribution of AgNPs on the fabrics. Visually the fabrics are less colored than the ultrasound spots. However, the exhaustion method displayed more intense coloration suggesting higher AgNPs deposition.

Reflectance measurements were made to estimate AgNPs concentration onto the fabrics. The decrease in specular reflection could be considered a direct measure of the AgNPs abundance in the topmost layer of the material since the NPs act as radiation absorbers on the entire visible spectra [18]. It is expected that a higher concentration of AgNPs on the fabrics would decrease reflectance values at 420 nm (figure 5). In the fabrics impregnated with AgNPs by dip coating or spray methods there are no significant differences in the plasma treated and untreated samples. In these cases due to the lower temperature and mechanical action of the system, the mass transfer is lower than ultrasound and exhaustion methods limiting the firmly AgNPs loading on the surface. Ultrasound tip and exhaustion at 70°C presented higher AgNPs loading due to the both mechanical action and higher temperature. Different exposition time was also tested (data not shown) and no significant difference was observed in the dip coating method. The best reflectance value was obtained with the exhaustion method at 70°C and 100 rpm for 60 minutes of treatment.

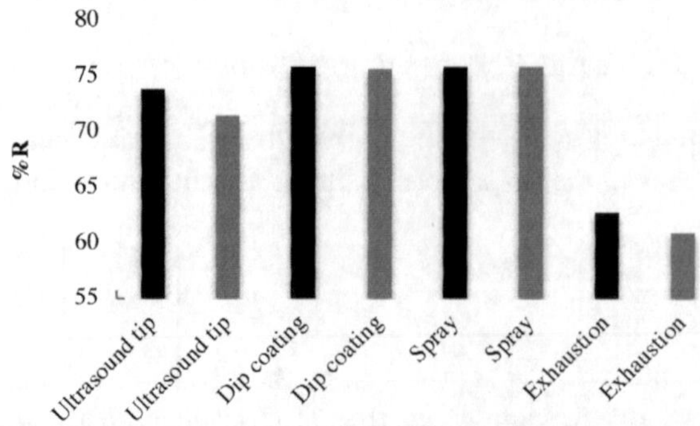

Figure 5. Reflectance measurements at 420 nm of PA66 samples impregnated with AgNPs using different deposition methods: Black bars: samples without plasma treatment. Gray bars: plasma treated samples.

The SEM images of PA66 untreated and DBD plasma treated samples shows AgNPs in all the tested methods (figure 6). However, better distribution, higher concentration and less agglomeration were observed in the plasma treated samples. The plasma treated fabrics with AgNPs deposited with the ultrasound method showed the highest concentration of nanoparticles onto the fabrics as well as the lowest NPs agglomeration due to the cavitation effect that is able to maintain the NPs well separated. However, this deposition is limited to certain areas of the fabric limiting its application as medical dispositive. The SEM images of the dip coating and spray methods showed very low amount of AgNPs for both plasma-treated and untreated fabrics. Moreover, the few AgNPs observed on the surface are present in the form of big agglomerated clusters. These methods are the most simple but also the worse in term of NPs application. The exhaustion method showed the best equilibrium between the NPs distribution and state of agglomeration. The SEM images showed AgNPs good distribution in the plasma treated sample. Some agglomeration is present but it is not so dramatic as in the DIP and spray methods. As previously observed at this relatively low concentration (~10 mg L^{-1}) the AgNPs tend to aggregate in clusters on the surface of the DBD-treated PA66 fibers due to the thermo-migration of the nanoparticles during the application of the exhaustion method and to the chemical changes induced by DBD plasma treatment [19, 20].

Despite the deep probe depth of EDX technique, the peak of silver was detected on the surface of the AgNPs-loaded fabrics (data not shown). Moreover, clear differences between the samples untreated and plasma treated as well as among the different methods can be noted in the silver peak intensity. The exhaustion method confirms as the best solution.

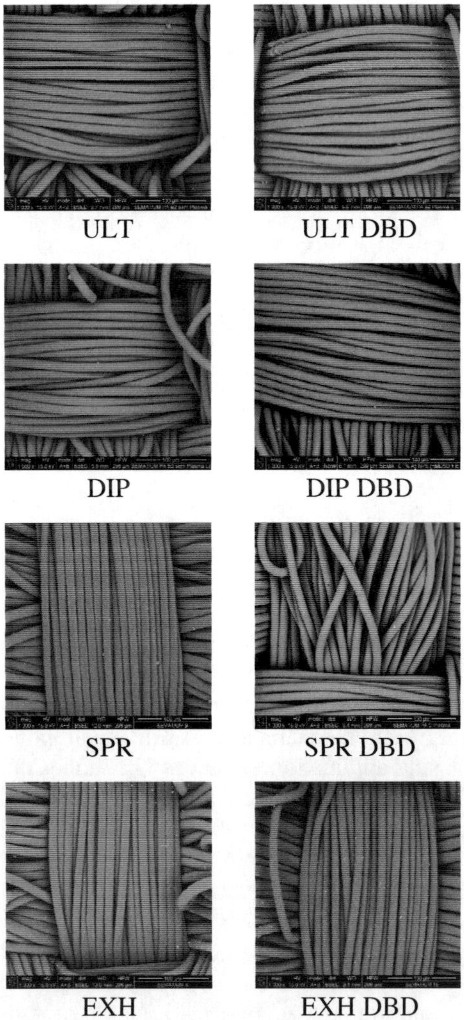

Figure 6. SEM images of the different deposition methods without and with (DBD) plasma treatment. ULT: Ultrasound; DIP: dip coating; SPR: Spray; EXH: Exhaustion at 70ºC.

AgNPs concentrations in the liquid dispersions were measured before and after the exhaustion deposition method by AAS (table 1). The initial concentration was 8.0 mg L^{-1}. After the deposition, the

IOP Conf. Series: Materials Science and Engineering **460** (2018) 012007 doi:10.1088/1757-899X/460/1/012007

AgNPs concentration in the dispersion was significantly lower in the plasma treated sample. These results confirm the higher AgNPs adsorption due to the DBD plasma treatment effect onto the fabric surface.

Table 1. AAE results

Dispersion	[AgNPs] (mg L^{-1})
Initial concentration	8.00
After deposition no DBD	3.50
After deposition with DBD	0.87

In order to estimate the kinetic of the release of AgNPs from the fabrics, the amount of silver released from AgNPs-loaded fabrics by exhaustion method at 70ºC was measured using the ICP-OES technique. AgNPs-loaded fabrics were immerged in water and liquid samples were collected after 1, 3, 5, 7, 10, 15, 30 days. From the collected data it is clear that the plasma treated samples are releasing more silver in the solutions (table 2). Moreover, it seems that plasma treated samples are able to better control the release of silver ions over time showing a higher stability in the released concentrations.

Table 2. ICP-OES results

Collect time (days)	[AgNPs] (mg L^{-1}) Samples no DBD	[AgNPs] (mg L^{-1}) Samples with DBD
1	0.10	0.12
3	0.00	0.20
5	0.08	0.20
7	0.05	0.20
10	0.06	0.15
15	0.05	0.08
30	0.14	0.20

4. Conclusion

The results obtained in this study showed that the mass transfer controls the AgNPs deposition onto PA66 fabric surface. Temperature and mechanical agitation are preponderant factors that influenced the loading of NPs. The DBD plasma application enhances NPs adhesion and adsorption for all the used methods demonstrating to be a promising waterless and chemically free surface modification technique. The ultrasound method displays an irregular distribution despite its local good deposition. The dip coating and spray methods did not reach the minimum amount of AgNPs on the fabric surface and showed high AgNPs agglomeration. The exhaustion method showed the best results for both NPs distribution and reduced agglomeration. The AAS and ICP method prove the good plasma-promoted adsorption of AgNPs and their controlled release during time. The next experiments will be conducted to test different temperature, plasma dosage and time of application in the exhaustion method including the use of polymeric barrier layers to tune the controlled release of AgNps.

5. Acknowledgment

Andrea Zille acknowledge FCT through the Investigator FCT Research contract (IF/00071/2015) and the project POCI-01-0145-FEDER-007136 and UID/CTM/00264 under the COMPETE and FCT/MCTES (PIDDAC) co-financed by European funds (FEDER) through the PT2020 program.

References

[1] Adlhart, C., Verran, J. et al., Journal of Hospital Infection 2018
[2] Lam, S. J., Wong, E. H. et al., Progress in Polymer Science 2017
[3] Zheng, K., Setyawati, M. I. et al., Coordination Chemistry Reviews 2018, 357, 1-17
[4] Jamal, M., Ahmad, W. et al., Journal of the Chinese Medical Association 2018, 81(1), 7-11
[5] Mitra, C., Gummadidala, P. et al., Environmental Science & Technology 2017, 51 (14), 8085–8093
[6] Surmeneva, M., Sharonova, A. et al., Colloids and Surfaces B: Biointerfaces 2017, 156, 104-113
[7] Wu, Y., Yang, Y. et al., Advanced Powder Technology 2018, 29(2), 407-415
[8] Losub, C. Ş., Olăreţ, E. et al, Nanostructures for Novel Therapy 2017, pp. 793-809
[9] Toh, H., Faure, R. et al, Nanotecn. Science and Applications 2017, 10, 147-162
[10] Yang, C., Yao, Y. et al., Advanced Materials 2017, 29(38)
[11] Liu, K. G., Abbasi, A. et al., Ultrasonics sonochemistry 2017, 34, 13-18
[12] Surmeneva, M., Sharonova, A., et. al, Colloids and Surfaces B: Biointerfaces 2017, 156, 104-113
[13] Zanini, S., Citterio, A. et al., Applied Surface Science 2018, 427, 90-96
[14] Petlin, D. G., Tverdokhlebov, S. I. et al, Journal of Controlled Release 2017
[15] Zanini, S., Citterio, A. et al., Applied Surface Science 2018, 427, 90-96
[16] Gorjanc, M., Gorenšek, M. et al., Eco-Friendly Textile Dyeing and Finishing 2013
[17] Oliveira, F. R., Souto, A. P. et al., Materials Science Forum 2010, 636-637, 846-852
[18] Zille, A., Fernandes, M. M. et al., ACS Applied Materials & Interfaces 2015, 7, 13731
[19] Radetić, M., Ilić, V. et al., Polymers for advanced technologies 2008, 19(12), 1816-1821
[20] Montazer, M., Shamei, A. et al., Progress in Organic Coatings 2012, 74(1), 270-276

Contribution for analyzing the intra-ply yarn sliding mechanism in preforming of the woven fabric

A. R. Labanieh[1], C. Garnier[2], P. Ouagne[2], O. Dalverny[2], D. Soulat[1]

[1]*Génie et Matériaux Textiles (GEMTEX), ENSAIT, 2 allée Louise et Victor Champier, 59056 Roubaix, France.*

[2] *Génie de Production (LGP), ENIT, 47 avenue d'Azereix, 65016 Tarbes, France.*

Corresponding author: ahmad.labanieh@ensait.fr

Abstract Manufacturing complex composite part via Liquid Composite Molding (LCM) processes involves preforming dry textile perform before injecting the liquid resin. The defect that may be encountered within the textile structure during preforming step decreases the expected mechanical properties of the composite part. The intra-ply yarn sliding is a defect frequently observed during preforming of a woven preform but its mechanism is far from being fully understood. In this paper, a contribution for analyzing the mechanism of this defect and the influencing process parameters is presented. This analysis is based on an experimental study performed for one ply carbon fabric using a hemispheric punch. The effect of the ply orientation and blank holder force on the fabric behavior during preforming is evaluated regarding the required preforming force and yarn slippage. For the performed tests, it was obvious the impact of these two factors on the fabric behavior. The slippage phenomenon occurred in zones with low shear angles, for specific ply orientation. The fabric behavior has been analyzed by considering the evolution of the yarn tension which is related to the contact shear stress induced by sliding the fabric across the die and blank holder during preforming. Using an analytical model for the yarn tension in friction sliding, the influencing ply geometry and process parameters have been identified. Based on this analysis a solution relative the ply geometry to reduce the yarn tension is examined.

***Keywords*—Woven Fabric, Composite manufacturing, Preforming, Defect**

I. Introduction

Liquid Composite Moulding (LCM) processes [1] offer a good compromise in terms of repeatability, production rates, low-energy consumption and low final cost [2] for manufacturing a complex composite parts. Preforming is the first step in this process. It is a difficult phase, and the physical mechanisms are complex because they depend on many parameters (shape of tools, characteristics of the preform, number and orientation of plies, loads applied, etc.). If the mechanical loadings (tension, shear, compression, bending, friction, etc.) to which the reinforcements are subjected during the preforming step have been well described in the literature [3]-[6], the generation and the control of defects are far from being entirely understood. Wrinkle is a frequent defect occurs at the macroscopic scale (preform scale). The origin of the wrinkle is related to the weak preform bending stiffness due to the possible inter-yarn slippage [8]. Several numerical [7] - [9] and experimental [10] - [12] studies were conducted to analyze the effect of the friction-based blank holder system on the wrinkles occurrence. This system induces high yarn tension on the preform that prevents wrinkles [13]. However, Nezami et al [14] underlined that this system or other technique may reduce wrinkles but induce other defects in the fabric such as parallel fiber distortions without gaps, fiber distortions with small/large gaps, filament damage, broken or pulledout roving. Except the filament damage, these defects are related to the lack of cohesion in the woven structure network and the possibility of intra-ply yarns slippage. Barbagello et al [15] noted the presence of measurable yarn slippages (up to a maximum that is around 10% of the total length of the yarns). In papers dedicated to the preforming of NCF fabrics [16], [17] intra-ply slippage between fibers has been observed, without solutions to prevent or reduce this defect qualified as irreversible in [17]. Except in these studies, few articles of the literature deal with this type of defect. In this paper, experimental preforming tests were conducted on a carbon twill weave fabric. The slippage phenomenon occurred in zones with low shear angles, for specific ply orientation. Yarn tensions were analytically computed relatively to tools load to understand this slippage defect. The influence of the ply orientation and the ply dimension were experimentally studied to reduce tension in yarns and reduce this slippage. Finally, solution based on the geometry of tools was proposed to prevent this type of defect.

II. Experimental condition

A. *Forming machine*

Experimental preforming machine developed at the GEMTEX laboratory [18], [19], which is illustrated in Fig 1, is used to conduct the preforming tests. The machine is composed of three basic parts: a punch and two square-shaped plates. The punch has hemisphere geometry (double curved surface) and its movement is controlled by a pneumatic actuator with a constant mounting speed (45 mm/sec). Both upper and lower plates are a square shaped plate with a circular hole in its center. The lower plate is fixed into position whereas the upper one is mobile, parallel to the punch axis. This allows the arrangement of the fabric ply between the two plates in the desired orientation. The upper plate is subjected to a normal pressure applied by four pneumatic actuators placed on the plate corners. This normal pressure is controlled by the pressure of the compressed air supplied to the actuators. The experiments were conducted with two operating pressures 0.025 and 0.075 MPa that correspond to normal forces by each cylinder equals to 77.9 and 233.7 N respectively, according to the actuator specifications. These operating pressures were chosen after a first preliminary set of test and are related to the appearance of the intra-ply yarn sliding defect. When the minimum operating pressure (0.025 MPa) is applied, no sliding defect occurs. From an applied operating pressure of 0.075MPa this defect takes place. The upper plate is considered as the blank-holder because it is the mobile plate and the normal force (P) is applied on its top surface, Fig 1. The force (P) is called the holder force all over this paper. The lower plate is considered as the die since it is fixed into position in this machine configuration.

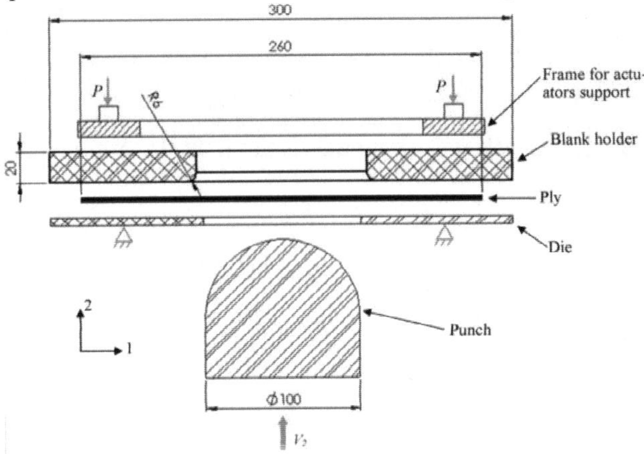

Fig 1: Forming machine scheme with characteristic dimensions. All dimensions are in mm.

For each experiment, the preforming force exerted by the punch is recorded as a function of the punch displacement. The force measurement is performed by means of a load cell mounted between the punch and the driving actuator. Also, the ply draw-in, border position and yarn arrangement after preforming are captured by a CCD camera. The camera is placed on the top side and its optical axis is aligned with the punch movement path.

B. *Material properties and ply geometry*

Commercial fabric "Hexcel HexForce 48600 U 1250" was used in this study to perform the experimental work. It is classified as a 2D woven fabric made with a twill 2/2 weaving architecture and constructed of 3.7 warps/cm and 3.7 wefts/cm with a nominal area density of 600 g/m² and a nominal thickness of 0.62 mm. The constituent yarns, both warp and weft, are "AS4 C GP 12K" high strength carbon fibres and they have a linear density of 0.8 g/m. Both yarns are not twisted and not powdered.

The forming tests were conducted using one ply of the woven fabric cut in a square shape with 260 mm side length. The tests were performed using two initial orientations (0° and 45°) of the fabric on the die as presented in Fig 2. The dashed lines sketched on the ply in Fig 2 represent the warp and weft yarns passing through the punch main axis and aligned with the radius of the hole. These yarns are called radial yarn in the next sections. On the 0° orientation ply, the warp and weft yarns are parallel to the plate side edges, Fig 2-a. For the 45° orientation, the warp and weft yarns are parallel to the plate diagonal and they create an angle of 45° with plate side edge, Fig 2-b. The plies with 0° and 45° orientations are denoted across this paper as ply-0 and ply-45 respectively. One can note that the length of the portion of the radial yarns which is in contact with the two upper and lower plates is different for both orientations. In ply-0, a half of this length is marked as L0 = 68.6 mm whereas in ply-45 it is marked as L45 = 122.5 mm as illustrated in Fig 2.

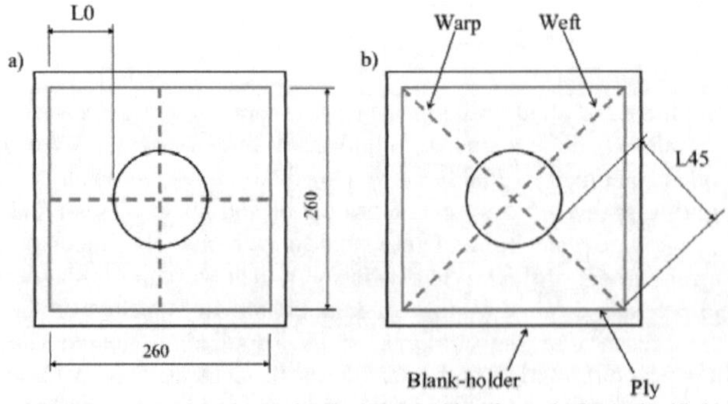

Fig 2: Dimensions and placement of the plies on the machine under the blank holder plate, ply-0 (a) ply-45 (b). Dashed lines represent the radial warp and weft yarns. All dimensions are in mm

III. RESULT

A. Preforming force

Fig 3 shows the preforming force as a function of the punch displacement for both ply-0 and ply-45 when preforming with the two operating pressures for the holder actuators. The experiments were repeated three times at the pressure (0.025 and 0.075 MPa) for both ply orientations. The mean value of the recorded force is presented on the graph with error bars pointing out the standard deviation of the measurements. It is clear from Fig 3 that the augmentation of the blank holder force causes the increase of the preforming force for both ply orientations. This is the results of a higher friction force standing against the ply sliding across the die and blank holder surfaces. The preforming of a ply with an initial 45° orientation requires higher effort in comparison to ply-0 in spite of similar blank holder pressure and identical initial ply dimension and shape as shown in Fig 2. The relative differences in the preforming force are about 56% and 63% for ply-0 at maximum displacement of the punch for the operating pressure of 0.025 MPa and 0.075 MPa respectively. The reason of this difference in the preforming force depending on the ply orientation for the performed test is investigated and discussed in the next sections.

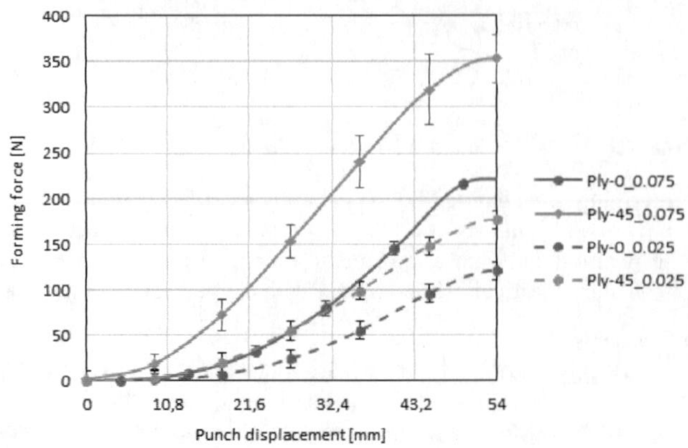

Fig 3: Preforming force for ply-0 and ply-45.

B. Intra-ply yarn sliding defect

Intra-ply yarn sliding defect is a loss of cohesion between the two orthogonal yarn sets constructing the woven ply [19]. It occurs when the applied efforts on a transversal yarn, such as axial tension or tangential friction force on the yarn exposed surface, overcomes the inter-yarn friction force, thus the transversal yarn is locked into position and the

longitudinal yarns slide across it. Length and width of the created gap between transversal yarns in addition to number of the involved yarns determine the amplitude of the defect.
For the preformed plies, the yarn sliding was observed when preforming ply-45 with 0.075 MPa operating pressure on the three tested samples. However, it was not remarked with lower pressure (0.025 MPa) nor for ply-0 with both applied operating pressures. The intra-ply yarn sliding occurred in the non-sheared zone of the ply on the rounded corner of the upper plate hole, Fig 4-a.
In order to characterize this defect, the yarns position after preforming and the local relative inter-yarn slippage have to be spotted. However, the carbon fibers are characterized by a light reflecting surface. Furthermore, the normal orientation and altitude of the useful part (3D deformed) of the ply surface change during the preforming process. Therefore white dots have been speckled on the apparent yarn surface with an alternative pattern (one with a dot and one with no dot), Fig 4-b. The dots have been speckled over a diagonal portion of the ply, as shown in Fig 4-a, where the defect is expected to take place. To distinguish between longitudinal and transversal yarns over this diagonal part, round (red) and cross (blue) marks were superimposed to the white dots on the treated photos.

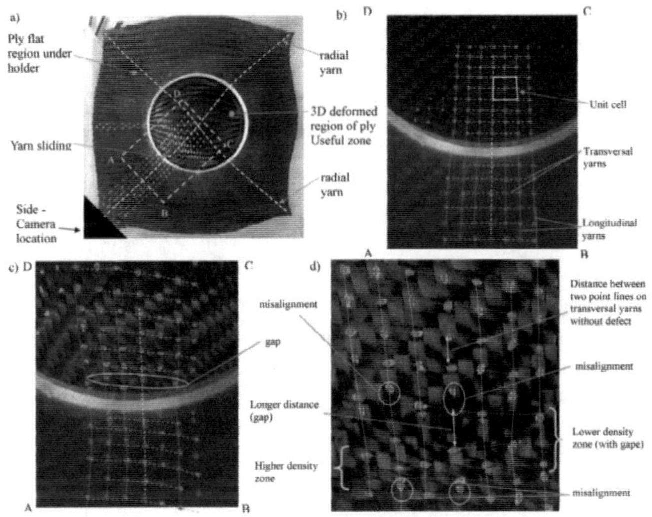

Fig 4: intra-ply yarn sliding on ply-45 after preforming with 0.075MPa operating pressure. Top view (a), zoomed portion (ABCD) before preforming (b), after preforming (c) and zoom on the defect zone after removal of the blank-holder plate (d).

On the zoomed rectangle zone ABCD defined in Fig 4-a, the yarn slippage defect is expected to take place. The yarns arrangements were observed before and after preforming. Before preforming, that corresponds to the initial planar state of the ply (Fig 4-b), the yarn count per unit length is uniform (same distance between dots). Furthermore, the dots belonging to the transversal yarn are aligned with the corresponding longitudinal yarns. The path of the longitudinal yarns was determined by spline connecting the dots. Similarly, the dots belonging to the longitudinal yarns are also aligned with corresponding transversal yarns.
After preforming, corresponding to the final state of the ply Fig 4-c and d, intra-ply yarn sliding defect occurs on the non-sheared region along the radial yarns on the holder corner when the ply exits from the die-holder zone. Out of the defect region (Fig 4-d), there is no disarrangement or misalignment between the two yarn sets. The longitudinal yarn dots remain aligned with transverse yarn indicating that no yarn sliding takes place. In the defect region, two zones are distinguished: high-density and low-density zone of the transversal yarns. On the first one (towards the 3D deformed zone), a gap is created between the last transverse yarn brought by the radial yarn and the next one. About 7 successive transverse yarns along the same radial yarns contribute to this defect and it is expanded across 11 longitudinal yarns. The first transversal yarn from the useful zone side shows the maximum bending. It is therefore associated to a large gap. Below the blank-holder, the transversal yarns accumulate creating a high density zone. This accumulation of the yarns increases the inter-yarn contact pressure between the two interlaced yarn sets on the crossing areas leading to stop the yarn slippage. Out of the defect region, a moderate misalignment is also observed but with a low amplitude. In the defect zone, no out of plane displacement of yarn is observed. This is even the case of yarns that bend in the ply plane. In those conditions one may have expected to observe the occurrence of tow buckles [21], but the load of the involved yarns as well as their in-plane bending angle magnitude did not favor the conditions of their occurrence. Otherwise, over the defect zones no transversal disarrangement of the longitudinal

yarn is observed. There is no gap between the longitudinal yarns and the dots placed on the transverse yarns remain aligned along the longitudinal yarns axis.

IV. Analysis and Discussion

Two strips may be distinguished on the deformed ply in relation to the dominated deformation mechanisms: tension strip and shear strip. The tension strip involves one radial yarn, pointed out on Fig 4-a, which intersects with the punch translating axis and it is aligned with the hole radius of the blank-holder plate. The radial yarn on this strip is driven by the punch that pulls it from the die-holder contact zone. During this slid an axial tension on the yarn due to the friction resistance on the contact surfaces (with the other yarns and with the die-holder system) takes place. The shear angle between the radial yarn and the transversal yarns was measured on the planar part of the strip. The measured angles between the radial and the transverse yarns are globally equal to zero. This indicates that no in-plane shear deformation takes place in this zone. Therefore, this strip is mainly submitted to tension deformation. The shear strip is characterised by a trellis of crossing yarns. In order to fit the punch 3D surface shape, the crossing yarns rotate relative to each other in the inter-yarn crossing area. The shear angles were measured on the planar part of this strip. The values vary along the strip and a maximum value is observed at the holder corner location near to punch. Therefore, the in-plane shear is considered as the dominating deformation mechanism that takes place in this strip and tension mechanisms can be neglected. The angle between the two strips is 45° on both ply-0 and ply-45. The ply behaviour between these two strips is seen as an interaction between these two deformation types.

The tension force on the tension strip is evaluated as a function of the punch displacement by implementing an analytical study of the fabric behaviour during preforming with a simple curved punch surface (hemi-cylindrical).

Based on this analysis, the friction force limit (once the strip slippage initiates) to pull one yarn of the tension strip across the machine plates is defined in the Equation (1), by considering the friction on the flat contact surface of the die and holder in addition to the friction on the curved contact surface of the die with the tension strip.

$$t^y = 2\,\mu\,(P - T\sin\alpha)\frac{A^p}{S}\,e^{\mu\alpha} \quad (1)$$

The yarn tension (t^y) depends on the friction coefficient between the tools and the fabric (μ), the contact angle (α) at the holder corner, the strip tension (T) and the holder force (P). The direct proportional relationship between the holder force and yarn tension in the equation coincides with experimental results, Fig 3. The yarn tension, Equation (1), depends also on the ratio (A^p/S) of the contact area of the sliding yarn on the flat part of the blank-holder (on the die) relative to the area of contact of the whole preform with the die.

Sliding the ply across the contact surfaces during the process with the punch motion is associated with normal contact and tangential contact shear stress generated on the contact surfaces and related directly to the yarn tension. These contact stresses increase gradually along the strip length over the flat area (A^p) from the ply border and it get its maximum value at the fabric entrance toward the punch on the curved die surface (A^r).

Back to Equation (1) the flat contact area (A^p) decreases whereas the contact angle (α) increases as the punch moves up. Consequently the contact area of the fabric with the curved die surface increase resulting in exponential augmentation of yarn tension, according to Capstan equation [22]. Thus highest normal contact and tangential contact shear stress are expected on this curved part. In case of higher contact shear stress than the inter yarn cohesion stress between the crossed yarns within the fabric, the generated contact shear stress causes sticking the transversal yarns of the tension strips on contact surface while the longitudinal radial yarns, which are driven by the punch, slid over them and create a gap between the successive transversal yarns. That conforms to the observation of the defect occurrence for ply-45°on the curved surface of the die along the radial yarns. However, on the shear strip, the yarns shear to fit the punch shape and the amount of the yarn movement and generated tension is too less than the yarn sliding on the tension strip as it is a structural deformation and the fabric has poor shearing resistance. Therefor less contact shear stress and no intra-ply yarn sliding is expected on the shear strip.

Otherwise, higher required yarn tension to pull the fabric out of the contact surfaces causes a higher preforming force, which is the sum of all required force to do the required yarn movement (sliding) and fabric deformation (shear and yarn bending). Therefore higher yarn tension generated by higher holder force (P) leads to higher forming force. Also the effect of the ply orientation (ply-0 and ply-45) on the preforming force which is illustrated in Fig 3can be explained through Equation (1). The length of radial yarn for the ply-45 (L45 = 121.42 mm) is longer than the ply-0 (L0=68.8 mm), as illustrated in Fig 2. That results in a higher (A^p/S) ratio for the ply-45 and consequently a higher yarn tension and preforming force on the tension strip.

V. SOLUTION

Based on the analytical analysis, a solution related to the ply geometry (radial yarn length) is proposed to inhibit the intra-ply yarn sliding. The yarn tension is directly proportional to the ratio (A^p/S) as shown in Equation (1). In order to explore into more depth the effect of the ply geometry, two different configurations for the ply-45 orientation were considered relative to the length of the radial yarn on the contact surface with the machine plates. The ply corners are cut into two different geometries to obtain two lengths for the radial yarn (L1 and L2) as illustrated in Fig 5. The length L2 is equal to the length of the radial yarn on the contact surface in the ply-0 configuration, ($L_2 = L_0 = 68.6$ mm). However, the fabric contact area (S) for ply-45_L2 is smaller than that for ply-0. The length L1 is defined by the following equation: $L_1 = \frac{1}{2}(L_{45} + L_0) = 95.5\ mm$.

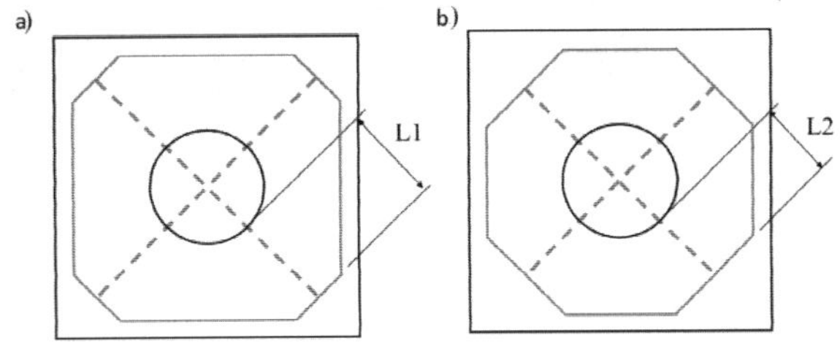

Fig 5: Modified ply-45-L1 and ply-45-L2

Preforming tests with a hemispheric shape punch and an operating pressure of 0.075 MPa for the holder actuators were conducted for both ply-45_L1 and ply-45_L2. The preforming punch force was recorded in both cases and plot as a function of its displacement (Fig 6). It clearly appears that a shorter radial yarn length is associated to a lower preforming force. This observation is completely consistent with Eq. (1). That reveals also the high impact of the yarn tension caused by the friction stress on the contact surfaces (die and blank-holder) and the preforming force. A difference in the preforming force is noted between the ply-0 and ply-45_L2, Fig 6, even if the radial yarn lengths are identical. This difference may be attributed to a higher fabric contact area (S) for ply-45_L2 resulting in a higher normal contact and friction contact force.

During the experimental tests, the intra-ply yarn sliding defect was only observed on ply-45_L1, Fig 7-a, but with lower amplitude than the one observed for ply-45 (Fig 4-d). Only a small gap appears between the transverse yarns on the tension strip in the useful zone. Furthermore, this defect occurs only along the radial yarns near the holder corner, as noticed earlier for ply-45. On the modified ply-45_L2 a slight misalignment can be observed on the non-sheared zone, Fig 7-b.

As a conclusion of this section, increasing the length of the radial yarns (or the longitudinal yarn of the non-sheared strip between the contact surfaces with the machine plates) has for consequence to raise the yarn tension resulting in higher preforming force. It also causes the rise of the normal contact force and the frictional force on that radial yarn at the holder corner that enhances the friction force between the transverse yarn and the machine tool at this zone. Consequently, the longitudinal yarns cannot bring the transversal yarns with them and the transversal yarns are locked into position on the holder corner causing the yarns sliding defect. On the contrary, reducing the length of the stretched yarn by cutting the samples corners may be a solution to prevent the occurrence of the yarn sliding defect.

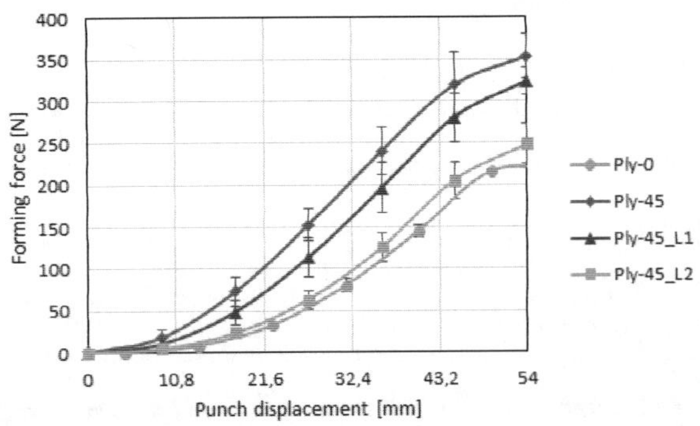

Fig 6: Preforming force of ply-0 and ply-45 with variation of the radial yarn length, ply-45-L1 and ply-45-L2.

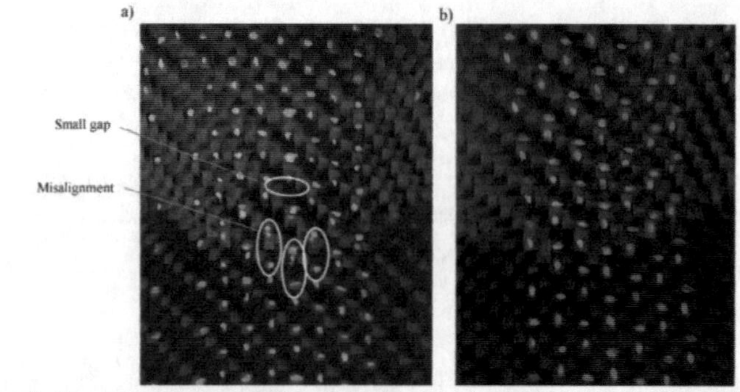

Fig 7: Zoomed photo for the ply/holder corner contact zone after preforming operation on ply-45-L1 (a) and ply-45-L2 (b).

VI. CONCLUSION

The behavior of the "HexForce 48600 U 1250" carbon woven fabric during the friction-based blank-holders preforming process with a hemispheric-shaped punch was conducted with a particular attention given to the observation of the occurrence of the intra-ply yarn sliding defect. The experiments were performed for two orientations of a mono-ply: ply-0 and ply-45. A higher preforming force is required to form the ply-45. Furthermore, this higher force is associated to the occurrence of the intra-ply yarn sliding defect on the holder corner along the radial yarns in the non-sheared zone.

Two strips from the ply were distinguished relative to the main deformation mechanism: tension strip and shear strip. The yarn tension on the tension strip was related to the friction on both ply/die and ply/holder contact surfaces based on analytical analysis developed previously. The analytical model shows that the longitudinal yarn tension, once the fabric slippage takes place across the die and the blank holder, depends only on the friction properties of the ply/tool contact surfaces and the applied holder force.

This work also showed that the yarn tension is related to its length between the machine plates contact surfaces. The effect of this length was investigated by testing two modified shapes for the ply-45. It was clear that a longer length of the radial yarn on the contact surface involves higher tension within the yarns and higher global preforming forces. This higher tension is associated to a higher normal contact and a higher tangential friction contact force on the ply/holder corner contact surface. When the friction between the transversal yarns and the holder corner surface becomes higher than the inter-yarns friction at the yarn crossing area, the radial yarns that are stretched and pulled by the punch cannot bring the transversal yarn with them. That has for consequence to initiate intra-ply yarn sliding that may be considered as a preforming defect as locally, the density of yarn may be severely reduced and therefore is a zone rich in resin and a potential zone of weakness for the composite part. Consequently, the yarn slippage is associated to the radial yarn tension induced by friction on the machine tool. Thus it is important to localize the zone with the high tangential friction on the ply during preforming and this work proposed it for the studied hemispherical shape.

The analysis of the evolution of process parameters and the fabric properties (weaving pattern, surface properties of the yarns) should be realized in future works.

REFERENCES

[1] Boisse P. Advances in composites manufacturing and Process Design. In Boisse P, editor. Woodhead Publishing Series in Composites Science and Engineering: Number 56. 2015.
[2] Allaoui S, Cellard C, Hivet G. Effect of inter-ply sliding on the quality of multilayer interlock dry fabric preforms. Composites: Part A 2015; 68 :336–345
[3] Boisse P. Hamila N, Guzman-Maldonad E, Madeo A. Hivet G. dell'Isola F. The bias-extension test for the analysis of in-plane shear properties of textile composite reinforcements and prepregs: a review. International Journal of Material Forming 2016 doi:10.1007/s12289-016-1294-7.
[4] Tephany C. Soulat D. Gillibert J. Ouagne P. Influence of the non-linearity of fabric tensile behavior for preforming modeling of a woven flax fabric. Textile Research Journal. 2016;86 (6):604-17.
[5] Harrison P. Modelling the forming mechanics of engineering fabrics using a mutually constrained pantographic beam and membrane mesh. Composites: Part A 2016;81:145–157
[6] Gereke T. Dobrich O. Hubner M. Cherif C. Experimental and computational composite textile reinforcement forming: A review. Composites: Part A 2013;46 :1–10.

[7] Hamila N. Boisse P. Sabourin F. Brunet M. A semi-discrete shell finite element for textile composite reinforcement forming simulation. International Journal for Numerical Methods in Engineering. 2009;79(12):1443–1466
[8] Charmetant A.Orliac J.G. Vidal-Sallé E.Boisse P.Hyperelastic model for large deformation analyses of 3D interlock composite preforms. Composites Science and Technology 2012;72(12):1352–1360
[9] BoisseP. Finite element analysis of composite forming.In Long A.C. editor. Composite Forming Technologies. Woodhead Publishing. Cambridge. 2007: 46–79
[10] Sjölander J. Hallander P. Åkermo M. Forming induced wrinkling of composite laminates: A numerical study on wrinkling mechanisms. Composites: Part A. 2016; 81: 41–51
[11] AllaouiS.BoisseP. ChatelS.Hamila N. HivetG.SoulatD. Vidal-Salle.Experimental and numerical analyses of textile reinforcement forming of a tetrahedral shape. Composites: Part A.2011; 42(6): 612–622
[12] Allaoui S. Hivet G. Soulat D. Wendling A. Ouagne P. Chatel S. Experimental preforming of highly double curved shapes with a case corner using an interlock reinforcement. International Journal of Material Forming. 2014; 7(2): 155–165
[13] Nezami F Nosrat. Gereke T. Cherif C. Analyses of interaction mechanisms during forming of multilayer carbon woven fabrics for composite applications. Composites: Part A 2016;84:406–16.
[14] Nezami F Nosrat.Gereke T.Cherif C. Active forming manipulation of composite reinforcements for the suppression of forming defects. Composites Part A 2017;99:94–101.
[15] Barbagallo G., Madeo A., Giorgio I., Morestin, F., Boisse P. Bias extension test on an unbalanced woven composite reinforcement: Experiments and modeling via a second-gradient continuum approach. Journal of Composite Materials, 51 (2), (2017), 153-170
[16] Margossian A. Bel S. Balvers J.M. Leutz D. Freitas R. Hinterhoelzl R. Finite element forming simulation of locally stitched non-crimp fabrics. Composites: Part A. 2014;61:152-162
[17] Schirmaier F. Weidenmann KA. Kaerger L. Henning F. Characterization of the draping behaviour of sewed unidirectional non-crimp fabrics (UD-NCF). Composites: Part A. 2016;80:28–38
[18] Wang P. Legrand X. Boisse P. Hamila N. Soulat D. Experimental and numerical analyses of manufacturing process of a composite square box part: Comparison between textile reinforcement forming and surface 3D weaving. Composites Part B 2015; 78: 26-34.
[19] Gatouillat S. Bareggi A. Vidal-Sallé E. Boisse P. Meso modelling for composite preform shaping simulation of the loss of cohesion of the woven fibre network. Composites: PartA 2013;54:135–44
[20] Jacquot P.B. Wang P. Soulat D. Legrand X. Analysis of the preforming behaviour of the braided and woven flax/polyamide fabrics. Journal of Industrial Textiles. September 2016;46(3):698-718.
[21] Ouagne P, Soulat D, Hivet G, Allaoui S, Duriatti D. Analysis of defects during the preforming of a woven flax. Adv Compos Lett 2011;20:105–8.
[22] W. Stephen. The mechanics of friction in rope rescue. International Technical Rescue Symposium (ITRS 99).

Application of Six Sigma in Clothing SMEs: A case study

N.ABBES[1], N.SEJRI[2], Y.CHAABOUNI[3], and M.CHEIKHROUHOU[4]

Textile Engineering Laboratory (LGTEX), Height Institute of Technical Studies (ISET) of Ksar Hellal, Monastir University, Tunisia

Corresponding author: Nedraabbes2@gmail.com

Abstract The Six Sigma approach aims to improve the quality of products in order to ensure the customers satisfaction. It focuses on the improvement of the quality by reducing of process variation in order to improve its efficiency. Small- and medium sized enterprises (SMEs) face several quality problems in progress or in the end of the work process. Few studies have been reported about the application of Six Sigma in SME. The purpose of this paper is to propose a new model based on the integration of Define, Measure, Analyze, Improve, Control (DMAIC) and Plan, Do, Check, Act (PDCA) approach in order to improve the effectiveness of the Six Sigma structure in solving quality problem in clothing SMEs .An example is given to illustrate the application of the proposed model to fix the parameters of a laying carriage in order to minimize tissue cutting defect

***Keywords*— Six Sigma, DMAIC, PDCA, Clothing SME.**

I. Introduction

THE Six Sigma methodology was developed by Motorola in USA in 1980. It aims to identify and eliminate defects or failures in business processes or systems by focusing on those process performance characteristics in order to assure the customers satisfaction [1]. With the statistical and non-statistical tools and techniques, Six Sigma aims to eliminate process variation and by this way it improves process performance and capability [2], and set this goal so that the process variability is kept in the range of ±6 Sigma [3]

Tunisian small- and medium sized enterprises (SMEs), precisely clothing SMEs face several problems of quality, in addition to other problems such as: Lack of adequate financial resources and manpower, lack of time to achieve the project, wrong choice of objectives, unacceptability of the culture of change and lack of knowledge about Six Sigma tools. For this reasons, the application of this Six Sigma approach still limited for large enterprises. In fact it requires investing in training and hiring of experts. In addition, the use of the complex tools remains difficult for small- and medium sized enterprises (SME) because it requires an adapted model to implement the approach successfully. The purpose of this paper is to present a framework to implement a Six Sigma approach in clothing SME. A simple model is developed to facilitate this application. This model is based on the combination of the DMAIC (Define, Measure, Improve, Analyze, and Control) approach and PDCA (Plan, Do, Check, Act) cycle to relate Six Sigma initiatives to process improvement.

II. LITERATURE

A. *Brief review of Six Sigma*

The Six Sigma methodology was developed by Motorola in USA in 1980. It aims to identify and eliminate defects or failures in business processes or systems by focusing on those process performance characteristics in order to assure the customers satisfaction [1]. With the statistical and non-statistical tools and techniques, Six Sigma aims to eliminate process variation and by this way it improves process performance and capability [2], and set this goal so that the process variability is kept in the range of ±6 Sigma [3]

Tunisian small- and medium sized enterprises (SMEs), precisely clothing SMEs face several problems of quality, in addition to other problems such as: Lack of adequate financial resources and manpower, lack of time to achieve the project, wrong choice of objectives, unacceptability of the culture of change and lack of knowledge about Six Sigma tools. For this reasons, the application of this Six Sigma approach still limited for

IOP Conf. Series: Materials Science and Engineering **460** (2018) 012009 doi:10.1088/1757-899X/460/1/012009

large enterprises. In fact it requires investing in training and hiring of experts. In addition, the use of the complex tools remains difficult for small- and medium sized enterprises (SME) because it requires an adapted model to implement the approach successfully.

The purpose of this paper is to present a framework to implement a Six Sigma approach in clothing SME. A simple model is developed to facilitate this application. This model is based on the combination of the DMAIC (Define, Measure, Improve, Analyze, and Control) approach and PDCA (Plan, Do, Check, Act) cycle to relate Six Sigma initiatives to process improvement.

B. Review on Six Sigma implementation in SMEs

Small and medium-sized enterprises are companies whose size is defined by the number of employees and turnover, which must not exceed certain limits. According to the European Union, the category of micro, small and medium-sized enterprises (SMEs) consists of companies with less than 250 employees and an annual turnover not exceeding 50 million euro [15].They are considered as the suppliers of large companies. Rowland (2004) noted that the traditional Six Sigma approach with Black Belt is not desirable in SMEs [16]. Snee and Hoerl (2003) presented the main obstacle to implementing of Six Sigma which consists in that the Six Sigma experts have formerly structured models that are not applicable for SME [1]. The result of the pilot survey presented in the work of Antony 2007 in UK manufacturing SMEs, which is primarily based on descriptive statistics, shows that many of the SMEs are not aware of Six Sigma .For its successful implementation in SME, management involvement and participation, and linking Six Sigma to customers and to business strategy are the most critical factors for the successful deployment of Six Sigma in SMEs [17]. In the same context, Kumar 2007 presented several critical success factors for the implementation in SMEs. The findings of his study revealed that management involvement and commitment, poor training and resource availability were ranked as the highest impeding factors during the implementation of the Six Sigma project [18]. Furthermore, DeRuntz and Meier (2010) show that the success of the Six Sigma implementation depends on adopting clear strategies, motivating employees and training stakeholders and involving the top management [19]. The result of the survey in our study in Tunisian Clothing SME shows that: The involvement and commitment of top management Performance measurement and communication are the most critical factors for the successful deployment of Six Sigma in these SMEs [20].

III. RESEARCH METHODOLOGY

The structure of the methodology is based on the continuous improvement model DMAIC and the PDCA quality. The PDCA (Plan-Do-Check-Act), cycle is the basic procedure of TQM. It is a problem-solving model in the context of quality management [12] [13]. A general description of the methodology and the process flow chart is given in figure1.

In particular the five phases in in Six Sigma methodology, representing (DMAIC) approach were deeply studied, with used tools.

In our recent study, we have determined the tools most used and known of Six Sigma in clothing SME by a survey. We have focused on the clothing industries in Tunisia with an employee's number between 10 and 300 people, with a turnover not exceeding 30,000 dinars and operating in 4 activities: Technical articles industries, Knitted articles industries, Denim articles industries and mixed articles industries (Table 1).

In our research we are interested in SMEs because they represent the major part of the Tunisian industrial and because large companies usually have more material resources.

The first step consists in identifying the most used techniques and tools, for implementation of Six Sigma in 85 Small and medium Tunisian clothing industries. A questionnaire was distributed to the selected companies situated in different regions in Tunisia. In order to improve the accuracy of the results, the questionnaire was followed by interviews with CEO of these companies.

To rate the application and the knowledge of methods, tools, techniques and indicators, the respondents were asked to rank on a Likert Scale of 1–5, were:

- ✓ 0 indicates: unknown
- ✓ 1 indicates: Known
- ✓ 3 indicates: known and partially used
- ✓ 5 indicates: known used continuously

The collected data was then analyzed using Microsoft Excel.

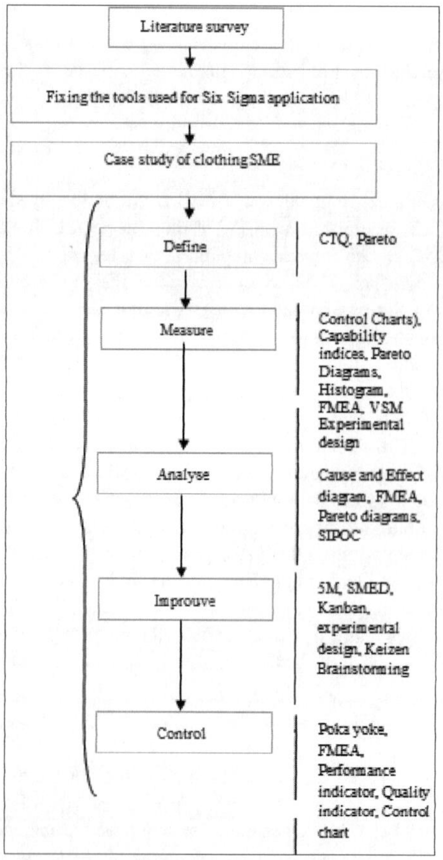

Fig. 1. Flow diagram of methodology adopted

TABLE I

DISTRIBUTION OF CLOTHING INDUSTRIES ACCORDING TO THEIR ACTIVITIES

Activity	number	Certified ISO 9001
Technical articles industries	77	2
Knitted articles industries	110	3
Denim articles industries	85	0
mixed articles industries	505	3
Total	777	8

Tools, techniques or indicators with highest average score means that they are the most known and used ones.

In Table 2, it can be seen the result of the most common unknown, known and used techniques by apparel manufacturing SMEs. Installation of balancing production line (4.2), the calculation of the cycle time (3.9), the Visual Management (2.9) and the 5S (2.8). They are essential and easy to use. All organizations have installed the balancing production line continuously or partially. Most of apparel manufacturing SMEs don't know the customer satisfaction evaluation methods such as Kano, CTQ, and VOC. Kai Zen, SMED, VSM, Six Sigma and

TPM, are well-known methods but not yet applied. These methods ensure continuous improvement of the production and the quality but require trainings to efficiently achieve their goals.

Our study shows that: 47% of companies that do not know the Six Sigma method, 43% of companies that know the method 5% of companies that applied this method partially and 5% of companies that applied this method continuously.

The structure of the methodology is based on the continuous improvement model DMAIC and the PDCA quality.

TABLEII

THE MOST COMMON UNKNOWN, KNOWN AND USED TECHNIQUES BY CLOTHING SMES

Tools, techniques and indicators	Average score
Installation of balancing production line	4.2
calculate the cycle time	3.9
Visual Management	2.9
5S	2.8
training	2.5
Acceptance of Six Sigma Culture	2.3
Just In Time	1.9
PDCA	1.9
KANBAN	1.6
Identification of bottlenecks	1.5
Kai Zen,Pareto	1.16
SMED	0.9
VSM	0.8
Six Sigma	0.7
TPM,FMEA	0.6
VOC, CTQ, KANO	0.4

The PDCA (Plan-Do-Check-Act), cycle is the basic procedure of TQM. It is a problem-solving model in the context of quality management [22] [23].In fact, it was considered as a well-established framework for process improvement where it focuses on continuous learning and knowledge creation [24]. We choice the PDCA cycle because it is similar to the steps in Six Sigma projects (DMAIC), [13] [24], and it adds the rigour of project life-cycle (PLC) to the implementation and close-out of Six Sigma projects. So that, we called it PDCA-applied-to-DMAIC. The principal is described by figure 2. The established framework PDCA is applied to each DMAIC steps to improve continuously the process, focusing on continuous learning and knowledge creation [24].

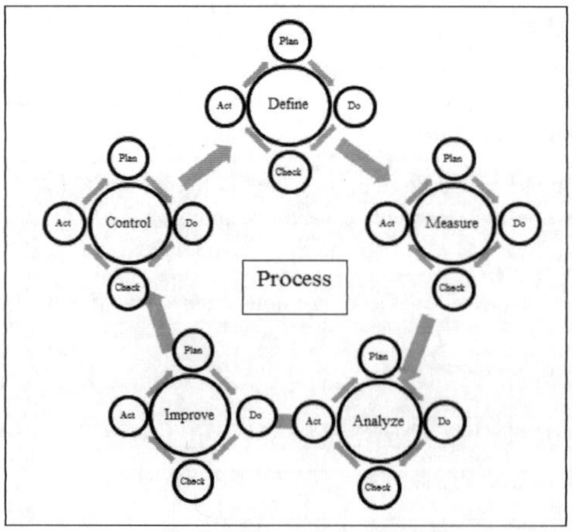

Fig. 2 .PDCA-applied-to-DMAIC

Present study was carried out in a clothing SME. The clothing SME contains a cut unit and an assembly unit. The problem is that cut pieces with the existing laying carriage parameters don't comply with the requested measures. In order to solve the former problem, a study was carried out to adjust of the laying carriage parameters (speed, tension) to ensure the correct superposition of the fabric ply mat. The unconformity rate of the cut pieces is huge due to the high variation in these parameters. That's why; it is of a great importance to need to fix these parameters in order to reduce the unconformity.

A. Define

This phase consists in defining the problem and the customer requirements. In order to precisely target the expectations and the requirement of the client .The Critical To Quality (CTQ) essential elements are determined. Figure 3 shows the results of the CTQ diagram of the wristband product .After conducting several brainstorming sessions; we have concluded that the measurement defect is the most significant issue.

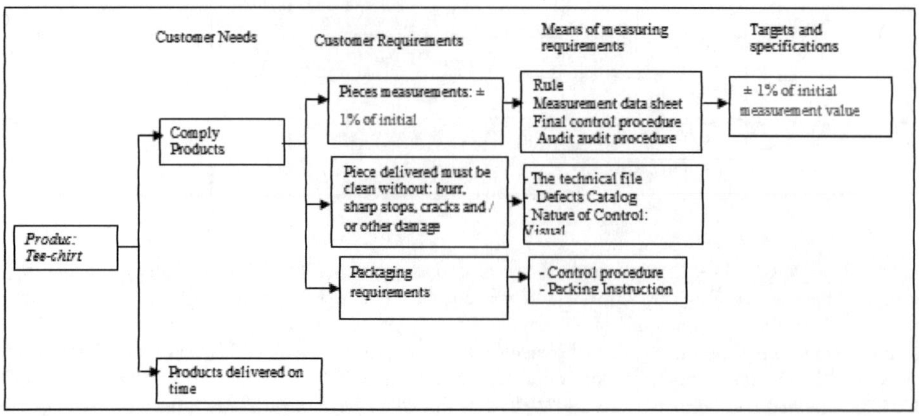

Fig. 3.Product CTQ diagram

B. Measure

In this step data on measurable indicators of production processes are collected. The objective is to determine this sigma of the process witch intern can give as an idea about the process capability index and hence its performances. The process capability analysis was performed to find out actual state of the process. The set of the possible values is divided into five states (Table 3) with the recommended actions relative to each state [25].

The variation of the process capability is shown in figure 4. The company was operating at a baseline capability of 0.2, the Z –sigma is equal to 0.7 and existing DPMO level of the process 780,000, which is remarkably high and this shows that there are lot of opportunities for improvement in the process.

TABLEIII

INTERPRETATION RULES OF THE CAPABILITY INDEX CP

Cp	Interpretation	Recommendation
Cp <1.67	Highly capable	No concerns
1.33 <Cp ≤1.67	Capable	Confortable situation
1.00 <Cp ≤1.33	Almost capable	Possibility of non-conforming products
0.67< Cp ≤1.00	Incapable	Existence of Possibility of non-conforming products
Cp ≤0.67	Chaotic	Shutdown, analysis of causes and corrections actions

C. Analyze

The analysis phase examines the collected data in order to generate a prioritized list of source of variation [26]. Using the result of the step ''Measure'', it's possible to determine the root causes and identify the significant process parameters of the defect. The cause and effect diagram (as shown in Figure 5) was drawn to find out the causes of '' Measurement defect''. Cause and effect diagram shows that the most important causes that affect the measurement defect are: the incorrect speed and tension setting, the change of these parameters by the operator, and the information lack of these parameters used for each fabric color. At this stage, it was essential to identify significant parameters so that they can be properly tuned for each fabric color.

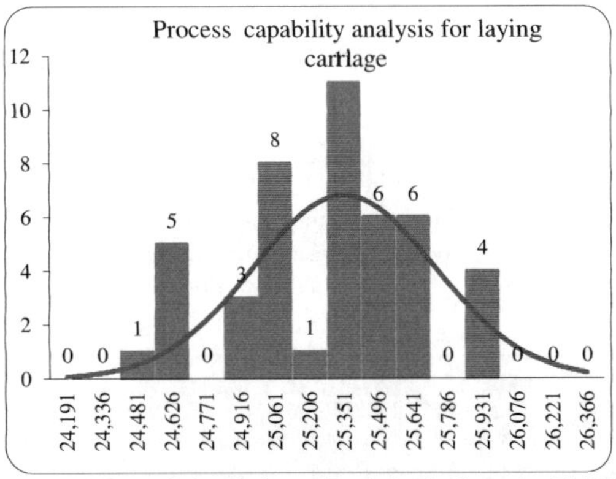

Figure. 4 Process capability analysis of laying carriage before implementing DMAIC methodology

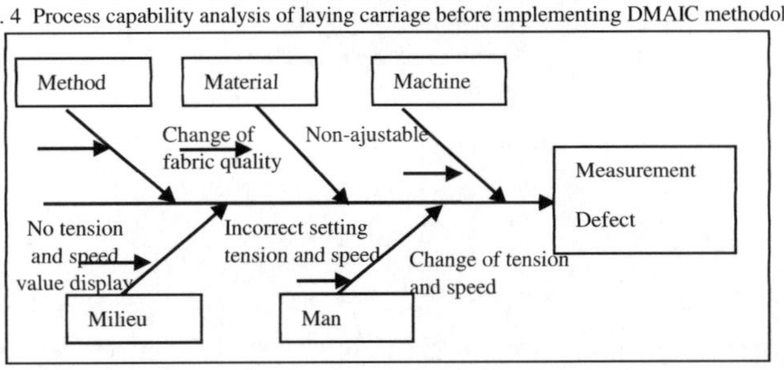

Fig. 5 .Cause and effect diagram

D. Improve

In the improve phase, we decided to carry out a designed experiment. The design of experiments was done to find out the optimum value of the vital few parameters found out after a two-sample t-test [27]. A 2*2 experiment was designed, i.e. an experiment with two levels (maximum value, minimum value) for each factors. Table 4 shows the four tests carried out to fix the values of the optimum carriage laying parameters. For each test, the Cp level given by these parameters values was calculated.

TABLE IV

THE DIFFERENT TESTS CARRIED OUT WITH THEIRS RESULTS

Test	Speed	Tension	Cp
1	100	80	0.4
2	100	40	0.2
3	25	80	0.31
4	25	40	0.35

Table 4 shows the four tests carried out for this objective. Minitab was used to determine the optimum combinations speed and tension as given in figure 6.The optimum values of the speed and tension various from 50 to 100, and 40 to 60 respectively. It is more practical and easier to work with equation than the flowchart of fig.5 to know the optimum of combination (tension speed). Using curve fitting technique, the function tension as a function of speed can be easily determined. Then, the user can manipulate only one variable (speed) to achieve the maximum Cp. The tension can be calculated using equation (1) when needed:

f(x) = a*x^b+c Coefficients (with 95% confidence bounds): x: speed ; f: tension

a = -4.043e+07 (-7.207e+07, -8.796e+06)

b = -3.705 (-3.905, -3.505)

c = 61.35 (61.03, 61.68)

Using equation 1 or figure 5,we have set the tension and speed values of the laying carriage for each fabric ply mat color. In order to validate the obtained results of the improve phase, a confirmatory test was implemented by using the optimal values of the combination speed and tension and the measuring again the cut pieces. In figure 6 shows, it can be seen the e new process capability of the laying carriage after.

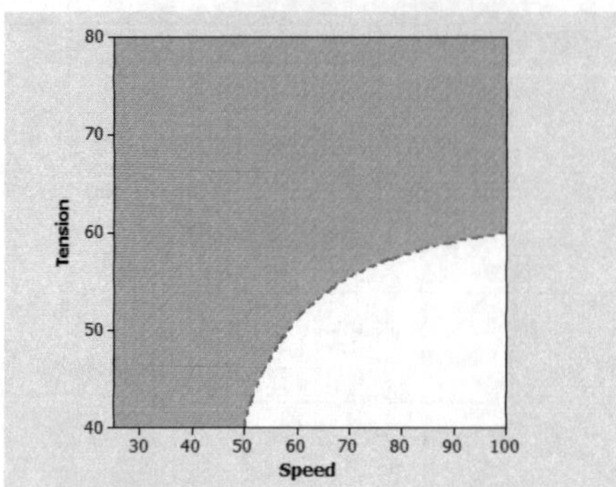

Fig. 6 .The optimum combinations speed and tension

Table 5 presents the significant improvements after implementation of Sigma methodology. The key metrics used for comparing the results after implementing the Six Sigma framework included Z-Sigma, Process Capability Cp, Defects per million opportunities (DPMO), and standard deviation of process.

TABLE V

COMPARISON BEFORE AND AFTER IMPROVEMENT BASED ON KEY METRICS

Key metrics used	Before improvement	After improvement
Z-Sigma	0.7	2
Cp	0.2	1.47
DPMO	780,000	308,000
Process standard deviation	0.38	0.056

Application of Six Sigma project brought up the sigma level for 0.7 to 2 with DPMO level of 308,000 and capability of 1.47, which is substantial for a clothing SME.

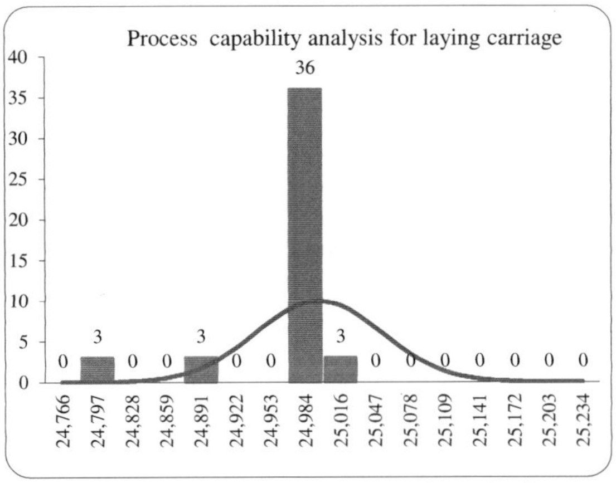

Figure.7 .Process capability analysis of laying carriage after improvement

E. Control

The main purpose of the Six Sigma methodology is not only improving the process performance but also having the improved results sustained in the long run. The standardization of the optimal parameters setting have been fully integrated into the training regime and the process documentation, the information related to company performance was sharing with its employees. A regular audit of parameters setting was carried out.

The cut unit process had been chosen in the chain manufacturing unit because it represents the first source of measurement defect. The success of Six Sigma application in this case study can definitely encourage the other clothing units to use Six Sigma as a quality tool to reduce the losses in their processes to ensure customer satisfaction.

IV. CONCLUSİON

In this paper, a new model was developed by integrating DMAIC and PDCA approach into the Six Sigma methodology in order to find out the sources of measurement defects cut unit and improve its index capability.

Analysis of the results shows that the speed and tension of laying carriage are the tow parameters responsible for this defect.

A 2*2 experiment was performed and the optimum combinations of these parameters were extracted. The new set of this optimum combination was applied for the laying carriage and yields to improve sigma from 0.7 to 2 and Cp from 0.2 to 1.47.

IOP Conf. Series: Materials Science and Engineering **460** (2018) 012009 doi:10.1088/1757-899X/460/1/012009

ACKNOWLEDGMENT

We would like to thank the director of the company in which we applied our work.

REFERENCES;

[1] Snee, R.D. and Hoerl, R.W., Leading Six Sigma – A Step by Step Guide Based on Experience at GE and Other Six Sigma Companies, 2003 (FT Prentice-Hall: NJ).
[2] Antony, J. and Banuelas, R. (2002), "Critical success factors for the successful implementation of Six Sigma projects", The TQM Magazine, Vol. 14 No. 2, pp. 92-9.
[3] Brady, J.E. and Allen, T.T. (2006): Six-sigma literature: a review and agenda for future research. Quality and Reliability Engineering International, Vol. 22, (2006), pp. 335-367. ISSN
[4] Linderman K, Schroeder RG, Zaheer S and Choo AS. (2003): Six Sigma: a goal-theoretic perspective. J Oper Mngt ,Vol 21, (2003), 10,pp,193–203. ISSN
[5] M.P.J. Pepper, T.A. Spedding. (2010): The evolution of lean Six Sigma. International Journal of Quality & Reliability Management, Vol., 27, (October 2010), 18, pp. 138 – 155, ISSN
[6] Roy Anderson, Henrik Eriksson, Håkan Torstensson, (2006): Similarities and differences between TQM, six sigma and lean .The TQM Magazine, Vol. 18 .,(March2006) , 15, pp. 282 – 296,ISSN
[7] Chiajou Lin and all 2013
[8] Starbird, D. 2002. "Business Excellence: Six Sigma as a Management System: A DMAIC Approach to Improving Six Sigma Management Processes." In ASQ World Conference on Quality and Improvement Proceedings, 47–55. American Society for Quality.
[9] Puga, R., Soler, A., Maximiano, A., & Wagner, J. A. (2005). Gerenciamento de Projetos Seis Sigma. V Seminário Internacional de Gestão de Projetos PMI São Paulo: Brasil Chapter, 1-8
[10] Rever, H. (2010). Six Sigma Can Help Project Managers Improve Results. International Institute for Learning, Inc. 1-5. http://www.microsoft.com/project/en/gb/six-sigma.aspx)
[11] Linderman, K.W., Schroeder, R.G., & Choo, A.S. (2006). Six Sigma: The role of goals in improve-ment teams. Journal of Operations Management, 24, 779–790.
[12] Mader, D.P. (2002). Design for Six Sigma. Quality Progress, 35(7), 82–86.
[13] Choo, A.S., Linderman, K.W., & Schroeder, R.G. (2007b). Method and psychological effects on learning behaviors and knowledge creation in quality improvement projects. Management Science, 53, 437–450.
[14] Naveh, E., & Erez, M. (2004). Innovation and attention to detail in the quality improvement para-digm. Management Science, 50, 1576–1586.
[15] European Union (consulted on 31 May 2011)}.
[16] Rowlands, H. 2004. "Implementation Issues of Six Sigma in an SME." Vol. 16, First International Conference on Six Sigma.
[17] Antony, J. 2006. "Six Sigma for Service Processes." Business Process Management Journal 12 (2): 234–248.
[18] Kumar, M. 2007. "Critical Success Factors and Hurdles to Six Sigma Implementation: The Case of a UK Manufacturing SME." International Journal of Six Sigma and Competitive Advantage 3 (4): 333–351.
[19] DeRuntz, B., and R. Meier. 2010. "An Evaluative Approach to Successfully Implementing Six Sigma." Technology Interface Journal 10 (3): 29–38.
[20] A.Nedra, S.Néjib, CH. Yassine, and CH. Morched,'' Critical success factors of Lean Six Sigma implementations within small and medium Tunisian clothing industries .International''.conference of Applied Research on Textile, CIRAT7, Hammamet, Tunisia, (2016).
[21] Michael L. George. "Lean Six Sigma: Combining Six Sigma Quality with Lean Speed". 2002. The McGrawHill Companies.
[22] Deming, W.E. (1986). Out of crisis. Cambridge, MA: MIT Press.
[23] Shewhart, W.A. (1931). Economic control of quality on manufactured product. New York: D. Van Nostrand.
[24] Deming, W.E. (1993). The new economics, for industry, government, education. Cambridge, MA: MIT Press.
[25] Taieb Ben Romdhane, Ahmed Badreddine & Manel Sansa,''A new model to implement Six Sigma in small- and medium-sized enterprises''. International Journal of Production Research
[26] Kapur, K.C. and Feng, Q. (2005), "Integrated optimisation models and strategies for the improvement of the Six Sigma process", International Journal of Six Sigma and competitive advantage, Vol. 1 No. 2, pp. 210-28.
[27] Kumar, M., J. Antony, R. K. Singh, M. K. Tiwari, and D. Perry. 2006. "Implementing the Lean Sigma Framework in An Indian SME: A Case Study." Production Planning and Control 17 (4): 407–423.

Analyzing the Visual Merchandising Elements for the Apparel Retailers

C Saricam[1], N Okur[1], D. Erdem[1], S. Akdag[1] and B.E. Kilikci[1]

[1]Istanbul Technical University, Faculty of Textile Technologies and Design, Department of Textile Engineering, Gumussuyu Campus, Gumussuyu, Beyoglu, Istanbul, Turkey

saricamc@itu.edu.tr

Abstract. Visual merchandising plays a key role, basically for pursuing the customer to enter the store by making the store attractive and changing the perceived image from the view point of consumers. The brand image and the retailers' image can be emphasized with the combination of these visual merchandising components. The aim of this study is to analyze and to compare the selection of design elements of different type of retailers such as fast fashion retailers, luxury apparel retailers, sportswear retailers or the retailers presenting products for women, men or both. To this aim, a content analysis was applied among 36 apparel retailers acting in Turkey. The findings showed that the visual merchandising elements were inarguably well employed by the luxury brand retailers, which was followed with the retailers selling high price fashion products. Among the visual merchandising elements special lightning and window displays were seen to be favored by all the retailers. Besides, the selection of different types of visual element were found to be based on the target customer group, while the apparel retailers selling products for females only are much more caring about the visual merchandising, it was vice versa for the apparel retailers selling only menswear.

Keywords— **Apparel retailer, Color Focus, Shop interior, Signage, Special Lighting, Visual merchandising, Window display.**

1. Introduction

In today's apparel retailing environment, visual merchandising plays a key role, basically for pursuing the customer to enter the store by making the store attractive and changing the perceived image from the view point of consumers [1]. Actually, there is a positive relationship between the pleasant store atmosphere and customer satisfaction and this causes to enhance the favorable purchasing experiment [2, 3]. Thus the concept of visual merchandising can be used as an effective strategy to strengthen the perceived image, to draw the attraction of the customers and to increase sales. Moreover, the retailers, which should develop unique positioning, create a clear differentiation and brand image enhancement can position themselves among their competitors and retain their market share by analyzing their visual merchandising concepts used by them.

Actually, including the exterior (the building, place, window display) and interior design components (store design), visual merchandising tries to create a pleasant store atmosphere. Among the elements of the visual merchandising, there are window displays, colors, lighting, signage and product information, and shop interiors which are actually addressing the visual senses [4]. Besides, there are other components of visual merchandising appealing the other senses such as music, smells, touch and even taste and technologies such as digital displays [5]. These components are usually brought together considering basically the contextual aspects such as type of store, brand and themes. The brand image is emphasized with the combination of these components such as the selection of prestigious and unique elements for a luxury brand and the selection of neat and plain elements for a brand identifying itself with simplicity. Besides, the target customer group and their socio demographic properties such as gender, age or education and the personal motivations are also considered to improve the attractiveness of the stores for them. Thus, the apparel retailers especially the successful ones include or highlight some of the visual merchandising elements more than the others. But, it is not known exactly, if the apparel retailers actually consider these points in the visual merchandising activities and if there are apparent differences between different types of retailers.

This study aims to analyze what kind of visual merchandising elements are used by the apparel retailers acting in Turkey and to compare the selection of design elements of different type of retailers such as fast fashion retailers, luxury apparel retailers, sportswear retailers or the retailers presenting products for women, men or both.

2. Literature review

Visual merchandising is described as an attractive and eye catching technique to present and display the products to the target and potential consumers [6]. It is actually an activity matching the effective assortment of the merchandise with the effective display of it, which means selecting the right merchandise and placing it effectively in the efficiently used retail space [7]. The major aim of using visual merchandising was stated to be increasing the sales by drawing the attention of the consumers to enter into the store and encouraging them to remain in the store and having positive experience in the retail atmosphere and even motivating the consumer to make an unplanned and impulse purchase. The reason why the visual merchandising is influential on the consumers is that, the people actually don't buy products just because of the products but they make purchases for various type of reasons such as cheering themselves up, beautifying their home, or adding value [6]. Being regarded as a 'silent salesman', visual merchandising was stated to be acting as a promotional tool by being an influential element in the marketing communication of the retailer which has the major role of appealing consumer senses [8-9].

Although it is a new concept, there are many studies regarding visual merchandising. Some of the studies are concentrated on describing the visual merchandising components, classifying the tools and making suggestions for improving the presentation of these tools. The other studies are focusing on analyzing the understanding of the consumers' perceptions about visual merchandising display, putting forward the effectiveness of the tools with some claims about the convenience of these tools for special type of customers or retailers.

The attributes associated with the retail environment were classified as tangibles and intangibles where the tangibles covered the fascia, overall store design, window display, layout and merchandising, fixtures and fitting, decoration and signage; the emotional and intangible attributes included the ambience, image, positioning, branding, sensory perceptions and emotional connection [10-12].

Soundhariya and Sathyan listed the visual merchandising techniques as presenting merchandise at the focal point, window display, colors and textures, exterior architecture, handling of the mannequins, lighting and music, display themes, props and fixtures, banners, signs and graphics, and the end caps which are at the end of the store [13].

Jain et al included the window displays, the signs, interior displays and special sales promotions and even the cosmetic promotions into the visual merchandising tools, which can be used by a fashion retailer [6]. Stating that there are actually two classes of tools as the interior and exterior components, the authors included any tools such as color combination, product placement, lighting arrangement, layout and highlighting of design, mannequin and props selection, fixtures and fittings selection in creating a favorable shopping atmosphere by Law, Wong and Yip [1].

IOP Conf. Series: Materials Science and Engineering **460** (2018) 012010 doi:10.1088/1757-899X/460/1/012010

Taskiran listed the dimensions of the visual merchandising as window display, color, lighting, signage and shop interior [4]. Regarding the window display, the author added that some of the objectives of the window display should be to create impact, image, theme and composition. On the other hand, it is claimed that digital signage can also be used as electronic display, which can be benefited to convey a message or information to the consumers. Specifically regarding the themes, Madhavi and Leelavati stated that the themes can be conveyed by the correct use of merchandise color, presentation style, awareness of fixtures, path finding and sensory qualities of materials and lighting [5].

Considering the dimensions and tools used as visual merchandising tools by different researchers, it can be stated that some dimensions of visual merchandising are highlighted.

The importance of the window display was underlined by many researchers since it is described as the initial way, a store would attract the attention of the consumer. Jain et al. claimed that, the window display should be attractive enough since only three to five seconds of attention is given by the consumers to window display [6]. Thus, it should create a visual appeal, relevance and interest value. Moreover, the window display of an apparel store was stated to be able to show the seasonal trends in fashion colors and looks. According to Madhavi and Leelavati, display themes, which are related with the retailers' imagination and creativity should be carefully selected and presented in good composition in window display creating an impact on the shopping mood of the customer [5].

Store layout and display as another visual merchandising element was claimed to be arranged in a way that the movement of the consumers is guided within the store. It was indicated that the front third of the store has 3 times more selling power when compared to back third of the store [11]. On the other hand, since a person entering into the room moves his/her eyes in Z pattern, the rear left of the store should be arranged in an appealing manner to guide the direction and keep the consumer visually interested [5]. Beside these, the usage of the empty spaces are declared to be significantly important within the perspective of visual merchandising.

Store design and shop interior covering the selection of the wall covering, carpets, furnishing, shelves and dividers are suggested to be designed to convey the image which is required to be projected to the customer [8].

Signages are referred to as providing a short cut for communication about a product. They were actually classified into two parts. The fixed signage was pointed that they give information about areas and facilities of the store such as fitting rooms and pay station and; the point of purchase signage such as speaker cards, posters and graphics was claimed to provide information about the product, institution or product and promote purchase. The proper signage was found to influence the sales up to 40%.

Colors were stated to be one of the powerful factor within the visual merchandising as they can be associated with emotions, gender or special occasion. Thus the selection of colors was emphasized to be matching with the theme of the items presented in the store [5] such as selecting light shades of pink and blue colors for baby accessorizes; red color with Valentines theme and so on.

Effective illumination and lightning was found to increase the sales up to 20% where the light blend with the interiors [5]. Besides, it was carried out that more items are handled by the consumer under bright lightning than soft lighting [14]. Actually, different lightings are proposed to be selected in different parts of the store considering the target consumer group. As a general concept it was added by some researcher that the main principles of design display should be to create balance, emphasis, proportion, rhythm, color, lighting and harmony where these principles should be applied for both window display and interior displays.

Although the most general dimensions and tools, which were investigated within the literature are as given above, some researchers also added some more components to the visual merchandising tools such as fragrances, music, and even store personnel. For instance, Fiore et al found out that sprayed fragrance, could stimulate one's affective state and boost mental imagination [15]. Appropriate music could be influential on the cognitive activity of the consumers [16].

Other than the analysis of the visual merchandising dimensions and tools, some researchers investigated the influence of these tools for specific type of retailers and consumers in order to find out the most influential elements of visual merchandising which well match with the target consumer group.

Pillai et al investigated the influence of the different visual merchandising tools including graphics and signage, trends and moot boards, space management, numeric and visuals, dressiforms and mannequins, synergy among different stalls, out of box thinking, merchandise planning and range planning, but the authors concluded that the most effective visual merchandising tool is to have proper lighting and attractive display themes [17].

A study established among a focus group of apparel customers revealed that the theme of the window and store display enabled the subconscious perception of the consumers to surface by creating an awareness and knowledge in consumers. Moreover, it was found that the attributes such as brightness, the size and the height of the display is important for them besides their suitability of to the brand image and their convenience to the store space. They revealed that they prefer the displays to be at eye-level, the use of mannequins or half mannequins instead of images and pictures but the displays should not prevent the customer to move freely within the store [18].

Krishnakumar investigated the attitude of the consumer towards visual merchandising in apparel purchase decision [8]. The authors found out that much consideration should be given into store front and store design adding that the proper and effective lightning (in terms of color, location and intensity) should be done to improve the effects of the display.

Hefer found out that the female consumers are much more attentive to the visual merchandising components whereas male consumers are showing attention only to the signages and directories which help them to find what they are looking for [19].

Oh and Petrie investigated the role of storefront window displays on the shoppers store entry decision and their perceptions about the brand or store image [20]. The authors concluded that visual perception of the store front display influenced the decision of the shoppers to enter the store. But this influence was stated to be depended on the inter-relationship among the type of the display, motivation and cognitive load effect. Moreover, they found that the merchandise focused window displays are easily understood by the consumers whereas the artistic window displays are promoting the wish for exploration within the consumers [20]. Nonetheless, the authors added that the artistic type of store display cannot be understood by the consumers having high cognitive load at that time.

Mehta and Chugan investigated the impact of visual merchandising on the buying behavior of customers visiting the central malls selling apparel products [21]. The authors analyzed four dimensions of the visual merchandising which are the window display, in store/mannequin display, floor merchandising and promotional signage and stated that except the in store mannequin display all the other merchandising dimensions are influential on the impulse buying behavior of the consumers.

Law, Wong and Yip investigated the effect of visual merchandising on the intimate apparel products for the women [1]. And they found that it should be taken care of the balance of the corporate images and perceived localness while preparing the visual merchandising strategies. Nonetheless, the authors added that the selection of the sexy mannequins as visuals could create negative feelings.

The studies investigating the effect of visual merchandising tools on consumers put forward that some of the tools are much more influential than the other. Besides, the studies emphasized that the selection of the tools should be made in a way that they should reflect the brand image and even personality. Last but not least, special type of merchandise or special type of stores are expected to use some visual merchandising tools more than the other.

Nonetheless, although the relation with the brand image and the usage of the visual merchandising tools are underlined, no study was come across in literature, which was putting forward if different type of apparel retailers are preferring different combination of these visual merchandising tool.

3. Method

Considering the lack of literature to analyze the selection of different combination of visual merchandising tools by different type of retailers, a content analysis was aimed to be made among the apparel retailers acting in Turkey. To this aim, 5 general visual merchandising dimensions with 7 sub dimensions were selected based on the literature. 39 apparel retailer were visited by an observation team of three people. The apparel retailers were assessed in terms of their usage of the predetermined

IOP Conf. Series: Materials Science and Engineering **460** (2018) 012010 doi:10.1088/1757-899X/460/1/012010

categories of the visual merchandising elements in their exterior and interior designs. All the data was recorded and then analyzed according to different type of retailers. Within this regard, these apparel retailers were classified in 8 categories according to type of the merchandise and price levels.

The statistical measures such as mode and average values were taken into account in order to make the analysis and make a comparison among different type of retailers.

Finally, a focus group of customers were asked about their approaches toward visual merchandising elements in their favorite apparel retailer store matching with the select apparel retailers to see if the findings are overlapping the results of the content analysis.

4. Results and Discussions

The dimensions of the visual merchandising tools were selected as window display, color focus, special lighting, signage and shop interior according to the study of Taskiran. But, considering that some tools are more important and some points should be especially covered during the observational study some sub dimensions were added within the dimensions of visual merchandising. Within this regard, 4 dimensions to the window display and 2 dimensions to the signage and 2 dimension to shop interior were added. Since, the window display is one of the most important exterior display as it is the first point that the customer makes his/her decision to enter the store 4 more dimensions, which are impact, image, theme and composition, were added. Considering that the signage easies the movement of the consumer within the store and it has the power of directly conveying the messages about the product, prices or promotions, two specific dimensions were added to signage, which were electronic display and lettering. Finally, covering many items, the shop interior was also handled by two more sub dimensions, which were named as the usage of empty space and focal point. The dimensions and sub-dimensions with their coding were given in Table 1.

Table1. Coding for dimensions and sub-dimensions

Dimension code	Dimension name
1	Window display
1a	Impact
1b	Image
1c	Theme
1d	Composition
2	Color focus
3	Special lighting
4	Signage
4a	Electronic display
4b	Lettering
5	Shop interior
5a	Usage of empty space
5b	Focal point

On the other hand, the 39 apparel retailers were divided into 8 categories considering the type of merchandise they are selling and average price levels. These categories, their explanations, and the number of apparel retailers falling in this category are given in Table 2 below. Moreover, the group names were stated within the Table 2, which will be reused in the presentation in Table 3.

Table 2. Categories of the retailers

Category name	Explanation	Number of retailer
Low price mass merchandise store (G1)	Huge capacity chain store selling low priced apparel for women, men, kids	1
Medium price level well known fast fashion stores (G2)	Mostly sell well known fast fashion brands at low, medium or medium to high price levels	9
Medium to high price specialty stores (G3)	Mostly specialty stores selling womenswear at medium or high price levels	6
Medium to high level sportswear retailers (G4)	They are selling younger age consumer mostly	2
High level sportswear retailers (G5)	They are selling older people including the smart casual apparel products	3
Moderate price level luxury retailers (G6)	They sell luxury items at moderate price levels compared to foreigners	4
High price luxury retailers (G7)	Well known luxury retailer	11
Menswear retailers(G8)	Selling only to male consumers	3

As seen from Table 2, not all the categories are involving the same number of retailers and even one category is involving only 1 retailer but since it is the largest apparel retailer in Turkey it was also included in the study. Besides, it should be mentioned that most of the retailers are predominantly selling women's wear products because of this reason a special category was added, which are only selling menswear products.

Table 3. The results of content analysis

		1	1a	1b	1c	1d	2	3	4	4a	4b	5	5a	5b
G1	Mode	0,0	0,0	0,0	0,0	0,0	1,0	1,0	1,0	1,0	1,0	0,0	0,0	0,0
	Average	0,0	0,0	0,0	0,0	0,0	1,0	1,0	1,0	1,0	1,0	0,0	0,0	0,0
G2	Mode	0,0	0,0	0,0	0,0	0,0	0,0	1,0	0,0	0,0	0,0	0,0	0,0	0,0
	Average	0,1	0,1	0,1	0,1	0,1	0,3	1,0	0,4	0,3	0,2	0,4	0,2	0,4
G3	Mode	0,0	0,0	0,0	0,0	0,0	1,0	1,0	0,0	0,0	0,0	1,0	1,0	1,0
	Average	0,3	0,3	0,3	0,3	0,3	0,7	1,0	0,3	0,0	0,3	0,7	0,7	0,5
G4	Mode	1,0	1,0	1,0	1,0	1,0	0,0	1,0	1,0	1,0	0,0	1,0	1,0	1,0
	Average	0,5	0,5	0,5	0,5	0,5	0,0	1,0	0,5	0,5	0,0	0,5	0,5	0,5
G5	Mode	1,0	1,0	1,0	1,0	1,0	0,0	1,0	0,0	0,0	0,0	1,0	1,0	1,0
	Average	0,7	0,7	0,7	0,7	0,7	0,3	1,0	0,0	0,0	0,0	0,7	0,7	0,3
G6	Mode	1,0	1,0	1,0	1,0	1,0	1,0	1,0	1,0	1,0	1,0	1,0	1,0	1,0
	Average	0,5	0,5	0,5	0,5	0,5	0,5	1,0	0,8	0,5	0,5	1,0	0,5	1,0

G7	Mode	1,0	1,0	1,0	1,0	1,0	0,0	1,0	0,0	0,0	0,0	1,0	1,0	1,0
	Average	0,8	0,8	0,8	0,8	0,8	0,4	1,0	1,0	0,0	0,1	0,9	0,6	0,8
G8	Mode	0,0	0,0	0,0	0,0	0,0	1,0	1,0	0,0	0,0	0,0	1,0	1,0	1,0
	Average	0,3	0,3	0,3	0,3	0,3	0,7	1,0	0,0	0,0	0,0	1,0	0,7	0,7

Table 3 gives the results of content analysis using the coding given in Tables 1 and 2. Regarding the assessment of mode values, it should be stated that the mode value was taken as the higher number (1) when the half of the retailers within the same category use this dimension of visual merchandising. From Table 3 general assessment of the all apparel retailers reveal that the mostly favored and used visual merchandising dimensions became special lighting, shop interior and focal point; the least favored ones became the lettering, electronic display and signage in the same order of magnitude.

Considering the categories, it can be stated that low price mass merchandise store (G1) was found to be emphasizing the color focus, special lighting and signage. Although it was not valid for all the retailers within the group medium price level well known fast fashion stores (G2), the highlighted visual merchandising dimensions were found to be shop interior, color focus and signage. Besides, it was observed most of them use special lighting. Similarly, for the third group, which is mostly specialty stores selling women's wear at medium or high price levels (G3), the visual merchandising elements were listed as special lighting, color focus and shop interior but it was observed that the number of retailers which are using the dimensions efficiently are higher than the (G2). For instance, all of the retailers within this group was observed to be using special lighting. Regarding the fourth group which are the retailers selling younger age consumer mostly (G4), it was interesting to see that almost all visual merchandising dimensions are used by these retailers, actually this was mostly because only two retailers have fallen in this category. Nonetheless, by looking at the retailers in this group it should be mentioned that the other interesting point about these retailers is that none of them use the color focus. In parallel with this group, the retailers falling in the fifth group, which are High level sportswear retailers (G5), only one retailer among them is using the color focus. Besides as in the case of G4, the retailers in G5 were found to be less frequently using the visual merchandising dimension regarding the signage. The two groups Moderate price level luxury retailers (G6) and High price luxury retailers (G7) became the retailer groups, which were benefiting most of the visual merchandising dimensions on the other hand. Nonetheless, regarding these two group it can be stated that whereas the G6 makes the balanced used of the visual merchandising dimensions, G7 make special emphasize on some of the dimensions such as window display and special lighting. It was observed that the sub dimensions of the window display are strongly used by the retailers that fall into this category. The final group of retailers, which is Menswear retailers (G8) was found to be emphasizing the shop interior, special lighting and color focus comparatively. Within this regard, it was observed that they are similar to G2 and G3.

Other than these findings, it was further studied with a focus group of consumer involving 16 people whose profile was given in Table 4 below.

Table 4. The profile of the focus group of consumers

	Gender	Age	Education	Income	Shopping frequency	Expenditure % of income
G1	Female	18-23	High school	Low	Once or twice a year	0-10 %
G1	Male	36-42	Undergraduate	High	Once or twice a month	20-30 %
G2	Female	18-23	Still undergraduate	Low to medium	Once or twice a month	30-40 %
G2	Female	30-35	Undergraduate	Low to medium	Once or twice three months	10-20 %
G3	Female	24-29	Undergraduate	Low to medium	Once a month	10-20 %
G3	Female	30-35	Undergraduate	High	Once a month	40-50 %

G4	Male	18-23	Still undergraduate	Low to medium	Once or twice three months	10-20 %
G4	Male	24-29	Still undergraduate	Medium	Once or twice three months	30-40 %
G5	Male	24-29	Undergraduate	Medium	Once or twice three months	10-20 %
G5	Male	36-42	Graduate	High	Often	10-20 %
G6	Male	30-35	Undergraduate	Low to medium	Once or twice a month	20-30 %
G6	Female	36-42	Undergraduate	High	Once a month	0-10 %
G7	Female	36-42	Undergraduate	High	Once a month	30-40 %
G7	Male	36-42	Undergraduate	High	Once a month	20-30 %
G8	Male	36-42	High school	High	Once or twice three months	40-50 %
G8	Male	36-42	Undergraduate	High	Once or twice a year	0-10 %

These participants were required to act as the consumer of one of these retailers according to their preference and rate these retailers using 7 point Likert scale in terms of 5 Visual Merchandising dimensions, which were quested with 10 questions. The findings were quite in parallel with the findings of the content analysis regarding the comparison of different type of retailers but there were some differences in terms of the most significant visual merchandising elements for this retailer within the consumers' minds. The difference may be related with the point that the consumers within the focus group were not the only consumer of this retailer but they are the ones that shop from this retailer as can be seen from Table 4 within the column regarding shopping frequency from the select retailer. Nonetheless, working with the focus group showed that the dimensions which are used in the visual merchandising environment can be easily detected by the consumers which meant that the improvement of the usage of these items will be certainly be recognized.

Table 5. The average rating of the consumers for retailers

	Window display	Color focus	Special lighting	Signage	Shop interior
G1	3.25	**5**	**4.75**	**5.5**	3.83
G2	6	**6.25**	5.5	**4**	**6**
G3	5.75	**5.75**	**5.5**	5.5	**5.5**
G4	4	6.25	5.5	4	2.83
G5	5.5	6	**6.5**	5.5	**6.5**
G6	4.5	5.5	4.25	3	4.2
G7	**5.5**	6	**5.75**	5.5	5.83
G8	5.5	**5.25**	4	4	**4.6**

5. Conclusion

In this study, it was analyzed to see if there is difference for the retailers in terms of usage of different type of visual merchandising dimensions. With the application of content analysis among 36 apparel retailers acting in Turkey, the findings revealed that the visual merchandising elements were inarguably well employed by the luxury brand retailers, which was followed with the retailers selling high price fashion products. It was interesting to see that the fast fashion retailers were not so much considering the application of visual merchandising components. Among the visual merchandising elements special lightning and window displays were seen to be favored by all the retailers. Besides, the selection of different types of visual element were based on the target customer group, while the apparel retailers selling products for females only are much more caring about the visual merchandising, it was vice versa for the apparel retailers selling only menswear. These results were also compared with findings obtained from focus group and it was seen that they were mostly confirmed but have small differences in details.

References

[1] D. Law, C. Wong C and J Yip, "How does visual merchandising affect consumer affective response?: An intimate apparel experience" European Journal of Marketing, vol.46, no. ½, pp.112-133, 2012.

[2] K. Spies, F. Hesse and K. Loesch, "Store atmosphere, mood and purchasing behaviour", International Journal of Research in Marketing, vol.14, no.1, pp 1-17, 1997.

[3] L. T. Wright, A. Newman and C. Dennis, "Enhancing consumer empowerment", European Journal of Marketing, vol.40, no.9/10, pp. 925-935, 2006.

[4] Z. Taskiran The elements of visual merchandising, Graduation thesis, Izmir University of Economics, 2012.

[5] S. Madhavi, T.S. Leelavati, "Impact of visual merchandising on consumer behaviour towards women apparel", International Journal of Management Research and Business Strategy, vol.2, no.4, pp.61-72, 2013.

[6] V. Jain, A. Sharme, P. Narwal, "Impact of Visual merchandising on Consumer Behavior towards Women's apparel", International Journal of Research in Management, vol.5, no.2, pp.106-117, 2012.

[7] J. Bell, K. Ternus, Silent Selling, 3rd ed., New York,:Fairchild Publications, (2006).

[8] M. Krishnakumar, "The role of visual merchandising in apparel purchase decision", IUP Journal of Management Research, vol.13, no.1, 37, 2014.

[9] L. Barnes, G. Lea-Greenwood, "Fast fashion in the retail store environment", International Journal of Retail & Distribution Management, vol. 38, no.10, pp.760-772, 2010.

[10] G. Lea-Greenwood, "Visual merchandising: a neglected area in UK fashion marketing?", International Journal of Retail Distribution and Management, vol. 26, no.8, pp.324-329, 1998.

[11] P.J. McGoldrick, Retail Marketing, 2nd ed., McGraw-Hill, Maidenhead, 2004

[12] R. Varley, M. Raffiq, Principles of Retail Management, Palgrave Macmillan, Basingstoke, 2004.

[13] S. Soundhariya, S. Sathyan, "Visual Merchandising and its Impact on Customer Buying Decisions", Indian journal of applied research, 5, 2015.

[14] C.S. Areni D. Kim, "The influence of in-store lighting on consumers' examination of merchandise in a wine store", Int J Res Mark, vol.11, no.2 ,pp.117-25, 1994.

[15] A.M. Fiore, X. Yah, E. Yoh, "Effect of a product display and environmental fragrancing on approach responses and pleasurable experiences", Psychology and Marketing, vol.17, no.1, pp.27-54, 2000.

[16] J.C. Chebat, C.G. Chebat, and D. Vaillant, "Environmental background music and in-store selling", Journal of Business Research, vol.54, 121-123, 2001.

[17] R. Pillai, A. Iqbal, H. Umer, A. Maqbool, A., and N. Sunil, Design, effectiveness and role of visual merchandising in creating customer appeal, 2011.

[18] M. C. Cant, Y. Hefer, "Visual merchandising displays-functional or a waste of space in apparel retail stores?.", Gender and Behaviour, vol.11, no.1, pp. 5336-5341, 2013.

[19] Y. Hefer, "Visual merchandising displays: wasted effort or strategic move? The dilemma faced by apparel retail stores", Journal of Applied Business Research, vol.28, no.6, pp.1489, 2012.

[20] H. Oh, J. Petrie, "How do storefront window displays influence entering decisions of clothing stores?", Journal of Retailing and Consumer Services, vol.19, 1, pp.27-35, 2012.

[21] N. Mehta, P. Chugan, Visual Merchandising: Impact on Consumer Behaviour (An Exploratory Study of Apparel Segment in Ahmedabad), 2012.

Antimicrobial activity of essential oils and their controlled release from the smart PLA fabric

D Štular[1], I Jerman[2], M Mihelčič[2], B Simončič[1] and B Tomšič[1]

[1]*University of Ljubljana, Faculty of Natural Sciences and Engineering, Department of Textiles, Graphic Arts and Design, Aškerčeva 12, 1000 Ljubljana, Slovenia*

[2]*National Institute of Chemistry, Hajdrihova 19, 1000 Ljubljana, Slovenia*

Corresponding author: brigita.tomsic@ntf.uni-lj.si

Abstract. In this study, "smart", stimuli responsive PLA fabrics with controlled antimicrobial activity were tailored, by the application of microgel based on temperature responsive poly(N-isopropylacrylamide) (poly-NiPAAm) and pH responsive chitosan, with incorporated β-cyclodextrine (PNCS/CD microgel), and subsequent embedment of essential oils (EO). Antimicrobial activity of lavender, thymol, savory and cinnamon bark EO was tested in 1 – 5 % concentrations. Based on the results, the 4% concentration of savory and cinnamon EO were chosen and prepared as emulsion, allowing further embedment into the PNCS/CD microgel using *in-situ* procedure on the PLA fabric. Morphological and chemical changes of the studied samples were assessed using scanning electron microscopy (SEM) and Fourier-transform infrared (FT-IR) spectroscopy. Temperature responsiveness was analyzed by determining moisture content (*MC*) at 20- and 40 °C, antimicrobial activity was determined by evaluating the reduction of gram-positive *Staphylococcus aureus* and gram-negative *Escherichia coli*, thus controlled release of the EO was tested by determining the zone of inhibition at 20- and 37 °C. The results show that the embedment of the EO emulsion did not affect the temperature responsiveness of the PNCS/CD microgel. The PNCS/CD microgel proved to be a suitable carrier of antimicrobial agents, assuring the effective controlled release of EO emulsion and excellent antimicrobial activity. Temperature related controlled release of the EO emulsion was proven by the formation of zone of inhibition, which occurred only at conditions dictating shrinkage of the microgel particles.

***Keywords*—Antibacterial activity, Cinnamon bark essential oil, Controlled release, Smart textiles, Savory essential oil, Stimuli responsive microgel**

I. Introduction

Essential oils (EO) which are distinguished by numerous advantages (i.e. anticancer, antiviral, antioxidant and antimicrobial properties), are indispensable in textile industry, to produce high value-added products which are non-toxic, biocompatible and environmentally friendly [1]. Their antimicrobial activity is based on small phenolic or terpenoid compounds, which hold an ability to invade the cell membrane, thus causing the proton gradient exhaustion, leading to the cell lysis or disruption of adenosine triphosphate synthesis [2]. In order to increase their lifetime on the surface of textile materials, EO can be incorporated into stimuli responsive hydrogels, which enables their release only at predetermined conditions. Namely, the ability of reversible volume change (i.e. swelling/de-swelling), triggered by the presence or absence of different external stimuli, makes stimuli responsive hydrogels ideal carriers of different active ingredients. Among stimuli responsive hydrogels, temperature and pH responsive microgels based on poly-(N-isopropylacrylamide) (poly-NIPAAm) and chitosan (PNCS microgel) has already been proven to be a suitable carrier of different antimicrobials, providing their controlled release only at required conditions [3]. PNCS microgel exists in the swollen phase when the temperature is below 32 °C and pH is below 6.5, thus undergoes transition in to shrunken and collapsed state when the temperature and/or pH rise above this values [4]. Nevertheless, embedding of EO into the structure of the microgel particles represents a challenge, due to the hydrophobic character of the EO. This challenge could be overcome by the addition of cyclodextrines into the microgel structure, as they are widely used as a hydrophobic guest molecules and have an ability to entrap substances in their hydrophobic cavity [5].

In the present research, we incorporated β-cyclodextrine into the structure of the PNCS microgel (PNCS/CD microgel). The suitability of the PNCS/CD microgel as a carrier of different EO was studied. For this purpose, antimicrobial activity of lavender, thyme, savory and cinnamon bark EO in various concentrations was assessed and selected EO were incorporated into the PNCS/CD microgel by the *in-situ* procedure on PLA fibers. With the use of adequate techniques, temperature responsiveness, antimicrobial activity and controlled release of EO from the functionalized PLA fabric were studied.

II. EXPERIMENTAL

A. Materials

In this study, 100 % poly(lactic acid) (PLA) fabric Revolution® (M+N Textiles, Nederland) was used. Lavender, thymol, cinnamon bark and savory essential oils (Florihana) were purchased from Magnolija, Nina Medved s.p, dimethyl sulfoxide (DMSO) (sigma Aldrich) was used as a solvent and Tween 20 (Sigma Aldrich) as a surfactant. For the preparation of the PNCS/CD microgel, chitosan (Chitoclear, Primex, Iceland; DD = 95%; η = 159 mPa), glacial acetic acid (Sigma Aldrich), N-isopropylacrylamide (NiPAAm) (Sigma Aldrich), N,N-methylenebisacrylamide (MBA) (Sigma Aldrich), ammonium persulfate (APS) (Sigma Aldrich), β-cyclodextrine (Sigma Aldrich) and sodium dodecyl sulphate (SDS) with high purity (GE Healthcare Life Sciences) were used.

B. Determination of antimicrobial activity of EO

Disk diffusion assay test was used to determine the antimicrobial activity of lavender, thymol, cinnamon bark and savory essential oils. EO were diluted in DMSO in 1, 2, 3, 4 and 5 wt% concentration. The 6 mm discs were impregnated with 20 μL of each diluted EO and air-dried. Blank disc and DMSO impregnated disc were used as controls. Dried discs were placed on agar plate infused with 200 μL of the test bacterium, where gram negative bacteria *Escherichia coli* (*E. coli*, ATCC 25922) and gram positive bacteria *Staphylococcus aureus* (*S. aureus*, ATCC 6538) were used. Samples were incubated on 37 °C for 24 hours and the zone of inhibition was observed.

C. PNCS/CD microgel synthesis, characterization and application

PNCS microgel, functionalised with β-cyclodextrin was prepared using in-situ synthesis, according to modified methods reported by Lee [6] and Yi [7]. The particle size of dispersed PNCS/CD microgel in correspondence with changes in temperature was determined with dynamic light scattering (DLS) analysis, on Zetasizer Nano S (Malvern, UK), equipped with 4mW He-Ne laser operating at wavelength 633 nm and an avalanche photodiode detector. Scattering light was detected at an angle of 173 °. Microgel was diluted 50 times. Particle size was determined at temperatures ranging from 20- to 40 °C in 5 °C intervals. 60 μL of sample was used for each measurement and results represent an average of 3 measurements. The PNCS/CD microgel was applied using a pad-dry method with a wet pick-up (WP) of 80±5 %.

D. Essential oil emulsion preparation, characterization and incorporation

Emulsions of cinnamon bark and savory essential oils were prepared by diluting 8 g of Tween 20 in 88 g distilled water and mixed for 20 min on room temperature with the speed of 100 rpm. Afterwards, the stirring speed was set on 1500 rpm, while 4 g of essential oil was added dropwise into the mixture. The mixing proceeded for 4 hours and was followed by mixing with lab homogenizer (Hielscher Ultrasonics GmbH) for 5 minutes, using 0.5 ultrasonic impulses. The particle size of EO emulsion was determined using dynamic light scattering (DLS) analysis in the same manner as the PNCS/CD microgel particles. The measurements were conducted at 20 °C and EO emulsion was diluted 10 times.

For the incorporation of EO emulsion into the microgel particles, the PNCS/CD microgel functionalized PLA samples were heated to 50 °C for 15 min, in order to achieve the shrinking of the microgel particles on the fiber surface. Next the EO emulsion was sprayed on the samples until the fabrics were completely wet. The used EO emulsion was previously stored in a refrigerator at 8 °C in order to advance the swelling of the shrunken microgel particles, leading to greater absorption of the EO emulsion. The wet samples were subsequently squeezed with foulard with the pressure of 3 bar in order to achieve homogenous distribution of the EO emulsion. The sample codes are presented in Table 1.

TABLE I
SAMPLE CODES ACCORDING TO THE CHEMICAL MODIFICATION OF THE PLA FABRIC.

Sample code	Description of chemical modification
PLA_UN	/[a]
PLA_M	PLA fabric treated with PNCS/CD microgel
PLA_M+S	PLA fabric treated with PNCS/CD microgel and savory EO emulsion
PLA_M+C	PLA fabric treated with PNCS/CD microgel and cinnamon bark EO emulsion

/[a] No finishing

G. Analysis and measurements

Add-on was calculated from the average mass of completely dried samples. The samples were placed in the moisture analyzer MLB-C (Kern & SOHN GmbH, Germany) and dried until a constant mass. For each sample, 10 measurements were obtained. The add-on values were calculated according to following equation:

$$Add - on = \left(\frac{m_f - m_{UN}}{m_{UN}}\right) * 100\ [\%] \qquad (1)$$

where m_f [g] is the average of the dry weights of the functionalized sample, and m_{UN} [g] represents an average of dry weights of the untreated sample.

Morphological changes of the studied PLA samples were determined using JEOL 6060 LV scanning electron microscope, operating at 10 kV. Before SEM images were taken, the samples were coated with a thin layer of mixture of platinum and gold to ensure sufficient electrical conductivity and to avoid charging effects.

Chemical changes were determined using Fourier-transform infrared (FT-IR) spectroscopy, obtained from a Spectrum GX I spectrophotometer (Perkin Elmer, Great Britain) equipped with an attenuated total reflection (ATR) cell and a diamond crystal (n = 2.0). The spectra were recorded over the range 4000–600 cm^{-1}, with a resolution of 4 cm^{-1} and averaged over 16 spectra.

Temperature responsiveness of the samples was assessed by measuring moisture content using a Moisture analyzer MLB-C (Kern & SOHN GmbH, Germany). Studied samples were pre-conditioned at 65±2% relative humidity at 20- and 40°C for 24 h, before they were put in a moisture analyzer and dried at 60 °C until the constant mass. Moisture content (*MC*) was determined by following equation:

$$MC = \left(\frac{m_o - m_f}{m_o}\right) * 100\ [\%] \qquad (2)$$

where m_o denote the initial mass of the pre-conditioned sample and mf represents the final mass of the sample after drying. *MC* was reported as mean values of ten measurements.
From results, the swelling degree of PNCS/CD microgel (SD_{MC}) on the studied PLA samples was calculated by following equation:

$$SD_{MC} = \left(\frac{MC_F - MC_{UN}}{MC_{UN}}\right) * 100\ [\%] \qquad (3)$$

where MC_F is the moisture content of the functionalized PLA samples, and MC_{UN} is the moisture content of the untreated PLA sample, determined under the same conditions.

Antibacterial activity of studied cotton samples was estimated by determination of bacterial reduction according to ASTM E 2149–01 standard method. Bacterial reduction of coated samples was evaluated against *E. coli* (ATCC 25922) and *S. aureus* (ATCC 6538). The reduction of bacteria (*R*) was calculated as follows:

$$R = \left(\frac{U - T}{U}\right) * 100\ [\%] \qquad (4)$$

where U is the number of CFU determined for the untreated sample after 24 h of incubation and T is the number of CFU determined for the treated samples under the same conditions. For each sample, the value R represent an average of eight counts.

Additionally, the zone of inhibition was determined referring to the ISO 20645 Agar diffusion plate test, where bacteria was let to grow at 20 °C and 40 °C for the purpose of determining the controlled release. According to the ISO 20645 Agar diffusion plate test, assessment of antibacterial activity was based on the absence or presence of an inhibition zone. All tests were performed in duplicate.

III. RESULTS AND DISCUSSION

A. Determination of antimicrobial activity of EO

Antimicrobial activity of lavender, thymol, cinnamon bark and savory essential oils in 1 – 5 % concentrations was determined against gram negative bacteria *E. coli* and gram positive bacteria *S. aureus,* using disk diffusion assay method. In general, the EO were more efficient against gram positive bacteria *S. aureus*, in comparison to gram negative bacteria *E. coli*. Lavender EO did not form any zone of inhibition in the tested concentrations, thus higher concentrations should be used. Furthermore, thymol, savory and cinnamon bark EO formed zone of inhibition in the concentrations ranging from 3 to 5 %, thus it increased with the rising concentrations (data not shown). Based on the results obtained for both bacteria, savory and cinnamon bark essential oils in 4 % concentration (Fig.1) were used for further incorporation into the PNCS/CD microgel.

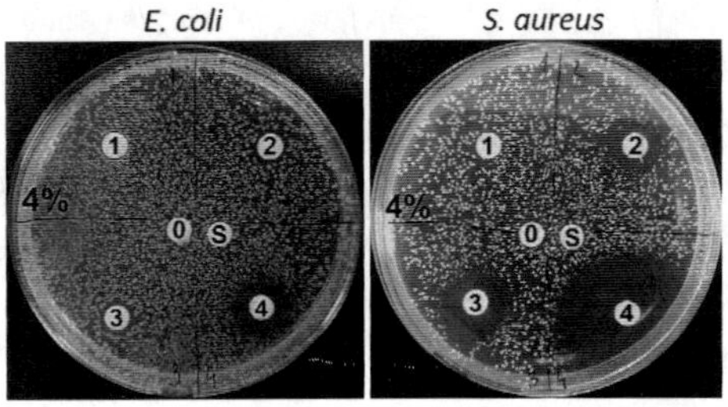

Fig. 1. Disk diffusion test against *E. coli and S. aureus*, of 1-Lavender, 2-Thymol, 3-Savory and 4-Cinnamon bark EO in 4% concentration, where 0-empty disk and S-DMSO impregnated disc were used as control.

B. PNCS/CD microgel and EO emulsion characterization

In order to investigate the hydrodynamic particle size of PNCS/CD microgel in dependence of changes in temperature, DLS analysis was used (Fig. 2). At 20 °C microgel particles were swollen and reached the size of 692 nm. Temperature responsiveness is seen as a decrease of the particle size with the rise of temperature, until the temperature reached 30 °C. Values gathered at 35- and 40 °C dramatically increased, which was contributed to fast agglomeration of the microgel particles. Such dramatic respond is in a good agreement with the literature, as Yi [7] reported that the incorporation of β-cyclodextryn into the microgel structure greatly improves its temperature responsiveness. In this case the formation of a complex between the temperature responsive poly-NIPAAm and β-cyclodextryn occurs, providing effective driving force, thus expelling hydrophilic poly-NIPAAm blocks at lower temperatures or pulling hydrophobic blocks of the temperature responsive polymer, upon heating.

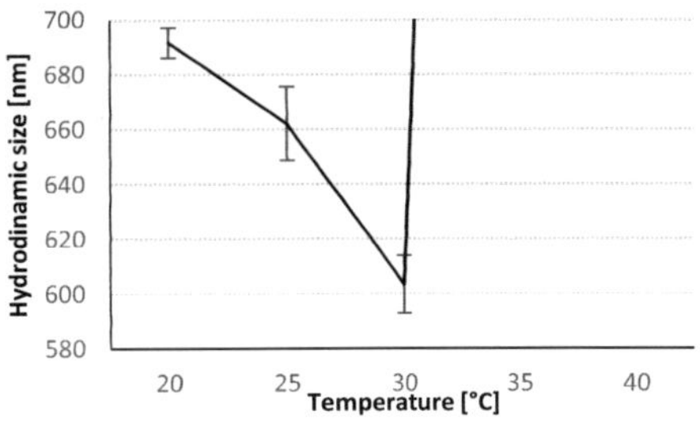

Fig. 2. Hydrodynamic diameter of the PNCS/CD microgel as a function of temperature.

DLS was also used to determine the size of EO emulsion particles which was 206.03±1.15 nm with PDI of 0.06 for the savory EO emulsion and 203.03±2.77 nm with 0.10 PDI for the cinnamon bark EO emulsion. Based on the results we can conclude that the type of EO used for preparation of the emulsion did not affect the size of the emulsion particles and that the size of the particles could allow for the emulsion to be embedded into the microgel structure.

C. Morphological and chemical changes of the functionalized PLA fibers

The PNCS/CS microgel was applied onto the PLA fabric in the first step and EO emulsion was embedded *in-situ* subsequently. The add-on determination (Table II) suggest that the incorporation of EO emulsion was successful, as the add-on values increased from approximately 3 % for the PLA_M sample, to greater than 10 % for the samples PLA_M+S and PLA_M+C.

TABLE II
ADD-ON OF THE SAMPLES FUNCTIONALISED WITH MICROGEL OR COMBINATION OF MICROGEL AND EO EMULSION.

Sample	Add-on [%]	SD
PLA_M	2.998	0.603
PLA_M+S	10.943	0.859
PLA_M+C	10.561	0.415

Next, the morphological changes were examined by SEM and are presented in Fig.3. PNCS/CD microgel formed round bulges on the fiber surface, with 350 - 500 nm in diameter. After the application of EO emulsion the bulges on the fiber surface are much more noticeable, due to the swelling of the microgel and the incorporation of EO emulsion. The diameter of the microgel bulges did not change significantly, yet the contrast is much more noticeable as the microgel particles rise in volume. By comparing the morphological changes on the PLA_M+S and PLA_M+C samples, we can conclude that the type of the EO used for the emulsion preparation did not influence the morphological changes of the studied samples.

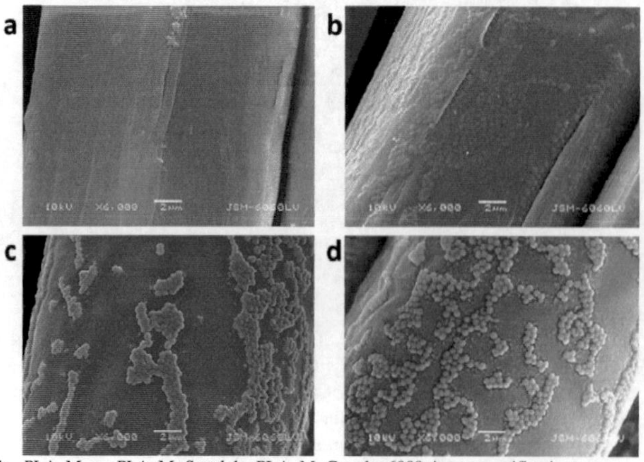

Fig. 3. SEM images of a - PLA_UN, b – PLA_M, c – PLA_M+S and d – PLA_M+C under 6000 times magnification.

Chemical properties of functionalized PLA fabric samples were studied by IR ATR spectroscopy (Fig. 4). The presence of PNCS/CD microgel in the IR ATR spectra was confirmed by the appearance of the absorption band at 1450 cm^{-1}, belonging to the N–H vibration of both poly-NiPAAm and chitosan, as well as by the absorption bands of amide I and amide II at 1645 and 1535 cm^{-1}, manifested by C=O stretching vibration of poly-NiPAAm [8]. The IR ATR spectra of the samples PLA_M+S and PLA_M+C, show the appearance of several additional peaks which can be contributed to the diverse composition of EO [9]. Savory and cinnamon bark essential oils have 20 common components, which was broadcasted in a form of similar additional bands in the area from 3000 to 2850 cm^{-1} belonging to C–H stretch of alkanes and between 900 and 675 cm^{-1}, belonging to aromatic C=C compounds. Furthermore, the sample PLA_M+C formed absorption band at 1667 cm^{-1} which can be ascribed to the C=O bond of aldehyde and the bands ranging from 1670 to 1620 cm^{-1}, belonging to C=C stretch of alkenes [10].

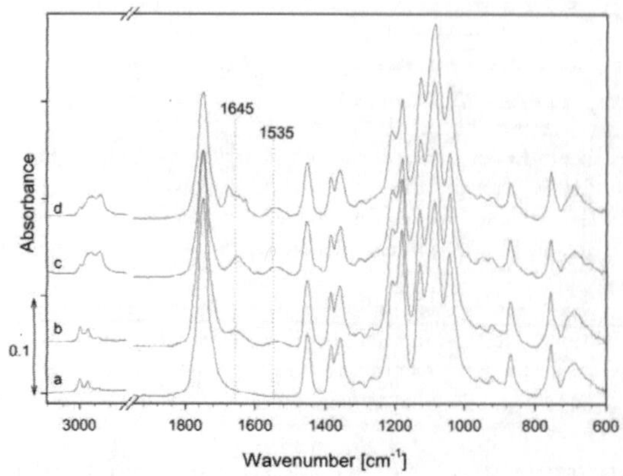

Fig. 4. FT-IR spectra a - PLA_UN, b – PLA_M, c – PLA_M+S and d – PLA_M+C.

D. Temperature responsiveness of the functionalized PLA fibers

Temperature responsiveness of the functionalized PLA samples was determined by moisture content (*MC*) analysis (Fig. 5a). With respect to the untreated PLA fabric, the application of temperature responsive PNCS/CD microgel improved the *MC* at 20 °C, when the microgel was in the swollen, hydrophilic state. Furthermore, when the EO emulsion was embedded into the microgel structure, the *MC* values gathered at 20 °C were even higher, as the microgel particles were already hydrated with the EO emulsion. On contrary, at 40 °C the microgel was in its hydrophobic state, therefore the *MC* of the functionalized samples decreased dramatically. In order to compare and further emphasize the stimuli responsiveness of the PNCS/CD microgel applied on PLA fabric, swelling degree (SD_{MC}) was calculated (Fig. 5b). It is evident that PNCS/CD microgel applied to the PLA fabric provided approximately 50% higher SD_{MC} compared to PLA_UN sample. Hence, the presence of the EO emulsion embedded in the microgel particles did not influence the temperature responsiveness, namely the SD_{MC} values lowered for only 1.6% and 2.6 % compared with the PLA_M sample.

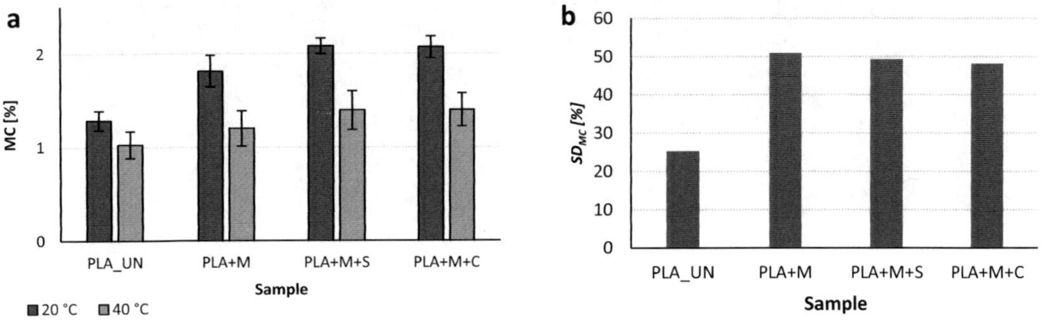

Fig. 5. a – Moisture content (*MC*) values and b – swelling degree of the PNCS/CD microgel (SD_{MC})applied on the studied samples, analyzed at 20- and 40 °C.

E. Antimicrobial activity of the functionalized PLA fibers

The antimicrobial activity of the studied samples was tested against Gram-negative *E. coli* and Gram-positive *S. aureus* (Table III). Microgel coated PLA fabric showed 26 % reduction of *E. coli* and 53 % reduction of *S. aureus*, which can be contributed to the presence of the chitosan in the microgel structure. Furthermore, excellent reduction was achieved for both PLA samples functionalized with the combination of PNCS/CD microgel and EO emulsion. Surprisingly, combination of PNCS/CD microgel and cinnamon bark EO showed lower bacterial reduction compared to savory EO, which contradicts the disk diffusion test. The reason for such behavior might be in the stableness of the EO emulsion. Namely, after few days some oil bubbles were seen in the cinnamon bark EO emulsion. In a view of that, further optimization of cinnamon bark EO emulsion is needed. Nevertheless, the antimicrobial reduction provided by the PLA_M+C sample is still efficient, namely 97±1.47 % for the gram negative bacteria and 87±12.5 % for the gram positive bacteria.

TABLE III
BACTERIAL REDUCTION, R, OF *E. coli* AND *S. aureus*, OF PLA SAMPLES FUNCTIONALISED BY PNCS/CD MICROGEL AND EO EMULSIONS

Sample code	***R* [%]**	
	E. coli	*S. aureus*
PLA_UN	/[a]	/[a]
PLA_M	26.4±7.5	53.0±10.7
PLA_M+S	>99.99	>99.99
PLA_M+C	97.82±1.47	87.02±12.5

/[a] bacterial reduction was not obtained

Controlled release of the EO emulsion was studied by the determination of zone of inhibition in bacterial growth. The samples were placed on the bacteria infused agar plate and incubated at 20 °C, i.e. conditions where microgel is swollen and 40 °C, when microgel is shrunken. Results gathered for the gram positive bacteria *S. aureus* are shown in Fig. 6. As expected, no zone of inhibition was seen for the samples PLA_UN and PLA_M at any given temperature. In contrast, samples PLA_M+S and PLA_M+C show the controlled release of the EO imposed by the shrinking of the microgel particles. At 20 °C the diameter of bacterial colonies that have grown on the agar plate is greatly reduced. At this temperature the microgel particles exist in the swollen state, which restricted the EO emulsion to be released from the particles on a larger scale. However, the swollen microgel particles are highly porous, thus the wetted agar plate allowed the leakage of the small amounts of EO emulsion. Despite that, bacterial growth next to the sample was still observed. The small amount of the released EO emulsion was not enough to provide the zone of inhibition, yet it was enough to interact with the growth of the bacterial colonies. At 40 °C the microgel particles shrunk and released some of the EO emulsion from their structure, which provided a noticeable zone of inhibition (i.e. > 1 mm) next to the studied samples.

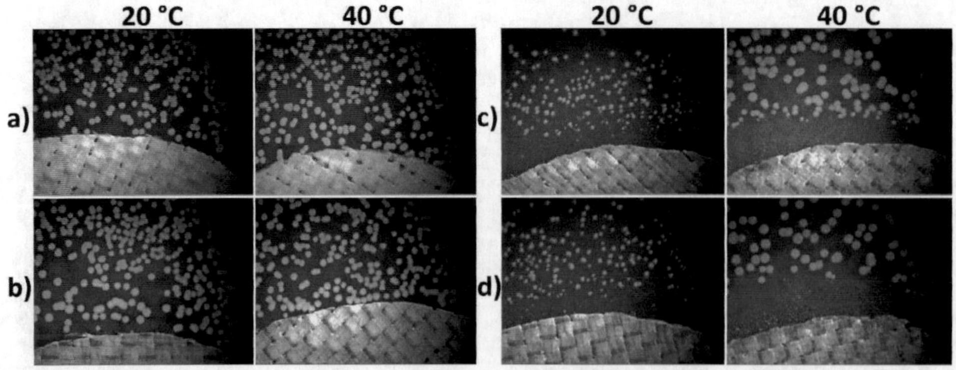

Fig. 6. Zone of inhibition against *S. aureus* for a - PLA_UN, b – PLA_M, c – PLA_M+S and d – PLA_M+C where bacteria was let to grow on 20 °C (left) or 40 °C (right).

IV. CONCLUSION

Based on the antimicrobial tests of the lavender, thymol, savory and cinnamon bark essential oils (EO), tested in 1 – 5 % concentrations, savory and cinnamon bark EO in 4 % concentration were chosen for the incorporation into the PNCS/CD microgel applied on the PLA fabric. The hydrodynamic microgel particles size at 20 °C was 692, thus the size of EO emulsions particles was approximately 200 nm, ensuring that the emulsion could be embedded into the microgel structure. All the samples padded with PNCS/CD microgel exhibited temperature responsiveness, which was preserved even after incorporation of EO emulsion. Accordingly, the type of EO emulsion (i.e. savory and cinnamon bark) did not influenced reversible responsive behavior of the studied microgel. Excellent bacterial reduction was obtained for both samples functionalized with EO emulsions. Temperature related release of the EO emulsion was proven by the determination of zone of inhibition, which was formed only at condition dictating shrinkage of the microgel particles, i.e. at 40 °C. This results prove the appropriateness of the functionalization process to create smart textiles with controlled moisture management properties and pro-active antimicrobial activity, triggered only at controlled conditions.

ACKNOWLEDGMENT

This work was supported by the Slovenian Research Agency—Slovenia (Program P2-0213, Infrastructural Centre RIC UL-NTF and a Grant for the Dr. Sc. student D. Š.).

REFERENCES

[1] S. Islam, M. Shahid and F. Mohammad, "Perspectives for natural product based agents derived from industrial plants in textile applications – a review," *Journal of Cleaner Production*, vol. 57, pp. 2-18, Jun. 2013.

[2] K. A. Rieger and J. D. Schiffman, "Electrospinning an essential oil: Cinnamaldehyde enhances the antimicrobial efficacy of chitosan/poly(ethylene oxide) nanofibers," *Carbohydrate Polymers*, vol. 113, pp. 561–568, Nov. 2014.

[3] D. Štular, I. Jerman, I. Naglič, B. Simončičc, B. Tomšič, "Embedment of silver into temperature- and pH-responsive microgel for the development of smart textiles with simultaneous moisture management and controlled antimicrobial activities," *Carbohydrate Polymers*, vol. 159, pp. 161–170, Dec. 2017.

[4] A. Tourrette, N. De Geyter, D. Jocić, R. Morent, M. M. C. G. Warmoeskerken, C. Leys, "Incorporation of poly(N-isopropylacrylamide)/chitosan microgel onto plasma functionalized cotton fibre surface," *Colloids and Surfaces A: Physicochem. Eng. Aspects*, vol. 352, pp. 126-135, 2009.

[5] U. Rashmi Bhaskara-Amrit, P. B. Agrawal and M. M. C. G. Warmoeskerken, "Applications of B -Cyclodextrins in Textiles," *AUTEX Research Journal*, vol. 11, no. 4, Dec. 2011

[6] C. F. Lee, C. J. Wen and W.Y. Chiu, "Synthesis of poly(chitosan-nisopropylacrylamide) complex particles with the method of soapless dispersion polymerization," *Journal of Polymer Science Part A: Polymer Chemistry*, vol. 41, pp. 2053–2063, May 2003.

[7] P. Yi, Y. Wang, S. Zhang, Y. Zhan, Y. Zhang, Z. Sun, Y. Li, P. He, "Stimulative nanogels with enhanced thermosensitivity for therapeutic delivery via β-cyclodextrin-induced formation of inclusion complexes," *Carbohydrate polymers*, vol. 166, pp. 219-227, Feb. 2017.

[8] G. Socrates, "Infrared and Raman characteristic group frequencies," *New York:John Wiley & Sons*, 2001.

[9] G.C. Feyzioglu, F. Tornuk, "Development of chitosan nanoparticles loaded with summer savory (*Satureja hortensis L.*) essential oil for antimicrobial and antioxidant delivery applications," *LWT – Food Science and Technology*, vol. 70, pp. 104-110, Jul. 2016

[10] B. Adinew, "GC-MC and FT-IR analysis of constituents of essential oil from *Cinnamon bark* growing in South-west of Ethiopia," *International Journal of Herbal Medicine*, vol. 1, no. 6, pp. 22-31, 2014.

Innovations in polymer nanocomposite coatings for EMI Shielding applications

M. Joshi[1], T. Bansala[1] and S. Mukhopadhyay[1]

[1] Department of Textile Technology, Indian Institute of Technology, Hauz Khas, New Delhi 110016, India

Corresponding author: mangala@textile.iitd.ac.in

Abstract. Start Excessive usage of electronic gadgets, telecommunication devices and electrical appliances has led to unwanted electromagnetic interference (EMI) as side effect. The EMI pollution interferes with the performance and functioning of the electrical equipments, which is generated through the conduction or radiation of emitted electro-magnetic (EM) waves. In addition to affecting the quality of electrical appliances, EMI has a huge negative impact on the health and life of living organisms. Moreover, microwave frequency operated functions such as radar surveillance systems; weather radar, military aviation, radar guns, wireless technology and satellite communication are more prone to EMI pollution and specifically with regard to stealth purposes. In this paper carbon nanomaterial based polyurethane nanocomposites have been synthesized and investigated for decreasing the EM wave pollution due to their light weight, flexibility, optimum conductivity and good dielectric properties. Three different types of graphene sheets such as thermally reduced and exfoliated graphene (TRG); chemically reduced graphene (CRG) and Polyvinyl-pyyrolidone stabilized silver nanoparticles based graphene nanohybrid (Ag-PVP-CRG) have been synthesized and characterized for EMI shielding properties. Further, TRG, CRG and Ag-PVP-CRG sheets have been dispersed in thermoplastic polyurethane (TPU) polymeric matrix to create multifunctional nanocomposite films with properties such as EMI shielding, electrically conductivity and dielectric behavior. The study presents for the first time a comparative evaluation of three different types of graphene sheets in both neat and nanocomposite form. The synthesis of Ag-PVP-CRG nanohybrids and its application for EMI shielding is novel and reported for the first time for such an application.

1. Introduction

The modern day electronic market is majorly governed by Electromagnetic waves and the major cause of unwanted electromagnetic interference (EMI) as side effect. The common examples of EMI in our daily routine are the flickering of picture and audio/video signal distortions in radio, television and computer screen. In addition to the quality degradation of electrical appliances, EMI has a huge negative impact on the health and life of living organisms. Moreover, microwave frequency operated functions such as radar surveillance systems; weather radar, military aviation, radar guns, wireless technology and satellite communication are more prone to EMI pollution and specifically with regard to stealth purposes.

In order to minimize the effect of EMI, emerging research has focused onto materials providing shielding against EMI pollution in all commercial and critical sectors. In general, the dissipation or absorption of incident electro-magnetic waves in the form of heat, conduction, dielectric or magnetic losses by using electrically conducting and magnetic material is known as electromagnetic shielding. EMI shielding materials either reflect or absorb the EMI radiations. Metal-based shielding materials are majorly reflection-based and suffer from processing difficulties, large density and sustainable to corrosion. Absorption based shielding materials such as polymer nanocomposites are preferred because of their camouflage nature as the incident EM signals will be absorbed instead of reflected which is desirable for defense systems. Literature review suggest that the electrical, mechanical, gas barrier, dielectric and EMI shielding properties of the polymeric nanocomposites are affected by a number of parameters viz. the aspect ratio, thermal or chemical treatment, dispersion (related to processing conditions) and loading level of graphene nanosheets. Therefore, a consolidated study dealing with the preparation of different types of graphene sheets such as chemically reduced graphene (CRG), thermally reduced and annealed graphene sheets (TRG) and silver nanoparticles decorated graphene sheets (Ag-PVPCRG) and their based thermoplastic polyurethane nanocomposites has been undertaken [1]. Furthermore, the effect of different types of graphene, their morphological aspects on the final tailored properties of TPU nanocomposites such as electrical, gas barrier, mechanical dielectric and EMI shielding properties has been discussed in detail. These nanocomposites show useful properties for decreasing the EM wave pollution along with light weight, flexibility, optimum conductivity and good dielectric properties. A new material based on silver nanoparticles and graphene sheets was synthesized and reported for first time for potential applications in the field of EMI shielding applications [2].

2. Methods and Materials

2.1. Materials

Natural graphite powder was acquired from Loba chemie, India (purity 99.5%, 60 mesh size). Concentrated sulphuric acid (H_2SO_4, 98%), potassium permanganate ($KMnO_4$), potassium per sulfate ($K_2S_2O_8$), di-phosphorus pentaoxide (P_2O_5), hydrochloric acid (HCl, 37%), silver nitrate (AgNO3 mol. weight 169.87 g/mol), hydrazine hydrate ($N_2H_4.H_2O$, 50.06 g/mol), N,N-dimethylformamide (density, 0.945 kg/L) and hydrogen peroxide (H_2O_2, 50%) were purchased from Merck, Mumbai and methanol from Fisher Scientific, India. Polyvinyl pyrrolidone PVP (K-40) was acquired from Sigma Aldrich. Procured chemicals and solvents required for synthesis were used directly.

2.2. Methods

Firstly, graphene oxide (GO) was synthesized through modified Hummer's method, which was reduced via chemical reduction method and thermal reduction routes [1-3]. The exfoliation by chemical reduction route is limited. Hence, thermal exfoliation of GO sheets was achieved via annealing in the thermal furnace. The increase in surface area and degree of exfoliation assist in effective dispersion in N, N-dimethylformamide (DMF). The polyvinyl pyrrolidone (PVP) stabilized silver nanoparticles decorated graphene sheets were produced through solvent assisted reduction of GO sheets without the addition of catalyst [2].

3. Results and discussion

3.1. Morphological characterization using SEM and HRTEM

The surface morphologies of CRG, TRG and Ag-PVPCRG nanohybrid were investigated using SEM and HRTEM analysis. Typical SEM images are shown in Figure 1. CRG is observed as aggregated stacked sheet like structures (Figure 1a). In comparison, TRG exhibits fluffy (exfoliated) sheets with lamellar structure consisting of randomly aggregated sheets (Figure 1b). Ag-PVP-CRG nanohybrid sheets morphology is shown in Figure 1c, which reveals agglomerated CRG sheets decorated uniformly with AgNPs.

Figure 1. SEM images at 5KX of (a) chemically reduced graphene (CRG), (b) thermally reduced graphene (TRG), inset shows exfoliated wrinkled morphology at high magnification and (c) Ag-PVP-CRG nanohybrid, inset shows AgNPs presence on graphene sheets.

HRTEM micrographs of Ag-PVP-CRG nanohybrid are shown in Figure 2. The distribution of PVP stabilized AgNPs over the CRG nanosheet is evident from Figure 2a. Graphene sheets are covered by spherical Ag nanoparticles with an average particle size of approximately 15–30 nm. The selected-area electron diffraction (SAED) pattern of single silver crystallite is shown in Figure 2b, and it confirms the presence of AgNP adherence on CRG sheets..

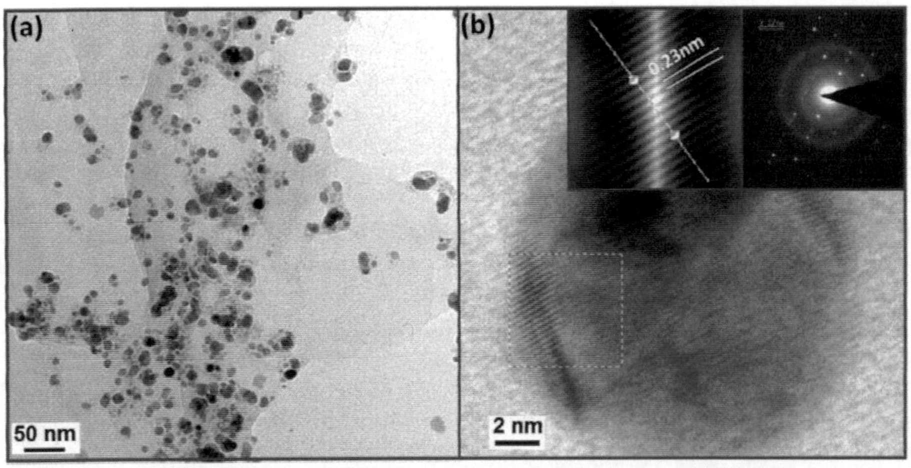

Figure 2. (a) TEM and (b) HRTEM image of Ag-PVP-CRG nanohybrid (inset shows lattice fringes and SAED pattern). The lattice-fringe spacing of silver

crystallites was measured to be 0.23 nm by HRTEM microscopy, which corresponds to the (111) crystal planes as shown in the inset of Figure 2b.

3.2. DC electrical conductivity analysis of Graphene nanosheets

Two-probe DC electrical conductivity was performed for GO, Ag-PVP-CRG nanohybrid, TRG and CRG at room temperature (30°C) as shown in Table 1. Rectangular shaped pellets measuring 2 mm in thickness were used for the measurement. The conductivity table shows the DC conductivity values as 1.2×10^{-4}, 0.25, 62.74 and 124.91 Scm^{-1} for GO, Ag-PVP-CRG, TRG and CRG respectively.

The rationale for low conductivity value in Ag-PVP-CRG nanohybrid relies upon the coating of non-conducting polymer (PVP) on the AgNP surface which hampers the conductivity of both AgNPs and graphene sheets. In the case of TRG, conductivity value is found be lower to that of CRG which is attributed to the presence of more residual OH functional groups in TRG, which are unable to restore the graphitic structure and its exfoliated morphology that leads to the lower stacking of the sheet structures in comparison to densely stacked sheets in CRG.

Table 1. Electrical conductivity values of Graphene nanosheets.

Graphene sheets	Electrical conductivity ($S\ cm^{-1}$)	BET Surface area (m^2/g)
CRG	124.91	18.20
TRG	62.74	159.61
Ag-PVP-CRG	0.25	n/a

3.3. EMI Shielding behavior of Graphene Nanosheets

The total EMI SE achieved for CRG, TRG and Ag-PVPCRG are ~ -80 dB, -45.1 dB and -57.8 dB at 15GHz, 2 mm thickness respectively as given in Table 2. The total EMI value (SE) for CRG, TRG and Ag-PVP-CRG majorly comes from the absorption losses, SEA. In general, the value of SER mainly depends upon the electrical conductivity of the material, whereas SEA depends on its thickness, conductivity and dielectric properties. The dielectric properties in terms of real and imaginary parts of the permittivity (ε', ε''), and tangent losses are given in Table 2 for all three forms.

The reason for highest EMI SE of CRG is attributed to (a) high conductivity of the order of 124.9 Scm^{-1} arising from increase in 3D conductive channels upon stacking of sheets, (b) the presence of clustered defects (residual functional groups) onto the sheet surface and (c) high values of dielectric permittivity. Thus, both high electrical conductivity and dielectric permittivity in terms of losses as well as stored energy plays a contributive role in attaining maximum SER and SEA value, yielding highest total SE. In comparison, TRG exhibits (a) low conductivity value (62.7 Scm^{-1}) due to highly exfoliated sheet structures, (b) lower residual bonds (oxygenic functional groups) and (c) low dielectric losses resulting in comparatively low SE value of -45dB. The dielectric permittivity in terms of stored energy plays a contributive role in attaining -45dB SE value for TRG. The high SE value (-58 dB) of Ag-PVP-CRG is a result of EM wave attenuation in the form of highest dielectric losses (Table 2) along with a moderate conductivity value of 0.25 Scm^{-1}. The dielectric losses ε''(ω) consist of both polarization losses and conductivity losses. The dielectric behavior of Ag-PVP-CRG comprises of polarization losses in terms of (a) interface polarization occurring at the interfaces between AgNPs and CRG sheet, (b) dipole polarization caused due to the presence of residual functional groups or defects in CRG sheets. The relatively high values of ε'' are attributed to the synergistic coordination between dielectric AgNPs and CRG sheets and effective conducting network formation of CRG sheets and PVP-AgNPs in Ag-PVP-CRG nanohybrid. The effective ε'' may also result from the quick dissipation of stored energy in the form of thermal energy. The high value of thermal conductivity of carbon and AgNPs in Ag-PVPCRG structures may add up to high dielectric losses resulting in effective dissipation of thermal energy. The

IOP Conf. Series: Materials Science and Engineering **460** (2018) 012012 doi:10.1088/1757-899X/460/1/012012

electrical conductivity of TRG is higher as compared to AgPVP-CRG, even though the EM wave losses due to reflection (SER) terms in case of Ag-PVP-CRG nanohybrid is higher which is attributed to the presence of metallic AgNPs in Ag-PVP-CRG.

Table 2. EMI Shielding Effectiveness summary of neat graphene nanosheets such as TRG, CRG and Ag-PVP-CRG @ 15 GHz

Material	SE_T	SE_A	SE_R	Real part of permittivity	Imaginary part of permittivity	Tangent losses
CRG	-80	-68	-11.9	73.4	56.2	0.76
TRG	-45.1	-40.1	-5	43.3	18.5	0.42
Ag-PVP-CRG	-57.8	-50.7	-7.06	88.5	139.9	1.58

3.4. SEM Fracture surface of nanocomposite films

The cross-sectional SEM micrographs of the cryofractured TPU/Graphene nanocomposites at 10 wt.% loading of different graphenes such as CRG, TRG and AgPVP-CRG is shown in Figure 3. The SEM morphology of TPU matrix exhibits smooth and continuous surfaces, while TPU/Graphene nanocomposites (TPU/TRG, TPU/CRG and TPU/Ag-PVP-CRG) appear to be rough, more likely as dendrite structures. The Ag-PVP-CRG nanohybrid in TPU matrix at 10 wt.% loading, appeared as scattered island morphology (Figure 3b), where PVP stabilized silver nanoparticles could be seen as droplets dispersed in TPU matrix. At high Ag-PVP-CRG loading, CRG sheets seem to be agglomerated and non-uniformly distributed in TPU matrix. Thus, TPU/Ag-PVP-CRG nanocomposite at higher content of Ag-PVP-CRG nanohybrid showed poor distribution of graphene sheets, revealing the presence of thick stacked CRG sheets. This agglomeration and aggregation of graphene sheets may be attributed to the solvent evaporation step in solution casting process during nanocomposite film preparation, leading to non-uniform distribution and dispersion of Ag-PVP-CRG in TPU matrix. In TPU/CRG nanocomposites, non-uniform and poor distribution of thick stacked CRG sheets was observed in Figure 3c .The SEM images of CRG revealed thick stacked sheets. The reason for CRG sheets stacking is the final stage water removal step during CRG synthesis from CRG/water.

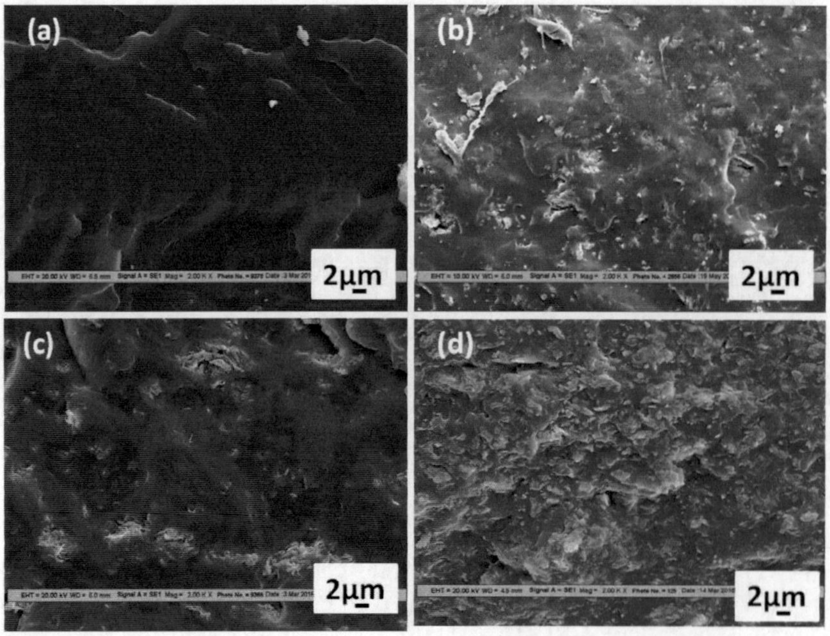

Figure 3. SEM micrographs of the fractured surfaces of (a) neat TPU, (b) 10 wt.% TPU/Ag-PVP-CRG, (c) 10 wt.% TPU/CRG, and (d) 10 wt.% TPU/TRG at 2KX magnification.

3.5. Electrical conductivity and EMI shielding property of Graphene/TPU nanocomposites

Inspite of highest electrical conductivity and EMI SE values of bare CRG sheets over TRG sheets, the TPU/TRG nanocomposite showed maximum EMI SE due to excellent exfoliation, fine dispersion and distribution of TRG sheets which form an effective 3-dimensional electrically conducting network inside the insulating polymer matrix (TPU) over CRG sheets which got agglomerated and aggregated within TPU matrix as shown in Figure 4 and 5. This has also been confirmed through DC conductivity values. The lowest EMI SE value obtained for TPU/Ag-PVP-CRG nanocomposite can be ascribed to the minimum conductivity values of Ag-PVP-CRG nanosheets as compared to CRG and TRG (Figure 5), due to the insulating coating of PVP onto AgNPs. Although the neat Ag-PVPCRG nanohybrid showed highest value ~ -58 dB EMI SE, but showed poor dispersion and aggregation when mixed in TPU matrix thus resulted in reduced EMI SE values when in nanocomposite with TPU. Therefore, TPU/TRG and TPU/CRG nanocomposite can acts as an effective shielding material, reducing the escape chances of incident EM wave more efficiently.

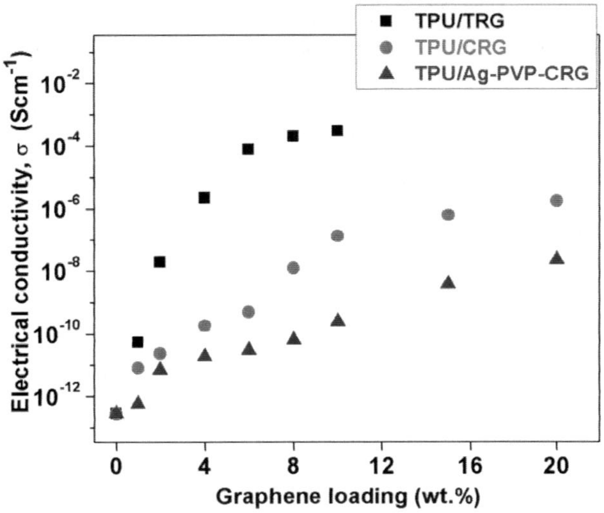

Figure 4. DC electrical conductivity TPU/Graphene nanocomposites vs. loading of graphene nanosheets.

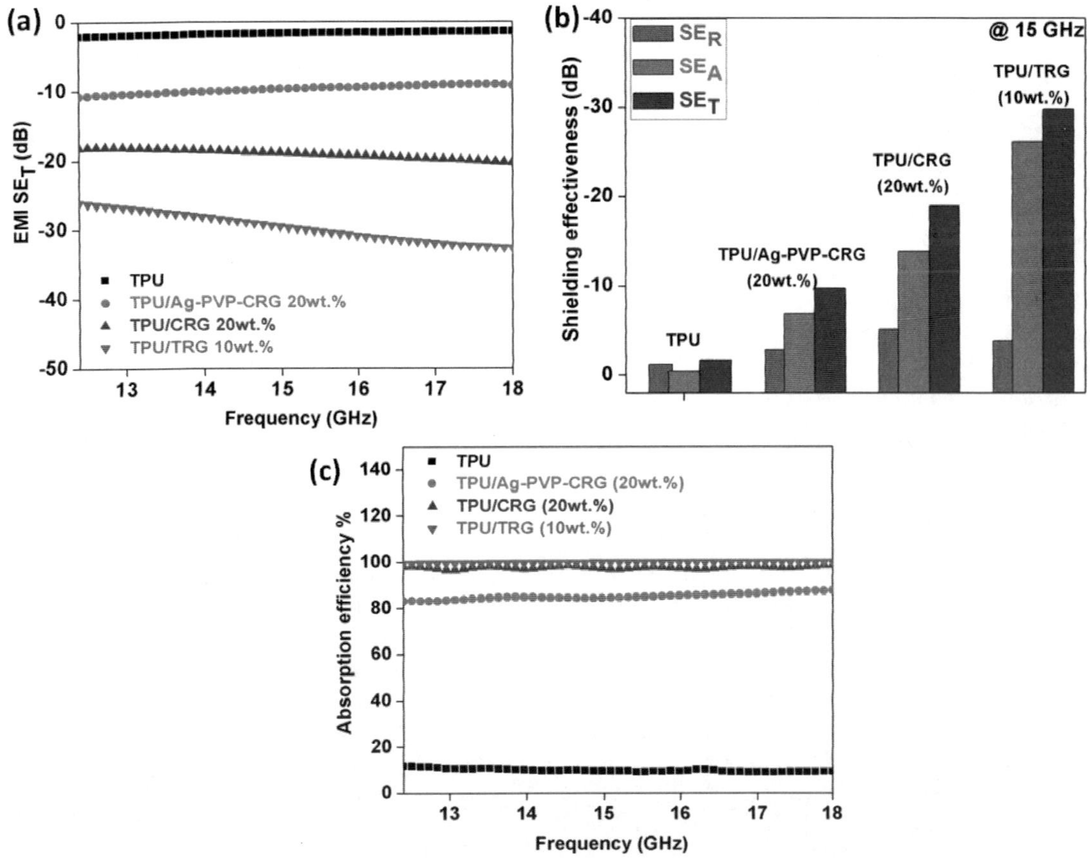

Figure 5. (a) EMI shielding effectiveness of TPU/graphene (TRG, CRG and Ag-PVP-CRG) nanocomposites with frequency (Ku Band region), (b) variation in SE_T, SE_A and SE_R with different graphene loading at 15 GHz (mid- frequency range), (c) variation in absorption efficiency.

3.6. EMI shielding behavior of graphene/TPU nanocomposites
The three different graphenes synthesized in the lab were then mixed with thermoplastic polyurethane by solution intercalation to prepare nanocomposite films and tested and characterized. A comparison of the conductivity, dielectric behavior and EMI shielding properties of TPU/graphene nanocomposites based on three different types of synthesized graphene i.e. CRG, TRG and Ag-PVP-CRG are presented in Table 3. Surface characterization via SEM of TPU/TRG showed fine dispersion of TRG sheets, due to excellent exfoliation of TRG sheets during synthesis as compared to TPU/CRG and TPU/Ag-PVP-CRG nanocomposite films which showed agglomerated CRG and Ag-PVP-CRG nanosheets. The highest DC conductivity value and lowest DC percolation threshold was obtained for TPU/TRG nanocomposite at 10 wt.% loading, comparatively at lower filler concentration as compared to TPU/CRG and TPU/Ag-PVP-CRG nanocomposites, where 20 wt.% loading could give similar property. The effective dielectric constant and tangent losses were obtained for 10 wt% TPU/TRG nanocomposites, as compared to TPU/CRG and TPU/Ag-PVP-CRG nanocomposites. TPU/TRG nanocomposite shows highest shielding effectiveness (SE) at 10 wt.% loading in comparison to TPU/CRG and TPU/Ag-PVP-CRG where SE achieved at higher loading ~20 wt.% (Table 3).

Table 3. Summary of EMI Shielding Effectiveness of different graphene (TRG, CRG and Ag-PVP-CRG) based TPU nanocomposites @ 15GHZ

Material	SE_T	SE_A	SE_R	Absorption efficiency %
TPU	-1.61	-0.43	-1.18	9.4
TPU/TRG 10wt.%	-29.8	-26.1	-3.7	99.9
TPU/CRG 20wt.%	-18.8	-13.8	-5.07	95
TPU/Ag-PVP-CRG 20wt.%	-9.6	-3.8	-5.8	82

4. Conclusion
The above comparative data analysis showed that amongst the three types of nanocomposites TPU/TRG nanocomposite show the most promising EMI shielding behavior at 10 wt.% loading and thus meet the basic requirements of the commercial electrical applications against electromagnetic pollution. The reason may be that it also shows highest electrical conductivity, dielectric constant and tangent losses amongst the three nanocomposites prepared. The maximum achieved exfoliation and uniform dispersion in TPU/TRG nanocomposites further helps in giving best EMI shielding parameter.

References
[1] T. Bansala, "Studies on Thermoplastic/graphene based nanocomposites", PhD Thesis, 2017, IIT Delhi, India.

[2] T. Bansala, M. Joshi, S. Mukhopadhyay, R A. Doong and M Chaudhary,"Synthesis and shielding properties of PVP-stabilized-AgNPs-based graphene nanohybrid in the Ku band" 2016, Synth. Met. vol. 221, pp. 86-94.

[3] T. Bansala, M. Joshi, S. Mukhopadhyay, R. A. Doong, M. Chaudhary, "Electrically conducting graphenebased polyurethane nanocomposites for microwave shielding applications in the Ku band" 2017, *J. Mater. Sci.*, vol. 52, no. 3, pp. 1546–60.

[4] T. Bansala, M. Joshi, S. Mukhopadhyay, R. A. Doong, M. Chaudhary, "Electrical and Dielectric Properties of Exfoliated Thermally Reduced Graphene Based Polyurethane Nanocomposites", 2017, *J. Nanosci. Nanotechnol.* vol. 17, no. 12, 8782-90.

IOP Conf. Series: Materials Science and Engineering **460** (2018) 012013 doi:10.1088/1757-899X/460/1/012013

Nanofibers in face masks and respirators to provide better protection

C. Akduman[1] and E. P. Akçakoca Kumbasar[2]

[1] *Pamukkale University, Denizli Vocational School of Technical Sciences, Department of Textile Technology, 20100, Denizli, Turkey*

[2] *Ege University, Faculty of Engineering, Department of Textile Engineering, 35100, Izmir, Turkey*

perrin.akcakoca@ege.edu.tr

Abstract A facemask is a loose-fitting, disposable device that creates a physical barrier between the mouth and nose of the wearer and potential contaminants in the immediate environment. They are generally labelled as surgical, isolation, dental or medical procedure masks. On the other hand, respirators are personal air purifiers. They are designed to protect the wearer from inhaling dangerous substances such as toxic chemicals and infectious particles. Respirators are designed to help reduce the wearer's respiratory exposure to airborne contaminants such as particles that are small enough to be inhaled - particles less than 100 microns (μm) in size. A face masks or a respirator consist entirely or substantially of filter material or comprises a face piece in which the main filter(s) form an inseparable part of the device. Nanofibers could be the key elements for filter materials in face masks or respirators. They have a very high surface area per unit mass that enhances capture efficiency and other surface area-dependent phenomena that may be engineered into the fiber surfaces (such as catalysis or ion exchange). They could enhance filter performance for capture of naturally occurring nanoparticles such as viruses, as well as micron-sized particles such as bacteria or man-made particles such as soot from diesel exhaust.

***Keywords*— face masks, filtration, nanofibers, respirators.**

I. Introduction

THIS study summarizes the general information about face masks and respiratory devices, face mask and respirator types, some related standards, and finally application of nanofibers in these devices.

Filtering face piece particulate respirators and facemasks are widely utilized for reducing inhalation exposure to airborne particles that may be associated with various health effects [1]. They are considered to prevent or slow down the transmission of airborne particles possibly causing adverse health effects. A simple definition for facemask is "a loose-fitting, disposable device that creates a physical barrier between the mouth and nose of the wearer and potential contaminants in the immediate environment". A face mask generally labelled as surgical, isolation, dental or medical procedure mask [2]. A conventional face mask consists of one, two or three layers of flat or pleated fabric, affixed to the head with ear loops [3]. A surgical mask is generally used as a physical barrier to body fluids and larger droplets in healthcare activities [4].

The most common facemasks are the surgical masks but the main differences are; simple face masks are not used to prevent infectious diseases but a surgical mask should have a minimum of 80% bacteria filtration efficiency, a conventional face mask consists of one, two or three layers of flat or pleated fabric [5], usually made from paper or cotton fabric but a "surgical mask" is made of at least a three-ply layer of material like polypropylene [3]. However surgical masks do not have either adequate filtering or fitting attributes to provide respiratory protection for the wearer. They are not intended to be used more than once [2] and are usually designed to help prevent contamination of the work environment or sterile field from large particles generated by the wearer (e.g. spit, mucous). Surgical masks may also be used to help reduce the risk of splashes or sprays of blood, body fluids, secretions and excretions from reaching the wearer's mouth and nose [6].

On the other hand, a respirator is a device designed to protect the wearer from inhaling dangerous substances such as toxic chemicals and infectious particles [5]. Respirators especially help to reduce the wearer's respiratory exposure to airborne contaminants such as particles that are small enough to be inhaled - particles less than 100 microns (μm) in size. Respiratory protective equipment (RPE) is a particular type of personal protective equipment (PPE) [7]. RPE is divided into two main types: 1. Respirator (filtering device), 2. Breathing apparatus (BA) [7]. Respirators use filters to remove contaminants in the workplace air and help reduce the wearer's respiratory exposure to airborne contaminants. This includes airborne particles that may contain biological material, e.g. mold, Bacillus anthracis, Mycobacterium tuberculosis, the virus that causes Severe Acute Respiratory Syndrome

(SARS), Avian Flu, Ebola Virus etc. or even PM2.5 (particulate matter with diameter of 2.5 micron or less) [6].

II. Standards And Classification

Surgical masks are cleared by the Food and Drug Administration (FDA) [2]. FDA does not do any testing; but reviews the information supplied by the manufacturers in their 510(k) premarket application. Manufacturers submit test results for fluid resistance, filtration efficiency for polystyrene latex and Staphylococcus aureus bacterial aerosol particles, differential pressure and flammability for Surgical masks clearance usually in accordance with ASTM F2100-11 [4]. ASTM F2100-11 specification covers testing and requirements for materials used in the construction of medical face masks that are used in providing health care services such as surgery and patient care. It provides performance classes for face mask materials as bacterial filtration efficiency (ASTM F2101), sub-micron particulate filtration efficiency (ASTM F2299), differential pressure (MIL-M-36954C), fluid penetration resistance (ASTM F1862) and flammability (16 CFR Part 1610) [8].

Medical face mask materials covered under this specification shall be designed as one or more of the following performance classes as based on the barrier performance properties of the materials used in medical face masks: Level 1 Barrier, Level 2 Barrier, and Level 3 Barrier. The properties of the medical face mask material shall conform to the specifications requirements in Table 1 [8].

TABLE I
MEDICAL FACE MASK MATERIAL REQUIREMENTS BY PERFORMANCE LEVEL (ASTM F2100-11) [8]

Characteristic	Level 1 Barrier	Level 2 Barrier	Level 3 Barrier
Bacterial filtration efficiency, %	≥95	≥98	≥98
Differential pressure, mm H_2O/cm^2	<4.0	<5.0	<5.0
Sub-micron particulate filtration efficiency at 0.1 micron, %	≥95	≥95	≥95
Resistance to penetration by synthetic blood,	80	120	160
Flame spread	Class 1	Class 1	Class 1

Besides, the respirators must provide protection at the highest concentration the person will experience. In the United States, the National Institute for Occupational Safety and Health (NIOSH) tests the filtration efficiency of particulate filtering, air-purifying respirators for certification purposes. NIOSH approves N-, R-, and P-series non-powered air-purifying respirators, each at 95, 99, and 99.97% filtration efficiency levels under 42 CFR Part 84 as shown in Table 2 [9].

TABLE II
NIOSH 42 CFR PART 84 PARTICLE FILTER CLASSIFICATIONS [6]

Minimum Efficiency	N-Series	R-Series	P-Series
95%	N95	R95	P95
99%	N99	R99	P99
99.97%	N100	R100	P100

According to NIOSH 42 CFR 84, the three efficiency levels are 95, 99, and 99.97%, tested at a flow rate of 85 L/min at the most penetrating particle size (generally about 0.1 to 0.3 μm). The three degradation resistance series were established by the choice of either charge-neutralized NaCl (sodium chloride salt), which is only mildly degrading to filter media (N series of filters), or DOP (dioctyl phthalate) liquid oil, which is highly degrading (R or P series) [10]. Then filters are rated as N, R and P series to indicate the filters are 'not oil resistant', 'resistant to oil' and 'oil proof' respectively [11]. Accordingly, N series filters tested with NaCl aerosol are recognized as not highly resistant to degradation and only appropriate for use with solid aerosol in the workplace. Among the respirators certified under 42 CFR Part 84, the N95 respirators are the most commonly used [6], [11]. Because the filter efficiency is based on the physical parameters of the particles to be filtered, any biologic particles can be expected to be filtered at no less efficiency than the test aerosol (ie, at least 95% efficient for an N95 filter [9]. The penetration, P, of such particles through a certified N95 respirator cannot exceed 5%; thus, the efficiency, E, of the respirator, which is calculated as E = 100% - P, must be at least 95% [12].

In Europe, the Personal Protective Equipment directive (89/686/EEC) requires that PPE placed within the European market should be certified by European Norm (EN) and marked with 'Conformité Européen'(CE), indicating European Community (EC) conformity. All respirators must be approved and tested to the performance requirements of the corresponding European Standard, which forms the following categories: filtering half masks, half masks and quarter masks, full face masks, powered air respirators, and supplied air respirators [12].

Legislation of European Standards for "filtering half masks" (also called filtering face pieces (FFP)) is covered by EN 149:2001 [13], [14]. They usually marked with EN 149:2001 and an additional sub category such as FFP1 (low efficiency; 80%), FFP2 (medium efficiency; 94%) and FFP3 (high efficiency; 99%) (Table 3) [6]. They are designed to protect against solid, non-volatile water-based, and oil-based aerosol particles. Therefore, all respirators have to meet both the solid and liquid filter performance requirements [14].

EN 149:2001 was followed by the amended version EN 149:2001+A1:2009 in July 2009. In this amendment two usability classes of disposable respirators: single shift only (non-reusable, marked 'NR') and reusable (marked 'R') were included. The US N95 standard is roughly equivalent to FFP2 or 3 as it is efficient up to 10 x the local occupational exposure limit [6]. The higher the FFP number, the more protection the respirator can provide if it is used properly [14]. EN 149 include filter penetration, extended exposure (loading), flammability, breathing resistance, total inward leakage (TIL), and dolomite dust clogging (optional). For testing filter penetration, EN 13274-7:2008 (Determination of particle filter penetration) is used and it dictates the use of a non-neutralized polydisperse 1% solution of sodium chloride (NaCl) against solid particles and paraffin oil against oil based aerosol at flow rate of 95 L min^{-1}. The NaCl aerosol is generated by a Collison atomizer with a particle size range of approximately 40–1200 nm and a mass median diameter (MMD) of $\approx$ 600 nm. The sodium chloride aerosol is analysed before and after the filtering device under test by flame photometry. Sodium chloride particles are vaporized while they are passing through, and the flame tube gives a characteristic sodium emission at 589 nm. The concentration of sodium in the air flow is calculated from the intensity of this emission [14].

TABLE III
EN 149:2001 PARTICLE FILTER CLASSIFICATIONS [13]

Filter Class	Efficiency
FFP1	80%
FFP2	94%
FFP3	99%

Beside EN 149, European standard EN 143 (Respiratory protective devices -Particle filters - Requirements, testing, marking) defines the P1, P2 and P3 classes of particle filters that can be attached to a face mask [16]. Based on their performance level; P1 filters have low filter performance (80% efficiency), P2 filters have medium filter performance (94% efficiency) and P3 filters have high filter performance (99.97% efficiency) [5].

III. FILTER MATERIALS

A filtering half mask is a face piece that consists entirely or substantially of filter material or comprises a face piece in which the main filter(s) form an inseparable part of the device [6]. Besides, respiratory protection offered by a particulate respirator is a function of the filter efficiency and face seal leakage [17].

The Elsevier Handbook of Filter Media has a precise definition of a filter medium: A filter medium is any material that, under the operating conditions of the filter, is permeable to one or more components of a mixture, solution or suspension, and is impermeable to the remaining components [18]. Fibrous materials are important for filtration. A significant feature of fibrous porous materials is that the fraction of solid material is often much lower than that for a granular material. The main driving force in the filtration industry is the requirements for finer degrees of filtration. In air filtration, the requirement for finer filtration is already being met by media with finer fibers. Typically, in filtration processes, the size of the items to be separated needs to be matched by the size of the fiber so that the removal of particles or microorganisms at diameters of less than 1 micrometer is possible.

More often, fiber blend is used to balance filtration efficiency and pressure drop. High pressure drop across a filter translates to the undesired high energy consumption to drive air flow through the filter. When fiber size is reduced the total filtration efficiency becomes higher. The increase in filtration efficiency is due to the large surface area per volume available for particle capture, especially for small submicron meter particles. This encouraged a wave of small fiber innovation and commercialization that boost the filtration industry starting from early 1980s [19] and leads to the development of nanofibrous filtering media.

IV. Nanofibers In Filtration

Nanofibers and nanofiber webs, as used in filter media, are an important part of that emerging technology to improve the efficiency of filter media [5]. The filtration and separation of sub-micron sized contaminants is a major concern of modern day nanotechnology. Because of their unique physical and chemical properties, engineered nanoparticles are synthesized for various applications. Nowadays in industrial workplaces the rapid growth of nanotechnology has increased the production of these engineered nanoparticles. Tasks and processes such as cleaning engineered nanomaterials from dust collection systems, spills, and disposal can present a potential for inhalation exposure and a high level of risk [17]. Nanofibrous media have low basis weight, high permeability and small pore size that make them appropriate for a wide range of filtration applications especially for smaller particles. In addition, nanofiber membranes offer unique properties such as high specific surface area (ranging from 1-100 m^2/g depending on the diameter of fibres and intrafiber porosity), good interconnectivity of pores and potential to incorporate active chemistry or functionality on a nanoscale [20]. Nanofibers do not exist independently in filtration, but are produced onto a substrate, typically a nonwoven fabric [5].

There are four main types of relevant filtration mechanism using nanofiber filter media. These are surface straining, depth straining, depth filtration and cake filtration [5]. In practice, the filtration processes often involve a combination of two or more mechanisms. In surface filtering, the surface of filter media plays an active role and the surface pore openings of filter media are precisely controlled in relation to the particulate size to be filtered. Depth straining type of filter media is relatively thick compare to surface filter media and the pores are variable in the flow path length. Depth filtration uses various physical mechanisms for removing a particle from a fluid even though the particle may be smaller than the pore diameter at any point in the pore structure. In cake filtration, the filter medium begins the filtration process; successive layers of particles deposit on the top of preceding layers to form a cake [20].

The technical requirements for filters are a balancing of the three major parameters of filters performance: filter efficiency, pressure drop and filter life time [20] and nanofibers provide a dramatic increase in particle capture efficiency as well as increased pressure drop [21] related to the nanofiber thickness on the base filter substrate.

The market for nanotechnology worldwide was estimated to grow at an annual rate of 19% during the years (2011-2013). Nanofiber-spun webs for nonwoven and filtration purposes are part of the growth [5]. Nanofiber filter media have enabled new levels of filtration performance in several diverse applications within a broad range of environments. Today, more people involved in the air filtration market are looking at nanofiber treated media in order to improve dust holding capacity and filtration efficiency. There are many applications of nanofibrous filters which are already commercialized as well as those still in development [20]. One of them is using nanofibers in facemasks and respirators.

Loesecke et al. indicated that a typical cellulose web with a nanofiber coating could catch particulates down to 0.1 microns, improving filtration efficiency with negligible penalty in permeability or resistance due to the significantly smaller fiber size compared to the substrate itself [22]. The most popular method for creating nanofibers for filtration applications has been electrospinning [5]. Li et al. produced polysulfone based nanofibers for mask filtration by electrospinning. They adjusted the thickness of the electrospun fiber matts with collecting period (15 min < 30min < 60min) and compared three nanofiber masks with disposable non-woven face mask, non-woven mask for operation room, Ito PM2.5, a N95 and a R95 repirator. They demonstrated that, produced nanofiber mask materials could efficiently filter out the PM2.5 particles and simultaneously preserve a good breathability [23]. Skaria et al. incorporate a nanofiber filter media (PT) into the surgical face masks. They observed that, PT with nanofiber filter was able to filter more efficiently than the commercial masks at 65.03% (60.47–69.58%) and when sealed 82.68% (80.57–84.78%). They indicated that, in contrary to the commercial face masks, produced nanofiber prototype retained its low resistance, even when sealed to the face, as well as its filtration performance when unsealed. They also found that this nanofiber media significantly reduced mask airflow resistance and resulted in more of the exhaled air from the manikin passing through the face mask as opposed to by passing the filter and going around the edges [24]. Wang et al. prepared an ultra-light weight (2.94 g/m^2) nanofiber-nets binary filter medium by the accumulation of nano-fibers/nets structured PA6-15 and PAN nanofibers via multi-jet spinning, which presented easy-to-use air filters for 300 nm aerosol particles [21]. In the study of Ramaseshan et al., they functionalized a polymer nanofiber membrane to capture chemical warfare agents. They used a catalyst for the detoxification of nerve agents and synthesized from β-cyclodextrin (β-CD) and o-iodosobenzoic acid (IBA). Then they produced PVC nanofiber membranes with β-CD, IBA, a blend of β-CD + IBA, and the synthesized catalyst. They observed that when tested against the decontamination of paraoxon, a nerve agent stimulant, the stimulant gets hydrolyzed. They indicated that, these functionalized membranes are suitable for use in filter media with the properties provided by a filter membrane with increased porosity and hydrophilicity in combination with improved catalytic performance given by catalyst material. Such functionalized nanofibers with different catalysts can also degrade other types of chemical and biological warfare agents and various nanofiber layers can be put together along with a substrate to form protective garments [25].

V. CONCLUSION

A filtering half mask or a respirator consists entirely or substantially of filter material to meet high protection. The filtration and separation of sub-micron sized contaminants is a major concern of today. Filter efficiency is known to improve as the size of the fibers decreases. The efficiency curves have a 'V' shape with the apex of the 'V' indicating the MPPS. MPPS for filters is an important factor to be considered in the respirator selection process. As the fiber diameter decreases, the most penetrating particle size decreases and the capture efficiency of the most penetrating particle size increases. In this point nanofibers could be the key elements for filter materials in face masks or respirators. They have very high surface area per unit mass that enhances capture efficiency and other surface area-dependent phenomena that may be engineered into the fiber surfaces (such as catalysis or ion exchange). They could enhance filter performance for capture of naturally occurring nanoparticles such as viruses, as well as micron-sized particles such as bacteria or man-made particles such as soot from diesel exhaust. We believe respiratory protective equipment developments require more research on advanced nanofibers filtration technology to provide necessary protection from airborne threats.

ACKNOWLEDGMENT

C.A. Author thanks to Pamukkale University 2018KRM002-173, HYFIBER-Abalıoğlu Teknoloji for the support given.

REFERENCES

[1] S. A. Grinshpun, H. Haruta, R. M. Eninger, T. Reponen, R. T. McKay & S. A. Lee, "Performance of an N95 filtering facepiece particulate respirator and a surgical mask during human breathing: two pathways for particle penetration," *J Occup Environ Hyg*, vol. 6 no.10, pp. 593-603, 2009.

[2] FDA, U.S. Food and Drug Administration, Medical Devices, (2017, December 10) [Online]. Available: https://www.fda.gov/MedicalDevices/ucm055977.htm

[3] HAS Health Sciences Authority, Health Products Regulation. (2017, December 10) [Online]. Available: http://www.hsa.gov.sg/content/dam/HSA/HPRG/Consumer_Information/Public%20Advisories/Influenza_A_H1N1_information/consumer_advisories/HSAConsumerGuide-MakingInformedDecisionsWhenBuyingMasks_Respirators-FINAL-Aug09.pdf

[4] S. Rengasamy, R. Shaffer, B. Williams & S. Smit, "A comparison of facemask and respirator filtration test methods," *J Occup Environ Hyg*, vol. 14, no.2, pp. 92-103, 2017.

[5] I. M. Hutten, *Handbook of nonwoven filter media*. Oxford: Elsevier, 2016, ch.8

[6] V. Pathak, (2016 Nov.) "Respirators for protection against PM2.5", 3M Personal Safety Division, [Online]. Available: https://multimedia.3m.com/mws/media/1313143O/respirators-for-protection-agains.pdf

[7] Respiratory protective equipment at work A practical guide, Health and Safety Executive, ISBN 978 0 7176 6454 2, [Online]. Available: http://www.hse.gov.uk/pUbns/priced/hsg53.pdf

[8] ASTM F2100-11 "Standard Specification for Performance of Materials Used in Medical Face Masks"

[9] A. Rengasamy, Z. Zhuang & R. Berry Ann, "Respiratory protection against bioaerosols: Literature review and research needs," *Am J Infect Control,* vol. 32, no.6, pp. 345-354, Oct. 2004

[10] Y.Qian, K.Willeke, S. A. Grinshpun, J. Donnelly & C. C. Coffey, "Performance of N95 respirators: filtration efficiency for airborne microbial and inert particles," *Am Ind Hyg Assoc J*, vol. 59, no.2, pp. 128-132, 1998.

[11] N. Mao, "*Methods for characterisation of nonwoven structure, property, and performance*", *Advances in Technical Nonwovens*, ed G Kellie (Amsterdam: Elsevier) 2016, ch 6.

[12] A. Balazy, M. Toivola, A. Adhikari, S. K. Sivasubramani, T. Reponen, & S. A. Grinshpun, "Do N95 respirators provide 95% protection level against airborne viruses, and how adequate are surgical masks?" Am J Infect Control, vol. 34, no.2, pp. 51-57, 2006.

[13] TS EN 149+A1, "Respiratory protective devices -Filtering half masks to protect against particles-Requirements, testing and marking"

[14] N. Serfozo, J. Ondráček, P. Otáhal, M. Lazaridis and N. Ždímal, "Size-Resolved Penetration of Filtering Materials from CE-Marked Filtering Facepiece Respirators," *J Occup Environ Hyg*, vol. 14, no.12, pp. 965-974, 2017.

[15] TS EN 13274-7: 2008-12," Respiratory protective devices - Methods of test - Part 7: Determination of particle filter penetration"

[16] TS EN 143, "Respiratory protective devices- Particle filters-requirements, testing and marking"

[17] S. Rengasamy, R. BerryAnn, & J. Szalajda, "Nanoparticle filtration performance of filtering facepiece respirators and canister/cartridge filters," *J Occup Environ Hyg*, vol. 10, no.9, pp. 519-525, 2013.

[18] K. Sutherland, "*Filter media*" in *Filters and Filtration Handbook*, 5th ed. K Sutherland Ed. Elsevier, 2008 pp. 41-95.

[19] Y. A. N. G. Chuanfang. "Aerosol filtration application using fibrous media—an industrial perspective" *Chinese J Chem Eng*, 20(1), 1-9.

[20] R S. Barhate, S. Sundarrajan, D. Pliszka and S. Ramakrishna, "Fine chemical processing: The potential of nanofibres in filtration" *Filtr Separat,* 45(4) 32-35, 2008.

[21] N. Wang, Y. Yang, S. S. Al-Deyab, M. El-Newehy, J. Yu & B. Ding, "Ultra-light 3D nanofibre-nets binary structured nylon 6–polyacrylonitrile membranes for efficient filtration of fine particulate matter" *J Mater Chem A*, 3(47) 23946-23954, 2015.

[22] D. Von Loesecke and B. Murphey "Ultra-light 3D nanofibre-nets binary structured nylon 6–polyacrylonitrile membranes for efficient filtration of fine particulate matter" *Filtr Separat*, 45(7) 17-19, 2008

[23] X. Li and Y. Gong "Design of Polymeric Nanofiber Gauze Mask to Prevent Inhaling PM2.5 Particles from Haze Pollution" *J Chem*, 1-5, 2015.

[24] S. D. Skaria & G. C. Smaldone "Respiratory Source Control Using Surgical Masks With Nanofiber Media" *Ann Occup Hyg*, 58(6) 771-7812014

[25] R. Ramaseshan, S. Sundarrajan, Y. Liu, R. S. Barhate, N. L. Lala & S. Ramakrishna, Functionalized polymer nanofibre membranes for protection from chemical warfare stimulants. *Nanotechnology*, 17(12), 2947, 2006.

IOP Conf. Series: Materials Science and Engineering **460** (2018) 012014 doi:10.1088/1757-899X/460/1/012014

Virtual technology of Made-to-Measure Men Shirt

J Q Yan[1] and V E Kuzmichev[1,2]

[1]Ivanovo State Polytechnic University, Department of Garment Design, No.21 Sheremetev Avenue, Ivanovo, 153000, Russian Federation
[2]Wuhan Textile University, Fashion Faculty, 1000 Talent Plan Laboratory, No.1 Sunshine Avenue, Jiangxia District, Wuhan, 430200, Hubei province, China

wkd37@list.ru

Abstract. The aim of this research is to propose the enhanced method of pattern block drafting with novel pattern-oriented body measurements and to optimize the fit of men shirts and customization process that demonstrating by the virtual system "body-garment". To investigate and devise this method, 119 young male bodies were scanned by 3D body scanner. New anthropometric indexes were proposed according to the theories and basic rules of sketching patterns[1]. Competitive neural networks and self-organized mapping neural networks[2] (*SOMNN*) were trained and tested to clustering the male body features into certain types. Thereby novel patterns for diverse types were employed as the database of customized men shirts. Compared with the traditional pattern blocks, the new ones are providing the better fit in virtual environment. Based on the proposed method, the individual pattern could be generated immediately from the existed database with facilitated precision and efficiency, by matching the body measurements of customers to certain anthropometric features.

1. Introduction

Made-to-Measure (*MTM*) garments have become pervasive especially since the flourish of 3D body scanning technologies in fashion industry[3]. While the final products are either hardly improved, or time-consuming comparing with conventional ones. One of the critical reason is the lack of applicable pattern-oriented anthropometric methods, which could be attained by 3D body scanner and directly utilized in pattern sketching[4]. To achieve the optimum fit and efficient procedure, new indexes of male bodies were extracted after body scanning[5], as the components for the individualized pattern blocks. Moreover, with *SOMNN*, the intricate indexes were automatically narrowed down into discernible clusters to express the features of body postures. Meanwhile, the database of men shirt pattern blocks was established, correlated with features of different body segments.

To achieve the fit and eliminate the misfit in individualized men's shirts, this research proposed the exemplary pattern block customization methods with novel scanned body measurements, and validated the feasibility in realistic virtual system "male body-shirt".

2. Methods and procedure

2.1. Attainment of pattern-oriented anthropometric indexes

119 males from China and Russia at the age of 18 to 30 were classified into Y, A, B, C type in advance according to Chinese standard sizing systems for garments (GB/T 1335.1-2008)[6]. Moreover, to acquire and validate new anthropometric indexes which are informative for personalized patterns, four pivotal

body segments were processed respectively: neck, area between torso and arm, shoulder, and torso. Hereby all subjects were scanned with 3D body scanner (Human Solutions GmbH) with naturally standing posture and uplifted arm. The generated 3D body mesh models were imported and processed in 3D modeling software Rhinoceros.

As Figure 1 shows, considering the protocol measurements and 3D models, the basic body anthropometrical points were marked as *FNP* (front neck point), *SNP* (shoulder neck point), *BNP* (back neck point), *SP* (shoulder point), *BP* (breast point), *ADP* (armpit depth point). Thus 16 new additional anthropometric indexes were procured and calculated, which were marked as *NGF* (neck girth front), *NGB* (neck girth back), *NDV* (neck depth in vertical), *HDH* (neck depth in horizontal), *D1* (distance between *SP* and *BP*), *D2* (distance between *SP* and *FNP*), *D3* (distance between *SP* and *BNP*), *D4* (distance between *SP* and back waist center), *SA* (shoulder angle), *DSFW* (distance between *SNP* and front waist), *DSBW* (distance between *SNP* and back waist), *DFPB* (distance between front center and profile center on bust level), *DBPB* (distance between back center and profile center on bust level), *DFPW* (distance between front center and profile center on waist level), *DBPW* (distance between back center and profile center on waist level), *AGA* (arm girth through *ADP*), *AD* (armpit depth) respectively in accordance with the segment located.

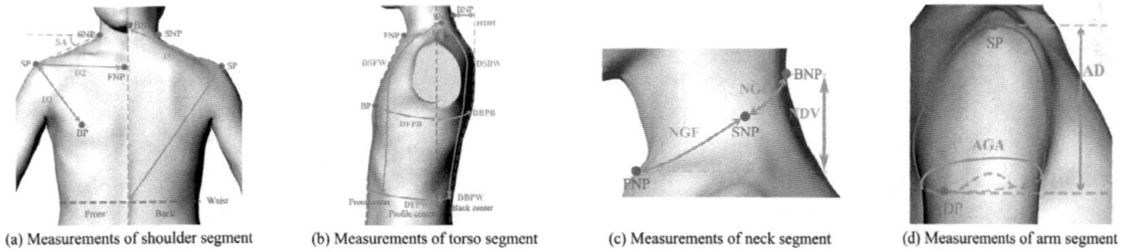

Figure 1. The new anthropometric indexes from 3D scanned bodies.

2.2. Attainment of pattern-oriented anthropometric indexes

According to the air volume between the shirt and the body, the styles of men shirts were generally classified as body fit, slim fit, regular, loose, etc., in accordance with E_{WG} (ease to half waist girth) ranging from 2 to 12 cm. Body fit, slim fit and regular shirts (E_{WG} from 2 to 10 cm) were primarily performed in this research.

At first, in terms of the features of different segments, the pattern prototype could be amended into certain categories accordingly. Besides, the fundamental frame was drawn from the initial prototype and basic body measurements such as chest girth, back length, etc. In consideration of these pattern-oriented body measurements, the shirt pattern prototype [1] was sketched as Figure 2 indicates.

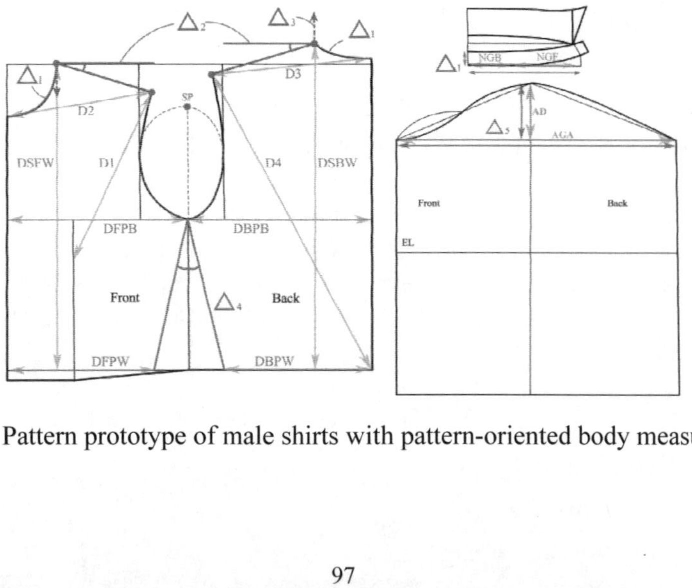

Figure 2. Pattern prototype of male shirts with pattern-oriented body measurements.

The patterns of torso, sleeve and collar were finalized in terms of corresponding new measurements with the pattern block parameters expressed as follow:

$$\Delta_1 = f(NGF,\ NGB,\ NDV) \quad (1)$$

$$\Delta_2 = f(D1,\ D2,\ D3,\ D4,\ SA) \quad (2)$$

$$\Delta_3 = f(DSFW,\ DSBW,\ HDH) \quad (3)$$

$$\Delta_4 = f(DFPB,\ DBPB,\ DFPW,\ DBPW) \quad (4)$$

$$\Delta_5 = f(AGA,\ AD) \quad (5)$$

where Δ_1 is the length and height of collar band, Δ_2 is the sloping angle and length of shoulder deployed by *SP*, Δ_3 is the vertical length of front and back torso deployed by *SNP*, Δ_4 is the proportion of front and back girth in bust and waist level deployed by side line, Δ_5 is the height and width of sleeve cap.

2.3. Classification of the new indexes with neural networks

The pattern blocks of *MTM* shirt are conventionally drafted with detailed measurements according to the basic pattern prototype whereas the postures and features of bodies are variable, sometimes particular and abnormal. Intricate adaption should be conducted to obtain the customized patterns deriving from the prototypes. To simplify the procedure of customization and to improve the fit of final products, the new anthropometric indexes responsible to Δ_1, Δ_2, Δ_3, Δ_4, Δ_5 were classified, determining the corresponding prototype pattern segments for different categories. Therefore, based on the body measurements and the database of categorized pattern segments, the unique customized pattern could be generated immediately without subsequent adaption.

As one of the most essential unsupervised learning methods in neural network, the competitive neural networks and *SOMNN* were utilized to clustering the new indexes. Figure 3 shows the models of these networks, two layers constitute the network: input layer and competitive layer. The indexes were imported in the input layer, after the calculation of the negative Euclidean distance of input matrix *P* and weight matrix *W* (plus bias matrix *b* in competitive neural networks), the results were exported in the competitive layer with winner neuron outputted 1, and other neurons outputted 0, which indicated the category of the inputs. Moreover, 36 neurons in the competitive layer of *SOMNN* were arranged as the topological structure (6 × 6 hexagon structure in this research), and the weight were modified with winner neurons (in competitive neural networks) or the neighborhood of winner neurons (in *SOMNN*) in every learning iteration.

(a) The structure of competitive neural networks (b) The structure of self-organized mapping neural networks (*SOMNN*)

Figure 3. The models of competitive neural network and *SOMNN*.

The samples were separated into training set with 84 objects for training and testing set with 35 for validating the models.

With configuration of necessary parameters (learning rate, weight, bias, learning function, transfer function, etc.), categories of different segments were clustered, and the results of test set proved the veracity of models. As table 1, 2, 3, 4, 5 demonstrates, the indexes were clustered into 4, 3, 3, 4, 2 categories respectively.

Table 1. The mean values of indexes responsible to Δ_1.

	NGF (cm)	*NGB* (cm)	*NDV* (cm)	Neck girth (cm)	Neck diameter (cm)
Cluster 1	14.1	6.9	7.7	42.0	12.4
Cluster 2	14.2	8.1	6.8	44.6	14.1
Cluster 3	12.6	7.5	5.6	40.1	12.9
Cluster 4	12.5	6.8	6.5	38.6	11.8

Table 2. The mean values of indexes responsible to Δ_2.

	SA (cm)	*D1* (cm)	*D2* (cm)	*D3* (cm)	*D4*(cm)
Cluster 1	21.3	24.9	21.9	21.1	46.7
Cluster 2	25.4	23.0	21.8	21.5	44.8
Cluster 3	26.9	22.0	20.8	20.0	42.2

Table 3. The mean values of indexes responsible to Δ_3.

	DSFW (cm)	*DSBW* (cm)	*HDH* (cm)
Cluster 1	43.5	42.3	6.3
Cluster 2	46.7	46.1	9.1
Cluster 3	47.1	44.6	6.5

Table 4. The mean values of indexes responsible to Δ_4.

	DFPB (cm)	*DBPB* (cm)	*DFPW* (cm)	*DBPW* (cm)
Cluster 1	54.1	50.4	49.4	37.0
Cluster 2	45.4	41.3	44.1	29.9
Cluster 3	46.7	51.5	40.4	38.9
Cluster 4	51.8	47.8	44.7	29.3

Table 5. The mean values of indexes responsible to Δ_5.

	ADP (cm)	*AGA* (cm)
Cluster 1	15.3	31.3
Cluster 2	13.2	28.2

Additionally, these categories revealed the certain feature of everybody segments. Taking neck indexes as an example, 4 categories of neck type were discovered after classification. The superposition of neck lines shows as Figure 4 obviously, with *SNP*s of all neck lines fixed at the same position. *NDV* of first type was much bigger than others, and the first type and second type had the similar girth while the first one had the biggest *NGB* and diameter. Moreover, the third type had the big *NGB* and small *NGF*, with the *NDV* smallest. The fourth had the smallest neck diameter and *NGB*, leading to the smallest girth. Thus, the classification of neck lines manifested the communality within each cluster, and the observable differentiations between clusters.

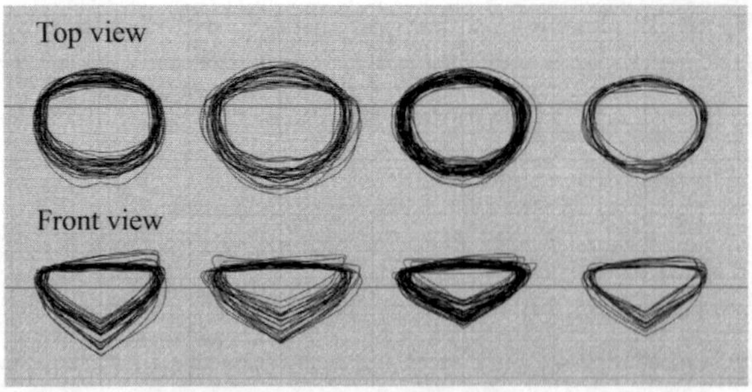

Figure 4. The superposition of different neck types in top and front views.

These classifications of the new anthropometric indexes, related to pattern block parameters, would contribute to the database of categorized pattern segments, which enabled the direct generation of customized patterns for unique people.

3. Results and discussion

According to the Chinese standard sizing systems for garments, the male type 170/92A was selected as an example. Bust girth, waist girth, back length and the ease allowances (ease to half bust girth = 5.6 cm, E_{WG} = 11 cm) were used to draw the shirt pattern block[9] with regular fit by concrete linear regression equations from conventional drafting methods as Figure 5 shows. Afterwards, the scanned male as avatar and the pattern block were imported into Marvelous Designer 6 (3D virtual clothing simulation software). As Figure 6 shows, the try-on results are unsatisfactory. The front piece is attached with the body with little wrinkles, showing proper fit with front torso. While the back piece is squeezed on the buttock area because the body has curved protruding buttock. Moreover, malignant protruding wrinkles are gathering around *SP* and back armhole, indicating the misfit of the shoulder, armhole and sleeve. The bust line and waist line aren't horizontal, indicating the improper proportion of front and back pieces.

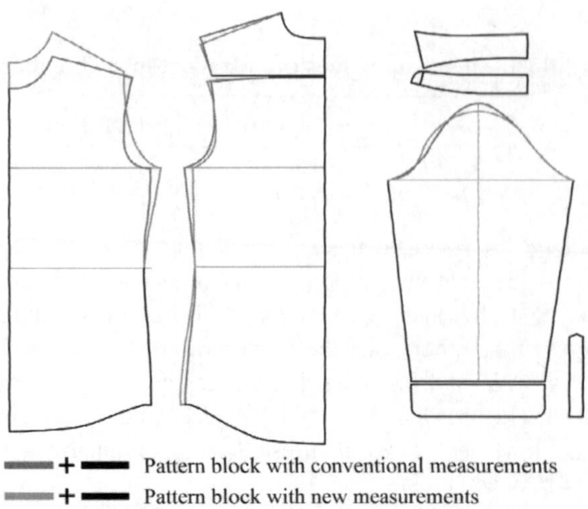

Figure 5. Pattern blocks sketched with conventional and new measurements.

The pattern block was drawn by applying the new pattern-oriented measurements and homologous clusters. The avatar is distinct as broad neck with large girth (42.2 cm) and small *NDV* (5.6 cm) in the second cluster of Δ_1, quite sloping shoulder with big *SA* (27.6 cm) and small *D1* (21.4 cm), *D2* (20.7 cm), *D3*(21.5 cm), *D4* (43.0 cm) in the third cluster of Δ_2. Moreover, the avatar has the prominent bust with *DSFW* (46.3 cm) much bigger than *DSBW* (41.7 cm) in the first cluster of Δ_3, and concave contour of back with similar *DFPB* (49.1 cm) and *DBPB* (43.7 cm) while *DFPW* (40.9 cm) is much bigger than *DBPW* (32.6 cm) in the second cluster of Δ_4. The armhole belongs to the second cluster in Δ_5 because of his small *ADP* (12.6 cm) and *AGA* (29.2 cm).

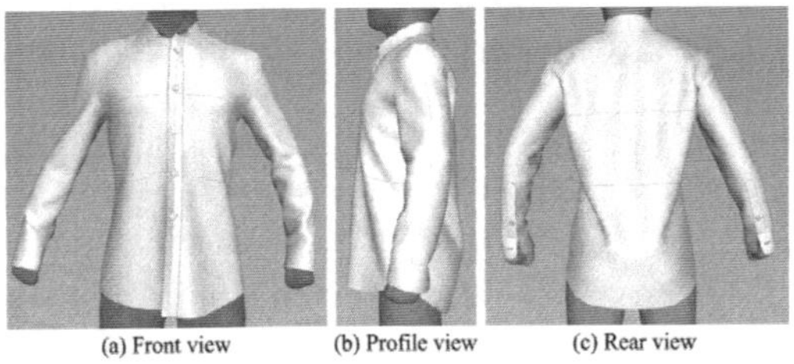

Figure 6. The try-on results of the conventional pattern block.

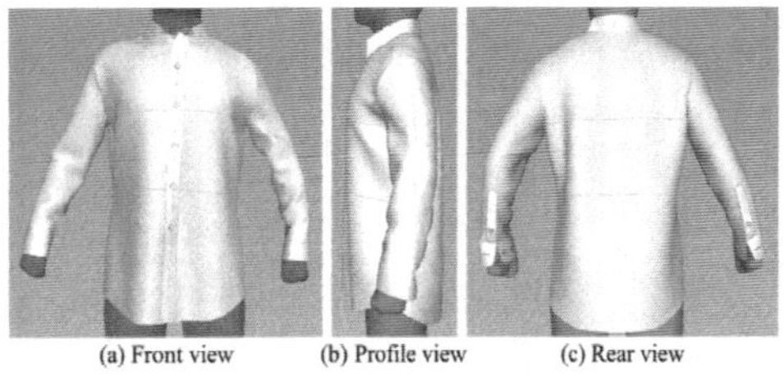

Figure 7. The try-on results of the pattern block utilizing the new measurements.

As shown in Figure 5, new pattern has been strongly modified to adapting the body feature. Firstly, to eliminate the wrinkles on the back piece, *SP* was moved down according to Δ_2. Secondly, to eliminate the folds and squeeze on the back piece, the bottom was increased and the bust girth was decreased considering Δ_4. *APD* of the front piece moved backwards accordingly. Then *SP* moved upwards to balance the front and the back pieces considering Δ_3, to keep the neck line and armhole length, *FNP* and *ADP* moved accordingly. Finally, because of the short armhole depth of this object according to Δ_5, the sleeve cap was decreased. As Figure 7 shows, the virtual shirt has the good fit. The surface of front piece is smooth with little wrinkle. And there's no squeezed folds around buttock. The bust line and waist line are more horizontal as well.

4. Conclusion

With male bodies scanned, new pattern-oriented anthropometric indexes were observed and measured. Hereby different bod segments were classified into distinct types through unsupervised learning neural networks, on the basis of which shirt pattern blocks were also accessible to be enormously individualized for a certain body. The optimized *MTM* men shirts enhanced by virtual technology revealed better fit than the conventional ones.

In the future, more objects and in-depth measurements will be involved, and the body posture classification model will be intensified gradually. The database of categorized pattern segments will be established accordingly, with which dedicated personalized shirt patterns will be generated immediately in virtual and real system for any customer.

IOP Conf. Series: Materials Science and Engineering **460** (2018) 012014 doi:10.1088/1757-899X/460/1/012014

References

[1] Miyoshi M 2008 *Costumes: theoretical papers* (Beijing: China Textile and Apparel Press) pp 143-148, 173-196, 321-325
[2] Zhou Z H 2016 *Machine learning* (Beijing: Tsinghua University Press) pp 108–109
[3] Kim K, Innami N, Takatera M, Narita T, Kanazawa M and Kitazawa Y 2017 Individualized male dress shirt adjustments using a novel method for measuring shoulder shape *Int Jnl of Clothing Sci & Tech* **29** 215–225
[4] Ashdown S P and Dunne L 2006 A study of automated custom fit: readiness of the technology for the apparel industry *Clothing & Textiles Research Journal* **24** 121–136
[5] Kozlova E S, Kuzmichev V E and Rennesson 2012 Development of anthropometric data base from scanned bodies for improving pattern block *Proc. 3th Int. Conf. 3D Body Scanning Technologies* 125–129
[6] General Administration of Quality Supervision, Inspection and Quarantine of the People's Republic of China 2008 *National Standard of People's Republic of China: standard sizing systems for garments – men* (China: Standard Press of China) pp 1–36
[7] Yan J Q, Zhang S C, Kuzmichev V E and Adolphe D C 2017 New database for improving virtual system "body-dress" *IOP Conf. Ser.: Mater. Sci. Eng.* **254** 172029–7
[8] Zhou K L and Kang Y H 2005 *MATLAB neural network model and simulation program design* (Beijing: Tsinghua University Press) pp 111–122
[9] Xiong N 2007 *World classic clothing design and pattern volume 5: men's clothing* (Nanchang: Jiangxi Fine Arts Publishing House) pp 121–123

Tensile Properties of a Novel Graphene Pattern Stitched Carbon/Epoxy 3D Composite

G. Kaya[1*], C. Soutis[2,3], and P. Potluri[3]

[1]Department of Textile Engineering, Kahramanmaras Sutcu Imam University, 46040, Kahramanmaras, Turkey
[2] Aerospace Research Institute, University of Manchester, Manchester M13 9PL, UK
[3] Northwest Composites Centre, University of Manchester, Manchester M13 9PL, UK

**Corresponding author: gkaya@ksu.edu.tr*

Abstract The focus of this study is on the tensile properties of a novel graphene pattern stitched carbon/epoxy 3D composite which could be considered in aerospace industry applications. Tensile modulus and strength are reduced by18-25% due to filament breakage during stitching; this reduction varies with stitching density. Fabrication induced voids and resin rich regions also contribute. However, extensive delamination observed in the tension loaded unstitched laminate is constrained by the graphene pattern stitching and the proposed 3D reinforced composite is expected to be more damage resistant under lateral impact loading.

***Keywords*— damage tolerance, stitching, tensile modulus, tensile strength, 3D composites.**

I. Introduction

Carbon fiber-reinforced composites have been utilized extensively by the aerospace industry. Their high specific stiffness and strength, resistance to cracking by fatigue loading, corrosion resistance and ability to shape and tailor make them produce more aerodynamically efficient structural configurations [1, 2]. 2D laminated composites, offering high in-plane stiffness and strength, have been widely used while these structures are generally weak in the thickness direction (z-axis) [3].

Some detailed studies were performed by researchers about the tensile properties of stitched 3D composites [4-13]. Yudhanto et. al. (2015) investigated the effect of various stitch parameters on damage characteristics of stitched composites under tensile load. It was stated that the certain combination of stitch density and diameter resulted as an increased tensile strength by 7% compared to unstitched composites. This was due to the effectiveness of stitching on impeding the growth of edge-delamination [4]. Pingkarawat and Mouritz (2015) studied on the healing performance of stitched composites. Thermoplastic stitching improved the interlaminar fracture toughness and healing efficiency while the mechanical properties reduced due to the microstructural damages as ply crimping and

fiber waviness [5]. Pappas et. al. (2018) stated that the delamination mechanisms in tufted composites are significantly affected by tuft geometry [6]. Ravandi et. al. (2016) investigated the effects of through-the-thickness stitching using natural fibers on the interlaminar fracture toughness and tensile properties of flax fiber/epoxy composites. It was concluded that the reduction in tensile properties of the composites due to imperfections caused by stitching was irrelevant from the material types of stitching yarn [7]. Effect of stitch density on tensile properties and damage mechanisms of Vectran® stitched carbon/epoxy composites was experimentally studied by Yudhanto et. al. (2013). It was found that low stitching density improved the tensile strength by 10.4%. The responsible of reduction in tensile stiffness was fiber misalignment in in-plane and out-of-plane directions. And also, the orthogonal binding effect of stitching architecture caused to decrease Poisson's ratio of composites [8]. Barile et. al. (2017) stated that the through-the-thickness reinforcement together with polar fiber arrangement improved tensile strength by 15.6% and tensile modulus by 16.5% due to a better distribution of the load on the laminas [9]. Goktas et. al. (2017) stated that the manual-type stitching and high stitch densities offer a significant improvement of 74.5% on Mode-I interlaminar delamination [10]. Joshi et. al. (2017) stated that the reinforcement density is one of the key factors affecting strength, stiffness and crack propagation in composite laminates. By suppressing the damage initiation, densely stitched laminates showed 15.2% higher in-plane stiffness than moderately stitched laminates [11]. Bilisik and Yolacan (2014) investigated the warp- weft directional and off-axis tensile properties of multistitched biaxial woven E-glass/polyester composites. It was stated that the stitching yarn type, the number of stitching directions, and the stitching density generally influenced the tensile properties of multistitched E-glass/polyester woven composites. Multistitching restricted the delamination in cross-sections of composites locally and the delamination did not propagate to the large areas [12, 13].

A stitched composite is among the impact tolerant design of stiffened composites and suitable for aircraft applications. Most investigators report a substantial deterioration of in-plane properties due to damage incurred as a result of stitching process depending on the type of laminate, the lamination technique, the stitching condition and the type of loading. However, any improvement in damage resistance and tolerance arising from stitching is more than compensate for a loss of in-plane strength [14, 15].

It can be seen from the literature that a lot of studies have been performed about the tensile properties of stitched 3D composites. In most of these studies, the influences of stitching parameters as stitching type (manual or machine), stitching density (dense or moderate), stitching direction (one, two or four directions) and stitching fiber types have been investigated. Stitching patterns are merely considered as using different stitching densities. However, the uniform distribution of z-fibers by stitching is critical for the performance of composites and constituted by stitching pattern. In addition, stitching is not only reinforcement to the out-of-direction of the composite but also the in-plane of the composite is reinforced by stitch loops which affect the damage formation under load. The novelty of this study is the used stitching pattern which is inspired by the unique molecular structure of graphene and called as 'graphene pattern stitching'. Tensile properties of unstitched (US) and graphene pattern stitched (GPS) composites are compared in this study. The array of the pattern is arranged in parallel series with unit-cell walls meeting at exactly 120°. By using this stitching pattern, not only the uniform distribution of z-fibers is provided through-the-thickness, but also the in-plane of the composite is reinforced by stitch loops directed at 0° and ±60°.

II. Methods and Procedures

A. *Fabric and Stitching Yarn*

Carbon woven fabric (Sigmatex, UK) sequenced as $[(0°/90°)/(90°/0°)]_3$ was used to produce US and GPS composites. Twisted carbon yarn (175 turns/m) was used for stitching. The linear density of stitching yarn is 430 tex. The weave of carbon fabric is 5 harness satin. Tow sizes of carbon fabric are 12K on both warp and weft directions. The warp/weft fabric densities are 18 ends/10 cm and warp/weft crimp ratios are 1.0%. The fabric weight is 285 g/m^2 and the fabric thickness is 0.42 mm.

B. *Graphene Pattern Stitching (GPS)*

The unit-cell view of GPS is shown in Fig. 1. The schematic views of the followed path of stitching on both front and back faces of the structure and the distribution of z-fibers in through-the-thickness are given in Fig. 2. The position of ±60° stitch-loops is the front-face and the position of 0° stitch-loops is the back-face while the z-fibers are uniformly distributed through-the-thickness.

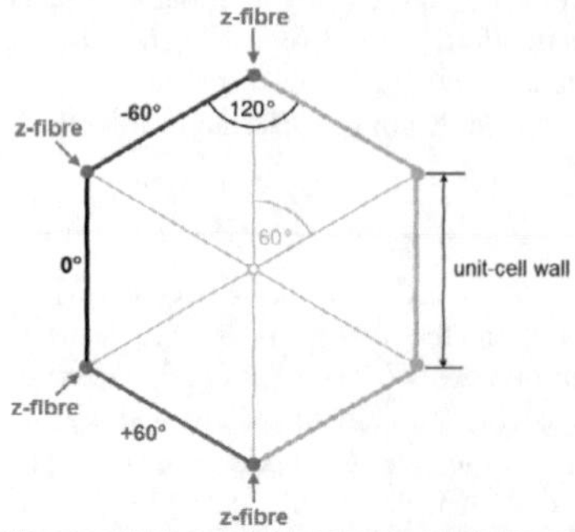

In one unit-cell:	Z-fibre	0°	+60°	-60°
Number	4	1	1	1
Position	out-of-plane	in-plane/back face	in-plane/front face	in-plane/front face

Fig. 1. The unit-cell of GPS.

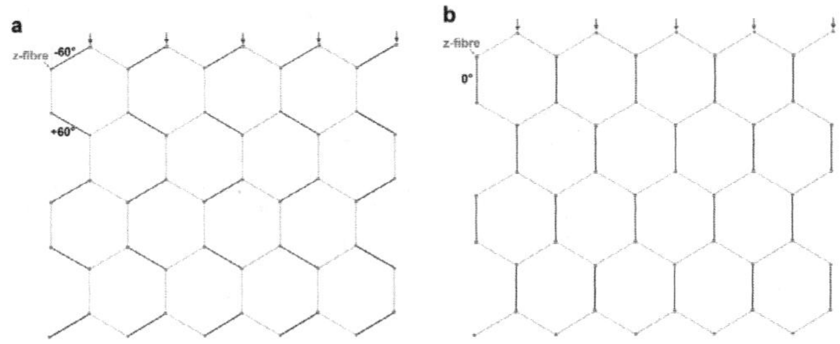

Fig. 2. GPS on the front face (a) back face and distribution of z-fibers.

The preforms were stitched manually using a darning-needle (size: 14, diameter: 1.0 mm, length: 90 mm). GPS was implemented as loosely and densely by two different dimensions of unit-cell walls as 10 mm (L-GPS) and 5 mm (D-GPS).

C. *Composite Manufacturing*

The US and GPS composites produced with a VARTM (vacuum assisted resin transfer molding) method. The meshes and peel plies were used on both sides to prevent dry regions as shown in Fig. 3. Epoxy resin system (Araldite LY 564 resin/Aradur 2954 hardener) was used as the matrix. The infused composite samples were kept in an oven at 2h 80°C for curing and 8h 140°C for post-curing of epoxy. Manufacturing and testing of composites are performed at Northwest Composites Centre, University of Manchester, UK.

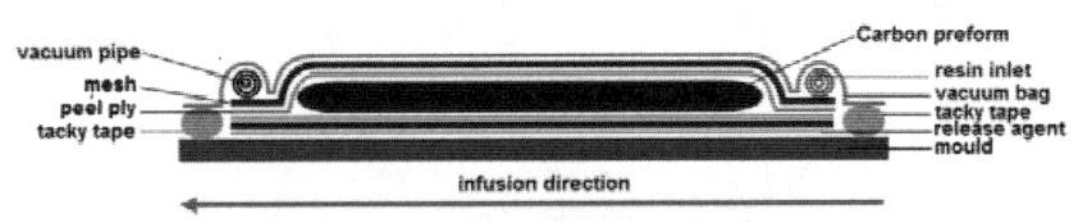

Fig. 3. The schematic view of VARTM.

D. *Measurement of Density and Fiber Fraction*

The density of the samples was calculated by measuring the weight of the specimens in the air and in the distilled water at room temperature according to ASTM D792-13. The composite fiber fraction (volume based, %) was determined by ASTM D3171-15 using matrix burn off method. Specimens were firstly weighted and placed in a furnace at a temperature of 600°C for 1 hour to burn epoxy portion. Then, the samples were re-weighted to measure the mass and volume percentages of the remained carbon fiber content.

E. *Tensile Test*

Tensile tests were conducted according to ASTM D3039/D3039M-12 using an INSTRON 5982 test machine fitted with 100 kN load cell and hydraulic grips at 2 mm/min constant displacement. A video extensometer was used to measure the strain of the samples. Table I shows the tensile test parameters. The dimensions of the specimens were 25 mm (width) × 250 mm (length) with a 50 mm gauge length for both the bottom/top of the specimen without tabbing. The test machine and fixed test specimen between grips are shown in Fig. 4. The ultimate tensile strength values were calculated automatically by the Bluehill® software.

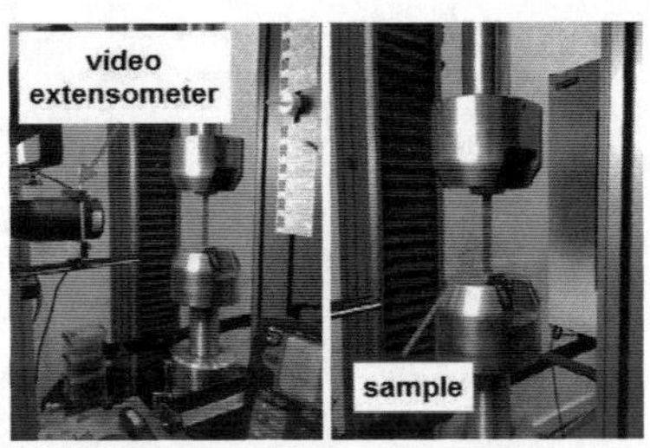

Fig. 4. The test machine and fixed test specimen between grips.

TABLE I
TENSILE TEST PARAMETERS

Sample width	25 mm
Sample length	250 mm
Load cell	100 kN
Testing length	150 mm
Testing speed	2.00 mm/min

III. Results

A. Density and Fiber Fraction Results

Table II shows the density, fiber fraction, thickness and void content results of US and GPS composites. The composite density decreased by GPS. This is due to the local gaps caused by the needle and resulted as resin rich regions. This was also the reason for the increased void contents from US to D-GPS. The positions of stitching loops on front/back faces were the responsible for increase in composite thickness.

TABLE II
DENSITY AND FIBRE FRACTION RESULTS

	Density (g/cm^3)	Thickness (mm)	Fiber fraction (V_f)	Void content (%)
US	1.522 (±0.004)	1.893 (±0.059)	53.76 (±0.16)	0.53 (±0.10)
L-GPS	1.493 (±0.004)	2.288 (±0.081)	49.30 (±0.65)	0.71 (±0.00)
D-GPS	1.470 (±0.006)	2.475 (±0.044)	47.86 (±0.69)	1.48 (±0.19)

B. Tensile Test Results

The tensile test results US and GPS composites are presented in Table III. Fig. 5 shows the stress-strain curves of US and GPS composites during tensile loading.

TABLE III
TENSILE TEST RESULTS

	Tensile strength (MPa)	Tensile modulus (GPa)	Tensile strain (%)
US	703.50 (±91.32)	36.76 (±1.00)	1.29 (±0.11)
L-GPS	526.50 (±82.03)	30.10 (±1.60)	1.20 (±0.18)
D-GPS	498.50 (±49.51)	28.40 (±0.98)	1.38 (±0.16)

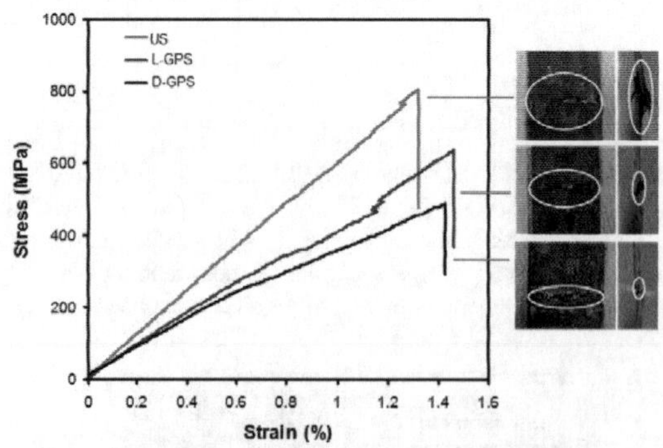

Fig. 5. Stress-strain curves of the US and GPS stitched composites.

As seen in Fig. 5, GPS slightly increased the ductility of composites. The damaged region of US was higher than those of L-GPS and D-GPS composites. A large fiber/matrix delamination was observed in US composites. The delamination of the GPS composite structures after the tensile load locally occurred.

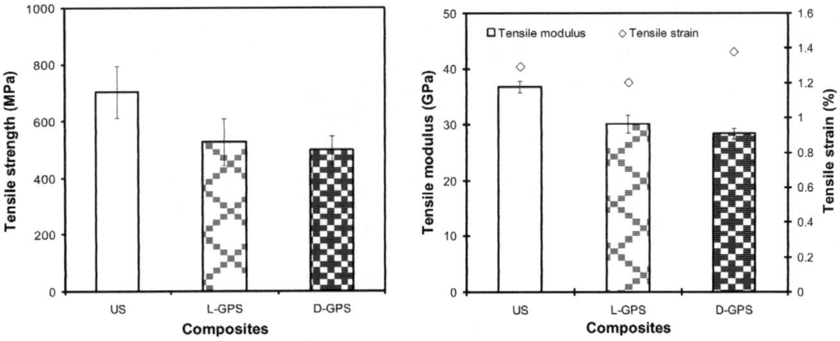

Fig. 6. Tensile test results.

As seen in Table III and Fig. 6, the tensile strength of US was 703.50 MPa while the tensile strengths of L-GPS and D-GPS composites were 526.50 and 498.50 MPa, respectively. These results indicated that tensile strengths of the L-GPS and D-GPS composites decreased by 25% and 29% compared to US composites since dense stitching caused more filament breakages. The tensile modulus of US was 36.76 GPa while the tensile strengths of L-GPS and D-GPS composites were 30.10 and 28.40 MPa, respectively. These results indicated that tensile modulus of the L-GPS and D-GPS composites decreased 18% and 23% compared to US composites as dense stitching caused more filament breakages. Tensile strains of US and L-GPS composites were similar while the tensile strain of D-GPS composite was the highest due to denser out-of-plane and in-plane fiber reinforcement by stitching.

C. *Tensile Failure Results*

Fig. 7 shows the views of composite samples before and after tensile load. As seen in Fig. 7, the tensile failure of US composite was due to delamination across the sample width, multiple matrix cracks and fiber breakages. D-GPS well confined the tensile failure in a small region, while matrix cracks were not observed as in US. Subjecting the D-GPS composite to uniaxial tension caused multiple fiber breakages, fiber/matrix delamination and stitching yarn breakages.

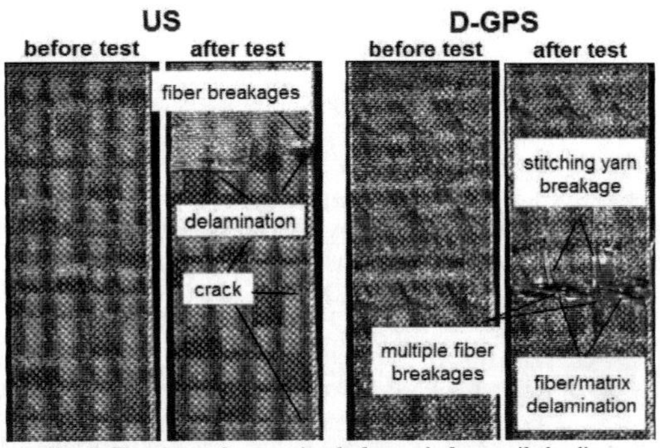

Fig. 7. The views of composites before and after tensile loading.

Tensile failures of US and GPS composites were also examined by an optical microscope as seen in Fig. 8. The cross-sectional view of US after tensile load was observed as a complete delamination of fabric layers. The delamination of L-GPS and D-GPS was locally occurred. Dense stitching confined the delamination more effectively than loose stitching. More severe multiple fiber breakages were occurred in GPS composites since the stitching inhibited the spread of delamination along the sample.

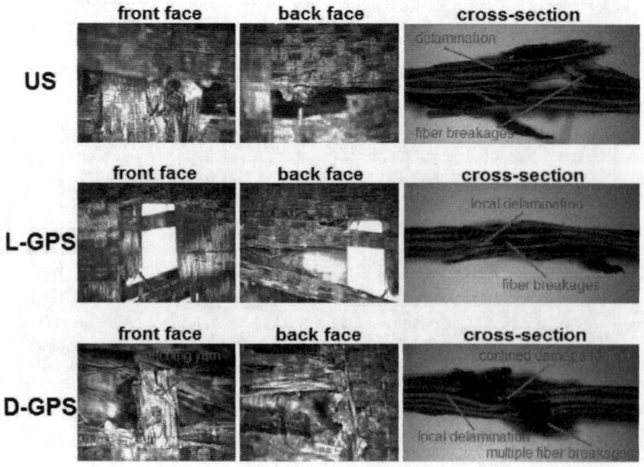

Fig. 8. Optical microscope views of composites.

IV. Conclusion

Tensile strength and modulus of the D-GPS laminate decreased by 29% and 23% when compared to US composite, respectively. These results indicated that dense stitching caused more filament breakages, fiber distortions, and resin-rich-regions created by the piercing of stitching needle. The tensile strengths of US and L-GPS were similar. However, D-GPPS showed higher tensile strain than those of the US and L-GPS. Dense stitching increased the ductility and hence toughness of the composite. In addition, dense stitching well confined the tensile failure in a small region. Multiple matrix cracks were not observed as in the US plate. More severe multiple fiber breakages were occurred in GPS composites since the stitching inhibited the spread of delamination along the sample.

It is expected GPS to provide more damage resistance under drop weight impact conditions and since delamination is constrained, the compression after impact strength will be affected less and lead to a more damage tolerant structural configuration. However, further work is required to support these arguments.

Acknowledgment

The Scientific and Technological Research Council of Turkey (TUBITAK) is acknowledged for granting Gaye Kaya postdoctoral study in the framework of TUBITAK BIDEB-2219 International Postdoctoral Research Scholarship Program.

References

[1] A. Kinsey, D. E. J. Saunders, C. Soutis, "Post-impact compressive behaviour of low temperature curing woven CFRP laminates", *Composites*, vol. 26, pp. 661-667, 1995.

[2] C. Soutis, "Fibre reinforced composites in aircraft construction", *Prog. Aerosp. Sci.*, vol. 41, pp. 143-151, 2005.

[3] M. N. Saleh, A. Yudhanto, P. Potluri, G. Lubineau, C. Soutis, "Characterising the loading direction sensitivity of 3D woven composites: Effect of z-binder architecture", *Compos. Part A Appl. Sci. Manuf.*, vol. 90, pp. 577-588, 2016.

[4] A. Yudhanto, G. Lubineau, I. A. Ventura, N. Watanabe, Y. Iwahori, H. Hoshi, "Damage characteristics in 3D stitched composites with various stitch parameters under in-plane tension", *Compos. Part A Appl. Sci. Manuf.*, vol. 71, pp. 17-31, 2015.

[5] K. Pingkarawat, A. P. Mouritz, "Stitched mendable composites: Balancing healing performance against mechanical performance", *Compos. Struct.*, vol. 123, pp. 54-64, 2015.

[6] G. Pappas, S. Joncas, V. Michaud, J. Botsis, "The influence of through-thickness reinforcement geometry and pattern on delamination of fiber-reinforced composites: Part I- Experimental results", *Compos. Struct.*, vol. 184, pp. 924-934, 2018.

[7] M. Ravandi, W. S. Teo, L. Q. N. Tran, M. S. Yong, T. E. Tay, "The effects of through-the-thickness stitching on the Mode I interlaminar fracture toughness of flax/epoxy composite laminates", *Mater. Des.* vol. 109, pp. 659-669, 2016.

[8] A. Yudhanto, N. Watanabe, Y. Iwahori, H. Hoshi, "Effect of stitch density on tensile properties and damage mechanisms of stitched carbon/epoxy composites", *Compos. Part B Eng.*, vol. 46, pp. 151-165, 2013.

[9] C. Barile, C. Casavola, C. Pappalettere, "The influence of stitching and unconventional fibres orientation on the tensile properties of CFRP laminates", *Compos. Part B Eng.*, vol. 110, pp. 248-254, 2017.

[10] D. Goktas, W. R. Kennon, P. Potluri, "Improvement of mode I interlaminar fracture toughness of stitched glass/epoxy composites, *Appl Compos Mater*, vol. 24, pp. 351-375, 2017.

[11] P. Joshi, A. Kondo, N. Watanabe, "Progressive failure analysis of carbon-fibre/epoxy composites laminates to study the effect of stitch densities under in-plane tensile loading", *Plast. Rubber Compos.*, vol. 46, pp. 147-154, 2017.

[12] K. Bilisik, G. Yolacan, "Warp and weft directional tensile properties of multistitched biaxial woven E-glass/polyester composites", *J. Text. Inst.*, vol. 105, pp. 1014-1028, 2014.

[13] K. Bilisik, G. Yolacan, "Off-axis tensile properties of multistitched plain woven E-glass/polyester composites, *Fib. Poly.*, vol. 15, pp. 589-598, 2014.
[14] E. Greenhalgh, M. Hiley, "The assessment of novel materials and processes for the impact tolerant design of stiffened composite aerospace structures," *Compos. Part A Appl. Sci. Manuf.* vol. 34 pp. 151-161, 2003.
[15] U. Beier, F. Fischer, J.K.W. Sandler, V. Altstädt, C. Weimer, W. Buchs, Mechanical performance of carbon fibre-reinforced composites based on stitched preforms, Compos. Part A Appl. Sci. Manuf., vol. 38, pp. 1655-1663, 2007.

IOP Conf. Series: Materials Science and Engineering **460** (2018) 012016 doi:10.1088/1757-899X/460/1/012016

Thermoregulating and hydrophobic nanofibers

Nuray Kizildag* and Nuray Ucar

Istanbul Technical University, Department of Textile Engineering, Gumussuyu, Istanbul, Turkey

E-mail: kizildagn@itu.edu.tr

Abstract. Phase changing materials are widely investigated in many different aspects due to the increasing interest in energy storage applications, thermal insulation and thermal comfort. Electrospinning is a simple technique for producing ultrafine (sub-micron) fibers from a wide variety of polymers. With PCMs, it is generally used as a method to obtain shape stabilized PCMs. In this study, microencapsulated phase change materials (MPCM) are incorporated into nanofiber structure. Morphology, chemical structure, thermal properties and water absorbency of the composite nanofibers are investigated. Uniform nanofibers displaying both thermoregulatory properties and hydrophobicity were obtained.

1. Introduction

Phase changing materials (PCMs), which include inorganic materials, organic materials and their blends, have the capability to absorb and release latent heat during a phase transition process over a narrow temperature range [1-2]. The storage or release of latent heat (thermal regulation) is based on the transition of a material (PCM) between phases. When a PCM is heated to its melting point, it absorbs heat as it changes from solid to liquid state and a temporary cooling effect is obtained. Conversely, when a PCM goes from liquid phase to a solid phase, it releases the latent heat to the environment and a temporary heating effect is obtained [3]. Paraffin waxes [4], polyethylene glycols [5], fatty acids and fatty acid derivatives [6] are types of PCMs which are very often used in research studies. Phase changing materials are widely investigated in many different aspects due to the increasing interest in energy storage applications, thermal insulation and thermal comfort [1-2]. Due to their high heat storage capacities, PCMs are applied to textiles in a variety of processes to improve thermal comfort of the products [7].

Electrospinning is a simple technique for producing ultrafine (sub-micron) fibers from a wide variety of polymers. It uses electrostatic forces to draw nanofibers. It provides a straightforward and cost-effective method to produce fibers with diameters ranging from nanometers to submicrons. The electrospinning setup consists of a high voltage power supply, a pump, and a grounded collector. The polymer solution, which is filled into a syringe, is fed by a pump through a needle. A high voltage power supply is used to form an electrostatic field between the droplet of polymer solution at the tip of a needle and the collector. As the voltage applied is increased, a force is built on the pendant drop and the drop is elongated forming a conical shape known as Taylor cone. When the voltage is further increased, the electrostatic force overcomes the surface tension of the solution droplet and a continuous jet forms, which is attracted by the grounded collector. During its fly from the tip of the needle to the collector, the jet undergoes instabilities (bending and whipping motions) during which the solvent evaporates and the jet diameter reduces. The jet rapidly thins and dries. Finally the nanofibers are collected on the collector in nanoweb form with random nanofiber orientation [8]. It is a widely used technique to obtain shape stabilized PCMs with the elimination of the need for the encapsulation process [1].

Chen et al. [1,9,10] conducted a series of studies on the incorporation of PCMs into nanofiber structure. They successfully produced ultrafine fibers of lauric acid/polyethylene terephthalate (LA/PET, 1:1, w/w) which showed smooth surfaces with acceptable tensile strength. The latent heat of fusion of the fibers was about 70.76 J/g. The composite nanofibers were suggested as suitable for the applications such as solar energy storage and thermo-regulating textiles [1]. They also produced composite nanofibers using stearyl stearate as additive. Increase was observed in average nanofiber diameter with addition of the additives. The morphology of the composite fibers had no obvious variation after thermal treatments, which confirmed that the developed material showed form stable phase changing property. DSC showed that the enthalpy of the composite fibers increased with the increase in additive content. The thermal performances of the fibers could be maintained after 100 heating–cooling thermal cycles, which indicated that the composite fibers had good thermal stability and reliability [9]. In another study, they produced composite PET nanofibers using a series of fatty acids such as lauric acid (LA), myristic acid (MA), palmitic acid (PA), and stearic acid (SA). The average diameter and morphology of the nanofibers was found to be affected by the amount of the fatty acid. While the latent heat of the composite fibers increased with the increase in the additive amount, the phase transition temperature of the fibers didn't varied compared to that of the fatty acid. The latent heat and phase transition temperature of composite fibers were dependent on the type of the fatty acids [10]. Producing composite PET nanofibers with a series of different fatty acid eutectics, Cai et al. [11] also reported that the latent heat and phase transition temperature of composite fibers were dependent on the type of the fatty acids and suggested that the composite nanofibers could be used in the field of building energy conservation. Ke et al. [12] electrospun composite PET nanofibers using a series of fatty acid eutectics for thermal energy storage. The composite nanofibers, which possessed smooth and cylindrical morphological structure with diameters of about 100-300 nm, showed different enthalpy depending on the content and the type of binary fatty acid eutectics. Chen et al. [5] produced cellulose acetate/ polyethylene glycol nanofibers, in which PEG acted as PCM. The heat capacities of the composite fibers with 10–70 wt% PEG content varied between 3.8 kJ/kg and 120.2 kJ/kg in DSC analyses. Cai et al. [6] electrospun composite nanofibers of polyethylene terephthalate (PET)/lauric acid (LA)/silica nanoparticles and investigated morphology, thermal storage, thermal stability, and combustion properties of the composite nanofibers. T. McCann et al. [4] used melt coaxial electrospinning for producing nanofibers consisting of n-octadecane and n-eicosane cores as phase change materials.

In this study, microencapsulated phase change materials (MPCM) are incorporated into nanofiber structure. The composite nanofibers are characterized in terms of morphology, chemical structure, thermal properties and water repellency.

2. Experimental

2.1. Materials

Polyacrylonitrile (PAN) (Sigma Aldrich, 181315, average Mw: 150.000g/mol), microencapsulated phase change materials (MPCM) (Microtek Laboratories, Inc., MPCM28D, Tm = 28°C), dimethylformamide (DMF) were used as received.

2.2. Solution Preparation

The required amount of MPCM (25 and 50 wt%) was added to the required amount of DMF and homogenized with ultrasonic tip for 10 minutes and with ultrasonic bath for 45 min. Then PAN was added to this dispersion and dissolved by stirring with a magnetic stirrer at 40°C, 300 rpm for 3 hours. Composite electrospinning solutions containing 25 and 50 wt% MPCM were prepared. The

IOP Conf. Series: Materials Science and Engineering **460** (2018) 012016 doi:10.1088/1757-899X/460/1/012016

concentration of PAN was 7 wt%. PAN/DMF solution was prepared for the production of the reference nanofibers.

2.3. Electrospinning
The nanofibers were produced using a horizontal electrospinning setup, the schematic of which is shown in figure 1. It contained a high voltage power supply (Matsusada, 0-50 kV), a precisely-controlled syringe pump (SINO MDT, SN-50C6), and a grounded rotating collector covered by aluminum foil and nonwoven. The nanofibers were collected on nonwoven to be able to easily peel off the nanowebs produced. Electrospinning solutions were filled into a syringe of 10 mL and fed through a needle (Ultradent Products Inc.) with a outer diameter of 1.25 mm. The flow rate of the electrospinning solution was set as 1 mL/h. The distance between the tip of the needle and the collector was 10 cm. The applied voltage was changed in order to obtain stable electrospinning.

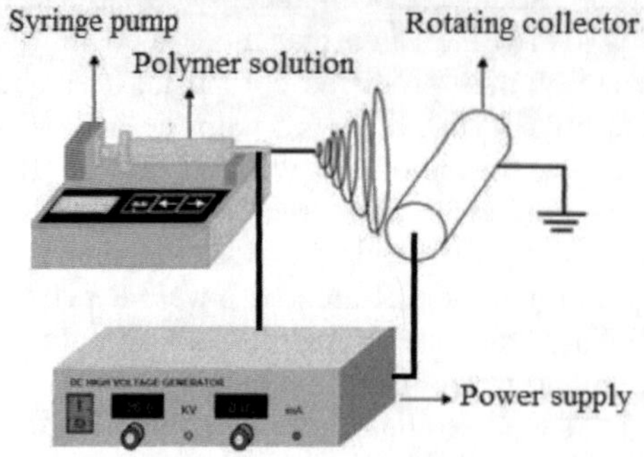

Figure 1. Schematic of the electrospinning setup.

2.4. Characterization
Composite nanofibers with MPCM were characterized in terms of morphology, chemical structure, thermal properties and hydrophobicity. The morphology of the nanofibers were analysed using scanning electron microscope (SEM, Jeol Quanta 200 FESEM). SEM images were taken at an accelerating voltage of 20 kV. Before SEM analysis, the samples were coated with gold using a sputter coater. The fiber diameters were measured using Image J. The average nanofiber diameters and the standard deviations were based 50 measurements per sample. Average nanofiber diameters were expressed as the mean±standard deviation. A Fourier transform infrared (FTIR) spectrophotometer from Perkin Elmer (Spectrum Two) was used to record absorption spectra of composite nanofibers in a range from 4000 to 400 cm^{-1} with a resolution of 4 cm^{-1}. 32 scans were taken for each experiment. The spectrum of MPCM was normalized to the largest peak observed, namely the $-CH_3$ stretchings observed at 2915 cm^{-1} and the spectrum of 100% PAN and composite nanofibers were normalized to the characteristic C≡N stretching band of PAN which appears around 2242 cm^{-1} to be able to compare them. Thermal properties were evaluated using differential scanning calorimetry (DSC, PerkinElmer DSC4000) in a temperature range from 0°C to 60°C, at a heating rate of 5°C/min in nitrogen environment. Hydrophobicity of the composite nanofibers were evaluated by water contact angle measurements.

3. Results and Discussion

3.1. Morphology
The reference and composite nanowebs were white in color. The SEM images taken with 25kX magnification and the average nanofiber diameters are presented in figure 2. The fibers were uniform in structure even at the MPCM content of 50 wt%. They were free of any beads. The average nanofiber diameter increased as a result of MPCM addition, likely due to the increased viscosity of the electrospinning solutions [13].

Figure 2. SEM images of reference and composite nanofibers: (a)100%PAN; (b)PAN/25%MPCM; (c)PAN/50%MPCM.

3.2. Fourier Transform Infrared Spectroscopy (FTIR)

FTIR spectroscopy analysis was performed to confirm the presence of MPCM in the composite fiber structure, and compare the spectral differences between the composite nanowebs and the reference 100% PAN nanoweb. The FTIR spectra obtained from MPCMs, 100% PAN nanoweb and composite nanowebs with MPCMs are shown in figure 3.

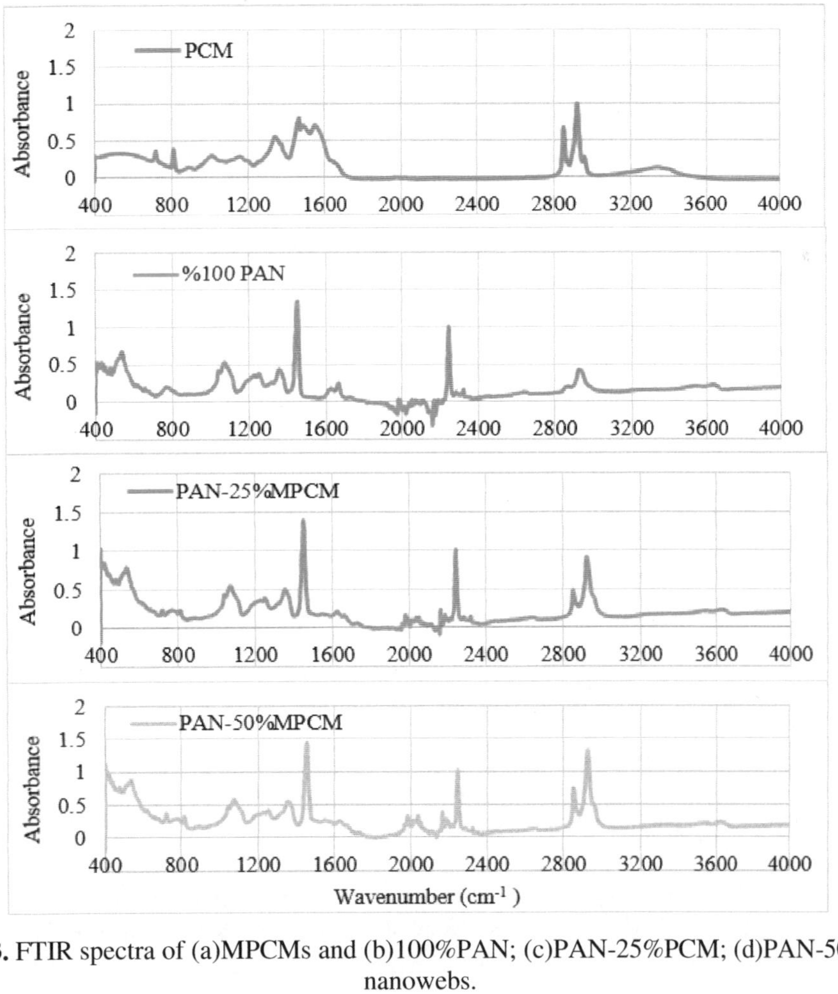

Figure 3. FTIR spectra of (a)MPCMs and (b)100%PAN; (c)PAN-25%PCM; (d)PAN-50%PCM nanowebs.

In the spectra of MPCMs (figure 3(a)), the bands located at 2915 cm^{-1} and 2849 cm^{-1} represented the stretching vibrations of $-CH_3$ and $-CH_2$ groups. The peak at 1470 cm^{-1} represented the $-CH_2$ bending vibration [14]. In the spectra of PAN nanofibers (figure 3(b)), the peaks are as follows: 3626 cm^{-1} (OH stretching), 2922 and 2870 cm^{-1} (C–H asymmetric and symmetric stretchings in CH, CH_2, and CH_3 groups), 2242 cm^{-1} (C≡N stretching), 1622 cm^{-1} (C=C stretching), 1452 cm^{-1} (CH_3 bending and CH_2 scissor vibration), 1356 cm^{-1} (CH_3 symmetric bending vibration in C-CH_3), and 1252 cm^{-1} (C–N stretching), 1070-1040 cm^{-1} (C=N bending), and 778 cm^{-1} (-C-CN-stretching) [15-19]. The FTIR spectra of composite nanowebs (figure 3(c) and 3(d)) showed characteristic peaks of MPCMs in addition to the characteristic peaks of PAN, which showed the successful incorporation of MPCMs into the PAN nanofiber structure. With the increase in the MPCM content, the intensities of the peaks related with MPCM (located between 1400 and 1600 cm^{-1}, 2915 and 2849 cm^{-1}) increased.

3.3. Differential Scanning Calorimetry Analysis (DSC)

DSC thermograms of 100% PAN nanoweb and composite samples are presented in figure 4. While 100% PAN nanoweb displayed no phase transitions between 0 and 60°C, composite samples showed phase transitions due to the presence of MPCMs in the nanofiber structure. The corresponding thermal properties of composite nanowebs are summarized in Table 1. Melting took place between the onset temperature of about 23°C and 28°C and crystallization occurred between 24°C and 13°C. Both melting and crystallization enthalpies increased as the MPCM content in the nanofibers increased. The melting enthalpy increased from 11.42 to 24.71 J/g as the MPCM content increased from 25% to 50%.

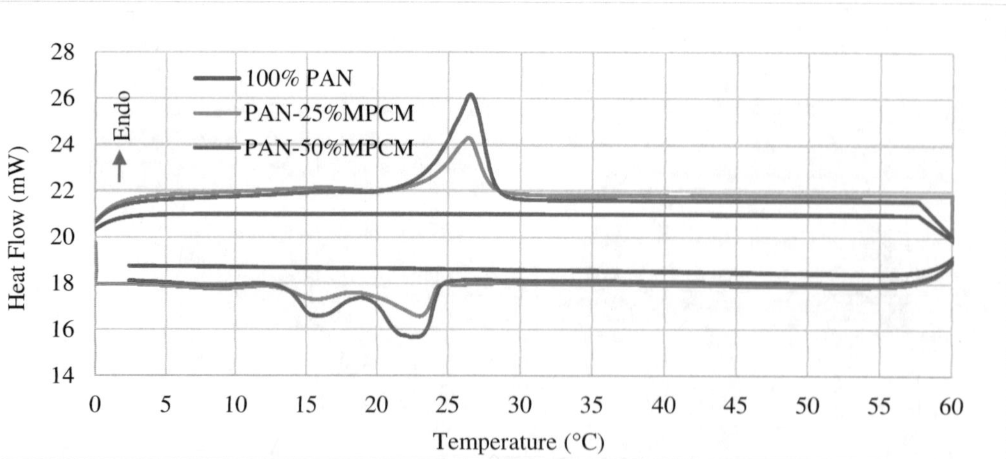

Figure 4. DSC thermograms of 100%PAN and composite nanowebs containing MPCMs.

Table 1. Thermal properties of composite nanowebs containing MPCMs.

Sample	Onset of melting (°C)	End of melting (°C)	Latent heat of melting (ΔH) (J/g)	Onset of crystallization (°C)	End of crystallization (°C)	Latent heat of crystallization (ΔH) (J/g)
PAN-25%PCM	23.53	27.81	11.42	24.26	13.38	11.53
PAN-50%PCM	23.85	28.13	24.71	24.45	13.58	26.45

3.4. Hydrophobicity

Water contact angle measurements were performed in order to be able to evaluate the hydrophobicity of the composite nanowebs. The basis of the contact angle measurement is to instill a drop of a liquid onto the surface and then record the drop using a camera. The video image is evaluated in order to determine the contact angle after a certain time. The larger the contact angle, the more hydrophobic the surface is considered. A surface is called "hydrophobic" if the water drop contact angle is higher than 90° and "hydrophilic" if the contact angle is smaller than 30°.

The images of drops taken by the camera are presented in figure 5. The composite nanofibers gained hydrophobic properties with the addition of MPCMs. While the water droplet was absorbed by 100% PAN nanoweb in less than a second, the water contact angle was measured as 107° for PAN nanoweb containing 25% MPCM.

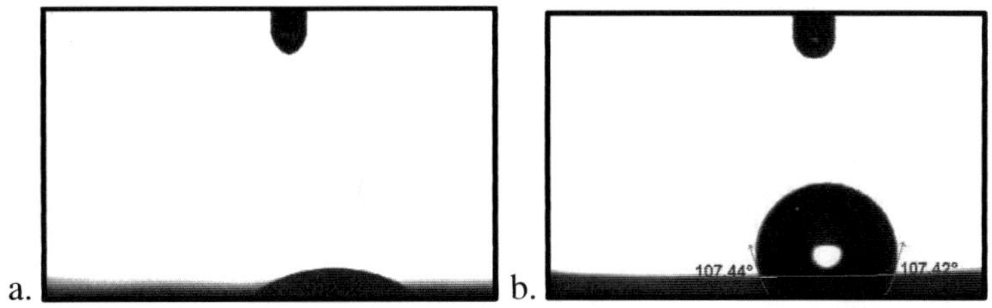

Figure 5. Contact angles obtained for a.100% PAN and b.PAN-25%MPCM.

4. Conclusions

This study showed the possibility of producing thermoregulating and hydrophobic nanofibers with the addition of phase change materials into the structure of nanofibers. Uniform nanofibers were produced which displayed thermoregulating and hydrophobic properties. The composite nanofibers with phase changing nanoparticles are suggested for use in active wear, undergarments, bedding, packaging materials, and insulation materials, etc.

Acknowledgments

We would like to thank ITU Textile Engineering Department Laboratories for FTIR and DSC measurements and ITU Global Textile Centre for water contact angle measurements.

References

[1] Chen C, Wang L and Huang Y 2008 *Mat. Lett.* **62** 3515.
[2] Sarier N and Onder E 2012 *Themochim. Acta* **540** 7.
[3] Erkan G 2004 Research Journal of Textile and Apparel **8(2)** 57.
[4] McCann J T, Marquez M and Xia Y N 2006 *Nano Lett.* **6** 2868.
[5] Chen C, Wang L and Huang Y 2011 *Appl. Energ.* **88(9)** 3133.
[6] Cai Y, Ke H, Dong J, Wei Q, Lin J, Zhao Y, Song L, Hu Y, Huang F, Gao W and Fong H 2011 *Appl. Energ.* **88(6)** 2106.
[7] Sarier N and Onder E 2007 *Thermochim. Acta,* **452** 149.
[8] Qiao B, Ding X, Hou X and Wu S 2011 J. *Nanomater,* **2011(839462)** 1.
[9] Chen C, Wang L and Huang Y 2009 *Chem. Eng. J.* **150** 269.
[10] Chen C, Wang L and Huang Y 2008 *Sol. Energ. Mat. Sol. C.* **92** 1382.
[11] Cai Y, Ke H, Lin L, Fei X, Wei Q, song L, Hu Y and Fong H 2012 *Energ. Convers. Manage.* **64** 245.
[12] Ke H., Li D, Zhang H, Wang X, Cai Y, Huang F and Wei Q 2013 *Fiber. Polym.* **14(1)** 89.
[13] De P, Sathyanarayana D N, Sadasivamurthy P and Sridhar S 2002 *Eur. Polym. J.* **38** 847.
[14] Cai Y, Sun G, Liu M, Zhang J, Wang Q and Wei Q 2015 *Sol. Energy* **118** 87.
[15] Deng S, Bai R and Chen J P 2003 *J. Colloid Interf. Sci.* **260** 265.
[16] Moafi H F, Shojaie A F and Zanjanchi M A 2011 *J. Chil. Chem. Soc.* **56(1)** 610.
[17] Minagawa M, Miyano K, Takahashi M and Yoshii F 1988 *Macromolecules* **21** 2387.
[18] Zhang D, Karki A B, Rutman D, Young D P, Wang A, Cocke D, Ho T H and Guo Z 2009 *Polymer* **50** 4189.
[19] Kizildag N, Ucar N, Onen A and Karacan I 2016 *J. Comp. Mater.* **50(28)** 3981.

PREPARATION OF ELECTROSPRAYED MICROPOROUS MEMBRANES

Mohanapriya Venkataraman[1], Rajesh Mishra[1], Kai Yang[1], Jiri Militky[1], Dana Kremenakova[1], Guocheng Zhu[2], Juming Yao[2]

[1]Faculty of Textile Engineering, Department of Material Engineering, Technical University of Liberec, Liberec 46117, Czech Republic.

[2]College of Materilas and Textiles, Zhejiang Sci-Tech University, Hangzhou 310018, P.R. China.

Abstract: Polytetrafluoroethylene (PTFE) which is also called as "TEFLON" is a synthetic fluoropolymer of tetrafluoroethylene that has wide applications due to its differentiating properties. In this study, different PTFE dispersions for preparation of polytetrafluoroethylene mesoporous active membranes doped by basalt and carbon particles which were created in high voltage electrostatic field has been studied. The adjusting of process parameters of common electrospinning system (Nanospider) for preparation of membranes with tunable porosity created by nanofibrous assembly (electrospinning) and interconnected particles (electrospraying) has also been investigated. The nanoparticles based on milled basalt and carbon was used for activation and achieving of special effects. Superhydrophobic Polytetrafluoroethylene (PTFE) microporous membranes with different surface structures were obtained by controlling operating parameters in the electrospinning process. The diameters and microstructure of the PTFE microporous membrane were characterized by scanning electron microscopy. The contact angles on the microporous membranes were evaluated by static micro-drop observation, and a modified Yang equation was applied to analyze the contact angles. The superhydrophobic PTFE microporous membranes were also tested for thermal properties. As a result of this study, optimized PTFE blend solutions were identified. The results also revealed that the specific surface area was the key factor affecting the contact angles. The thermal properties revealed that thermal conductivity was higher and thermal resistance was lower for carbon and basalt doped membranes.

Key words: Teflon, Hydrophobic, Electrospraying, Microporous membrane

I.INTRODUCTION

Polytetrafluoroethylene (PTFE) which is also called Teflon is a synthetic fluoropolymer of tetrafluoroethylene that has numerous applications. The major application of PTFE is due to the fact that PTFE has excellent dielectric property [1]. It can be stretched to contain small pores of varying sizes and is then placed between fabric layers to make a waterproof, breathable fabric in outdoor apparel [2]. It is used widely as a fabric protector to repel stains on formal school-wear, like uniform blazers [3]. It is also used as a film interface patch for sports and medical applications, featuring a pressure-sensitive adhesive backing, which is installed in strategic high friction areas of footwear, insoles, ankle-foot orthosis, and other medical devices to prevent and relieve friction-induced blisters, calluses and foot ulceration [4]. Electrospinning has been established as a simple and effective method to produce polymer fibers with diameters at nanometer or submicron scale. Continuous fibers can be collected in the form of nonwoven fibrous membranes or as aligned yarns [6]. Nanofibers prepared by electrospinning have many superior properties, such as good pore structure and high surface porosity [6, 7]. Recently, several research groups have proved that the surface energy of materials was closely related to the surface properties, especially surface roughness [8], which presented a new way to control material surface wettability. Many theoretical researches on electrospinning process have been done by many teams and groups [9-11]. At the end of the 1500s, Sir William Gilbert explained the behavior of electrostatic and magnetic emanation. He found that by affecting the water droplet by electrostatic field, the water gets cone and hopper shape, and a droplet extrudes from the head of the hopper. This formed the first process of electro-spraying. Electrospinning can be viewed as a kind of electro-spraying. As with electrospraying, the raw material of electrospinning is linked to a high-voltage power supply to enhance the liquid electrostatic potential [12]. The parameter that has affected morphology, structure, physical and chemical properties of electrospinning fibers is the distance between nozzle and collector. It has the direct effect on final fiber properties based on evaporation rate, deposition time and inconsistency interval. The studies show that by decreasing the distance between nozzle and collector, we have wet electrospinning fiber which has beaded structure. Also in some fibers the morphology of final fibers has changed from circular shape to flat shape [13]. On the other hand, the study shows that for aqueous polymer dispersion, more distance is needed to dry the fiber [14].

In this study, optimization of PTFE blend solution was carried out. The surface energy of microporous PTFE membranes was investigated by static contact angle measurements. The effect of the fibrous structures on the contact angle of microporous membranes was analyzed. The thermal properties of the samples were also evaluated and statistically analyzed.

II.METHODOLOGY

A. *Materials*

Polytetrafluoroethylene (PTFE) was used as-received without further purification. The PTFE microporous membranes were prepared based on various compositions of PTFE, organic salt Tetraethylammonium bromide (TEAB), water, surfactant and doping of carbon/basalt nanofibers. The detailed description of the samples is shown in Table 1.

TABLE 1
DESCRIPTION OF SAMPLES

Sample No.	Sample Description
S002	PTFE +TEAB without water
S003	PTFE +TEAB+Carbon without water
S004	PTFE +TEAB+Basalt without water
S005	PTFE +TEAB+Carbon with water
S006	PTFE +TEAB with water
S007	PTFE +TEAB+Basalt+ with water

B. *Methods*

Electrostatic spinning is a method of producing superfine fibers with diameters ranging from 10 μm down to 10 nm by forcing a polymer melt or a solution through a spinneret by electric field and subsequently drawing the resulting filaments as they solidify or coagulate [19, 20]. Electrospinning exerts voltage on polymer solution to get nanofiber which is different from traditional methods. At the beginning, the polymer solution or melt bring thousands of high voltage. The charged droplet in the capillary was accelerated under the force of electrostatic force. When the electric force is large enough, the droplet can overcome surface tension to formation of a thin stream. Due to solvent evaporation in the electric field, nanofibers which are like non-woven mats are collected in the receiver [21].

Polymer concentration plays a major role in the electrospinning process. Under the same electrospinning conditions, an increasing polymer concentration usually increases the diameter of the electrospun fibers. However, Deitzel et al. found that there is often a nonlinear relationship between the solution concentration and fiber diameter [22].The reason for this nonlinear relationship can be attributed to the nonlinear relationship between the polymer concentration and solution viscosity [23].

Electrospraying (electrohydrodynamic spraying) is a method of liquid atomization by means of electrical forces. In electrospraying, the liquid at the outlet of a nozzle is subjected to an electrical shear stress by maintaining the nozzle at high electric potential [24]. The advantage of electrospraying is that droplets can be extremely small, in special cases down to nanometers, and the charge and size of the droplets can be controlled to some extent by electrical means, i.e., by adjusting the flow rate and voltage applied to the nozzle. Due to its properties, electrospraying is considered as an effective route to nanotechnology. The paper considers the latest achievements in micro and nano thin-film production, including self-assembled nanostructures, in solid nano-particle generation, and in the formation of micro- and nanocapsules.

Figure 1 Rotating roller electrode, Single wire electrode and Rotating wire electrode (from left to right).

"Nanospider" instrument was used to spin the membranes by electrospraying. The Nanospider electrodes which have been developed through last ten years of optimization can give big portfolio for electrospraying as well. Three different electrodes were used to spin the membranes out of which the rotating wire electrode was suitable for preparing the membranes by electrospraying. Since Polytetrafluoroethylene (PTFE) was insoluble in water, it was introduced in "particle" form in the solution. The solution prepared was optimized with the change in concentration of the solution, PTFE composition, organic salt, surfactant, carbon/basalt doping, relative humidity, conditional temperature, distance of the electrodes and substrate speed. The prepared solution was taken in the bath and electrosprayed by using rotating wire electrode. The electrosprayed particles were collected in the polyprolylene spun bond nonwoven fabric. The details of spinning parameters are shown in Table 3.

TABLE 2
SPINNING PARAMETERS

Sample No.	Substrate speed (mm/min)	Voltage (kV)	Speed of electrode (ot/min)	In	Tent	Air
S002	static	-10/30	5	43.1%/22.3°C	43.3%/22.5°C	48.2%/22°C
S003	static	-10/30	5	44%/22.7°C	48%/23.3°C	50.1%/23.1°C
S004	static	-10/30	5	42%/23,4°C	47%/22.7°C	50%/22.1°C
S005	static	-10/30	5	41.4%/22.6°C	47.7%/22.6°C	51.1%/22.4°C
S006	15	-10/30	5	43.3%/22.4°C	48.9%/22.5°C	49.9%/22.5°C
S007	static	-10/30	6	40.8%/22.9°C	45.8%/23°C	50.1%/22.3°C

The microporous membranes with different concentrations were characterized using SEM (VEGA TESCAN Inc. USA) at 30 kV. The contact angle was measured on a See System E is a portable computer-based instrument for contact angle measurement and surface energy determination. Deionized water was dropped onto the sample from a needle on a micro syringe (5 μl) during the test. A picture of the drop was taken a few seconds after the drop set onto the sample. The contact angles could be calculated by analyzing the shape of the drop. For thermal measurements, Alambeta was used. All results were statistically analyzed.

III.RESULTS AND DISCUSSION

A. Analysis of Microstructures

SEM images of PTFE membranes with different compositions are shown in Figure 5. The electrospinning parameters have a strong influence on fiber morphology. The concentration of the polymer solution and the spinning voltage were found to be especially significant. In this study, the effects of concentration and spinning voltage on fiber morphology for PTFE microporous membrane with different blend ratio were observed.

Electrospraying influences the microstructure of the microporous membrane where the technique utilizes the electrical forces for liquid atomization [25]. Droplets obtained by this method are highly charged. The advantage of electrospraying is that the droplets can be extremely small (tens nanometers), and the charge and size of the droplets can be controlled to some extent by electrical means. Motion of the charged droplets can be controlled by electric field. The deposition efficiency of the charged spray on an object is usually higher than that for uncharged droplets. The membrane morphologies can be categorized into two main groups: dense and porous. The dense layer can be amorphous, crystalline (of different structures) or amorphous with incorporated particles (intrusions).

The porous layer can be reticular, grainy, or fractal like. The membrane morphology depends, on the temperature, the solvent used for spraying, voltage, physical properties of liquid phase (emulsion, dispersion, solution, density and surface tension), distance of collector from source, doping agents and the time of solvent evaporation [26].

It can be seen that the viscosity of the solution and the spinnability increased with the increase of concentration of the solution, which is consistent with other published results [27-30]. When the concentration of the polymer solution is lower, the fibers tend to form beadlike structures. The change in fiber diameter cane attributed to the presence of surfactant in the PTFE emulsion. The existence of surfactant in the solution tends to increase the solution conductivity and decrease the surface tension, thus facilitating the formation of finer fibers [31].

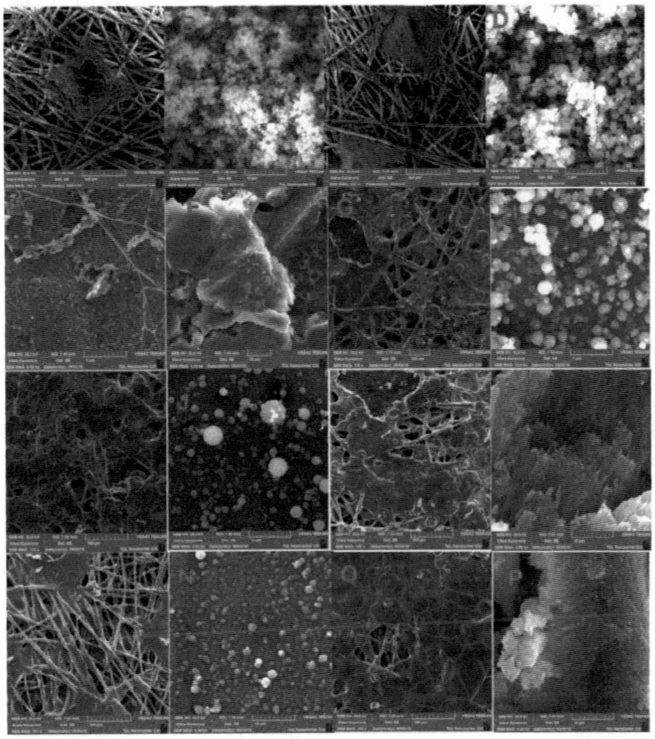

Figure 2 SEM images of PTFE membranes.

Since surfactant is a part of the PTFE emulsion, the surfactant content changes with solution concentration. At higher concentration, the effect of surfactant tends to be stable, whereas polymer concentration remains as the main influencing factor on fiber diameter, with fiber diameter increasing rapidly as polymer concentration increases. The membranes S002, S003 and S004 had dispersion of 60% concentration (original solution) with the range of different relative humidity. Lowering moisture leads to a slightly larger deposit (higher coverage), but the difference is not substantial.

The membranes S005, S006 and S007 had dispersion of 50% concentration (diluted with water) with the range of different relative humidity. The samples S003, S004, S005 and S007 were spun with carbon and basalt nanofibers. The output range of relative humidity i.e., the decreasing RH leads to a bit higher surface layer, and the difference between 25% a 10% RH is negligible. Diluting the solution does not bring visible change in process or change in surface layer.

B. Contact Angle Analysis

The results of the static contact angle measurements are shown in Figure 3. It clearly reveals that the contact angles of all the microporous membranes are over 90°. The image in Figure 3 presents a water droplet formed on the PTFE microporous membrane electrospun at 10 to 30 kV. The contact angle increases to 143.2° when the voltage for the electrospinning increases to 30 KV, as illustrated in Figure 3(c). The high contact angle in the image clearly reveals the superhydrophobic behavior of the microporous membrane. The curve in Figure 3 indicates that the contact angle increases with the specific surface area formed by different voltages. It also, indicates the contact angle for three different wash times (2 mins, 4 mins and 6 mins). 2 mins wash time showed a better contact angle than the 4 and 6 mins wash time. This phenomenon can be interpreted in terms of the Young equation and its modification.

(a) (b) (c)

Figure 3 Contact angles of the samples (a) 125.7o (b) 135.3o (c) 143.2o

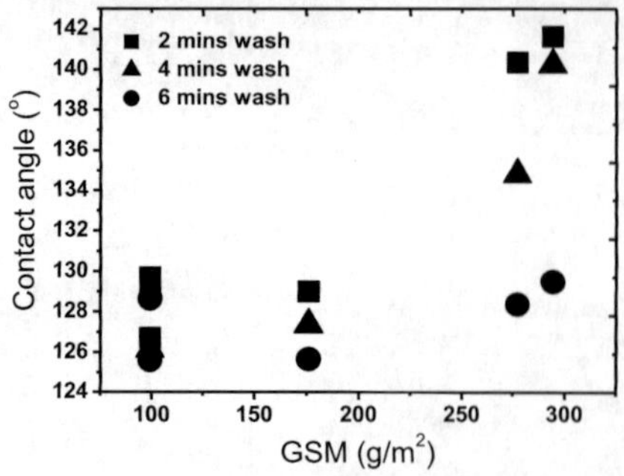

Figure 4 Contact angle of PTFE microporous membranes.

The Young equation is the common model to describe the contact angle on a surface:

$$\cos\theta = \frac{\gamma sv - \gamma sl}{\gamma lv} \qquad (1)$$

Where h is the contact angle, and γsv , γsl and γlv are surface energies for liquid-vapor, vapor-solid, and solid-liquid interfaces, respectively. Most practical surfaces, however, are rough and heterogeneous to some extent. Accordingly, different modifications of the Young equation have been proposed to apply to rough surfaces [32, 33]. In this study, it is assumed that all of the nanofibers under the water drop have contact with the water. A roughness factor r is the relation between the real area and the projected area. The model of contact angle for microporous membrane is shown in Figure 5. Thus, the contact angle of a microporous membrane can be denoted by:

$$\cos\theta' = \frac{r(\gamma sv - \gamma sl)}{\gamma lv} = r\cos\theta \qquad (2)$$

The increase in contact angle with voltage is attributed to the change in roughness factor r, as described by Equation (2). Roughness, however, is so complicated that it is difficult to develop a general method for roughness measurement.

$$B = \frac{S_r}{m} \qquad (3)$$

$$r = \frac{S_r}{S_p} = \frac{mB}{S_p} \qquad (4)$$

Where S_r is the real surface area of microporous membrane, S_p is the projected area, B is the specific surface area, and m represents the mass of a membrane. For a given microporous membrane, the mass m and projected area S_p are constant and can be easily measured. Consequently, the modified Young Equation (2) can be derived:

$$\cos\theta' = r\cos\theta = \frac{mB}{S_p}\cos\theta \qquad (5)$$

It can be seen from Equation (5) that the contact angle, h0, of the membrane depends on the ideal contact angle, h, and surface roughness, and the specific surface area plays an important role in the material's contact angle. Larger specific surface area leads to higher contact angles.

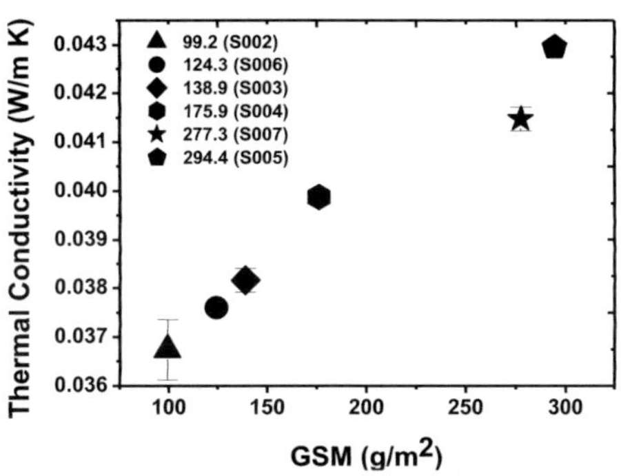

Figure 5 Thermal conductivity of the microporous membranes.

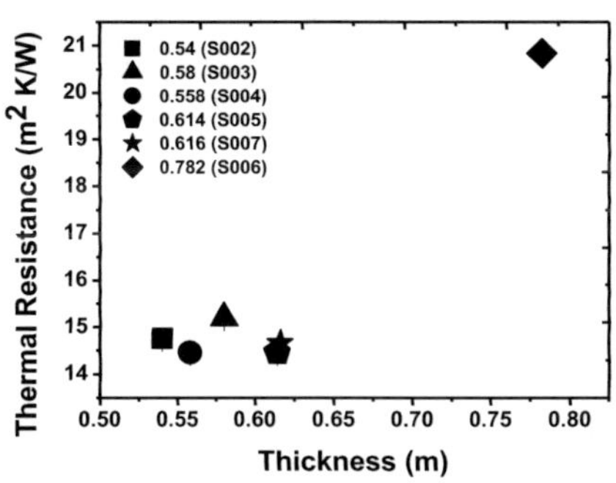

Figure 6 Thermal resistance of the microporous membranes.

Figure 5 and 6 contains the thermal properties of the samples. We can see from the Figure 5 that S002 has the lowest thermal conductivity and S005 has the highest thermal conductivity. The thermal conductivity of the carbon doped membrane (S005) is higher than other membranes. From the Figure 6, it can be seen that the membrane S006 has highest thermal resistance than other membranes. The thermal resistance of the membrane S006 is higher due to the higher thickness and also it can be seen from the sample description (Table 1) that the membrane is diluted with water and concentration is reduced to 50%. This can also attribute to the higher thermal resistance of the membrane. Compared to other membranes, carbon and basalt doped membranes have higher thermal conductivity and lower resistance. .

IV.CONCLUSION

PTFE microporous membranes were electrospun by optimizing the concentration and processing parameters. By adjusting the process parameters of common electrospinning system, preparation of membranes with tunable porosity was prepared by nanofibrous assembly and interconnected particles (electrospraying). The surface structures, diameters, and specific surface areas of the electrospun nanofibers were successfully controlled by altering the electrostatic voltage. SEM observations revealed that when the concentration of the polymer solution was lower, the fibers tend to form beadlike structure and also the change in fiber diameter was attributed to the presence of surfactant. The experimental results also proved that the static force applied on the jet might be the key factor affecting the surface structure, as well as the specific surface area. It was concluded by the analysis of the modified the Young equation that specific surface area played the most important role in contact angles of the microporous membrane. The larger specific surface areas resulted in high contact angles. The thermal properties results reveal that thermal conductivity was higher and thermal resistance was lower for carbon and basalt doped membranes as compared to other membranes.

ACKNOWLEDGEMENT

This work was supported by the Ministry of Education, Youth and Sports of the Czech Republic and the European Union - European Structural and Investment Funds in the frames of Operational Programme Research, Development and Education - project Hybrid Materials for Hierarchical Structures (HyHi, Reg. No. CZ.02.1.01/0.0/0.0/16_019/0000843), project "Design optimization and application of smart heat-insulating nano-layers" [LTACH-17014, 18301] and project "Sophisticated hybrid tapes for fabrication of composites by precise winding" [Project no: TJ01000292, 14014/136].

REFERENCES

[1] https://en.wikipedia.org/wiki/Polytetrafluoroethylene, *Accessed:*2017-05-01

[2] R. Rhodes, "*The Making of the Atomic Bomb*." Simon and Schuster, New York. 1986, pp. 494.

[3] *"Teflon PTFE Properties"*,http://www.rjchase.com/ptfe_handbook.pdf, Dupont, USA, Retrieved 11 October 2012.

[4] http://www.plastechcoatings.com, D.T.C, *Accessed:*2017-05-01

[5] H. Chen, G. Jiang, Y. Liu, L. Li, & Q. Huang. "Template Synthesis of Mesoporous Bi4(GeO4)3 Microspheres From Na4Ge9O20 Nanospheres via Ion-Exchange Strategy for Photo-degradation of Reactive Brilliant Red X-3B. Acta Metallurgica Sinica (English Letters)", 2015, vol.28, no.9, pp. 1156-1161.

[6] H. Pan, L. Li, L. Hu & X. J. Cui, "*Continuous aligned polymer fibers produced by a modified electrospinning method*". Polymer, 2006, vol. 47, pp. 4901–4904.

[7] M. Bognitzki, W. Czado & T. Frese, "*Nanostructured fibers via electrospinning*". Advanced Materials, 2001, vol.13, pp.70–73.

[8] T. S. Meiron, A. Marmur, & I. S. Saguy. "*Contact angle measurement on rough surfaces*". Journal of Colloid and Interface Science, 2004, vol. 274, pp. 637–644.

[9] Y.M. Shin, M.M. Hohman, M.P. Brenner, & G.C. Rutledge, "*Experimental characterization of electrospinning: the electrically forced jet and instabilities*". Polymer, 2001, vol. 42, no. 25, pp.09955–09967.

[10] A. L. Yarin, S. Koombhongse, & D. H. Reneker, "*Bending instability in electrospinning of nanofibers*". Journal of Applied Physics, 2001, vol. 89, no. 5, pp. 3018.

[11] Q. Long, M. Cai, J. Li, H. Rong, & L. Jiang, "*Improving the electrical catalytic activity of Pt/TiO2 nanocomposites by a combination of electrospinning and microwave irradiation*". Journal of Nanoparticle Research, 2010, vol. 13, no. 4, pp.1655–1662.

[12] G. Taylor, "*Disintegration of water drops in an electric field*". In Proceddings of the Royal Society of London A, Mathematical, Physical and Engineering Science, 1964, vol. 280, no. 1382, pp. 383-397.

[13] S. Megelski, J.S. Stephens, D.B. Chase, & J.F. Rabolt, "*Micro- and nanostructured surface morphology on electrospun polymer fibers*". Macromolecules, 2002, vol. 35, no. 22, pp.8456–8466.

[14] C.J. Buchko, L.C. Chen, Y. Shen, & D.C. Martin, "*Processing and microstructural characterization of porous biocompatible protein polymer thin films*". Polymer, 1999, vol. 40, no. 26, pp. 7397–7407.

[15] G. Eda, & S. Shivkumar, "*Bead structure variations during electro- spinning of polystyrene*". Journal of Material Science, 2006, vol. 41, no. 17, pp. 5704–5708.

[16] J. Tao, & S. Shivkumar, "*Molecular weight dependent structural regimes during the electrospinning of PVA.* Materials". Letters, 2007, vol. 61, no.11–12, pp. 2325–2328.

[17] S.V. Fridrikh, J.H. Yu, M.P. Brenner, & G.C. Rutledge, "*Controlling the fiber diameter during electrospinning*". Physical Review Letters, 2003, vol. 90, no. 14, pp.144502.

[18] J.M. Deitzel, J. Kleinmeyer, D. Harris, & N.C.B. Tan, "*The effect of processing variables on the morphology of electrospun nanofibers and textiles*". Polymer, 2001, vol. 42, no. 1, pp. 261–272.

[19] N.S.B Raymond Scheffler, & W. Sigmund, "*Electrospun Teflon AF fibers for superhydrophobic membranes*". Journal of Materials Research, 2010, vol. 25, pp.1595-1600.

[20] S. Zhang, F. Zhou, H. Peng, T. Liu, L. Ding, & Y. Fang. "*Fabrication and humidity sensing performance studies of a fluorescent film based on a cholesteryl derivative of perylene bisimide*". Spectrochimica Acta Part A: Molecular and Biomolecular Spectroscopy, 2016, vol. 165, pp.145-149 .

[21] F. Xia, & J. Lai, "*Design and creation of superwetting/antiwetting surfaces*". Advance Materials, 2006, vol. 18, pp. 3063-3078.

[22] T. Lin, & X.G. Wang, "*Controlling the Morphologies of Electrospun Nanofibres*", 2007, CRC Press.

[23] F. Cengiz, & O. Jirsak. "*The effect of salt on the roller electrospinning of polyurethane nanofibers*". Fibers and Polymers, 2009, vol. 10, no. 2, pp. 177–184.

[24] B. Ding, X. Wang, J. Yu, & M. Wang. "*Polyamide 6 composite nano-fiber/net functionalized by polyethyleneimine on quartz crystal microbalance for highly sensitive formaldehyde sensors*". Journal of Materials Chemistry, 2011, vol. 21, pp.12784–92. .

[25] N. Bock, T.R. Dargaville, & M.A. Woodruff, "*Electrospraying of polymers with therapeutic molecules: State of the art*", Progress in Polymer Science, *2012*, pp.1510 - 1551.

[26] A. Jaworek. "*Electrospray droplet sources for thin film deposition*", Journal of Materials Sicence, 2007, pp. 266-297.

[27] S. Sukigara, M. Gandhi, J. Ayutsede, M. Micklus, & F. Ko. "*Regeneration of Bombyxmori silk by electrospinning—Part 1: Processing parameters and geometric properties*", Polymer, 2003, vol. 44, pp. 5721.

[28] Y.C. Ahn, S.K. Park, G.T. Kim, Y.J. Hwang, C.G. Lee, H.S. Shin, & J.K Lee. "*Development of high efficiency nanofilters made of nanofibers*", Current Applied Physics, 2006, vol. 6, pp.1030.

[29] H. Fong, I. Chun, & D.H. Reneker, "*Beaded nanofibers formed during electrospinning*", Polymer, 1999, vol. 40, pp. 4585.

[30] K.H. Lee, H.Y. Kim, H.J. Bang, Y.H. Jung, & S.G. Lee, "*The change of bead morphology formed on electrospun polystyrene fibers*", Polymer, 2003, vol. 44, pp. 4029.

[31] K. Lin, K.N. Chua, G.T. Christopherson, S. Lim, & H.Q. Mao, "*Reducing electrospun nanofiber diameter and variability using cationicamphiphiles*". Polymer, 2007, vol. 48, pp. 6384.

[32] R.N. Wenzel, "*Resistance of solid surfaces to wetting by water*". Industrial and Engineering Chemistry, 1936, vol. 28, pp. 7426–7431.

[33] A.B.D. Cassie, & S. Baxter. "*Wettability of porous surfaces*". Transactions of the Faraday Society, 1944, vol. 40, pp. 546.

Yarn Tension Change with Respect to Bobbin Diameter in Yarns Including Lycra Component

Özge Çelik[1] and Recep Eren[2]

[1,2] *Uludag University, Faculty of Engineering, Textile Engineering Department, Gorukle Campus 16059 Nilüfer-Bursa, Turkey*

Corresponding author: celikozge@uludag.edu.tr

Abstract The yarn is transferred from one bobbin to another in many textile processes. In these kinds of processes like warping, weaving, doubling and twisting, yarn tension change with respect to bobbin diameter has a significant role because it affects product quality and machine efficiency. Yarns including lycra component are widely used in fabric production which are sensible to tension changes. This paper presents an experimental research aiming at determining the relationship between yarn tension and bobbin diameter during unwinding for continuous filament polyester-lycra yarns. For this aim, an experimental research set-up was established representing unwinding from bobbins in many textile processes. Experimental research set-up consists of a winding machine, a creel with single bobbin holder, a laser sensor measuring bobbin diameter, a tension sensor for measuring yarn tension, a PC and a DAQ card. A software program was developed in C programming language to read and record the tension and bobbin diameter simultaneously as a data file. Based on the recorded data, average values of warp tension and bobbin diameter were calculated and relationship between yarn tension and bobbin diameter was obtained in this way. Experimental work was carried out for 3 different yarn numbers (75, 150 and 300 denier) at 5 different unwinding speeds (100, 200, 400, 600 and 800 m/min). The results were compared with the yarns without lycra component and found to be in agreement with theoretical findings in the literature expressing that high elasticity of yarns limited the tension change from full to empty bobbin.

***Keywords*—At least four keywords or phrases in alphabetical order, separated by commas. Bobbin diameter, tension measurement, unwinding tension, yarn elasticity, yarn tension**

I. Introduction

Before using in weaving, warping, knitting etc. departments, yarns sometimes undergo different processes like dyeing, doubling, twisting. While going through all these processes, yarn tension changes with respect to the bobbin diameter. Due to the aim to achieve higher processing speeds, tension changes become more pronounced. Tension change from full to empty bobbin diameter has a significant effect on the product quality as well as process efficiency. Hence, optimization the parameters affecting this tension change plays an important role. Yarn tension tends to increase towards to empty bobbin diameter inherently. In warping process, yarn tension change causes warp sections of the same length to be wound at different diameters and this leads tension differences between sections on the warp and causes irregular appearance on the fabric surface. Besides, in bobbins to be dyed tension change with respect to bobbin diameter affects hardness variations in radial direction. Although today many warping and winding machines have their own tension controlling systems to keep the tension at an adjusted value, machines without tension control are still in use in industry. Yarns including lycra component are widely employed in weaving sector which are sensible to tension changes. This paper investigates experimentally yarn tension change with respect to bobbin diameter during unwinding in yarns including lycra component.

There are numerous theoretical and experimental research investigating the unwinding behavior of yarns from bobbins. Without considering centrifugal and Coriolis forces Mack [1] formulated the differential equations for spinning balloons. Padfield [2] investigated the motion and tension of an unwinding thread and modified Mack's equations by adding effect of Coriolis forces. She suggested an empirical quadratic relation between yarn tension during unwinding and balloon length/bobbin radius ratio. She didn't consider the effects of gravitational force and the internal elastic forces within the yarn. Booth [3] studied the sliding region of over-end unwinding and investigated the dynamics of the yarn motion over the bobbin surface. Kothari and Leaf [4-8] improved Padfield's [2] theory by adding the effects of gravity and tangential air drag. Fraser, Ghosh, and Batra [9] presented an integrated model for both balloon and sliding regions for the first time. Ghosh, Murthy, and Batra [10] analyzed

Fraser, Ghosh, and Batra's [9] study and recommended the third region for unwinding process (stationary region). Popova and Efremov [11] investigated the yarn tension bobbin diameter relation during unwinding from cone bobbins. Kurilenko, Matyushev and Goncharenko et al. [12] investigated yarn tension during unwinding from a cylindrical bobbin. They measured yarn tension at 3 different bobbin diameters, 3 different yarn guide-bobbin front distances and 3 different unwinding speeds. Cooray and Fernando [13] studied the conditions both mathematically and experimentally for a uniform unwinding tension from bobbins and developing a device for this aim. Godawat [14] studied the experimental verification of non-linear behavior of over-end yarn unwinding from cylindrical packages. He designed and built an experimental set up measuring yarn tension, bobbin diameter and balloon shape by a high-speed camera. Apart from analyzing the effects of bobbin diameter, bobbin front-yarn guide distance and unwinding speeds on yarn tension during unwinding, he also intended to test the theory of Pan's [15] mathematical model which investigates and deduces relationships between balloon shape and tension distribution in the end region of unwinding. Kong, Rahn and Goswami [16] investigated steady-state yarn motion during unwinding from nonzero winding angle packages. Pracek, Pusnik and Simoncic, et al. [17] studied a model for simulating yarn unwinding. Çelik [18] investigated yarn tension change with respect to bobbin diameter during unwinding experimentally for both continuous filament polyester (twisted and intermingled) yarn of 7 different numbers and cotton yarns of 5 different numbers. For each yarn or bobbin, experiments were carried out with 3 different bobbin front surface-yarn guide distances (120 mm, 240 mm, 480 mm) and 5 different unwinding speeds (100 m/min, 200 m/min, 400 m/min, 600 m/min and 800 m/min). In addition, the effect of winding types (random winding, step precision winding and precision winding for 2 different winding ratios) on the relationship between the yarn tension and bobbin diameter was investigated.

When literature was reviewed it is realized that most of the studies which reveal models for unwinding and balloon formation made the same simplifying assumption that the yarn is inextensible [2, 3, 5, 6, 9, 16]. In 1992, Fraser [19] published a framework for unwinding of non-linear elastic or viscoelastic yarns and investigated the relation between the elastic or viscoelastic properties of the yarn and the yarn tension during unwinding. He presented balloon equations which included linear elasticity term and made numerical solutions of these equations for small values of elasticity. He assumed that yarn inextensibility is acceptable for typical yarns. He concluded that the inclusion of elasticity in the theoretical model caused a decrease in balloon tension and also in balloon radius. Nevertheless, he indicated that this effect is very slight for real yarns (yarns not including lycra or similar elastic component). So, the inextensibility assumption is reasonable for these types of yarns and the extensibility of these yarns could be neglected. But for highly extensible yarns, yarn elasticity must be considered.

Godawat [14] referred a study which was done by Ma, Ghosh and Batra and investigating unwinding an elastic yarn from a cylindrical bobbin. They presented an integrated model for elastic yarn unwinding process obtained for balloon, sliding and stationary regions. According to Godawat [14] they concluded that when the balloon was formed as a single loop for the yarns which had an initial modulus higher than 0.1 N/Tex (almost all man- made fibers and natural fibers), the effect of elasticity was negligible, and these types of yarns could be assumed as inextensible. However, for the yarns which had an initial modulus less than 0.1 N/Tex (elastomeric yarns) both the balloon radius and yarn tension decreased with decreasing initial modulus.

Available research in the literature provides no experimental result for tension change with respect to bobbin diameter for yarns having high elasticity. This paper presents an experimental work investigating yarn tension change during unwinding for textured polyester with lycra component.

II. Methods and Procedures

An experimental set-up was built to investigate the relationship between yarn tension and bobbin diameter during unwinding for continuous filament polyester-lycra yarns. Experimental set up had a laser sensor measuring bobbin diameter, a tension sensor for measuring yarn tension, a bobbin winding unit and a PC and a DAQ card. A software program was developed in C programming language to read and record the tension and bobbin diameter simultaneously.

Experimental set up had a single unit creel as shown in Fig. 1. The single unit creel had a bobbin holder, yarn guides as well as laser bobbin diameter measurement sensor. Laser sensor was mounted on the creel to measure bobbin diameters of maximum 300 mm.

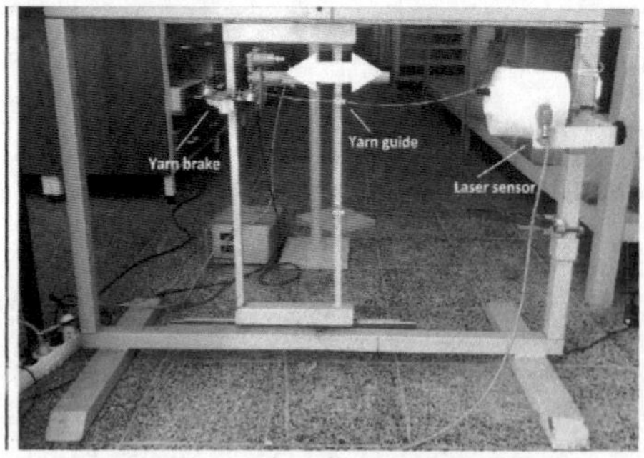

Fig. 1. Single unit creel

After the yarn guide and other guide roller, yarn went through the tension sensor as shown in Fig. 2. The tension sensor is of Schmidt made and has measuring interval of 0-200 cN [20].

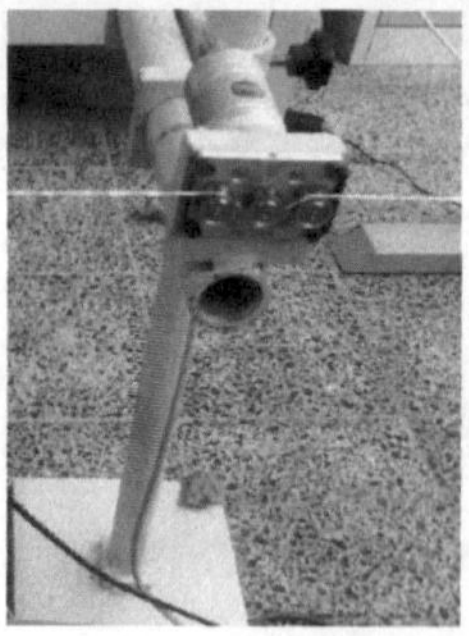

Fig. 2. Tension sensor

Output signals of laser sensor and tension sensor are connected to a PC by a DAQ card. DAQ card has 12 bits bipolar ADC converters, therefore 11 bits are used for reading the signals and last bit is reserved as a sign bit. Hence, 10 volts are converted to 2047 numbers for diameter and tension measurements. A software program was developed in C programming language to read and record 2 sensor signals simultaneously.

In the experimental work, 75 denier, 150 denier and 300 denier yarns with lycra were unwound from bobbins at 100, 200, 400, 600 and 800 m/min speeds. Yarn tension was recorded at 13 different diameters and the relationship between yarn tension and bobbin diameter was obtained by averaging yarn tension for each bobbin diameter. Distance between yarn guide and bobbin front was kept constant at 240 mm during all the experiments.

Table I shows the bobbin dimensions and yarn properties used in the experimental work.

TABLE I
BOBBIN DIMENSIONS AND YARN PROPERTIES USED IN THE EXPERIMENTAL WORK

Yarn number/ Yarn and bobbin properties	75 denier	150 denier	300 denier
Lycra yarn number	36 dtex	44 dtex	78 dtex
Lycra drawing ratio	3,4	3,4	3,6
Full bobbin diameter (mm)	230	250	240
Empty bobbin diameter (mm)	76	75	75
Full bobbin length (mm)	215	215	220
Empty bobbin length (mm)	255	245	255
Winding angle (degree)	31	31	31

III. Results and Discussion

Figure 3 shows unwinding tension change with respect to bobbin diameter for 75 denier polyester-lycra yarn at 100, 200, 400, 600 and 800 m/min unwinding speeds. Changing unwinding speed from 100 m/min to 800 m/min increased yarn tension around 4 cN at full bobbin diameter. At all speeds during unwinding tension tended to decrease up to half of the full bobbin diameter and then it increased up to empty bobbin diameter. Tension change from full to empty bobbin diameter increased with increasing unwinding speeds. A less than 1 cN tension fluctuation occurred up to the empty bobbin diameter with 100 m/min, 200 m/min, 400 m/dak and 600 m/dak unwinding speeds. At 800 m/min unwinding speed, tension fluctuation occurred around 1,5 cN from full to empty bobbin diameter. The reason for the higher initial tension is thought to be due to the friction between unwound yarn and bobbin surface. Friction between bobbin surface and unwound yarn decreases with a decrease in bobbin diameter and therefore yarn tension decreases. On the other hand, angular rotation of yarn around bobbin axis increases with decreasing bobbin diameter and this causes centrifugal force to increase. After half bobbin diameter, yarn tension continuously increased because of the effect of centrifugal force with decreasing bobbin diameter. As the yarn is too thin and has a small mass forming the balloon, the effect of centrifugal force on yarn tension showed a limited effect even at 800 m/min unwinding speed.

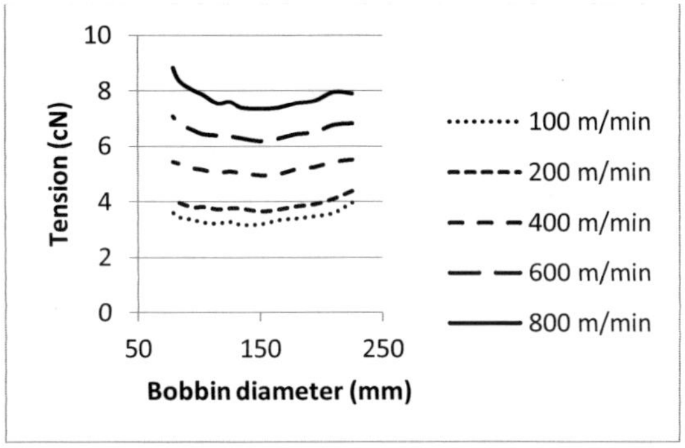

Fig. 3. Yarn tension bobbin diameter relation for 75 denier continuous filament polyester- lycra yarn.

Figure 4 shows unwinding tension change with respect to bobbin diameter for 150 denier polyester-lycra yarn at 100, 200, 400, 600 and 800 m/min unwinding speeds. At 100 m/min and 200 m/min unwinding speeds, tension change showed a slight decrease at around 0,5 cN from full to empty bobbin. At 400 m/min, 600 m/min and 800 m/min unwinding speeds, tension decreased very slightly up to half bobbin diameter, then it increased up to empty bobbin diameter. Tension change increased as unwinding speed increased. Tension increased less than 1 cN at 400 m/min unwinding speed whereas it occurred around 1,5 cN at 600 m/min unwinding speed and 3 cN at 800 m/ min unwinding speed.

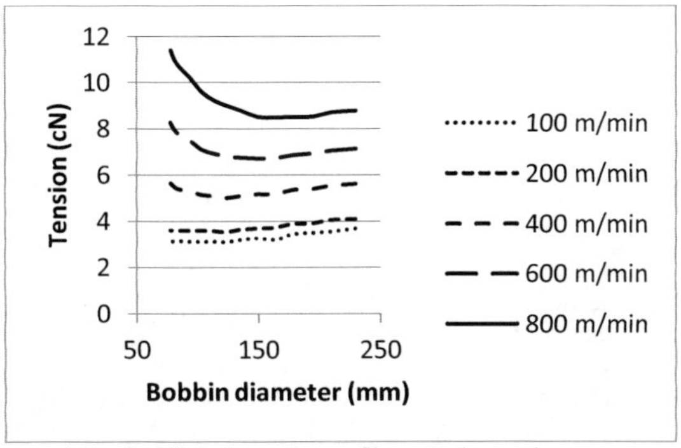

Fig. 4. Yarn tension bobbin diameter relation for 150 denier continuous filament polyester- lycra yarn.

Figure 5 shows tension change with respect to bobbin diameter for 300 denier polyester-lycra yarn. No significant tension change occurred with respect to bobbin diameter at 100 m/min and 200 m/min unwinding speeds. At 400 m/min, tension increased very slightly towards empty bobbin diameter. At 600 m/min and 800 m/min unwinding speeds, tension increased about 3 cN and 7 cN respectively. Tension changed slightly at higher bobbin diameters and it increased at a higher rate towards empty bobbin diameter. In fact, the relationship between unwinding tension and bobbin diameter shows a change like quadratic curve at 600 m/min and 800 m/min unwinding speeds.

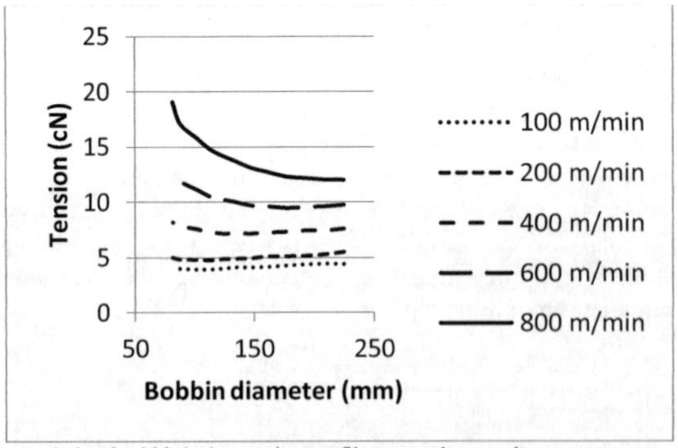

Fig. 5. Yarn tension bobbin diameter relation for 300 denier continuous filament polyester- lycra yarn.

Table II shows the maximum tension (max tension), minimum tension (min tension), average tension and difference between maximum and minimum tension values from full to empty bobbin diameter for each yarn number and unwinding speed used in the experimental work.

TABLE II
MAXIMUM TENSION (MAX TENSION), MINIMUM TENSION (MIN TENSION), AVERAGE TENSION AND DIFFERENCE BETWEEN MAXIMUM AND MINIMUM TENSION VALUES FROM FULL TO BOBBIN DIAMETER FOR EACH YARN NUMBERS AND UNWINDING SPEEDS

Yarn number (denier)	Tension (cN)	Unwinding speed (m/min)				
		100	200	400	600	800
75	**Max tension**	3,97	4,38	5,50	7,07	8,82
	Min tension	3,16	3,64	4,94	6,17	7,34
	Average tension	3,40	3,88	5,19	6,52	7,79
	Difference between max and min tension	0,80	0,74	0,56	0,90	1,48
150	**Max tension**	3,68	4,09	5,64	8,27	11,40
	Min tension	3,08	3,53	5,00	6,71	8,48
	Average tension	3,26	3,73	5,30	7,11	9,26
	Difference between max and min tension	0,60	0,56	0,64	1,56	2,92
300	**Max tension**	4,44	5,50	8,18	12,60	19,07
	Min tension	3,93	4,76	7,15	9,47	12,00
	Average tension	4,16	5,02	7,45	10,34	14,07
	Difference between max and min tension	0,51	0,73	1,03	3,13	7,07

To analyze the effect of yarn elasticity on tension change during unwinding, results of this paper (polyester yarn with lycra component) were compared with Çelik's [18] results of 75 denier, 150 denier and 300 denier intermingled polyester textured yarn without lycra component (distance between yarn guide and bobbin front and the unwinding speeds are equal). When the graphs showing yarn tension change from full to empty bobbin diameter were examined, it was realized that same yarn numbers with and without lycra component showed the same characteristics.

Table III shows the difference between maximum and minimum tension values from full to empty bobbin diameter for 75 denier, 150 denier and 300 denier polyester yarns without lycra component [18]. Although they showed the same yarn tension change characteristics, as seen from the Table II and Table III, yarn elasticity showed a decreasing effect on yarn tension change from full to empty bobbin especially for higher unwinding speeds. Tension fluctuation occurred as about 2,5 cN at 400 m/min, 6 cN at 600 m/min and 11 cN at 800 m/min unwinding speeds for 300 denier polyester yarn without lycra component, whereas it occurred about 1 cN, 3 cN and 7 cN respectively with lycra component. For 150 denier yarn, without lycra component this tension difference was about 1,5 cN, 2,5 cN and 4 cN respectively, with lycra component it was about 0,5 cN, 1,5 cN and 3 cN. Lycra component showed a very limited effect on 75 denier polyester yarn. At even 600 m/min unwinding speed the difference of this value between with and without lycra component was 0,05 cN, at 800 m/min unwinding speed this difference tension value was just 0,32 cN.

TABLE III
DIFFERENCE BETWEEN MAXIMUM AND MINIMUM TENSION FROM FULL TO EMPTY BOBBIN DIAMETER FOR 75 DENIER, 150 DENIER AND 300 DENIER POLYESTER YARNS WITHOUT LYCRA COMPONENT [18]

Yarn number (denier)	**Unwinding speed (m/min)**				
	100	**200**	**400**	**600**	**800**
75 denier	0,67	0,54	0,36	0,95	1,80
150 denier	1,08	1,18	1,73	2,55	3,77
300 denier	0,84	0,98	2,37	5,87	11,01

IV. CONCLUSION

An experimental research was carried out to investigate the relationship between yarn tension and bobbin diameter during unwinding from bobbins for polyester yarns including lycra component. Following conclusions can be drawn regarding the experimental research.

-In polyester yarns with lycra component, unwinding tension-bobbin diameter relation showed almost the same characteristic with polyester yarns without lycra component.

-In accordance with the literature, the amount of tension increase from full to empty bobbin decreased significantly with polyester yarns with lycra component. Tension change with respect to bobbin diameter became more pronounced as the yarn got thicker and unwinding speed increased like 600 m/min and 800 m/min.

-Although tension increase decreased with polyester yarns having lycra component, it is still important from practical and processing point of view because this tension change causes more yarn elongation therefore mass change in unit area of a fabric.

ACKNOWLEDGMENT

This research has been done as a part of research project supported by The Scientific and Technological Research Council of Turkey (TÜBİTAK). We would like to express our sincere thanks to Technological Research Council of Turkey (TÜBİTAK) for supporting this project (Project Number: 215M372). Also, we would like to thanks to administration of Polyteks Tekstil San. Araştırma ve Eğitim A.Ş. for providing yarns used in the experiments.

REFERENCES

[1] C. Mack, "Theoretical Study of Ring and Cap spinning Balloon Curves (With and Without Air Drag)", J. Text. Inst., 44, pp. 483-498, 1953.
[2] D.G. Padfield, "The Motion and Tension of an Unwinding Thread I" Proceedings of the Royal Society of London. Series A, Mathematical and Physical Sciences; 245(1242), pp. 382-407, 1958.
[3] K. H. V. Booth, "Vibration in Tension of An Unwinding Thread" British J. of Appl. Physics, 8, pp. 142-144, 1957.
[4] V.K. Kothari and G.A.V. Leaf, "The Unwinding of Yarns From Packages, Part I: The Theory of Yarn Unwinding", J. Text. Inst., 70, pp. 89-95, 1979a.
[5] V.K. Kothari and G.A.V. Leaf, "The Unwinding of Yarns From Packages, Part II: Unwinding From Cylindrical Packages", J. Text. Inst., 70, pp. 95-105, 1979b.
[6] V.K. Kothari and G.A.V. Leaf, "The Unwinding of Yarns From Packages, Part III: Unwinding From Conical Packages", J. Text. Inst., 70, pp. 172-183, 1979c.
[7] V.K. Kothari and G.A.V. Leaf, "The Unwinding of Yarns From Packages, Part IV: Two-for-One Twisting: The Storage Disc and Balloon", J. Text. Inst., 70, pp. 184-192, 1979d.
[8] V.K. Kothari and G.A.V. Leaf, "The Unwinding of Yarns From Packages, Part V: Two-for-One Twisting: Two-for-One Twisting: Before the Storage Disc", J. Text. Inst., 70, pp. 193-198, 1979e.
[9] W. B. Fraser, T. K. Ghosh and S. K. Batra, "On Unwinding Yarn From A Cylindrical Package", Proc. R. Soc. London, A436, pp. 479-498, 1992.
[10] T.K. Ghosh, A.S. Murthy and S.K. Batra, "Dynamic Analysis of Yarn Unwinding From Cylindrical Packages, Part I: Parametric Studies of the Two-Region Problem", Text. Res. J., 71(9), pp. 771-778, 2001.
[11] G.K. Popova and E.D. Efmerov, "The Yarn Tension in Unwinding From a Cone Bobbin Under Warping Conditions", Tech. Of Textile Industry U.S.S.R., 1, pp. 46-50, 1970.
[12] Z.N. Kurinlenko, I.I. Matyushev, A.E. Goncaranko and A.F. Mel'nikov, "Yarn Tension During Unwinding From a Package", Khimicheskie Volokna, 3, pp. 43-44, 1980.
[13] T. Cooray and E. Fernando, "Mathematical Modeling of Over-end Yarn Withdrawl, and the Development/ Design of a Device For Uniform Unwinding Tension" in 85th Textile Conference, Colombo, Srilanka, 1-3 March 2007, pp. 798-812.
[14] P. Godawat, "Experimental Verification of Non-Linear Behavior of Over-end Yarn Unwinding from Cylindrical Packages", MSc Thesis, North Carolina State University, USA, 2003.
[15] Z. Pan, "Dynamic Analysis of Over-end Unwinding of Yarn", PhD Thesis, North Carolina State University, USA, 2001.
[16] X.M. Kong, C.D. Rahn and B.C. Goswami, "Steady-State Unwinding of Yarn from Cylindrical Packages", Textile Research Journal, 69(4), pp. 292-306, 1999.

IOP Conf. Series: Materials Science and Engineering **460** (2018) 012018 doi:10.1088/1757-899X/460/1/012018

[17] S. Pracek, N. Pusnik, B. Simoncic and P.E.F. Tavcer, "Model for Simulating Yarn Unwinding from Packages", Fibres & Textiles in Eastern Europe, 23 2(110), pp. 25-32, 2015.

[18] Ö. Çelik, "Experimental Investigation Of The Factors Affecting Yarn Tension During Unwinding From Bobbin", PhD Thesis, Uludag University, Turkey, 2018.

[19] W.B. Fraser, "The Effect of Yarn Elasticity on An Unwinding Balloon", J. Text. Inst., 83(4), pp. 603-613, 1992.

[20] Schmidt Control Instruments Tension Sensor. Stationary Electronic Tension Meter, Single Place Systems3Rollers (accessed 2017, December 24), Available: http://www.hansschmidt.com/EN/products/tension_meter/stationary_electronic/single_place_systems_3_rollers/ts_series/model_ts1.

The essential conditions for developing high electrical conductivity capability in carbon nanofibers

S. Arbab[1], A. Teimoury[2], D. Adolphe[3], B. Noroozi[2] and P. Nourpanah[4]

[1]*ATMT Research Institute, Department of Textile Engineering, Amirkabir University of Technology, P.O. Box 15875-4413, Tehran, Iran*

[2]*Department of Textile Engineering, University of Guilan, Rasht, Iran*

[3]*Université de Haute-Alsace - ENSISA- Laboratoire de Physique et Mécanique Textiles, Mulhouse CEDEX, France*

[4]*Department of Textile Engineering, Amirkabir University of Technology, P.O. Box 15875-4413, Tehran, Iran*

Corresponding author: shahram.arbab@aut.ac.ir

Abstract

Carbon nanofiber webs have high electrical and thermal conductivity, porosity, surface area and good mechanical properties promising great potential for different applications. In this paper, three types of chemically different polyacrylonitrile (PAN) copolymers were electrospun in various concentrations and nanofibers with average diameter between 220 and 530 nm were produced. After stabilization and carbonization of PAN nanofibers, carbon nanofibers with diameter in the range of 110 to 300 nm were produced. The effects of chemical composition and processing parameters on the formation of the sponge-bond interconnected morphology and electrical conductivity of the carbon nanofibers were studied. The results revealed that the progress of stabilization reactions higher than 98% is inappropriate, whereas the stabilization progress in the range of 87% was considered adequate for the development of a proper structure for obtaining high electrical conductivity during carbonization process. Formation of nanofiber mats in shape of a network interconnected sponge-like structure was believed to be necessary for obtaining much higher electrical conductivity (16.70 S/cm compared to 1.10 and 2.33 S/cm in the case of nanofiber mats without interconnections and sponge-like structure).

Keywords— **Carbon nanofibers, Electrical conductivity, Morphology, PAN nanofibers, Stabilization.**

I. Introduction

Carbon microfibers have been widely used for numerous applications particularly for the development of large load-bearing composites. Obtaining carbon fibers with smaller diameter is one of the important objectives in carbon fiber industry, because reduction of fiber diameter results in increased surface area per unit mass, which facilitates the stabilization of precursor fiber and prevents from formation of core-shell structures [1-3]. Carbon nanofibers are receiving increasing attention because of their large length to diameter ratio, high strength, elastic modulus, and relatively low density [1].

PAN copolymers are the best precursor for the production of carbon nanofibers [4-7]. Incorporation of comonomers into the PAN chains reduces the nitrile – nitrile interactions, increases the solubility of the polymer, facilitates the thermochemical reactions during oxidative stabilization by reducing the activation energy and the initiation temperature of the reactions. Moreover, the temperature range of exothermic reactions is broadened, which improves the uniformity and mechanical properties of precursor fibers and resulting carbon microfibers [2, 8-12]. During the stabilization, which is the most complicated and time-consuming process in the production of carbon nanofibers and carried out in air atmosphere in the temperature range of 200 – 350 °C, chemical and physical changes such as cyclization, dehydrogenation, oxidation, crosslinking and chain fragmentation take place. The linear structure of PAN is converted into an infusible ladder-like structure which can tolerate higher temperatures in carbonization process [1, 7, 10, 13, 14]. If PAN fibers are not properly stabilized, either obtaining carbon fibers will be impossible or the produced carbon fibers will have low quality [15]. High performance carbon fibers are usually made from PAN copolymers. Acidic comonomers such as itaconic acid are used to decrease initiation temperature of the cyclization reactions in the stabilization process, whereas esteric comonomers such as methyl acrylate are used to improve solubility, spinning and stretchability of the fibers [2].

After stabilization, carbonization process is carried out in an inert atmosphere at temperatures as high as 800-2000 °C. During this process, structural changes including crosslinking, and integration of cyclized segments converts the ladder-like structure of the stabilized fibers to the graphitic structure of carbon fibers [9, 16]. A number of researchers showed that the conversion of disordered carbonaceous (turbostratic) structures to ordered graphitic structures as well as graphite crystal size are increased in carbon nanofibers carbonized at higher temperatures [17-20]. Electrical conductivity is one of the important properties of carbon nanofibers, which is usually lower than 20 S/cm [21]. Several factors can increase the electrical conductivity of carbon nanofibers, such as higher proportion of graphitic structure, higher chain orientation, more connection points between nanofibers, bigger size of graphitic crystallites, less structural defects (like holes and fractures) as well as increased compression between nanofibers [9, 17, 20, 22-24]. Some researchers showed that electrical conductivity of carbon nanofibers increases with temperature and duration of the carbonization process. They both result in a more graphitic and ordered structure and increased connections between carbon nanofibers [17, 22, 25, 26].

Less research has been done on the production of carbon nanofibers from various PAN copolymers with different chemical compositions. In this work, three types of PAN copolymers were used to obtain PAN nanofibers by electrospinning and stabilization and carbonization thermal treatments were employed to produce carbon nanofibers. Different analysis techniques including Differential Scanning Calorimetry (DSC), Field Emission Scanning Electron Microscope (FE-SEM) and electrical conductivity were used to investigate carbon nanofibers properties and the effect of processing parameters on PAN nanofibers during thermal treatments.

II. Methods and Procedures

A. *Materials*

Three types of polyacrylonitrile copolymers with different chemical composition were used to produce PAN nanofibers (Table I). N, N-dimethylformamide solvent (99.5%, Merck) was used to prepare electrospinning solutions.

TABLE I
CHARACTERISTICS OF PAN COPOLYMERS

Polymer Code	Manufacturer company	Chemical composition	Monomer feed in polymerization reactor (%)	[η] (dL/g)	M_V (g/mol)
P_1	Polyacryl Co.	AN-MA-SMS	95-4.5-0.5	1.441	112452
P_2	Courtaulds Co. Ltd	AN-MA-IA	95-4-1	2.216	202000
P_3	Jilin Chemical Fiber Group, Co. Ltd	AN-MAA-AM	95-1.5-3.5	2.174	197000

P: PAN Copolymer, AN: Acrylonitrile, MA: Methyl Acrylate, IA: Itaconic Acid, MAA: Methacrylic Acid, AM: Acrylamide, SMS: Sodium Methallyl Solfunate.

B. *Nanofiber production*

After preparation of the electrospinning solutions (10 wt%) from each copolymer, electrospinning process was carried out at room temperature with a positive high voltage of 11 kV, solution flow rate of 0.25 ml/h. The distance between the spinneret and the fixed metal collector was 15 cm. The electrospun PAN nanofibers were stabilized using two different procedures in air atmosphere from the room temperature to 300 °C with the heating rate of 5 °C/min, followed by holding the temperature at 300 °C for 90 minutes to allow the stabilization to complete. Then stabilized nanofibers were carbonized in pure nitrogen atmosphere at 1250 °C with the heating rate set at 40 °C/min and the samples were held at the final temperature for 5 minutes. Nomenclature of PAN nanofibers (PF), stabilized PAN nanofibers (SF) and carbon nanofibers (CNF) are listed in Table II.

TABLE II
NOMENCLATURE OF INITIAL PAN NANOFIBERS, STABILIZED PAN NANOFIBERS AND CARBON NANOFIBERS

Polymer Code	PAN nanofibers	Stabilized nanofibers	Carbon nanofibers
P_1	PF_1	SF_1	CNF_1
P_2	PF_2	SF_2	CNF_2
P_3	PF_3	SF_3	CNF_3

C. Characterization

Thermal behavior of PAN nanofibers (temperature range of 50- 450 °C) and stabilized PAN nanofiber (temperature range of 50- 600 °C) was studied using a Mettler Toledo DSC 802 in air and at a heating rate of 5 °C/min. In order to estimate the progress of the stabilization reactions, cyclization index (CI) was calculated using the following equation [27-31]:

$$CI = \frac{\Delta H_{PF} - \Delta H_{SF}}{\Delta H_{PF}} \times 100 \qquad (1)$$

Where ΔH_{PF} and ΔH_{SF} are the reactions enthalpy of initial PAN and stabilized PAN nanofibers, respectively.

A Hitachi U-4080 field-emission scanning electron microscope (FE-SEM) was employed in order to examine the surface morphology of PAN and carbon nanofibers. The average diameter of nanofibers was calculated using Image J software. Diameter measurements were done in 100 different points randomly selected in the micrographs of each sample. The average diameter ± standard deviation of each sample is presented [2, 32].

Electrical conductivity of carbon nanofiber webs was measured by standard four-point probe method using the Keithley 196 System DMM_2:

$$\sigma = \frac{L}{RA} \qquad (2)$$

Where L is distance between two electrodes in “cm”, R is the web resistance in “Ω” and A is the area of carbon nanofibers sheet in “cm^2”.

III. RESULTS

A. DSC study of PAN and stabilized nanofiber Webs

DSC curves of the initial and stabilized PAN nanofibers are shown in Fig. 1. The presence of acidic comonomers in PF_2 and PF_3 nanofibers leads to initiation of cyclization reactions through ionic mechanism [10, 33-36]. As a result, the exothermic peak becomes broader and its initiation temperature decreases compared to PF_1 nanofibers (Fig. 1). Unlike PF_2 and PF_3, PF_1 nanofiber does not contain acidic comonomers. Despite similar stabilization conditions, the progress of stabilization reactions in PF_1 nanofibers is approximately 87%, whereas the progress of stabilization reactions in PF_2 and PF_3 nanofibers containing acidic comonomers is more than 98%, demonstrating maximum progress of stabilization reactions in these fibers (Table III).

B. Morphology study of PAN and carbon nanofiber webs

Field emission scanning electron microscopy (FE-SEM) images of PAN and carbon nanofiber webs are shown in Fig. 2 and their statistical data are reported in Table III. The produced initial PAN nanofiber webs were uniform and bead-free.

TABLE III
CHARACTERIZATION OF PAN AND CARBON NANOFIBERS.

PAN nanofibers	Average diameter of PAN nanofibers (nm)	Cyclization Index (%)	Carbon nanofibers	Average diameter of carbon nanofibers (nm)	Electrical conductivity of carbon nanofibers (S/cm)
PF_1	223	87.25	CNF_1	110	16.70
PF_2	240	98.59	CNF_2	178	2.33
PF_3	529	99.17	CNF_3	312	1.10

In similar spinning conditions, the diameter of PF_1 nanofibers is lower than those of PF_2 and PF_3 nanofibers in various concentrations of electrospinning solution. This can be attributed to the lower molecular weight of PF_1 polymer compared to PF_2 and PF_3 (Table I). Lower diameter of PF_1 nanofibers may lead to a better diffusion of oxygen into nanofiber structure during stabilization process. Oxygen has an important role in initiation and progress of stabilization reactions. Despite the absence of acidic comonomers in PF_1 nanofibers, better diffusion of oxygen results in proper oxidation of these nanofibers during stabilization process.

The morphology of CNF_1 nanofibers is completely different compared to CNF_2 and CNF_3 nanofibers. Most nanofiber strands in CNF_1 are merged and formed a sponge-like network structure. However, the morphology of CNF_2 and CNF_3 has not changed considerably. Only a number of fractures has been introduced in the structure due to the thermal treatment.

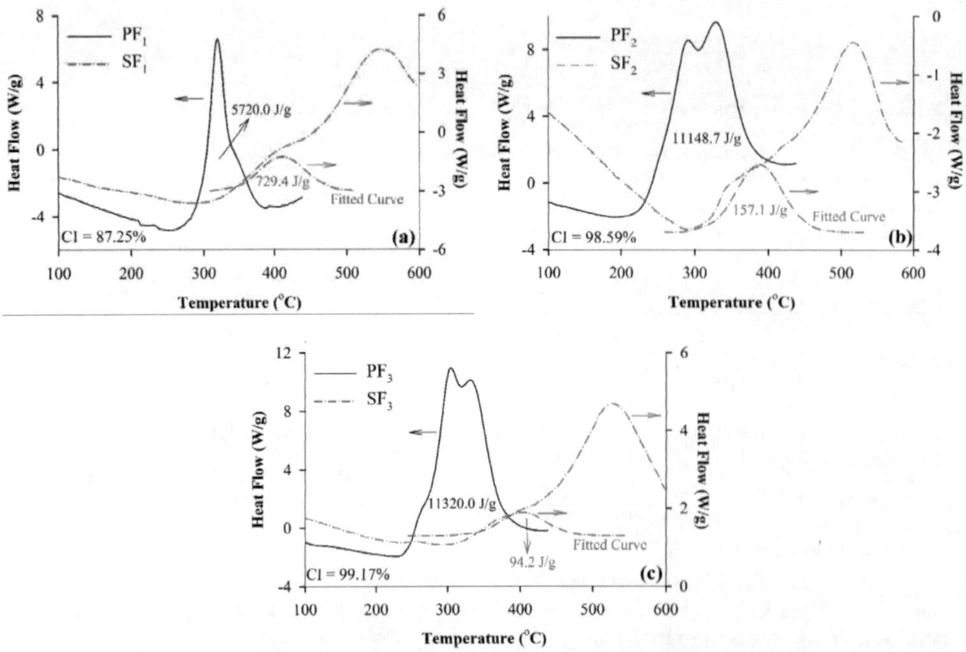

Fig. 1. DSC curves of PAN and stabilized PAN nanofibers. (a) PF_1 and SF_1, (b) PF_2 and SF_2, (c) PF_3 and SF_3.

C. Electrical conductivity of carbon nanofiber webs

Electrical conductivity of CNF_1 nanofiber webs is significantly higher than CNF_2 and CNF_3 nanofiber webs (Table III). The significantly higher electrical conductivity of CNF_1 nanofibers compared to CNF_2 and CNF_3 can be attributed to formation of connection points between nanofibers during thermal treatment, resulting in a songe-like compact network structure of carbon nanofibers mats (Fig. 2).

The electrical conductivity of carbon nanofiber webs are mainly influenced from the compactness of structure and formation of connection points due to thermal treatment. The more compact nanofiber mats increases the probability of presence of connection points and enhancement of carbon nanofiber webs electrical conductivity. In CNF_1 nanofibers (Fig. 2), a number of nanofiber strands are merged together and a sponge-like network structure is created.

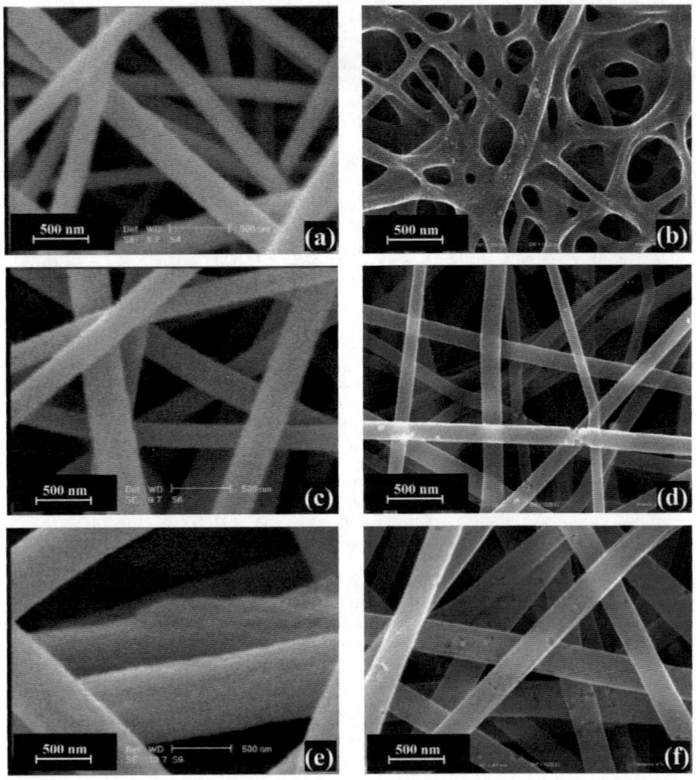

Fig. 2. SEM images of PAN and carbon nanofibers, (a) PF_1, (b) CNF_1, (c) PF_2, (d) CNF_2, (e) PF_3 and (f) CNF_3.

PF_2 and PF_3 nanofibers show higher progress in stabilization reactions due to the presence of acidic comonomers. However, the electrical conductivity of their carbon nanofibers is lower than CNF_1 nanofibers. Therefore, it can be concluded that obtaining proper carbon structure, specially higher electrical conductivity, requires an optimum stabilization process with optimum progress in stabilization reactions and formation of a sponge-like interconnected network of carbon nanofibers.

IV. Conclusion

Carbon nanofiber webs were produced from three different types of PAN precursor through the stabilization and carbonization of electrospun PAN nanofiber webs. The results showed that despite the common viewpoint in production technology of carbon fibers in micro scale, presence of acidic comonomers in the copolymer stucture is not necessary for production of carbon fibers at nanoscale. The results revealed that progress of stabilization reactions higher than 98% is inappropriate, whereas the stabilization progress in the range of 87% leads to development of proper graphitic structure during carbonization process and formation of nanofiber mats in shape of an interconnected sponge-like network structure with much higher electrical conductivity. Therefore, it can be concluded that the formation of interconnected structure in carbon nanofiber web is essential for obtaining high electrical conductivity.

References

[1] Arshad, S.N., Naraghi, M., and Chasiotis, I., *Strong carbon nanofibers from electrospun polyacrylonitrile.* Carbon, 2011. **49**(5): p. 1710-1719.

[2] Liu, J., He, L., Ma, S., Liang, J., Zhao, Y., and Fong, H., *Effects of chemical composition and post-spinning stretching process on the morphological, structural, and thermo-chemical properties of electrospun polyacrylonitrile copolymer precursor nanofibers.* Polymer, 2015. **61**: p. 20-28.

[3] Ali, A.A. and El-Hamid, M.A., *Electro-spinning optimization for precursor carbon nanofibers.* Composites, 2006. **37**(10): p. 1681-1687.

[4] Zhang, B., Kang, F., Tarascon, J.-M., and Kim, J.-K., *Recent advances in electrospun carbon nanofibers and their application in electrochemical energy storage.* Progress in Materials Science, 2016. **76**: p. 319-380.

[5] Kim, J.H., Ganapathy, H.S., Hong, S.-S., Gal, Y.-S., and Lim, K.T., *Preparation of polyacrylonitrile nanofibers as a precursor of carbon nanofibers by supercritical fluid process.* The Journal of Supercritical Fluids, 2008. **47**(1): p. 103-107.

[6] Lee, H.-M., Kim, H.-G., Kang, S.-J., Park, S.-J., An, K.-H., and Kim, B.-J., *Effects of pore structures on electrochemical behaviors of polyacrylonitrile (PAN)-based activated carbon nanofibers.* Journal of Industrial and Engineering Chemistry, 2015. **21**: p. 736-740.

[7] Sabet, E.N., Nourpanah, P., and Arbab, S., *Quantitative analysis of entropic stress effect on the structural rearrangement during pre-stabilization of PAN precursor fibers.* Polymer, 2016. **90**: p. 138-146.

[8] Morgan, P., *Carbon fibers and their composites.* Boca Raton, FL: CRC Press, 2005.

[9] Liu, J., Zhou, P., Zhang, L., Ma, Z., Liang, J., and Fong, H., *Thermo-chemical reactions occurring during the oxidative stabilization of electrospun polyacrylonitrile precursor nanofibers and the resulting structural conversions.* Carbon, 2009. **47**(4): p. 1087-1095.

[10] Ouyang, Q., Cheng, L., Wang, H., and Li, K., *Mechanism and kinetics of the stabilization reactions of itaconic acid-modified polyacrylonitrile.* Polymer Degradation and Stability, 2008. **93**(8): p. 1415-1421.

[11] Wangxi, Z., Jie, L., and Gang, W., *Evolution of structure and properties of PAN precursors during their conversion to carbon fibers.* Carbon, 2003. **41**(14): p. 2805-2812.

[12] Zhang, L., Aboagye, A., Kelkar, A., Lai, C., and Fong, H., *A review: carbon nanofibers from electrospun polyacrylonitrile and their applications.* Journal of Materials Science, 2014. **49**(2): p. 463-480.

[13] Arbab, S., Mirbaha, H., Zeinolebadi, A., and Nourpanah, P., *Indicators for evaluation of progress in thermal stabilization reactions of polyacrylonitrile fibers.* Journal of Applied Polymer Science, 2014. **131**(11): p. n/a-n/a.

[14] Rafiei, S., Noroozi, B., Arbab, S., and Haghi, A., *Characteristic assessment of stabilized polyacrylonitrile nanowebs for the production of activated carbon nano-sorbents.* Chinese Journal of Polymer Science, 2014. **32**(4): p. 449-457.

[15] Simitzis, J.C. and Georgiou, P.C., *Functional group changes of polyacrylonitrile fibres during their oxidative, carbonization and electrochemical treatment.* Journal of Materials Science, 2015. **50**(13): p. 4547-4564.

[16] Rahaman, M.S.A., Ismail, A.F., and Mustafa, A., *A review of heat treatment on polyacrylonitrile fiber.* Polymer Degradation and Stability, 2007. **92**(8): p. 1421-1432.

[17] Zhou, Z., Lai, C., Zhang, L., Qian, Y., Hou, H., Reneker, D.H., and Fong, H., *Development of carbon nanofibers from aligned electrospun polyacrylonitrile nanofiber bundles and characterization of their microstructural, electrical, and mechanical properties.* Polymer, 2009. **50**(13): p. 2999-3006.

[18] Zhou, Z., Liu, K., Lai, C., Zhang, L., Li, J., Hou, H., Reneker, D.H., and Fong, H., *Graphitic carbon nanofibers developed from bundles of aligned electrospun polyacrylonitrile nanofibers containing phosphoric acid.* Polymer, 2010. **51**(11): p. 2360-2367.

[19] Zussman, E., Chen, X., Ding, W., Calabri, L., Dikin, D.A., Quintana, J.P., and Ruoff, R.S., *Mechanical and structural characterization of electrospun PAN-derived carbon nanofibers.* Carbon, 2005. **43**(10): p. 2175-2185.

[20] Kim, C., Yang, K.S., Kojima, M., Yoshida, K., Kim, Y.J., Kim, Y.A., and Endo, M., *Fabrication of Electrospinning-Derived Carbon Nanofiber Webs for the Anode Material of Lithium-Ion Secondary Batteries.* Advanced Functional Materials, 2006. **16**(18): p. 2393-2397.

[21] Yang, Y., Simeon, F., Hatton, T.A., and Rutledge, G.C., *Polyacrylonitrile-based electrospun carbon paper for electrode applications.* Journal of Applied Polymer Science, 2011. **124**(5): p. 3861-3870.

[22] Wang, Y., Santiago-Aviles, J.J., Furlan, R., and Ramos, I., *Pyrolysis Temperature and Time Dependence of Electrical Conductivity Evolution for Electrostatically Generated Carbon Nanofibers.* IEEE Transactions On Nanotechnology, 2003. **2**(1): p. 39-43.

[23] Sharma, C.S., Katepalli, H., Sharma, A., and Madou, M., *Fabrication and electrical conductivity of suspended carbon nanofiber arrays.* Carbon, 2011. **49**(5): p. 1727-1732.

[24] Li, M., Han, G., and Yang, B., *Fabrication of the catalytic electrodes for methanol oxidation on electrospinning-derived carbon fibrous mats.* Electrochemistry Communications, 2008. **10**(6): p. 880-883.

[25] Hedin, N., Sobolev, V., Zhang, L., Zhu, Z., and Fong, H., *Electrical properties of electrospun carbon nanofibers.* Journal of Materials Science, 2011. **46**(19): p. 6453-6456.

[26] Prilutsky, S., Schechner, P., Bubis, E., Makarov, V., Zussman, E., and Cohen, Y., *Anodes for glucose fuel cells based on carbonized nanofibers with embedded carbon nanotubes.* Electrochimica Acta, 2010. **55**(11): p. 3694-3702.

[27] Arbab, S., Mirbaha, H., Zeinolebadi, A., and Nourpanah, P., *Indicators for evaluation of progress in thermal stabilization reactions of polyacrylonitrile fibers.* Journal of Applied Polymer Science, 2014. **131**(11).

[28] Qin, X.H., *Structure and property of electrospinning PAN nanofibers by different preoxidation temperatures.* Thermal Analysis and Calorimetry, 2010. **99**(2): p. 571-575.

[29] Tsai, J.-S. and Hsu, H.-N., *Determination of the aromatization index for oxidized polyacrylonitrile fibre by the differential scanning calorimetry method.* Journal of materials science letters, 1992. **11**(21): p. 1403-1405.

[30] Tsai, J.S. and Lin, C.H., *The effect of molecular weight on the cross section and properties of polyacrylonitrile precursor and resulting carbon fiber.* Journal of applied polymer science, 1991. **42**(11): p. 3045-3050.

[31] Bang, Y., Lee, S., and Cho, H., *Effect of methyl acrylate composition on the microstructure changes of high molecular weight polyacrylonitrile for heat treatment.* Journal of applied polymer science, 1998. **68**(13): p. 2205-2213.

[32] Moon, S.C. and Farris, R.J., *Strong electrospun nanometer-diameter polyacrylonitrile carbon fiber yarns.* Carbon, 2009. **47**(12): p. 2829-2839.

[33] Devasia, R., Reghunadhan Nair, C.P., Sivadasan, P., Katherine, B.K., and Ninan, K.N., *Cyclization Reaction in Poly(acrylonitrile/Itaconic Acid) Copolymer: An Isothermal Differential Scanning Calorimetry Kinetic Study.* Applied Polymer Science, 2003. **88**(4): p. 915-920.

[34] Arbab, S. and Zeinolebadi, A., *A procedure for precise determination of thermal stabilization reactions in carbon fiber precursors.* Polymer Degradation and Stability, 2013. **98**(12): p. 2537-2545.

[35] Arbab, S., Teimoury, A., Mirbaha, H., Adolphe, D.C., Noroozi, B., and Nourpanah, P., *Optimum stabilization processing parameters for polyacrylonitrile-based carbon nanofibers and their difference with carbon (micro) fibers.* Polymer Degradation and Stability, 2017.

[36] Mirbaha, H., Arbab, S., Zeinolebadi, A., and Nourpanah, P., *An investigation on actuation behavior of polyacrylonitrile gel fibers as a function of microstructure and stabilization temperature.* Smart Materials and Structures, 2013. **22**(4): p. 045019.

Investigation of a Natural pH-Indicator Dye for Nanofibrous Wound Dressings

A. Pakolpakçıl[1], E. Karaca[2] and B. Becerir[2]

[1]*Bursa Uludağ University, Graduate School of Natural and Applied Sciences, Görükle Campus 16059 Nilüfer/Bursa, Turkey*
[2] *Bursa Uludağ University, Textile Engineering Department, Görükle Campus 16059 Nilüfer/Bursa, Turkey*

Corresponding author: ayben_p@yahoo.com

Abstract In this study, fast responding and user-friendly biocompatible, halochromic nanofibrous mat was successfully produced for the purpose of wound healing monitoring, including a natural pH-indicator dye which was extracted from Red cabbage (*B. oleracea L.*) inside an alginate (NaAlg) and polyvinyl alcohol (PVA) mixture solution. The morphological characterization of the halochromic nanofibrous mat was examined by scanning electron microscope. Nanofibrous mats exhibited a pink color when exposed to pH 4-6, simulating the wound-milieu salubrious to the wound healing, and a green-blue color when exposed to pH 7-10, simulating the wound-milieu that obstacle wound healing.

Keywords— **pH-sensitive dyes, polyvinyl alcohol, red cabbage, smart textile, sodium alginate, wound dressing .**

I. Introduction

Color changing textiles are intelligent textile materials which change color upon an external stimulus. These materials can change their colors due to the change of several triggers, such as temperature (thermochromism), light (photochromism) and pH (halochromism)[1]. The usage of halochromic textiles in the determination of acidity of sweat, urine, and coryza is possible. The pH degree of a wound gives information about its healing property and pH of the wound leakage changes according to the healing or worsening of the wound [2]. For this reason, the pursuit of the change of wound pH is of vital importance in wound healing [3].

In this study, natural pH-indicator anthocyanins which are compatible with human health and environment were extracted from red cabbage (RC) plants. Nanofiber wound dressings (mats) were produced by electrospinning method by adding this extract in sodium alginate (NaAlg) and polyvinyl alcohol (PVA) mixture solution. Morphological characteristics of electrospun PVA/NaAlg/RC nanofiber mats were investigated in terms of extract concentrations. Further, PVA/NaAlg/RC nanofiber mats were tested against simulated conditions having different pHs for the purpose of wound healing monitoring.

II. Methods and Procedures

Red cabbage (*B. oleracea L.*) leaves (300 g) were smashed by using a domestic blender and soaked in 500 ml of alcohol solution (80% ethanol+20% pure water) in a magnetic stirrer for 24 h at room temperature. The product solution was filtered through a filter paper and the remaining solid was extracted two times with 500 ml of fresh solvent. The extract was lyophilized (0.188 mBar pressure and 72 h at -45°C). At the end of the process, the solid extract was dissolved in 100 ml of pure water.

The NaAlg polymer $(NaC_6H_7O_6)_n$ which was used for the production of nanofibrous mats had a viscosity of 700-900 cp. It was a commercial product, Cecalgum® S1300, and it was purchased from Cargill. It was dissolved in distilled water by mixing on magnetic stirrer at 60°C for 6 h with a concentration of 1% (w/w). PVA polymer $(CH_2CHOH)_n$ with Mw 85.000-124.000 g/mol (%87-89 hydrolyze) was purchased from Sigma-Aldrich. It was dissolved in distilled water by mixing on magnetic stirrer at 80°C for 12 h with a concentration of 12% (w/w). PVA solution with a concentration of 12% and NaAlg solution with a concentration of 1% were mixed in the volume ratio of 2/1 PVA/NaAlg. After that different extract (RC) concentrations were (2-3%) added the solution.

The pH, conductivity, surface tension and viscosity properties of the electrospun solutions were measured. Viscosity was measured by using a Brookfield RV-DV II viscometer. Conductivity and pH were determined by HANNA HI-98129 pH-meter and surface tension was determined by using Attension Theta surface tensiometer.

Electrospinning of the prepared solutions was carried out by a set-up consisting of a high-voltage DC power supply. The feed rate of the solution was controlled at about 0.7 ml/h by using a syringe pump and a tip-to collector distance of 12 cm.

In order to stabilize the PVA/NaAlg/RC nanofibers, electrospun mats were cross-linked by immersing them in acetone solution of 39.3 ml including 0.5 ml glutaraldehyde (50%) and 0.2 ml hydrochloric acid (37%) at room temperature for 10 min. The cross-linked nanofibrous mats were neutralized in ethyl alcohol (99.5%) for 5 min. The cross-linked and neutralized mats were washed with phosphate buffer solution (1%) three times and then dried.

The following buffer solutions were used to test the color schemes of the PVA/NaAlg/RC nanofibrous mats: sodium acetate (CH_3COONa) and acetic acid (CH_3COOH) buffer for pH 4–5; sodium phosphate dibasic dehydrate ($Na_2HPO_4.2H_2O$) and sodium dihydrogen orthophosphate ($NaH_2PO_4.2H_2O$) buffer for pH 6–9; sodium phosphate dibasic dehydrate ($Na_2HPO4.2H_2O$), sodium dihydrogen orthophosphate ($NaH_2PO4.2H_2O$) and NaOH buffer for pH 9–10.

The morphological appearance of the electrospun PVA/NaAlg/RC mats was observed under CARL ZEISS EVO 40 scanning electron microscope. The mean diameters of the resultant fibers were calculated from the measurements of fiber diameters on the images of X 10.000 magnification by using Image J program. Approximately 100 measurements were carried out on different parts of each sample.

III. RESULTS AND DISCUSSION

Solution properties are very important for electrospinning process and nanofiber surface structure. The properties of the solutions were given in Table I. When the concentration of red cabbage extract added to the PVA/NaAlg solutions increased, the viscosity of the solutions gradually decreased. The reason was the decrease in PVA / NaAlg polymer concentration per unit volume after mixing the aqueous extract solutions with the PVA / NaAlg solution. On the other hand, increases in conductivity of solutions with increasing concentrations of red cabbage extract were observed. It was possible for the conductivity of the solution to increase because of the presence of calcium, iron, and sodium ions in extract solutions [4]. The nanofiber surfaces obtained by electrospinning are affected by the pH of the solutions. When the pH of the solution is basic, smooth and fine structured nanofibers are obtained. When nanofibers are obtained from acidic solutions, beads are observed due to the presence of protons [5]. After the addition of red cabbage extract into PVA/NaAlg solutions, the pH slightly increased. The results of surface tension measurements did not significantly change, however a slight decrease was observed.

TABLE I
PROPERTIES OF ELECTROSPUN SOLUTIONS

Solutions	Viscosity (cp)	Conductivity (μS/cm)	Surface tension (mN/m)	pH
2/1 12 %PVA /1% NaAlg	880	1344	48.48	5.62
2/1 12 %PVA /1% NaAlg with 2% RC	649.60	2564	44.56	5.75
2/1 12 %PVA /1% NaAlg with 3% RC	499.20	3212	45.69	5.86

Figs. 1-3 illustrate the SEM images and diameter distributions of electrospun PVA/NaAlg and PVA/NaAlg/RC mats with different RC concentrations. Based on these images, the calculated mean diameters of fibers, standard deviation, minimum and maximum values were given in Table II.

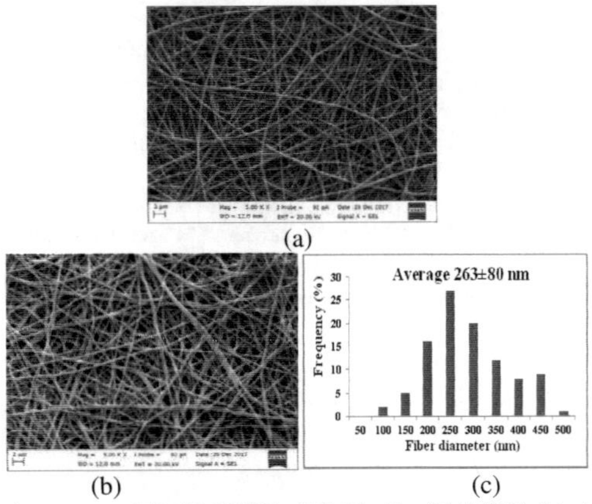

Fig. 1. SEM image of (a) the electrospun 2/1 12 %PVA /1% NaAlg (X 5.000) (b) the cross-linked electrospun 2/1 12 %PVA /1% NaAlg (X 5.000) (c) cross-linked electrospun diameter distribution

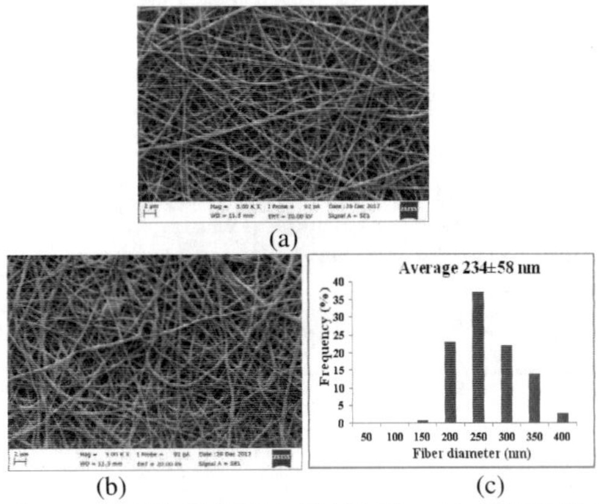

Fig. 2. SEM image of (a) the electrospun 2/1 12 %PVA /1% NaAlg with 2%RC (X 5.000) (b) the cross-linked electrospun 2/1 12 %PVA /1% NaAlg with 2%RC (X 5.000) (c) cross-linked electrospun diameter distribution

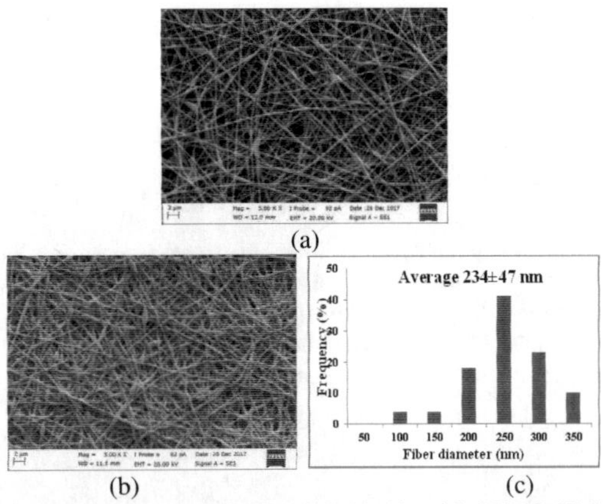

Fig. 3. SEM image of (a) the electrospun 2/1 12 %PVA /1% NaAlg with 3%RC (X 5.000) (b) the cross-linked electrospun 2/1 12 %PVA /1% NaAlg with 3%RC (X 5.000) (c) cross-linked electrospun diameter distribution

SEM images (Fig. 1) showed that smooth, beadless, uniform and continuous fibers were successfully produced from PVA/NaAlg polymer solutions. Fiber diameter average (263±80nm) value was a little bit higher than that of PVA/NaAlg/RC nanofibers mats because of high viscosity and low conductivity properties of PVA/NaAlg. It has been found that the fiber diameter distribution was close to the normal distribution and it was obtained between 200-350 nm.

SEM images of the 2/1 12 %PVA /1% NaAlg with 2-3% of RC extract mats (Figs. 2 and 3) showed that uniform and continuous fibers were produced with fewer beads. This could stem from the decrease in solution viscosity. When there is low viscosity, there is low chain complexity and the influence of surface tension forces on the jet during electrospinning is dominant [4]. After the addition of 2-3% red cabbage extract to the PVA / NaAlg solution, the conductivity values increased as the viscosity values decreased. As a result, fine fibers with lower diameters (234 nm) were produced. SEM images showed that the morphology of electrospun nanofibers progressively changed from a uniform structure to a beaded-like structure as the viscosity of the blended solutions decreased.

It was seen from SEM images (Figs. 1b, 2b and 3b) of cross-linked nanofibrous mats that the fibers appeared flattened, and bundle structures were obtained so that average fiber diameters increased. In the cross-linking process, acetal bonds are formed between the aldehyde groups of glutaraldehyde (which was used as crosslinker) and the hydroxyl groups present in the structure of polyvinyl alcohol and alginate [6]. This flattened structure of nanofibrous mats could be due to formation of new crosslinking bonds which leaded to squeeze the nanofibrous structure [7]. Furthermore, the rapid evaporation of the solvent from the nanofiber surfaces causes the formation of flattened structure. When a quantity of solvent evaporates which is cramped in the fibers, the fibers become flat [8].After cross-linking process, it is possible that the flattening may happen due to the compression of the nanofiber structure which could stem from the formation of new-crosslinking bonds. This could be the result of the separation of water molecules presented in the structure. It was observed that the average diameters of cross-linked electrospun nanofibers increased by the decreasing viscosity of the blended solutions.

TABLE II
MEAN DIAMETER, STANDARD DEVIATION, MINIMUM AND MAXIMUM VALUES OF ELECTROSPUN AND CROSS-LINKED ELECTROSPUN PVA/NAALG AND PVA/NAALG/RC NANOFIBERS

Nanofibrous mats	Mean Diameter (nm)	Standard Deviation (SD)	Minimum (nm)	Maximum (nm)
2/1 12 %PVA /1% NaAlg	242.29	82.74	68.72	416.67
cross-linked electrospun 2/1 12 %PVA /1% NaAlg	263.08	79.99	104.42	479.66
2/1 12 %PVA /1% NaAlg with 2% RC	221.24	61.78	79.94	371.67
cross-linked electrospun 2/1 12 %PVA /1% NaAlg with 2% RC	234.14	57.61	111.17	388.68
2/1 12 %PVA /1% NaAlg with 3% RC	203.12	55.48	59.51	333.49
cross-linked electrospun 2/1 12 %PVA /1% NaAlg with 3 %RC	233.71	46.79	132.64	390.67

The PVA/NaAlg/RC nanofibers took approximately 10 s to develop color in each pH solution (pH 4-10). The PVA/NaAlg/RC nanofiber mats exhibited rather visibly distinguishable colors as seen in Fig. 4, especially at pHs of 4-6 and 7-10. When compared with the PVA/NaAlg /RC solutions, the PVA/NaAlg /RC nanofibrous mats exhibited very similar color patterns. Consequently, the use of the nanofibrous mat as a pH sensor is advantageous; being an alternative and a simple tool for pursuing the wound healing process [9-11].

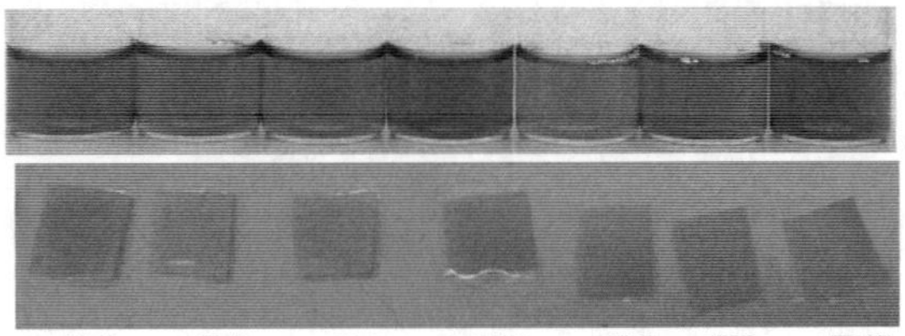

Fig. 4. The color schemes of **(a)** 2/1 12 %PVA /1% NaAlg with 3% RC electrospun solution **(b)** 2/1 12 %PVA /1% NaAlg with 3% RC nanofibers mats at pH 4-10

IV. CONCLUSION

In this research, the preparation of a nanofibrous mat including a natural halochromic dye (extracted from RC) for detecting the changes in wound environments' pH values was presented. Nanofibrous mat was successfully prepared by electrospinning using sodium alginate (NaAlg) and polyvinyl alcohol (PVA). Uniform and continuous fibers were obtained from PVA/NaAlg/RC solutions, with diameters of 200-250 nm. It is clear that PVA/NaAlg/RC nanofibrous mats exhibit different properties depending on pH of the surrounding environment. It could be concluded that natural pH-indicator dye (RC) was successfully incorporated into nanofibrous mats and the obtained nanofibrous structures could be used as pH-biosensors.

ACKNOWLEDGMENT

The authors thank TUBITAK (The Scientific and Technological Research Council of Turkey) for its financial support in the research project of (No. 116M540).

REFERENCES

[1] H. Mattila.,P.Talvenmaa, *Intelligent Textile and Clothing*, Woodhead Publishing London:, 2006 , ch 3.
[2] L. V. Schueren, K. , Clerck , "Coloration and Application of pH-Sensitive Dyes on Textile Materials", *Coloration Technology*, vol.128, no.2, pp. 82-90, 2012.
[3] A. Pakolpakçıl "Examination of Wet Fastness Halochromic Dyed Polyamide 6,6 Fabrics for Using Smart Textiles", *MSc. Thesis*, Namık Kemal University, 2013.
[4] S. Ramakrishna, K. Fujihara, W Teo. , T Lim., Z Ma. , *An Introduction to Electrospinning and Nanofibers*. Singapore: World Scientific Publishing Co., 2005, pp.80-81.
[5] W. K. Son, J. H. Youk, T. S. Lee, W. H. Park, "Effect of pH on Electrospinning of Poly(vinyl alcohol)", *Materials Letters*, vol. 59, no. 12, pp.1571-1575, 2005.
[6] M. C. Hassan, N.A. Peppas, "Structure and applications of poly(vinyl alcohol) hydrogels produced by conventional crosslinking or by freezing/thawing methods", *Advances in Polymer Science*, 153, pp.37-65, 2000.
[7] M. K.Sinha, B. R. Das, A. Srivastava, A. K. Saxena , "Needleless Electrospinning and Coating of Poly Vinyl Alcohol with Cross-Linking Agent via in-Situ Technique", *International Journal of Textile and Fashion Technology*, vol.3, no.5, pp. 29-38, 2013.
[8] A. Baji, Y. W. Mai, S. C. Wong, M. Abtahi, P. Chen, "Electrospinning of Polymer Nanofibers: Effects on Oriented Morphology, Structures and Tensile Properties", *Composites Science and Technology*, vol. 70, pp. 703–718, 2010.
[9] K. Devarayan, B. S. Kim, "Reversible and Universal pH Sensing Cellulose Nanofibers for Health Monitor", *Sensors and Actuators B:Chemical*, vol. 209, pp. 281-286, 2015.
[10] A. Agarwal, A. Raheja, T. S Natarajan, T. S.Chandra, "Development of Universal pH Sensing Electrospun Nanofibers", *Sensors and Actuators B: Chemical*, vol. 161, no. 1, pp. 1097-1101, 2012.
[11] L. V. Schueren, T. Mollet, Ö. Ceylan, K. Clerck, "The Development of Polyamide 6.6 Nanofibres with a pH-Sensitive Function by Electrospinning", *European Polymer Journal*, vol. 46, no. 12, pp. 2229-2239, 2010.

3D woven fabric with cross rib as a composite reinforcement

A. Kaminska[1,2], M. Barburski[1]

[1] *Lodz University of Technology, Faculty of Material Technologies and Textile Design, 116 Zeromskiego Street, Lodz 90–924, Poland.*
[2] *Institute of Security Technologies "MORATEX", M. Skłodowskiej-Curie 3 Street, Lodz 90-505, Poland*

Corresponding author: marcin.barburski@p.lodz.pl

Abstract A woven 3D composite is a new type of an advanced construction material. A prototype of three dimensional (3D) textile based on combination of a flat textile with two intersecting tunnels with a rectangular cross-section was constructed and the feasibility of this type of weaving was studied. Prototype of 3D woven fabric was made.
Based on 3D textiles preforms composites for the mechanical trial was made. Composites used for strength tests were made using a vacuum infusion method. The basic elements of the composite, that is the flat part and the element with the rib, were selected for testing. This selection resulted from a complicated process of making a composite based on the entire fabric. A 3D composite, created through the lamination of two flat fabric, was also made for comparison. The study compared the material strength with and without rib reinforcement. Tensile and bending test was made. The analysis of the third composite was aimed at demonstrating the purpose of implementing the complicated weaving process.
The laminate reinforced with 3D fabrics made from a unified weaving process is characterised by the best mechanical properties of the tested variants. The results can be considered to be encouraging.

***Keywords*— bendning test, composite, flax fibres, 3D textile.**

I. Introduction

Three-dimensional (3D) fabrics play an important role in the development of advanced composites reinforced with textiles. These fabrics are used as shaped elements ready for impregnation with resins, or as thicker materials with structural durability, which - following impregnation - gain a high durability against interlaminar shearing — that is better than in conventional laminated products [6].

Three-dimensional woven fabrics possess a third dimension — thickness (the dimension relative to the Z axis), which is significant in relation to dimensions X and Y. The yarns are interwoven with each other in all three directions X (longitudinal), Y (transverse), and Z (vertical) [1]. According to Chen, structures that have a fundamental dimension in their thickness direction, that is formed by a layer of fabric or yarn layering, are generally referred to as three-dimensional (3D) textiles [6]. 3D fabrics can be understood as thick multilayer fabric with a simple regular form, o as structure made of more complex shapes — that is, multilayer fabrics containing hollow spaces [9]. Khodar defined woven 3D products as fabrics, whose constituting yarns are to be arranged in a tripartite-perpendicular relation [5,10,11].

What we currently define as three-dimensional fabrics may have a generic 3D shape, a more complex internal structure, or both of these features. Besides weaving methods, 3D textiles can be made via means of knitting and breading or by obtaining non-woven fabrics [5].

Strengthening a composite with 3D fabrics can overcome problems related with thickness and low interlaminar strength, which are typical of traditional laminates using 2D fabrics, and have been known for many years. At present, the application of 3D textiles in composites is not only aimed at improving their mechanical properties, but also at reducing production costs and integrating a greater amount of functions in the component [2].

A woven 3D composite is a new type of an advanced construction material. In the case of technical textiles, most of them are used in specialised markets (i.e. aviation and automotive industries) where the requirements concerning durability and efficiency of textile composites are high. The most important applications of fabric-based composites are as stiffeners for Joint Strike Fighter air inlet panels, Beech Starship aircraft wing joints and carrier's cones. Reinforced layered fabrics are also used as floor plates for trains, and hard skeletons for delivery vans and fishing boats. Composites based on 3D fabrics have found application as tapes, fan blades in gas turbines or helicopter blades; while on the military market they are used - among others – as reinforcement of vehicle armours. In addition, they are used in the construction industry or in the sports goods market [5].

A potential market for the introduction of fibre composites is the urban furniture industry, whose goods are expected to have high durability. Urban elements are subjected to intensive usage and are exposed to high and low temperatures. Urban furniture, left unattended, is often vandalised.

3D fabrics were designed in order to increase the mechanical strength of seats and tops of urban furniture, allowing to withstand greater forces than traditional products, thus improving the lifespan of the furniture. The proposed shape is a flat fabric reinforced with two intersecting tunnels with a rectangular cross-section.

II. Design of a 3D fabric

A. *Combination of a flat textile with two intersecting tunnels with a rectangular cross-section*

The proposed 3D fabric envisages the combination of a flat textile and two intersecting tunnels with rectangular cross-sections. The proposed design's notion sees the span-reinforced composite panel as able to carry higher loads, thus allowing the elimination of certain steel or wood (frames and joints) elements in the production of furniture. By employing a composite product as a seat/table top, the weight of the product is lowered and its mechanical strength increases, which could potentially favour the assembly of such furniture within urban spaces, bus stops or stadiums. The shape of the seats and tops can be any, though they are most frequently square or rectangular. As a result, the furniture is easy to arrange and set up in both a closed and open space. For the purpose of creating a prototype, a square shape was adopted.

The virtual model of a 3D fabric depicted below, has been reinforced with two intersecting ribs (Fig. 1). The designed ribs have the form of tunnels with square cross-sections. In principle, the rib's inner canal should be empty, to reduce - among others - the mass of the composite. 3D fabric with a tunnel will full structural filling would be easier to produce (i.e. using an orthogonal loom). Such a structure would, however, lead to problems at the fabric laminating stage — the resin could have difficulty infiltrating into the depths of the fabric, thus not all fibres would be impregnated. Moreover, the occurrence of large areas of resin only would be a danger, which would have a weakening effect onto the structure. The rods inserted into the tunnels will allow to preserve the rectangular rib shape. In the case of filled tunnels within a woven structure, the use of a void could deform the originally established shape. The designed fabric is asymmetrical — the left side is a flat fabric, while on its right side there are convex ribs in the aforementioned cross shape (Fig. 2). It is a homogeneous product characterised by a differing thickness in specific places.

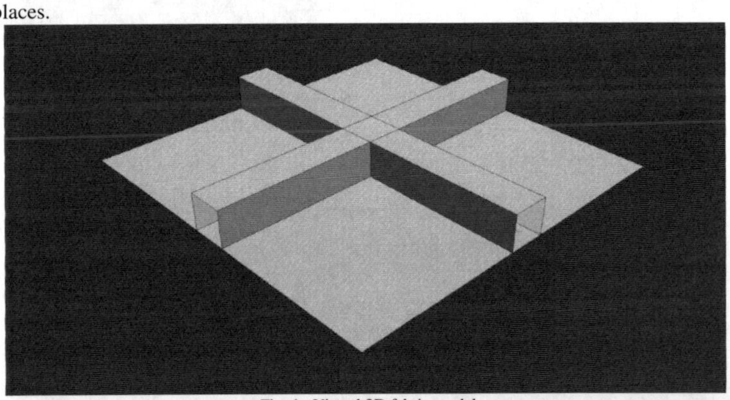

Fig. 1. Virtual 3D fabric model

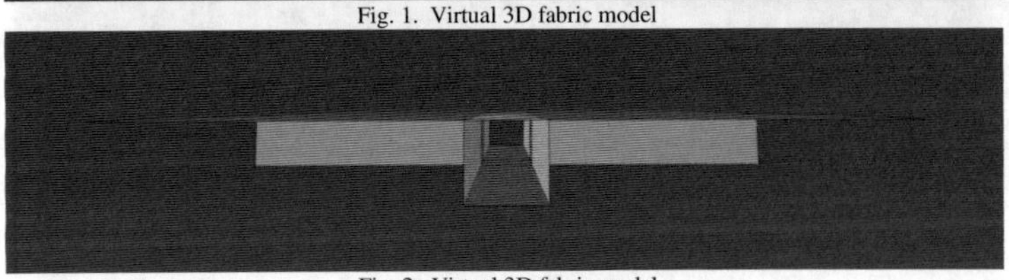

Fig. 2. Virtual 3D fabric model

B. *Material*

In the case of trials for strength tests, based on the conducted analysis of flax yarns, yarns of 100 tex linear mass were used as a warp, and 2000 tex roving as a weft. Roving of a mass of 2000 tex is characterised by a high breaking force fluctuate at around 120 N. The 100 tex yearn was considerably thinner and had a larger twist, allowing for it to be properly tensed and stretched over the frame. Its strength was about 4N.Both yarns were produced by the Safilin company.

C. Manufacture of a 3D fabric

With the currently known technologies it is not possible to manufacture such a fabric on a mechanical loom. To be able tomake ribs in both directions, three warp systems would be necessary: 1) lower layer warp — co-creating both the flat element and the tunnel walls located on the left side of the fabric; 2) upper layer warp — participating in the weaving process of the flat element and the convex part of the horizontal tunnel (creating an overlay equal to the width of three tunnel walls); 3) a warp creating the tunnel perpendicular to the direction of the warp. A fabric made on a mechanical loom equipped in a three warp system would still fail to meet the criteria. At the intersection of tunnels, inside the ribs, there would be non-woven threads ofweft and warp, that would impede the introduction of moulds. As a result of this, it was necessary to develop a method offabricating the fabric, that would reproduce the project as faithfully as possible. The design required the mounting of three warps systems onto the frame: the lower layer warp, the upper layer warp and the warp making a tunnel in a parallel direction tothe warp. The warp forming the vertical tunnel should be mounted so that the threads lay parallel to the walls of the mould. The creation of a parallel tunnel in regard to the weft is possible by making an overlay in the upper layer and then joining the upper and lower layers. The size of the overlay in the upper layer should equal to three widths of the rods wall, a fragment equal to the width of one wall of the rod should be woven in the lower layer. The weaving of the flat part with the vertical tunnel should be carried out in accordance with the study report (Fig. 3). By retaining the set order, the fabric's growth is the same forthe flat textile and the tunnel.

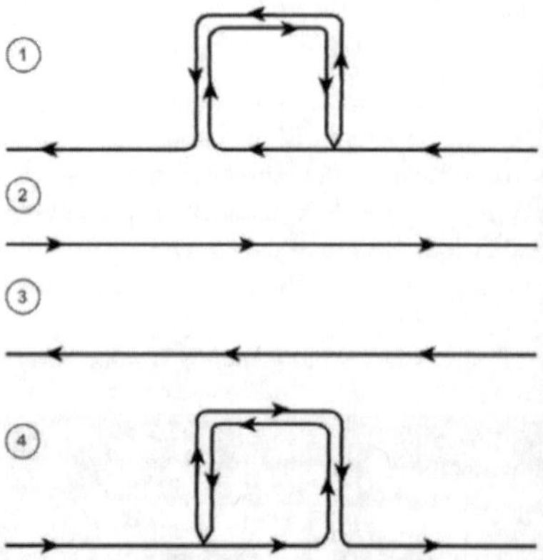

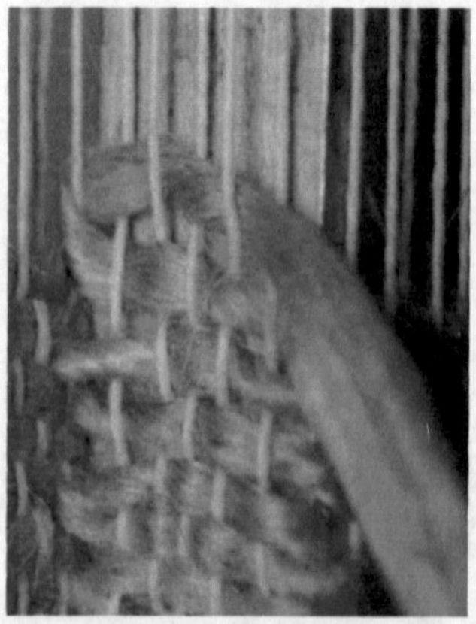

Fig. 3. Weft intromitting cycle for a flat textile with a vertical tunnel. Numbers 1 to 4 refer to the next steps in weaving the vertical tunnel. The left photo shows weft intromitting –step 1.

Fig. 4. Flaxen prototype of the 3D fabric based on a combination of a flat textile and two intersecting tunnels with rectangular cross-sections

D. Manufacture of a composite

Composites used for strength tests were made using a vacuum infusion method, employing a vacuum pressure to press in the resin for the lamination of a textile product. The dry materials are laid on a previously prepared surface and covered with a rigid or flexible film membrane, which is sealed around the perimeter of the mould. A vacuum is produced before the resin is introduced. Only after obtaining a vacuum equal to 1 Bar the resin is sucked into the laminate via inserted tubes. Infusion is considered as a process of closed forms. Vacuum lamination significantly improves the relation of fibre to resin, allowing to obtain a stronger and lighter product. When making the composite, epoxy resins were used.

III. EXPERIMENT

A. Three-point bending of flat laminates of flax fibres

The samples were evaluated for determining the flexural properties under three-point loading (Method A). This is a test dedicated for products made of fibre-reinforced thermoplastic and thermosetting plastic composites. It was carried out in accordance with the norm: PN-EN ISO 14125:2001 Fibre-reinforced plastic composites — Determination of flexural properties.

During the test, the sample - fixed in the form of a supported beam - is subject to bending at constant speed until fracture or until the deformation value reaches the desired quantity. As part of the test, maximum tension stress, modulus of elasticity and deformation on the external surface of the sample are determined. The tested samples were classified into class II materials, and in accordance with the recommendations regarding the types of moulds tested with method A (bending in a three-point system) their length was > 80mm and width 15 ± 0,5mm. The average thickness of the samples is 1,702mm. The tests were carried out on the INSTRON 4485 strength machine in the Laboratory of Strengths of Materials and Structure of the Łódź University of Technology.

During the tests spacing between supports was equal to 80 mm, initial force amounted to 1N and speed of deformation increment was 5mm/min.

The fabric for testing was made of yarns of various linear masses — 100tex warp, 2000tex weft. By studying flat samples in both directions it is possible to analyse the effect of the yarns linear mass on the composite's strength. The experiment was aimed at demonstrating the differences in bending stiffness of the composite along the warp and weft resulting from different linear masses of the used yarns.

B. Three-point bending of 3D laminates

The designed 3D fabric would serve as a reinforcement for seats and table tops of urban furniture. Thus, the ribbed fabric has to be analysed for its mechanical properties. The basic elements of the fabric, that is the flat part and the element with the rib, were selected for testing. This selection resulted from a complicated process of making a composite based on the entire fabric. A 3D fabric, created through the lamination of two flat fabric, was also made for comparison.

The three-point bending test was carried out on three laminates, all 9,5cm wide and 18cm long: flat (P), shaped - reinforcement woven in one single process (K1), shaped - created from two singular fabrics (K2).

(P) — flat laminate of single-cloth flaxen reinforcement with plain weave (Fig. 5);

(K1) — flaxen fabric laminate with a square cross-section tunnel. The reinforcement was made in a uniform weaving process in accordance with the afore-described production method;

(K2) — composite based on two flaxen fabrics (single fabrics with plain weave), which, during the laminating process, were formed into a shape analogous to the reinforcement of composite K1 (Fig. 6).

The study compared the material strength with and without rib reinforcement. The analysis of the third composite (K2) was aimed at demonstrating the purpose of implementing the complicated weaving process.

During the tests spacing between supports was equal to 120 mm, initial force amounted to 10N and speed of deformation increment was 10 mm/min.

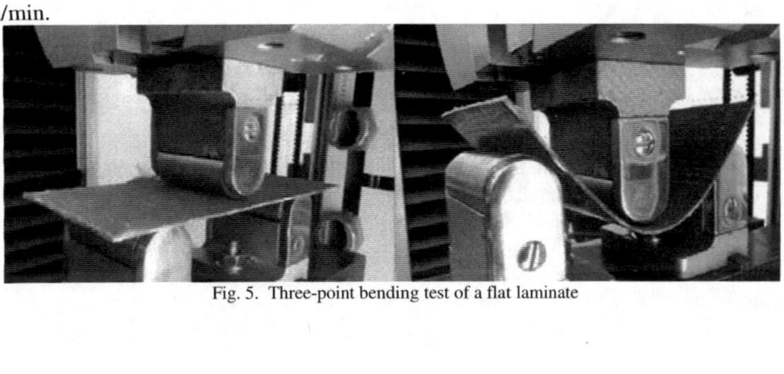

Fig. 5. Three-point bending test of a flat laminate

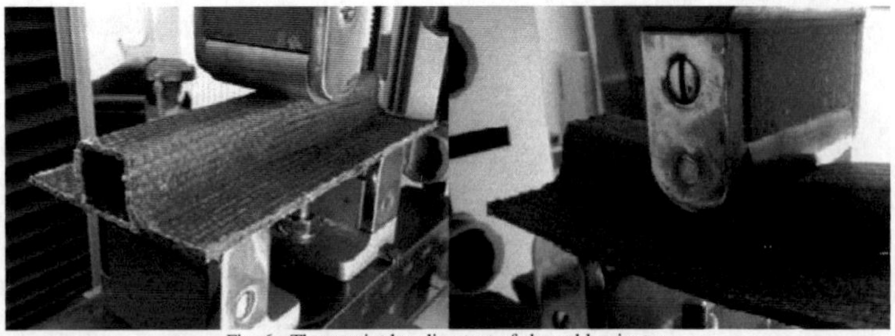

Fig. 6. Three-point bending test of shaped laminates

IV. RESULTS

A. Three-point bending of flat laminates

Lower flexural strength is characterised by samples tested for yarns with 100 tex linear mass, warp (Fig. 7). Laminate is characterised by a greater flexural stress along the weft. (63,52 MPa). It is more than three times higher than for the tested samples along the warp line (19,55 MPa). The modulus of elasticity along the weft (5131,52 MPa) is likewise three times greater (Fig. 8). The strength of a material is dependent on the testing direction and the internal positioning of the fabric threads. This is due to the use of yarns with different linear weights — weft (2000 tex) and warp (100 tex). Yarn with a higher linear mass influences better strength properties. In order to obtain uniform results for both axes, the same warp and weft should be used.

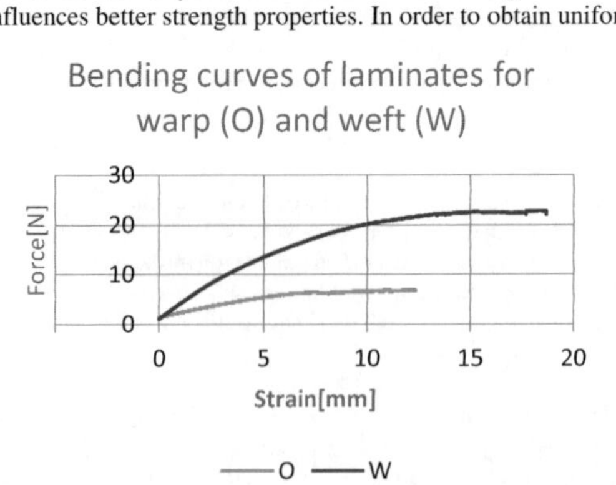

Fig. 7. Graph of bending curves of laminates for average values in both directions

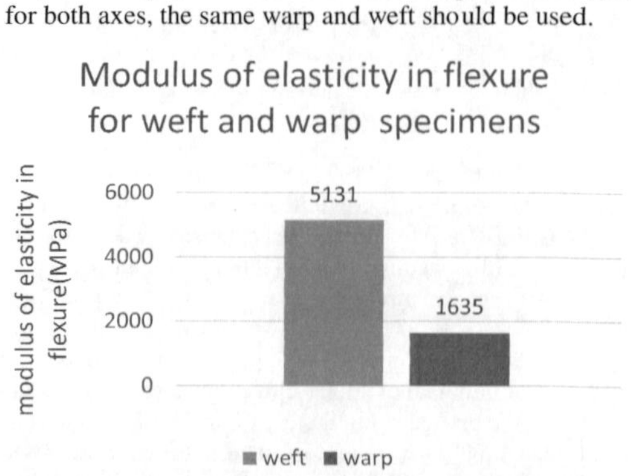

Fig. 8. Average modulus of elasticity in flexure trials along weft and warp

B. Three-point bending of shaped laminates

The laminate reinforced with 3D fabrics made from a unified weaving process is characterised by the best mechanical properties of the tested variants (Fmax = 1122.25 N). The laminate formed from two single fabrics was characterised by a significantly lower flexural strength (809,25 N) (Fig. 9).

In the process of 3D weaving the weft threads are interwoven from the beginning so that they are arranged according to the premise. This results in a greater percentage of micro-structural yarn defects, but provides less internal stresses. The fabric formed into a suitable shape at the stage of lamination is subject to an unnatural arrangement of it. In the critical spots, that is at the bends, stresses occur which lower the mechanical strength of the final material. Strengthening in the form of a tunnel leads to an increased stiffness of the composite. The highest load force for the 3D fabric (1112,25 N and 809, 25N) is even several times greater than for a flat one (86,24N). After exceeding the limit of flexural strength, the samples did not break, however the upper support began to dent into the sample. Few cracks in the structure around the working load appeared in both the structures of the shaped laminates and the flat product.

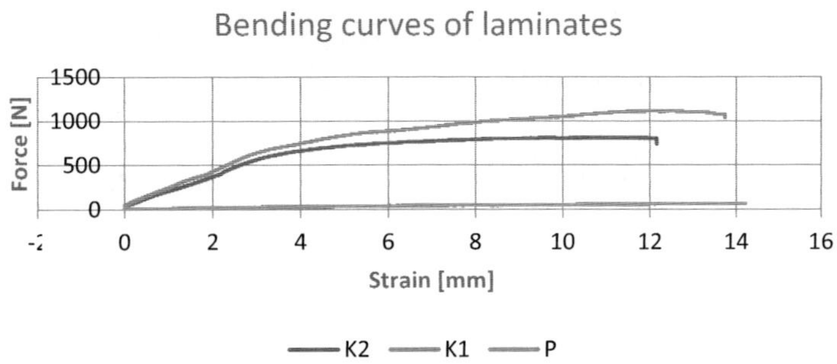

Fig. 9. Graph of bending curves of laminates

V. Conclusion

The bending stiffness of the material along the roving with a larger linear mass (2000tex) is greater than for threads with a lower mass (100tex). With the increasing of linear yarn mass the stiffness of the product increases. In order to obtain similar characteristics in both directions of the study (along the weft and the along the warp) the same yarn should be used in both arrangements.

Higher flexural strength was demonstrated by a laminate reinforces with a 3D fabric made in a unified weaving process. In a 3D laminate made of two fabrics, higher internal stresses occurred in the critical places (wall fractions) which influenced the decrease of the product's stiffness.

Laminate made of a single textile without ribbing demonstrates very low stiffness and poor strength properties compared to both 3D composites. The bending stiffness of a laminate reinforced with a 3D fabric is even several times higher than for a flat laminate. This fact proves that reinforcement in the form of a rectangular tunnel significantly increases the flexural strength of the composite.

Little cracks appeared in the tested composites in the places with the highest load. At the critical moment, when F>F max, the tested 3D composites did not break — the upper support began however to dent into their structure. Composite based on two flaxen fabrics formed in the laminating process is characterized by a lower bending stiffness than a composite based on a shaped fabric made in one weaving process.

The research of polymer laminates reinforced with flaxen fabric described above gives hope for the potential use of these shaped composites in the technical industry. To make this possible, further mechanical tests should be carried out.

References

[1] Badawi, S.S. 2007. Development of the Weaving Machine and 3D Woven Spacer Fabric Structures for Lightweight Composites Materials. PhD Thesis, Technical University of Dresden, Dresden, Germany.
[2] Barburski M, Weigert L, Fernández I, Pouplier S, Roth S, Huurnink G "Woven Reinforced Composites for Improving the Design of the Hyperextension Brace" J Fashion Technol Textile Eng 2017, S3 DOI:10.4172/2329-9568.S3-002
[3] Barburski M., Masajtis J., „Modelling of the change of structure of woven fabric under mechanical loading" Fibres and Textiles Easter Europe No 1(72)2009, p. 39-45 (ISSN 1230-3666)
[4] Barburski M., Straumit I., Zhang X,. Wevers M,. Lomov S.V. „Micro-CT analysis of internal structure of sheared textile composite reinforcement" Composites: Part A 73 (March 2015) 45–54, http://dx.doi.org/10.1016/j.compositesa.2015.03.008;
[5] Chen X. Red., Advances in 3D Textiles, Woodhead Publishing, Cambridge 2015, s. 1-120.
[6] Chen X. Waterton Taylor L., Tsai L., An overview on fabrication of three-dimensional woven textile preforms for composites, „Textile Research Journal", 2011, nr 81 (9).
[7] Chen, X.; Taylor, L.W.; Tsai L.J. . 2011. An Overview on Fabrication of Three-Dimensional Woven Textile Preforms for Composites. Textile Research Journal 81 (9):932-944.
[8] Czub K. and Barburski M. „Mechanical properties of flax roving composites reinforcement" IOP Conf. Series: Materials Science and Engineering 254 (2017) 042004 doi:10.1088/1757-899X/254/4/042004
[9] Hearle, J.W.S.; Chen, X. 2009. 3D Woven Preforms and Properties for Textile Composites. Seventeenth International Conference on Composite Materials, Edinburgh.
[10] Khokar, N. 2001. 3D-Weaving: Theory and Practice. Journal of the Textile Institute 92 (2):193-207.
[11] Long A.C., Design and Manufacture of Textile Composites, wydaw. Woodhead Publishing, Cambridge 2005
[12] Vanleeuw B., Carvelli V., Barburski M., Lomov S.V., Aart W. van Vuure "Quasi-unidirectional flax composite reinforcement: deformability and complex shape forming" Composites Science and Technology (110), February 2015, 76–86

AgNW coated on poplar fibres for flexible capacitors

Muhammed E. Narinc[1], Alper Gurarslan[1]

[1]Istanbul Technical University, Faculty of Textile Technologies and Design, Textile Engineering Department, İnönü Caddesi, No: 65Gümüssuyu (34437) - Beyoğlu, İSTANBUL, Turkey

Corresponding author: gurarslan@itu.edu.tr

Abstract Recently there has been great interest in flexible electronics and wearables. Traditional electronics fail to satisfy that demand due to being rigid and uncomfortable to wear. As an important element of electronic systems, flexible capacitors are needed in the application of wearables. Poplar fibres, which do not have any usage are in industry, perfectly suit with the demand of market owing to their unique properties. Silver nanowire coated non-woven poplar fibre webs exhibited very good electrical properties and they were quite sensitive to loads.

***Keywords*—Poplar fibres, Silver Nanowires, coating, capacitor, flexible.**

I. Introduction

FLEXIBILITY lexibility of traditional electronics are quite poor, which restraints feasibility of them in numerous applications so that lately, there has been great interest in flexible electronics [1–5]. Flexible capacitor, which are one of the most essence part of flexible electronics, have attracted attention from many researchers because of their potential employment in wearable electronics, robotics and many more applications. [6–8] Researches showed that capacitors [9], field-effect transistors [10], piezoresistive materials [11] and triboelectric nanogenerators [12] can be employed as flexible capacitors. Besides, carbon nanotubes [13,14], silver nanowires [15,16], gold nanowires [17] and graphene [18,19], whose main task is to increase electrical conductivity of matrix materials, are used to create flexible capacitors. However, when cost and efficiency parameters are taken into consideration, difficulties, expensiveness and complexity of many applications are faced. Moreover, sustainability of some of aforementioned methods are rather controversial. Since usage of hazardous chemicals, nonbiodegradable materials and high energy consumption heavily undermine ecology and depletes natural resources. Hence, developing flexible and sustainable capacitor at low cost still stands as a big challenge. The fibres produced from poplar seeds are short, hydrophobic, porous, lightweight and abundant in nature [20]. Additively, they possess a similar morphological structure to kapok fibre [21]. Poplar fibre that consists mainly of lignocellulose exhibits favourable physical properties, which are highly suitable to create flexible capacitor. Despite having very good physical characteristics, every year tons of poplar fibres are treated as waste material due to not being employed in any industrial application [21]. Herein, we demonstrate a simple method to create flexible capacitors by coating non-woven raw poplar fibre web with AgNWs. Non-woven raw poplar fibre webs were prepared via spray coating a polymer solution on aforementioned fibres, where poplar fibres serve as a skeleton and spandex fibres as binder. Coating porous non-woven raw poplar fibres with AgNWs formed a conductive network on the surface of poplar fibres. This study shows that AgNW coated nonwoven raw poplar fibre webs exhibit good electrical conductivity and flexibility, besides capacitors made of them are quite sensitive to loads.

Do not change the font sizes or line spacing to squeeze more text into a limited number of pages. The paper should not exceed 6 pages.

II. Experimental

A. Fibre preparation

The raw poplar used in the experiments was harvested during the middle of May in Istanbul. On account of, raw poplar fibres contain some amount of undesired foreign particles, mechanical cleaning was applied prior to following stages in order to prevent undesirable results. The poplar fibres were freed from the remnants of seeds, leaves and small particles with the help of steel tweezers and then placed in a glass container. After all, clean raw poplar fibres obtained to be used in the experiments.

B. Preparation of non-woven web samples

B1. Preparation of solution for spray coating

Commercial spandex fibres and Dimethylacetamide (DMAc), which were obtained from Hyosung and Sigma Aldrich Company, respectively, were put into the Erlenmeyer flask at the concentration of 0.04 g/ml, where 2.7 gram of spandex and 65 ml of Dimethylacetamide were used. Immediately after, the solution was placed on the magnetic stirrer whose rotation speed was set to 250 rpm at 50 °C. The Spandex/DMAc solution stirred on the magnetic stirrer for 2 hours to completely dissolve spandex fibres in Dimethylacetamide. The colour of the Spandex/DMAc solution became creamy white at the end of 2 hours and the whole spandex fibres were fully dissolved in DMAc.

B2. Spray coating

MagicBrush CP-101 Airbrush air compressor, which has ability to blow 350 mL air per second, and MagicBrush Airbrush Kit AB-125A gun, whose hose nozzle is 0.33 mm and reservoir capacity is 5 cc were used to carry out spray coating process. The reservoir of air gun was fully filled with of the Spandex/DMAc solution. After weighting each poplar specimen, the cleaned raw poplar fibres were finely placed in a long glass beaker for the purpose of giving smooth shape to non-woven webs. The air gun was set to a constant height of 50 cm and the Spandex/DMAc solution was sprayed at constant speed onto poplar fibre specimens. Poplar, which serves as a base fibre, covered with the Spandex/DMAc solution. The front and back surfaces of poplar specimens were held in the range of the gun of the air compressor for 15 minutes, evenly. Non-woven raw poplar webs formed after the spray-coating process were rested for 1 hour in an attempt to dry up the remnant DMAc.

C. Coating with AgNWs

The non-woven raw poplar webs were soaked in a glass container filled with AgNW/Ethanol solution at the concentration rate of 0.0175, where 0.35 gram of AgNW and 20 mL of C2H6O were used. Subsequently, the glass container was gently located in sonicator with the intention of increasing interaction of AgNWs and non-woven raw poplar web truly.

Fig. 1. SEM image of non-woven poplar capacitor under magnification rate of 5000X.

III. Results and Discussion

As clearly seen in Fig. 1, the surface of poplar fibres covered with AgNWs via aforementioned continuous experimental methods. The conductivity of non-woven poplar fibre webs range from 6-15 Ω which indicates the success of methodology. In Table 1, the electrical conductivity range of five nonwoven poplar fibre web sample demonstrates the good electrical property of non-woven poplar fibre webs. Besides, standard deviation of average electrical conductivity of non-woven poplar fibre web is 1.78.

TABLE I
Electrical conductivity of non-woven poplar fibre webs

Sample	Average Resistance (Ω)
Non-woven poplar fibre web-I.	6.80

Non-woven poplar fibre web-II	11.45
Non-woven poplar fibre web-III	8.80
Non-woven poplar fibre web-IV	10.35
Non-woven poplar fibre web-V	8.75

Fig. 2. An image of non-woven poplar capacitor during determination of capacitance values under precise weight.

As seen in the Fig. 2, several precise load are put onto non-woven poplar fibre webs in order to collect data about sensitivity of them. In the Fig. 3, the collected data about non-woven poplar fibre capacitors turned into weight versus time graph which indicates sensitivity of non-woven poplar fibre capacitors. Moreover, they can exhibit rather well properties by being sensitive even to small amount of loads such as 0.1 gram.

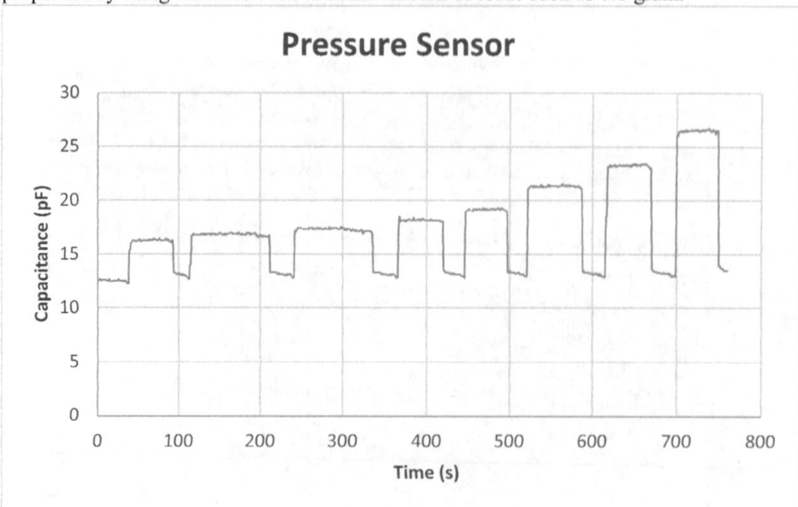

Fig. 3. Capacitance versus time graph of responds provided by non-woven poplar fibre web.

IV. Conclusion

In conclusion, non-woven poplar webs treated with AgNWs can be used in wearable and many more applications thanks to their flexibility, good electrical conductivity and hydrophobicity. Utilization of populus fibres has significant favourable impacts on the environment and wearable electronic industry because carbon emission is utmost minimized and biodegradability of product is achieved. Hopefully, this study can bring a bio-waste into the world of wearable electronics industry.e either SI (MKS) or CGS as primary units. (SI units are strongly encouraged.)

References

[1] S. Park, M. Vosguerichian, Z. Bao, A review of fabrication and applications of carbon nanotube film-based flexible electronics, Nanoscale. 5 (2013) 1727. doi:10.1039/c3nr33560g.

[2] W.S. Wong, A. Salleo, Flexible Electronics: Materials and Applications, 2009. doi:10.1007/978-0-387-74363-9.

[3] A. Manekkathodi, M.Y. Lu, C.W. Wang, L.J. Chen, Direct growth of aligned zinc oxide nanorods on paper substrates for low-cost flexible electronics, Adv. Mater. 22 (2010) 4059–4063. doi:10.1002/adma.201001289.

[4] I.C. Cheng, S. Wagner, Overview of flexible electronics technology, in: Flex. Electron., 2009: pp. 1–28. doi:10.1007/978-0-387-74363-9.

[5] T. Agcayazi, K. Chatterjee, A. Bozkurt, T.K. Ghosh, Flexible Interconnects for Electronic Textiles, Adv. Mater. Technol. 1700277 (2018) 1700277. doi:10.1002/admt.201700277.

[6] S.C.B. Mannsfeld, B.C.K. Tee, R.M. Stoltenberg, C.V.H.H. Chen, S. Barman, B.V.O. Muir, A.N. Sokolov, C. Reese, Z. Bao,Highly sensitive flexible pressure sensors with microstructured rubber dielectric layers, Nat. Mater. 9 (2010) 859–864. doi:10.1038/nmat2834.

[7] G. Ren, F. Cai, B. Li, J. Zheng, C. Xu, Flexible pressure sensor based on a poly(VDF-TrFE) nanofiber web, Macromol. Mater. Eng. 298 (2013) 541–546. doi:10.1002/mame.201200218.

[8] B.Y. Lee, J. Kim, H. Kim, C. Kim, S.D. Lee, Low-cost flexible pressure sensor based on dielectric elastomer film with micropores, Sensors Actuators, A Phys. 240 (2016) 103–109. doi:10.1016/j.sna.2016.01.037.

[9] J. Bae, M.K. Song, Y.J. Park, J.M. Kim, M. Liu, Z.L. Wang, Fiber supercapacitors made of nanowire-fiber hybrid structures for wearable/flexible energy storage, Angew. Chemie - Int. Ed. 50 (2011) 1683–1687. doi:10.1002/anie.201006062.

[10] T. Someya, T. Sekitani, S. Iba, Y. Kato, H. Kawaguchi, T. Sakurai, A large-area, flexible pressure sensor matrix withorganic field-effect transistors for artificial skin applications, Proc. Natl. Acad. Sci. 101 (2004) 9966–9970. doi:10.1073/pnas.0401918101.

[11] S.E. Zhu, M. Krishna Ghatkesar, C. Zhang, G.C.A.M. Janssen, Graphene based piezoresistive pressure sensor, Appl. Phys. Lett. 102 (2013) 2014–2017. doi:10.1063/1.4802799.

[12] F.R. Fan, L. Lin, G. Zhu, W. Wu, R. Zhang, Z.L. Wang, Transparent triboelectric nanogenerators and self-powered pressure sensors based on micropatterned plastic films, Nano Lett. 12 (2012) 3109–3114. doi:10.1021/nl300988z.

[13] D.J. Lipomi, M. Vosgueritchian, B.C.K. Tee, S.L. Hellstrom, J.A. Lee, C.H. Fox, Z. Bao, Skin-like pressure and strain sensors based on transparent elastic films of carbon nanotubes, Nat. Nanotechnol. 6 (2011) 788–792. doi:10.1038/nnano.2011.184.

[14] N. Sinha, J. Ma, J.T.W. Yeow, Carbon Nanotube-Based Sensors, J. Nanosci. Nanotechnol. 6 (2006) 573–590. doi:10.1166/jnn.2006.121.

[15] Y. Joo, J. Byun, N. Seong, J. Ha, H. Kim, S. Kim, T. Kim, H. Im, D. Kim, Y. Hong, Silver nanowire-embedded PDMS with a multiscale structure for a highly sensitive and robust flexible pressure sensor, Nanoscale. 7 (2015) 6208–6215.doi:10.1039/C5NR00313J.

[16] J. Wang, J. Jiu, M. Nogi, T. Sugahara, S. Nagao, H. Koga, P. He, K. Suganuma, A highly sensitive and flexible pressure sensor with electrodes and elastomeric interlayer containing silver nanowires, Nanoscale. 7 (2015) 2926–2932. doi:10.1039/C4NR06494A.

[17] S. Gong, W. Schwalb, Y. Wang, Y. Chen, Y. Tang, J. Si, B. Shirinzadeh, W. Cheng, A wearable and highly sensitive pressure sensor with ultrathin gold nanowires, Nat. Commun. 5 (2014). doi:10.1038/ncomms4132.

[18] H. Bin Yao, J. Ge, C.F. Wang, X. Wang, W. Hu, Z.J. Zheng, Y. Ni, S.H. Yu, A flexible and highly pressure-sensitive graphenepolyurethane sponge based on fractured microstructure design, Adv. Mater. 25 (2013) 6692–6698. doi:10.1002/adma.201303041.

[19] H. Tian, Y. Shu, X.F. Wang, M.A. Mohammad, Z. Bie, Q.Y. Xie, C. Li, W.T. Mi, Y. Yang, T.L. Ren, A graphene-based resistive pressure sensor with record-high sensitivity in a wide pressure range, Sci. Rep. 5 (2015). doi:10.1038/srep08603.

[20] H.L. Chen, B. Cluver, Assessment of Poplar Seed Hair Fibers as a Potential Bulk Textile Thermal Insulation Material, Cloth. Text. Res. J. 28 (2010) 255–262. doi:10.1177/0887302X10375845.

[21] M. Likon, M. Remškar, V. Ducman, F. Švegl, Populus seed fibers as a natural source for production of oil super absorbents, J. Environ. Manage. 114 (2013) 158–167. doi:10.1016/j.jenvman.2012.03.047.

Comparing the Influence of the Drop Fill and Overflow Rinsing On the Reactive Dyeing Process in a Textile Dye House

F. Filiz Yıldırım[1], Esra Gelgec[1], Saban Yumru[1] and Sema Palamutcu [2], Mustafa Çörekcioğlu[1]

[1]*Ozanteks Tekstil San ve Tic A.S R&D Center, Bozburun Mah.7042 Sokak No:6 Denizli / TURKEY*

[2] *Pamukkale University, Textile Engineering Department, TURKEY*

Corresponding author: filiz@ozanteks.com.tr

Abstract Drop/fill rinsing method was used for reducing the high water consumption of overflow rinsing process in textile dye house. However, it is not known whether this drop/fill method has adverse effects on the final fabric properties or not. In this study, the effects of using drop/fill method instead of overflow rinsing process on colorimetric and fastness properties of 100% cotton towel fabrics are investigated. It is found that, skipping to the drop fill rinsing method from the over flow rinsing process and obtained fresh water saving has no negative effect on colorimetric and color fastness values of the processed fabric samples and can be used conveniently in the industrial sized production.

Keywords; **drop/fill rinsing, overflow rinsing, reactive dyeing, water consumption**

I. Introduction

Fresh water has become one of the most valuable natural source. In the past water resources was abundant and cheap. Therefore, high amount of fresh water consumption and consecutive high amount of textile wastewater discharge were acceptable both by textile companies, municipalities and also governments without any restrictions [1]. Nowadays, water is expensive and the increased costs of clean water and wastewater disposal have increased the social awareness and economic incentives for environmental friendly technologies [1, 2]. Thanks to these environmental friendly applications the natural water resources can be used efficiently [1]. Water management model, an integrated water resource management (IWRM), is offered as a useful tool for complex water research using production systems in industrial companies. The best available techniques (BAT) are introduced to present for a number of industrial processes in order to remove or reduce waste water discharge and emissions.

The textile industry is known as its large amount of water consumption in various processes, especially wet processing. In wet processing, water consumption is extremely high compared to the spinning or weaving processes. According USEPA (United States Environmental Protection Agency), 36,000 litres of water should be consumed for producing 20,000 lb fabric per day during the regular dye house processing. In the textile mills, millions of gallons of wastewater are created [2, 3]. Additionally, percentage of the water consumption of the dyeing process is approximately 16% of the total water consumed during the whole wet processes. In wet processing, water is used primarily for chemical processing and secondly for washing and rinsing [2]. Water conservation or reuse processes are development starts with the determination of how and where the water being used. Furthermore, it would be helpful to develop a diagram of water consumption with specific details such as, quantity and location of water usage, water quality or temperature requirements [2].

In this industrial sized research work, water consumption ratios of each processes were studied to determine water consumption amounts and Pareto analysis were performed. As result of Pareto analysis, data are processed on Minitab and HT fabric rope dyeing and cone dyeing departments were identified as the highest water consuming processes in the dye house.

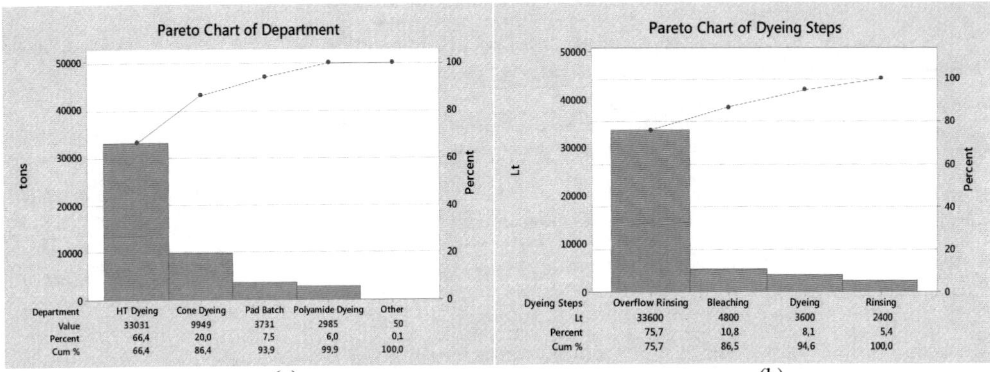

(a) (b)

Fig 1. Minitab Pareto analysis results (a) for water consumption ratios of dye house departments,(b) for water consumption ratios of reactive dyeing process

As seen on Figure 1, the highest water consumption ratios of the reactive dyeing process are belongs to rinsing processes, especially overflow rinsing. Similarly, in literature, washing/rinsing processes were known to be responsible for the relatively high water consumptions in the companies [3]. These processes have significant potential for sustainable production. The overflow rinsing is found to be most water consuming processes in the reactive dyeing process. It has reported that avoidance of overflow rinsing, water consumption could be reduced by 25% [2, 3]. It has also reported that instead of overflow rinsing, drop/fill method should be preferred to manage water saving in the reactive dyeing of cotton fabric [2, 4-16].

In this study, the water consumption ratios of a textile dye house are evaluated to reduce water consumption by replacing the overflow rinsing with the the drop/fill method. Drop/fill rinsing method is applied gradually. The effect of drop/fill method utilization on the color and fastness properties of the fabric is also examined to clarify not to cause any deterioration, since it is very important removal of the hydrolyzed dyes completely removed from the fabric surface with the application of intensive rinse and soaping operations. Drop/fill rinsing process was applied in four stages for light shades and five stages in medium and dark shades. At each stage, color and fastness values of samples were examined to explain whether a negative effect on the processed fabric has occurred or not.

II. MATERIALS AND METHODS

100% cotton towels were used in this study. Drop/fill rinsing method is applied gradually. Gradually processes applied in this study are given in Figure 2. In light shades, 3 processes changed, in medium and dark shades 4 processes changed.

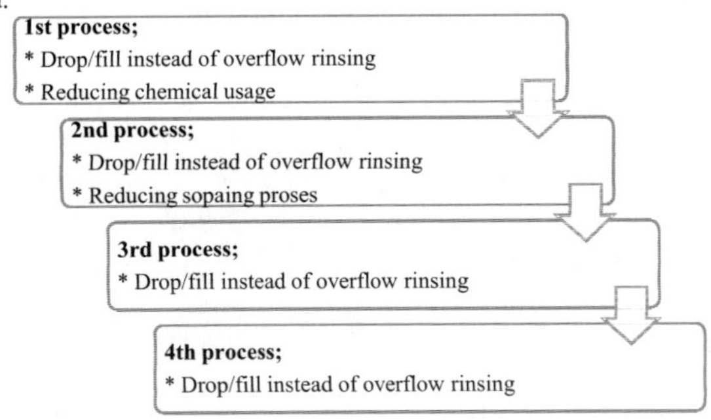

Fig 2. Processes applied to obtain water savings

Fabrics were dyed with 3 different color shades of light, medium, and dark shade. Three reactive dyes with light, medium and dark shades were selected for experiments. The CIE Lab values (L^*, a^*, b^*, C^*, and h^o) *and* ΔE were measured from the reflectance values with using a spectrophotometer under illuminant D65, using 10° Standard observer for each samples. Both dry and wet rub fastness testing were performed according to the ISO

105: X12 protocol and wash-fastness test was carried out according to ISO 105:C06 C1S test standard. Overall fastness properties were evaluated by using ISO grey scales in the light box.

III. RESULTS AND DISCUSSION

A. Colorimetric Properties

It is important to note that reactive dyed samples have the same yarn count and fabric structures. For each dye shades four or five color measurements and washing and rub fastness examinations were done.

The colorimetric data of the samples dyed in light color shades obtained from each process steps are given in Table 1 and Figure 3-5.

TABLE 1
COLORIMETRIC PROPERTIES OF SAMPLES DYED IN LIGHT COLOR SHADES

Samples	L*	a*	b*	C*	h	dE*
11171Dyed fabrics without processes	77,96	7,69	-4,67	9,00	328,73	
11171-After 1st process step	78,42	7,75	-5,04	9,25	326,92	0,603
11171-After 2nd process step	78,46	7,62	-5,18	9,22	325,77	0,725
11171-After 3rd process step	78,33	7,02	-5,02	8,64	324,43	0,843

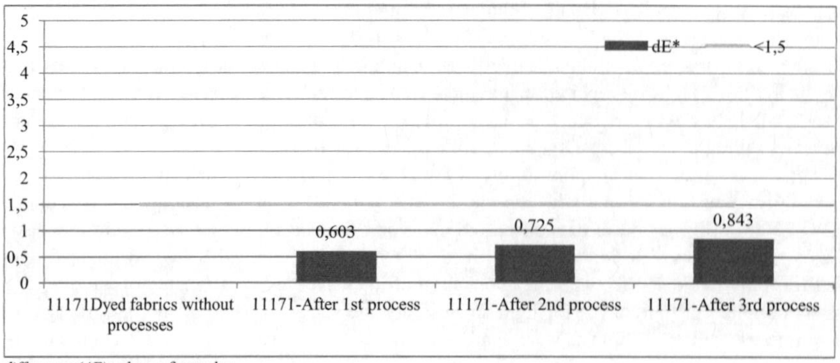

Fig 3. *Color differencs (ΔE)* values of samples

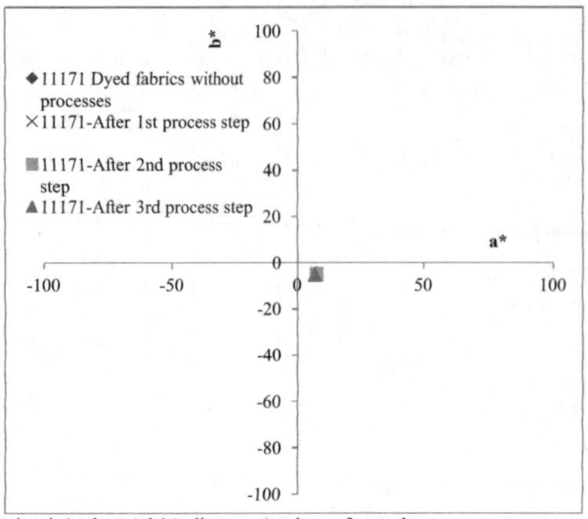

Fig 4. *a* (redness)-b*(yellowness)* values of samples

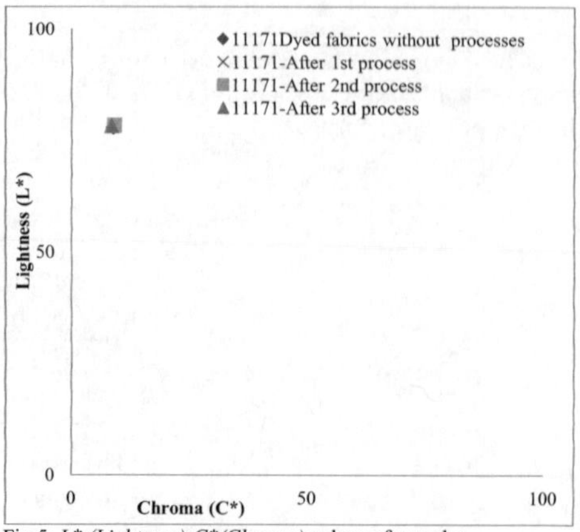

Fig 5. *L* (Lightness)-C*(Chroma)* values of samples

For each stages, the CIE Lab values (*L*, a*, b*, C**, and *h°) and ΔE* of samples are found to be nearly the same. There have been no significant differences between Lightness (*L*), a*, b**,Chroma *(C*)*, and hue angle (*h°*) values of the samples.

The colorimetric data of the samples dyed in medium color shades obtained from each process steps are given in Table 2 and Figure 6-8.

TABLE 2
COLORIMETRIC PROPERTIES OF SAMPLES DYED IN MEDIUM COLOR SHADES

Samples	L*	a*	b*	C*	h	dE*
2952 Dyed fabrics without processes	68,68	-18,1	27,08	32,57	123,7	
2952-After 1st process	68,23	-18,1	26,91	32,41	123,9	0,484
2952-After 2nd process	68,43	-18,1	26,69	32,25	124,1	0,458
2952-After 3rd process	68,21	-18,2	27,14	32,14	123,9	0,49
2952-After 4th process	68,28	-18	27,31	32,73	123,4	0,46

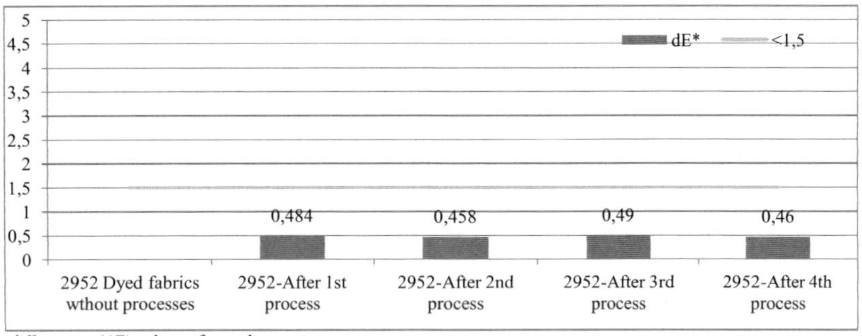

Fig 6. *Color differences (ΔE)* values of samples

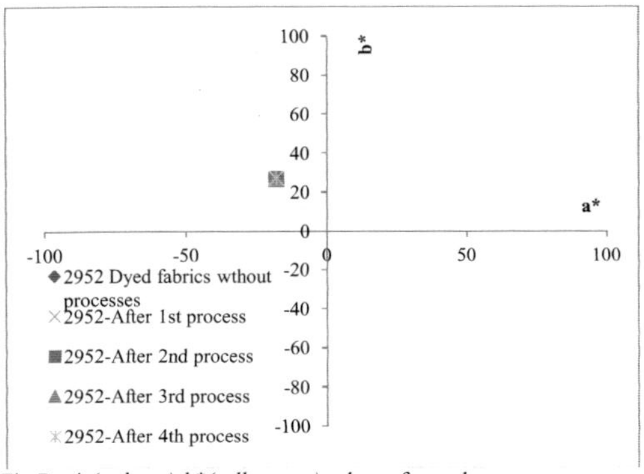

Fig 7. *a* (redness)-b*(yellowness)* values of samples

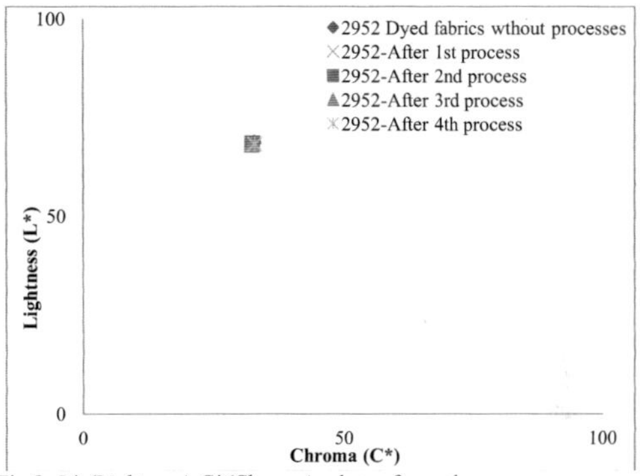

Fig 8. *L* (Lightness)-C*(Chroma)* values of samples

For each stages, the CIE Lab values (*L*, a*, b*, C**, and h^o) *and ΔE* of samples are found to be nearly the same. There have been no significant differences between colorimetric values of the samples.

TABLE 3
COLORIMETRIC PROPERTIES OF SAMPLES DYED IN DARK COLOR SHADES

Samples	L*	a*	b*	C*	h	dE*
1946 Dyed fabrics before processes	37,4	50,46	12,72	52,04	14,15	
1946-After 1st process	37,4	50,46	12,73	52,04	14,15	0,011
1946-After 2nd process	37,58	50,73	12,97	52,37	14,34	0,414
1946-After 3rd process	37,66	50,08	12,67	51,66	14,2	0,46
1946-After 4th process	38,22	50,68	12,58	52,22	13,94	0,858

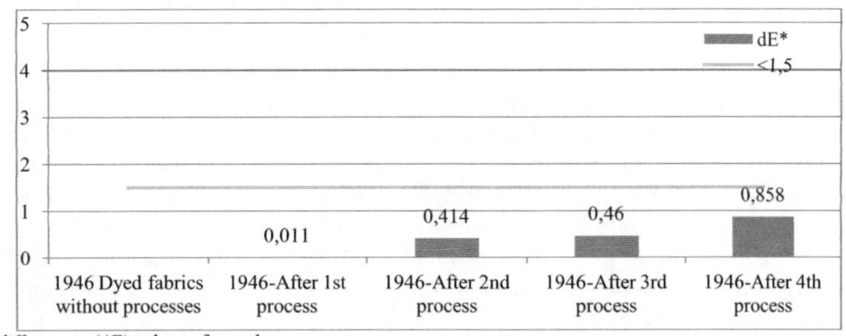

Fig 9. *Color differences (ΔE)* values of samples

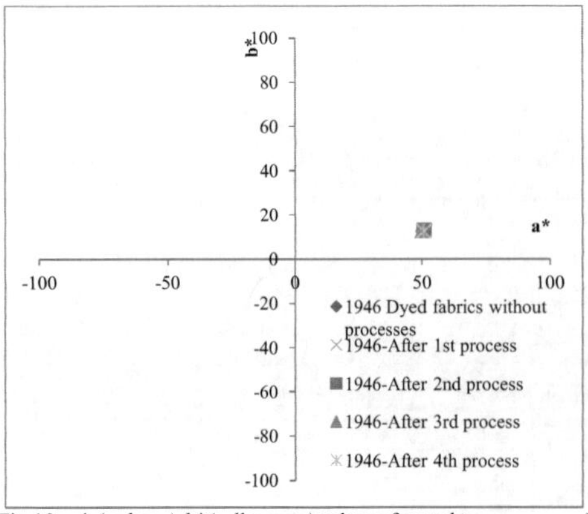

Fig 10. *a* (redness)-b*(yellowness)* values of samples

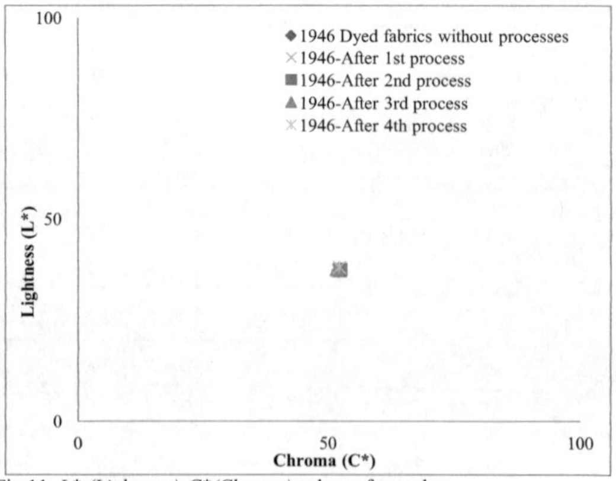

Fig 11. *L* (Lightness)-C*(Chroma)* values of samples

For each stages, the CIE Lab values (L^*, a^*, b^*, C^*, and h^o) and ΔE of samples are found to be nearly the same. There have been no significant differences between colorimetric values of the samples.

In this case, it can be stated that use of drop/fill rinsing method instead of overflow rinsing method have no negative effect on color properties of fabrics and the water consumption reduction method can be used in the dyeing process conveniently.

B. Wash and Rub Fastness

The rub and wash fastness values of samples were given on Table 4.

TABLE 4
RUB AND WASH FASTNESS PROPERTIES OF SAMPLES

Samples	Rub Fastness		Wash fastness					
	Wet	Dry	WO	PC	PES	PA6.6	CO	AC
11171Dyed fabrics without processes	5	5	5	5	5	5	5	5
11171-After 1st process step	5	5	5	5	5	5	5	5
11171-After 2nd process step	5	5	5	5	5	5	5	5
11171-After 3rd process step	5	5	5	5	5	5	5	5
2952 Dyed fabrics without processes	4/5	5	5	5	5	5	4/5	5
2952-After 1st process	4/5	5	5	5	5	5	4/5	5
2952-After 2nd process	4/5	5	5	5	5	5	4/5	5
2952-After 3rd process	4/5	5	5	5	5	5	4/5	5
2952-After 4th process	4/5	5	5	5	5	5	4/5	5
1946 Dyed fabrics before processes	3/4	4/5	5	4/5	5	5	4	5
1946-After 1st process	3/4	4/5	5	4/5	5	5	4	5
1946-After 2nd process	3/4	4/5	5	4/5	5	5	4	5
1946-After 3rd process	3/4	4/5	5	4/5	5	5	4	5
1946-After 4th process	3/4	4/5	5	4/5	5	5	4	5

All dyed samples in each stages exhibited moderate to good rub and wash fastness values according to 5 grey scale ratings. There were no differences between the fastness values of samples in each step. It can be stated that use of drop/fill rinsing method instead of overflow rinsing method have no negative effects on washing and rubbing fastness values of fabrics and the water saving approach can be used conveniently.

IV. CONCLUSION

Drop/fill rinsing method was used for reducing the water consumption in textile dye house. In one textile dye house, %61,8 reduction in water consumption and %20,4 reduction in process time were achieved by using drop/fill rinsing method instead of overflow rinsing. However, it is not known whether this drop/fill method has adverse effects on fabric or not. In this study, the effects of using drop/fill method instead of overflow rinsing, on colorimetric and fastness properties of fabrics were investigated. The colorimetric and fastness properties of samples exhibited almost the same values. Therefore, it is shown that this water consumption reduction method has no negative effect on colorimetric and color fastness values of the fabric samples and can be used conveniently in the dye house.

Acknowledgment

This research was supported by the Ozanteks Textile R&D center and related equity project number is 17S02.

References

[1] Dvarioniene, J., & Stasiskiene, Z. "Integrated water resource management model for process industry in Lithuania". *Journal of Cleaner Production*, 2007, 15(10), 950-957.
[2] Shaikh, M. A. "Water conservation in textile industry". *Pakistan Textile Journal*, 2009, 48-51.
[3] Alkaya, E., & Demirer, G. N. "Sustainable textile production: a case study from a woven fabric manufacturing mill in Turkey". *Journal of Cleaner Production*, 2014, 65, 595-603.
[4] Environmental Technology Best Practice Programme (ETBPP), 1997. Water and Chemical Use inTextile Dyeing and Finishing (No. GG62). Retrieved fromhttp:// www.wrap.org.uk/sites/files/wrap/GG062.pdf.
[5] Shaikh, M.A., "Water conservation in textile industry". *PTJ* 9, 2009, 48e51.
[6] N.C. Division of Pollution Prevention and Environmental Assistance (NCDENR), 2009. Water Efficiency Industry Specific Processes: Textiles. North Carolina. Retrieved from http://infohouse.p2ric.org/ref/01/0069203.pdf.
[7] Lai B., Chen W., Yeh J., "A novel reactive dyestuff range with excellent color yield in continous and cold pad-batch dyeing",
[8] Kar, A., Keane, S. E., & Greer, L."Best Practices for Textile Mills to Save Money and Reduce Pollution in Bangladesh. Natural Resources Defense Council" (NRDC), Dhaka, 28, 2012.
[9] http://www.lebanese-cpc.net/CP_Textile.pdf
[10] Welham, A. C., & Charlotte, S. A. 1994 Development of More Efficient Textile Rinsing Systems. American Dyestuff Reporter, 83, 59-59.
[11] Schlaeppi, F., "Optimizing textile wet processes to reduce environmental impact." *Text Chem Color*, 1998, 30(4), 19-26.
[12] http://www.temizuretim.gov.tr/Files/haberfiles/d120215/Prof.%20Dr.%20Mehmet%20KANIK-Tekstil%20Terbiye%20Sekt%C3%B6r%C3%BCnde%20Temiz%20%C3%9Cretim%20%C4%B0%C3%A7in%20%C3%96mek%20Modellerin%20Olu%C5%9Fturulmas%C4%B1.pdf
[13] http://www.greenpoint.com.tr/uploads/docs/192015Ig3Lc4.pdf
[14] http://www.gpcb-kp.in/live/hrdpmp/hrdpmaster/igep/content/e48745/e49028/e58164/e58173/reductionofwaterconsumption.pdf
[15] Alper, F. "Sürdürülebilirlik Kavramı içerisinde Su Ayak İzi: Tekstil Sektörü Örneği", 2015, (Doctoral dissertation, Fen Bilimleri Enstitüsü).
[16] Defrawy, N. E. "Water management in textile industry: Technical and economic aspects", *International journal of environmental studies*, 2002, 59(5), 573-587.

Development of pressure sensors for smart textiles

P Gomes[1], D Tama[1,2], Y Yao[3], M J Abreu[1], A P Souto[1], H Carvalho[1]

[1] University of Minho, Center for Textile Science and Technology, Campus de Azurem, Guimaraes, Portugal
[2] Ege University, Department of Textile Engineering, 35100 Bornova / Izmir, Turkey
[3]University of Minho, School of Engineering, Dep. Industrial Electronics, Campus de Azurém 4800-058 Guimarães, Portugal

Corresponding author: vpcgomes@gmail.com

Abstract. The investigation of textile sensors has increased and has gained an important role in the creation of new solutions for the most varied products, from monitorization sensors, actuators and controllers, implemented in garments, house textile items, or textiles structures used for a specific purpose e.g. bridge structural monitoring, soil proprieties, among others. This paper presents the development of pressure sensors base in a conductive silicone with piezoresistive conductive properties. The objective is to explore the combination of different types of conductive material (conductive silicon, nonconductive silicon, conductive ink and conductive fabric) and test the interactions between them. In the production methods the conductive ink, and conductive fabric were fixed to the silicone in the curing phase. Conductive and nonconductive silicon were mixed to obtain better control of sample conductivity. Samples with different thicknesses were developed, it was study the influence on the voltage variation and conductivity for each sample.

Keywords. Pressure sensors, piezoresistive materials, conductive fabric, conductive ink, conductive silicone, e-textiles.

1. Introduction

Smart textiles using flexible integrated pressure sensors have many potential applications. In the medical field, these products may be used for measuring the pressure applied to the body via garments, namely, hosiery or bandages for varicose veins or leg ulcers [1,2] determining the suitability of a wheelchair cushion [3], monitoring the wearer's respiration [4]. In sports science, there are applications in martial arts to measure the impact [5-7] or monitoring the performance of muscles during exercise [8]. Also, in home automation objects with a textile base, is appearing an increase interest in creating new sensor and controller products [9].

Although there are different kinds of smart materials for sensor applications, the present study focused on building pressure sensors using piezoresistive materials.

Piezoresistive materials exhibit a change of electrical resistance when pressure is applied, enabling them to work as transducer material for sensors [3]. The development of flexible force sensors based on piezoresistive films was presented before. A copper tape, a ripstop conductive fabric and conductive knitted stretch fabric were used as electrodes and it has been found out that the different electrodes affect the sensor behavior [10].

In a previous study [11], Linqstat film was used as piezoresistive conductive polyethylene film. Conductive ink was used as electrodes and the adhesion of ink on plasma-treated piezoresistive film was analyzed, and the functionality of achieved sensors was evaluated. Pressure sensors were constructed as shown in Figure 1, using conductive fabrics as electrodes.

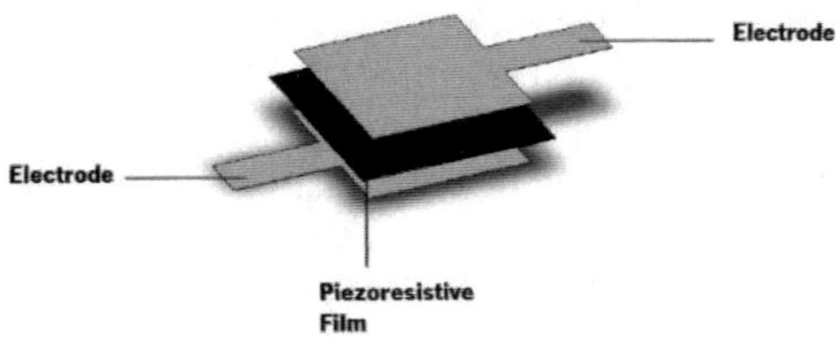

Figure 1. The construction of pressure sensors.

The current study aims at developing optimum pressure sensors for integrating into smart textile products.

2. Materials and Methods

2.1. Materials

In this study, pressure sensors were made by combining different materials. Conductive silicon (Wacker ELASTOSIL® LR 3162 A/B) was used as piezoresistive material. Linqstat film (Linqstat MVCF), conductive fabric (Silver plated polyamide fabric: Shieldex® Bremen), conductive ink (DuPont PE 825 Silver Composite Conductor, DuPont PE 828 Silver Conductor) and aluminum foil were used as electrodes. Wacker® primer G 790 was used to improve the adhesion between layers. Nonconductive silicone (Wacker Elastosil® LR 3003/50 A/B were also used.

2.2. Methods

Three different methods were used to produce the pressure sensors; high temperature curing, hot press and painting. In order to cure the conductive silicon, it has used a laboratory oven, and the temperature of 160°C was applied for 10 minutes. The hot press technique, was used to create prototypes using Linqstat film and conductive silicon.

The sensor dimensions were (30 x 30) mm and the electrode dimensions were (25 x 25) mm. In order to improve adhesion of Linqstat on the silicon, a primer was added between the layers.The conductive inks were applied on the conductive silicon by brush painting technique. Two different conductive inks were used and applied in a (25 x 25) mm area of the sensor. The curing conditions of PE 825 was 120°C, 10minutes and the curing conditions of PE 828, which was more flexible, was 80°C, 20 minutes in the oven.

The conductive fabric and conductive silicon sensor was put together during the curing process at 160°C for 10 minutes. Plasma treatment was applied in the conductive silicon samples to increase the adhesion properties of the material. All the sensors developed were tested using a Hounsfield dynamometer producing 10 cycles of compression between 2 and 100 N at a speed set at 5 mm/min. Figure 2 shows the setup for the cyclic compression test. Encapsulation is achieved using 2mm-thick EPDM foam, which conveys mechanical protection to the sensing element.

The sensor was connected to a Fluke 45 multimeter that acquires resistance values and transmits them to a PC via RS-232. In a second trial, the sensor was connected to a signal conditioning circuit that produces a voltage signal according to the sensor's resistance, as depicted in Figure 2.

Figure 2. Compression test setup.

3. Results

3.1. Piezoresistive Silicon

To create the pressure sensor, firstly the piezoresistive material must be obtained. Therefore, the inner part was produced, conductive silicone was prepared using "ELASTOSIL-LR-3162, A/B". It was prepared using A and B compounds according to the mixing ratio 1:1. The curing condition of this silicone is 160°C, 10 minutes.

Firstly, bare silicone samples were prepared in order to determine conductivity. After preparing the mixture, it was used a coating machine to obtain an even sample with 2 mm thickness (Figure 3).

Figure 3. Silicone sample from coating machine.

During curing process, the silicone swells and bubbles emerged on one side's surface of sample, which contacts with air. In order to overcome that problem, it was decided to use a different method, the mixture was compressed between 2 glasses and cured (Figure 4).

Figure 4. Silicone sample by squeezing between 2 glasses.

In order to obtain a specific form, a mask (Figure 5) was used between the glasses to shape the silicon in different forms.

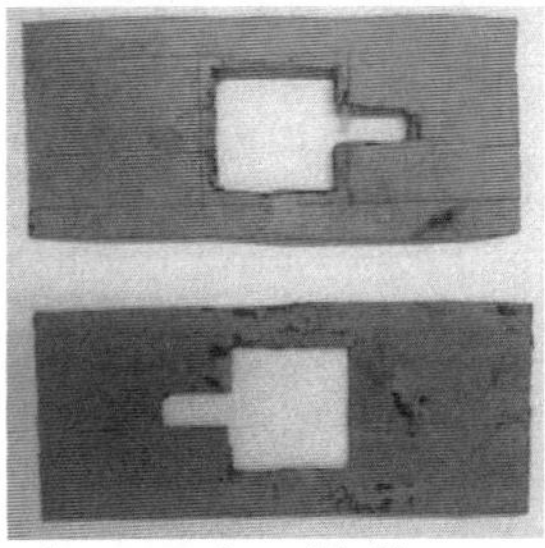

Figure 5. Mask used for the sample with linqstat and conductive silicone.

By this method, an even surface was obtained. When the conductivity was analyzed, we determined that the silicone has high conductivity and behaves like a piezoresistive material.

3.2. Linqstat and conductive silicon sample

In order to make the silicone and linqstat film connection, it was used the hot press method. Firstly, it was used the hot press at 120°C, for 10 seconds, and the pressure of the plates were set to 3,5 bars. The sample were checked, but the silicone didn't show adhesion to the linqstat film. Therefore, the condition was changed to 125°C, for 30 seconds. After that process, the linqstat film was melted and destroyed (Figure 6).

Figure 6. Destroyed sample prepared by hot press.

To obtain a better adhesion, it was used a primer (WACKER® PRIMER G 790. The primer serves as adhesion promoter between silicone elastomers and other substrates like metals, glass and thermoplasts. The primer was applied on the linqstat by spraying method. The same conditions were applied in the hot press machine (120ºC, 10 minutes) but the conductive silicone doesn't stick on the velostat even when the primer is used. By these tests, is possible to ensure that, isn't possible to obtain a working sample with linqstat and conductive silicone, so other methodologies had to be tested.

3.3. Conductive ink and conductive silicon sample

The samples were prepared using both PE 825 (Figure 7) and PE 828 (Figure 8), the conductive inks were applied on silicone by brush. The curing conditions were 120°C, 10 minutes for PE 825 and 80°C, 20 minutes for PE 828. The samples presented cracks and bad adhesion between the inks and the silicon.

Figure 7. Sample with conductive silicone and PE 825 conductive ink.

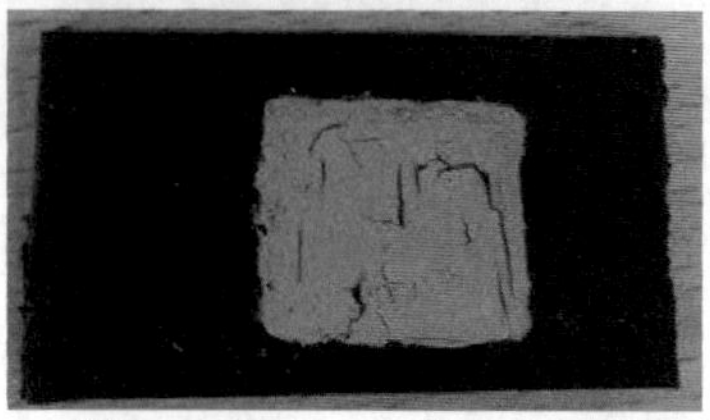

Figure 8. The sample with conductive silicone and PE 828 conductive ink.

In order to improve the bonding properties plasma treatment can be used to induce surface texturization (creation of microroughness) increasing the possibility to get a better absorption and adhesion of finishing agents, stamping, inks. The process may induce secondary reactions such as crosslinking thus allowing graft polymerization. Therefore, plasma treatment was applied on the conductive silicone. This treatment was inefficient due to the high conductivity of the silicon material, that didn't allow the generation of plasma in the plasma machine. It was concluded that, the conductive ink isn't appropriate for using with conductive silicone.

3.4. Conductive fabric/aluminium foil and conductive silicon sample

As determined before the conductive silicone sample created behaves like piezoresistive material, it was prepared a sensor with conductive silicone and conductive fabric. We used the glasses to press the silicon and a mask in order to have the sample in desired dimensions (Figure 9).

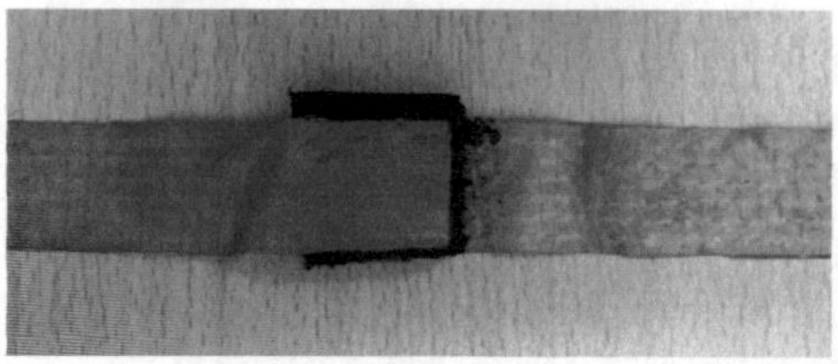

Figure 9. Sensor with conductive silicone and conductive fabric.

The sample didn't work as a sensor, it seems it has a short circuit. For that reason, we decided to prepare a sample using aluminum foil instead of conductive fabric, which is well known high conductive material (Figure 10).

Figure 10. Sensor with conductive silicone and aluminum foil.

The same process for preparing the sample was used. After the tests the results showed that the sample works as a sensor.

3.5. Different conductivity silicon samples

In order to improve the signals inputs and the construction of the sensors, different mixtures, and different thicknesses were made to determine the possibility to change the conductivity in the silicon, and the influence of each parameter in the final results. The conductive silicone was mixed in 90%, 80% and 50% rate with nonconductive silicon, they cured at 160ºC for 10 minutes.

After preparing the samples they were tested in 10 cycles of compression between 2 and 100 N force in the Hounsfield dynamometer and the data were collected. It was analysed the force vs extension behaviour of the sample (Tables 1, 2 and 3).

The graphics present a small deviation for all the samples, in an ideal graphic for force and extension, the lines should overlap in every cycle. These results present a loss in recovering capacity in every cycle. The samples of 80% conductive silicone and 20% nonconductive in all thicknesses had the worst performance presenting the greatest dispersibility.

The samples were also analysed for changes in the resistivity during the 10 cycles test (Tables 4, 5 and 6).

As it could be seen the samples conductivity vary with the changes in the concentrations and thickness. The resistance increases with the increases of nonconductive material. In the 1mm samples (Table 4) the variation in resistivity is to low, turning it not suitable for sensor application.

IOP Conf. Series: Materials Science and Engineering **460** (2018) 012024 doi:10.1088/1757-899X/460/1/012024

Table 1. Silicon samples 1mm thickness - force vs extension.

	1 mm
90% Cond. Silc - 10% Non-cond. Silc.	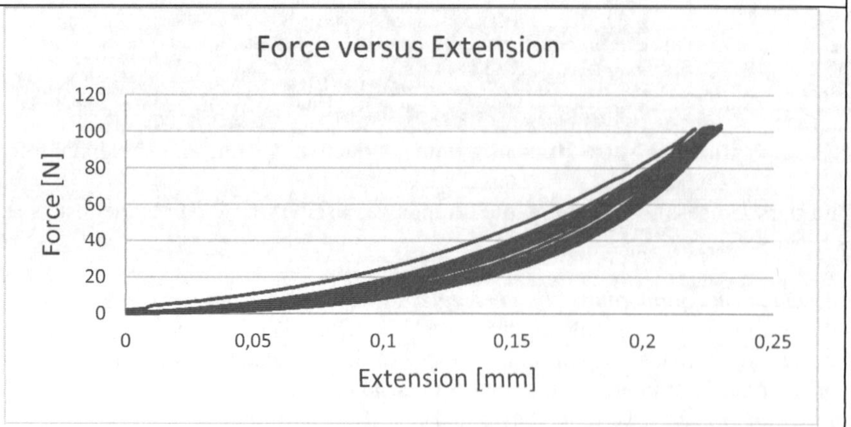
80% Cond. Silc - 20% Non-cond. Silc.	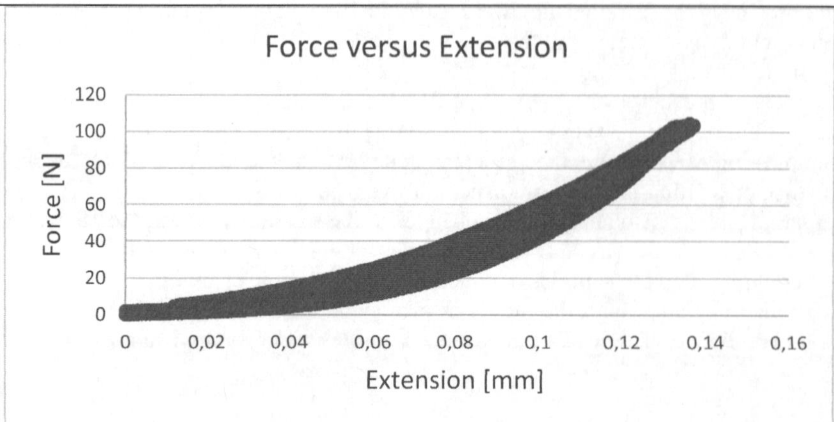
50% Cond. Silc - 50% Non-cond. Silc.	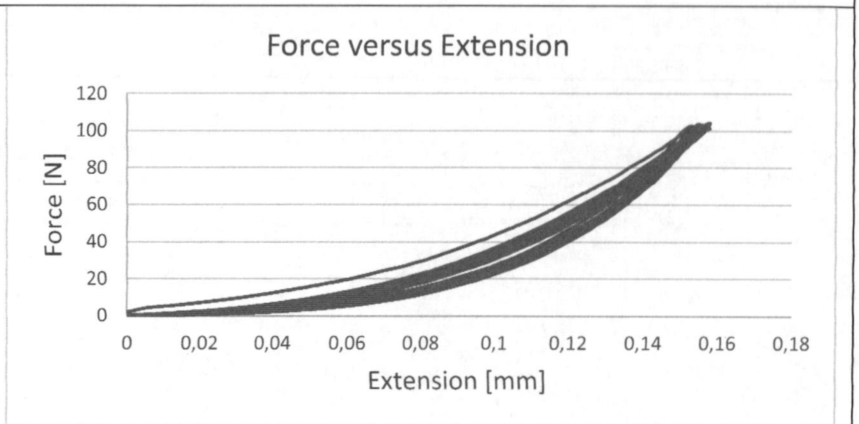

IOP Conf. Series: Materials Science and Engineering **460** (2018) 012024 doi:10.1088/1757-899X/460/1/012024

Table 2. Silicon samples 2mm thickness - force vs extension.

<table>
<tr><td></td><td>2 mm</td></tr>
<tr><td>90% Cond. Silc - 10% Non-cond. Silc.</td><td>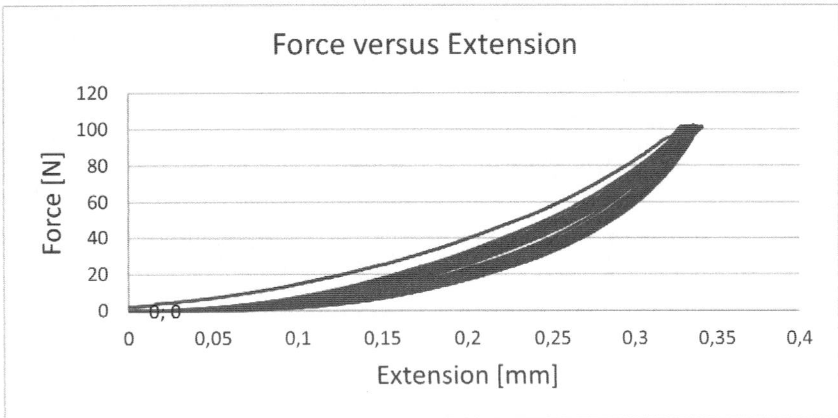
</td></tr>
<tr><td>80% Cond. Silc - 20% Non-cond. Silc.</td><td>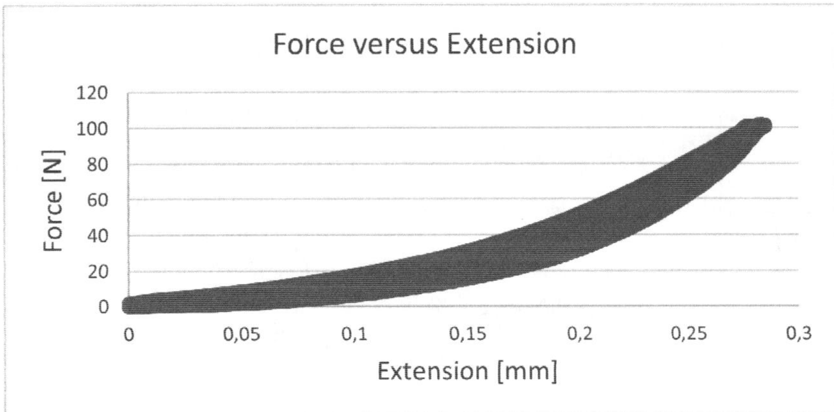
</td></tr>
<tr><td>50% Cond. Silc - 50% Non-cond. Silc.</td><td>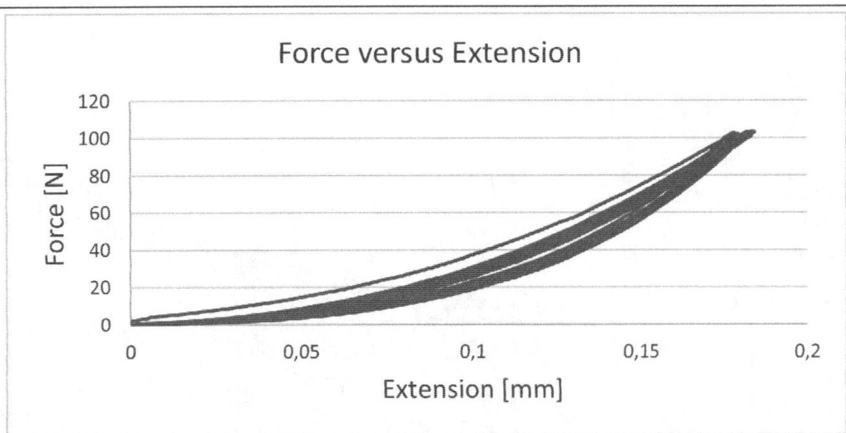
</td></tr>
</table>

IOP Conf. Series: Materials Science and Engineering **460** (2018) 012024 doi:10.1088/1757-899X/460/1/012024

Table 3. Silicon samples 3mm thickness - force vs extension.

	3 mm
90% Cond. Silc - 10% Non-cond. Silc.	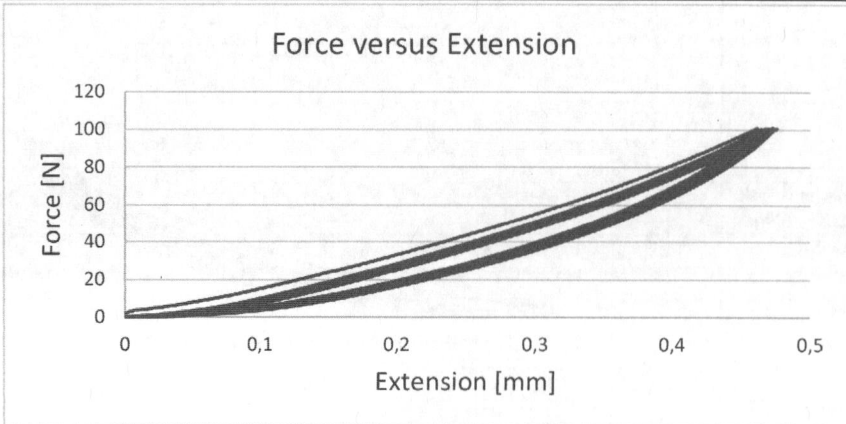
80% Cond. Silc - 20% Non-cond. Silc.	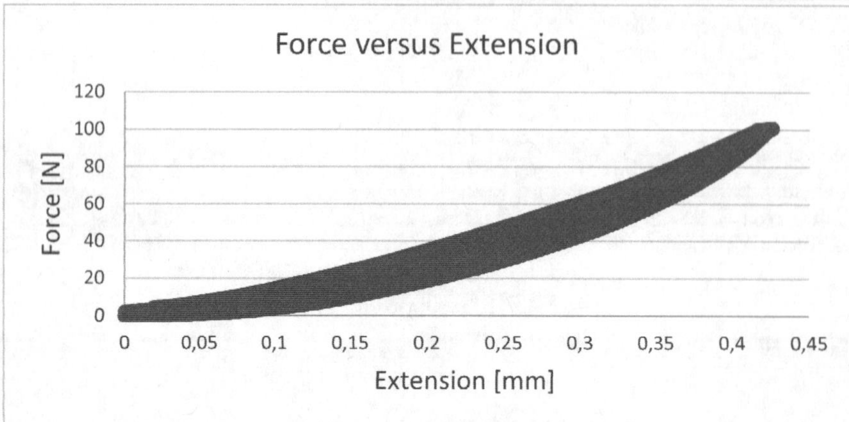
50% Cond. Silc - 50% Non-cond. Silc.	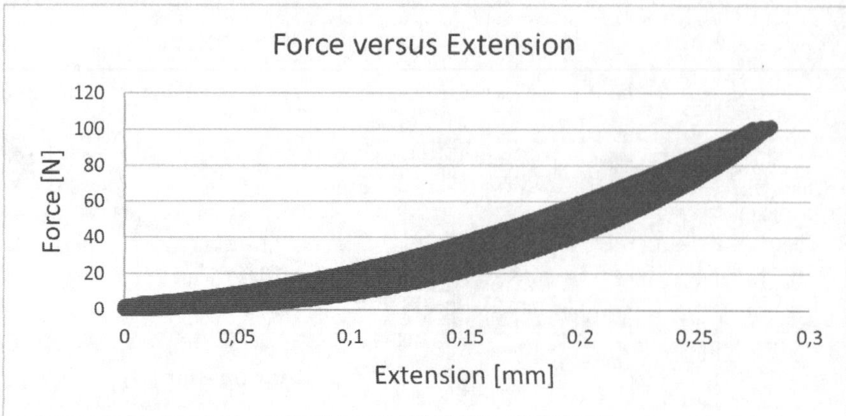

IOP Conf. Series: Materials Science and Engineering **460** (2018) 012024 doi:10.1088/1757-899X/460/1/012024

Table 4. Silicon samples 1mm thickness - resistance vs time.

<table>
<tr><td></td><td>1 mm</td></tr>
<tr><td>90% Cond. Silc - 10% Non-cond. Silc.</td><td>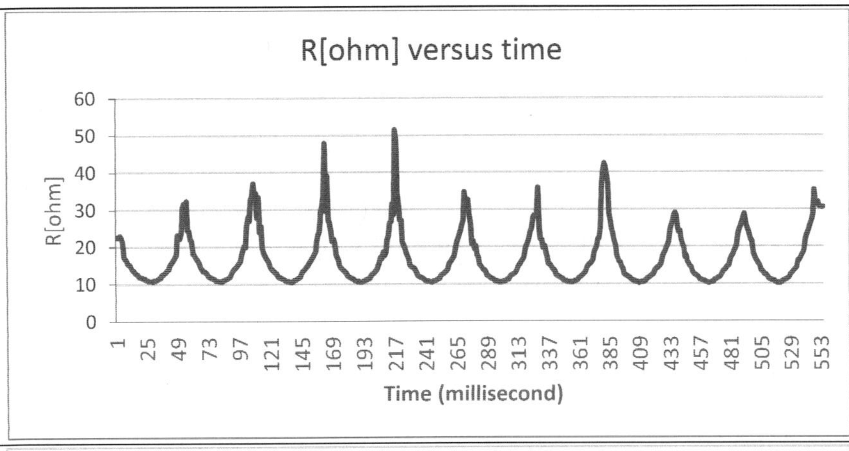
</td></tr>
<tr><td>80% Cond. Silc - 20% Non-cond. Silc.</td><td>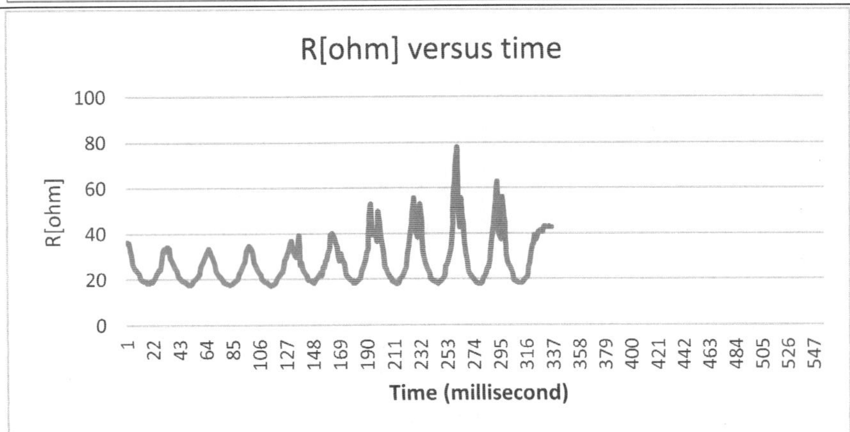
</td></tr>
<tr><td>50% Cond. Silc - 50% Non-cond. Silc.</td><td>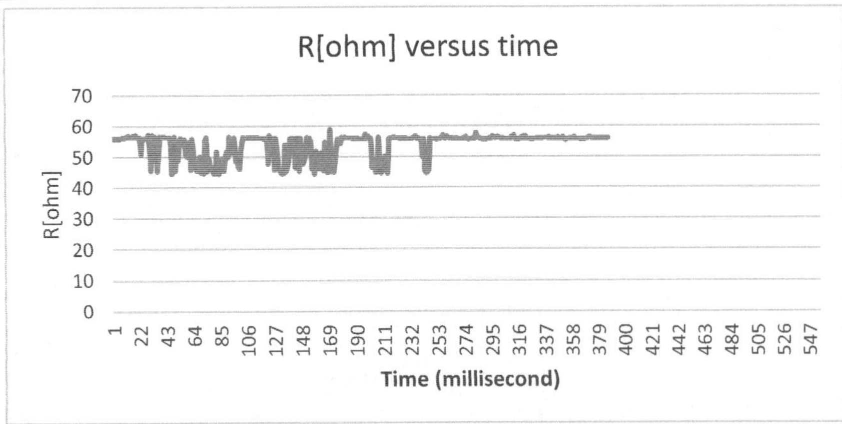
</td></tr>
</table>

Table 5. Silicon samples 2mm thickness - resistance vs time.

	2 mm
90% Cond. Silc - 10% Non-cond. Silc.	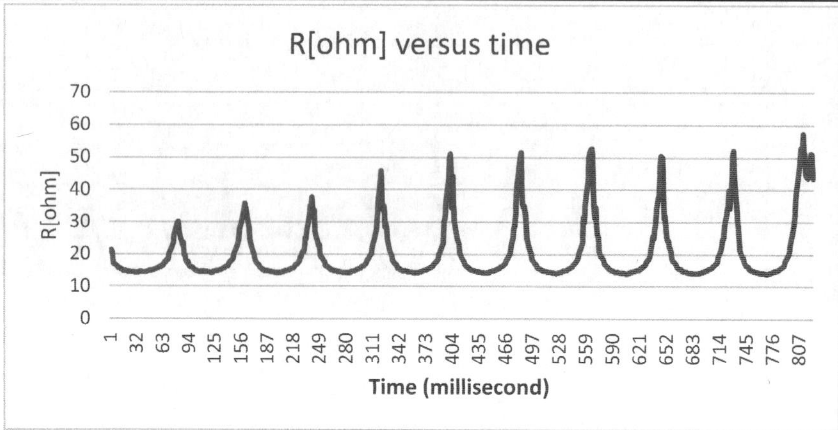
80% Cond. Silc - 20% Non-cond. Silc.	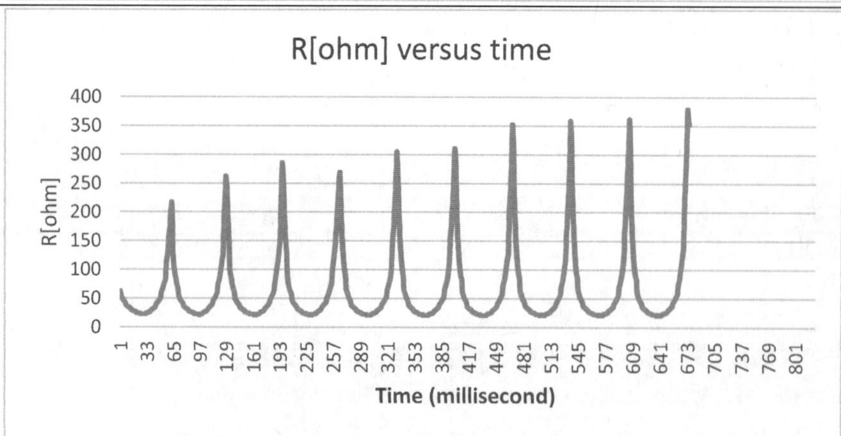
50% Cond. Silc - 50% Non-cond. Silc.	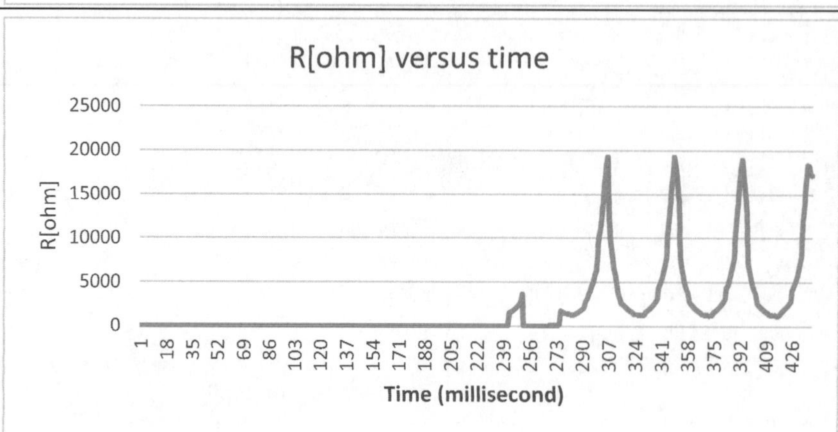

Table 6. Silicon samples 3mm thickness - resistance vs time.

	3 mm
90% Cond. Silc - 10% Non-cond. Silc.	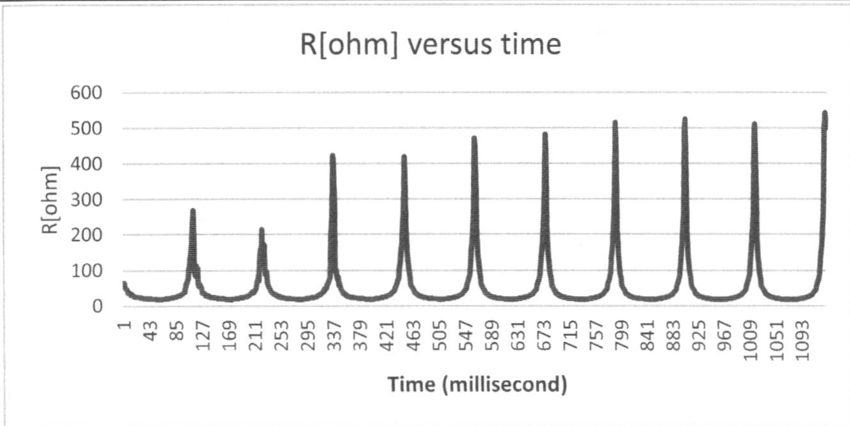
80% Cond. Silc - 20% Non-cond. Silc.	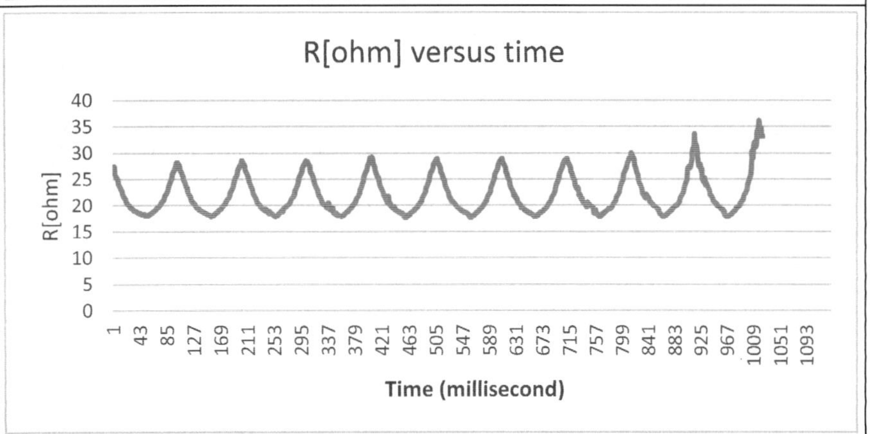
50% Cond. Silc - 50% Non-cond. Silc.	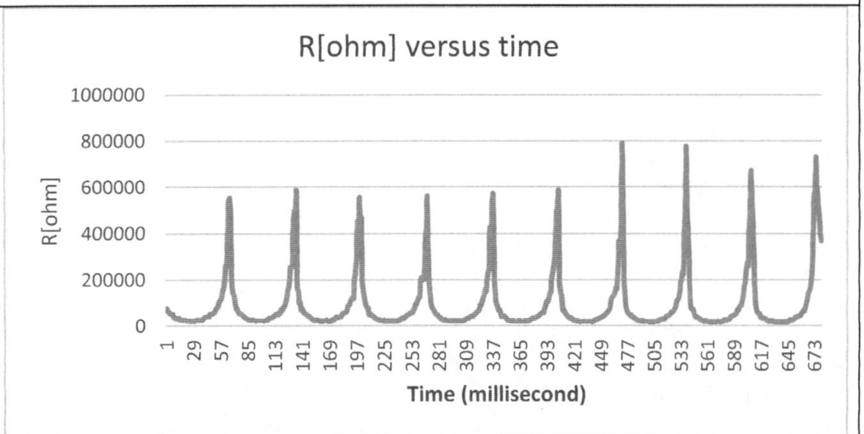

The 2mm samples (Table 5) presented a uniform behaviour increasing the resistivity values with the increase of nonconductive material.

The 3mm samples (Table 6) presents an increase in resistivity when compared with 2mm samples for the same values (90% and 50% conductive material). The 80% conductive sample presented a low value of resistance and low oscillation in the values.

4. Conclusion

A piezoresistive material could be obtained by using conductive silicon. The thickness can be adjusted in the process of high temperature curing using a compressing technique. This technique prevents the expansion of the silicon and the formation of bubbles inside the sample. Silicon did not present good adhesion properties to conductive ink and the linqstat, turning it impossible to use them as electrodes, even with the application of primer. A linking was established with conductive fabric and aluminum foil during the curing phase. Although the adhesion the conductive fabric prevents the silicon to act as a piezoresistive material, the aluminum foil was a god electrode. Mixing nonconductive material to the conductive silicon is possible in order to control the resistance values. In further studies new samples will be prepared and tested with new material to improve the adhesion and prevent the short cutting with the conductive fabric. Different thickness samples and different concentration of conductive and nonconductive material will be tested in order to determine the values of resistance that is possible to obtain for the creation of pressure sensors.

Acknowledgment

This work is financed by FEDER funds through the Competitivity Factors Operational Program - COMPETE and by national funds through FCT – Foundation for Science and Technology within the scope of the project POCI-01-0145-FEDER-007136. Derya Tama thanks FCT for fellowship 2C2T-BPD-08-2017.

References

[1] Macintyre, L., 2011, New calibration method for I-scan sensors to enable the precise measurement of pressures delivered by 'pressure garments', Burns, Vol:37, pp. 1117 –1181.

[2] Baldoli, I., Mazzocchi, T., Paoletti, C., Ricotti, L., Salvo, P., Dini, V., Laschi, C., Francesco F.D. and Menciassi, A., 2016, Pressure mapping with textile sensors for compression therapy monitoring, Proceedings of the Institution of Mechanical Engineers, Part H: Journal of Engineering in Medicine, Vol 230, Issue 8, pp. 795 – 808, https://doi.org/10.1177/0954411916655184

[3] Bosowski, P., Hoerr, M., Mecnika, V., Gries, T. and Jockenhövel, S., 2015, Electronic Textiles, Smart Fabrics and Wearable Technology: Design and manufacture of textile-based sensors, Woodhead Publishing, England, 307 p.

[4] Brady, S., Dunne, L. E., Tynan, R., Diamond, D., Smyth, B. and O'Hare G. M. P., 2005, Garment-Based Monitoring of Respiration Rate Using a Foam Pressure Sensor, Ninth IEEE International Symposium on Wearable Computers (ISWC'05), 2005, pp. 214-215, doi: 10.1109/ISWC.2005.23

[5] Cowie, J. I., Flint, J. Z. and Harland, A. R., 2009, The Engineering of Sport: Wireless Impact Measurement for Martial Arts, Springer, France, 713p.

[6] Del Vecchio, F.B., Franchini, E., Del Vecchio, A.H.M. and Pieter, W., Energy absorbed by electronic body protectors from kicks in a taekwondo competition, Biology of Sport, 2011,Vol. 28(1).

[7] Mcloughlin, J., Cashman, P., Power, J. and Ledbury, J., A Smart Textile for Measuring the Impact of Energy used in Competitive Contact Sports', The Textile Institute International Conference: Advances in Functional Textiles, 25 26th July 2013, Manchester, UK. (Unpublished), retrieved 10/06/2017 fromhttp://eprints.hud.ac.uk/id/eprint/18533/3/28_A_Smart_textile_for_measuring.pdf

[8] Zhou, B., Sundholm, M., Cheng, J., Cruz, H. and Lukowicz, P., 2016, Never skip leg day: A novel wearable approach to monitoring gym leg exercises, 2016 IEEE International Conference on Pervasive Computing and Communications (PerCom), Sydney, NSW, pp. 1-9, doi: 10.1109/PERCOM.2016.7456520

[9] Ojuroye O., Torah R., Beeby S., Wilde A. (2017) Smart Textiles for Smart Home Control and Enriching Future Wireless Sensor Network Data. In: Postolache O., Mukhopadhyay S., Jayasundera K., Swain A. (eds) Sensors for Everyday Life. Smart Sensors, Measurement and Instrumentation, vol 22. Springer, Cham, DOI; https://doi.org/10.1007/978-3-319-47319-2_9

[10] Carvalho H., Yao Y. and Gonçalves M., 2017, Flexible force sensors for e-textiles, 17th World Textile Conference AUTEX 2017- Textiles - Shaping the Future, Conf. Series: Materials Science and Engineering 254 (2017) 072007 doi:10.1088/1757-899X/254/7/072007

[11] Tama, D., Gomes, P., Yao, Y., Abreu, M.J., Souto, A.P. and Carvalho, H., 2017, Developing Smart Body Protector for Sports Karate, XIVth International Izmir Textile & Apparel Symposium (IITAS 2017), Izmir, October 26-28.

IOP Conf. Series: Materials Science and Engineering **460** (2018) 012025 doi:10.1088/1757-899X/460/1/012025

Preparation of magnetite nanoparticle and fatty acid incorporated poly(methacrylic acid-ethyl acrylate) nanowebs via electrospinning for magnetic hyperthermia application

N Sarier[1,6], E Onder[2], M D Carvalho[3], L P Ferreira[4,5], M M Cruz[5] and R Arat[1]

[1] Istanbul Kultur University, the Faculty of Engineering, 34156 Bakirkoy, Istanbul, Turkey
2 Istanbul Technical University, the Faculty of Textile Technologies and Design, 34437 Gumussuyu, Istanbul, Turkey
[3] CQB, Centro de Química e Bioquímica, Faculdade de Ciências, Universidade de Lisboa, Campo Grande, 1749-016, Lisboa, Portugal
[4] Department of Physics, University of Coimbra, 3004-516, Coimbra, Portugal
[5]BioISI, Biosystems and Integrative Sciences, Faculdade de Ciências, Universidade de Lisboa, Campo Grande, 1749-016, Lisboa, Portugal
[6] Corresponding Author: n.sarier@iku.edu.tr

Abstract. Magnetic hyperthermia has been arising as a promising approach for treatment of cancer. When magnetic nanoparticles (MNPs) are locally injected through cancerous tissues and subjected to an appropriate alternating magnetic field, they generate heat due to the rotation of the nanomagnets, causing the destruction of the cancer cells. For repeated applications of magnetic hyperthermia, it is highly preferred to keep the temperature constant at about 41–46°C while preventing the leakage of MNPs, to minimize secondary effects on surrounding healthy tissues. In our study, we produced novel nanowebs provided with magnetic and thermal buffering properties by encapsulating MNPs and Lauric Acid in poly(methyl ethyl acrylate) matrix via uniaxial electrospinning. For the oleic acid functionalized magnetic nanoparticles, the magnetization for the corresponding nanowebs is higher by a factor of two for all concentrations than the ones with non-functionalized magnetic nanoparticles, indicating a more effective integration of the functionalized nanoparticles through the nanoweb. The heat absorption and release capacities of the nanowebs, incorporated with 1.25–2.50 % functionalized MNPs, vary between 70–75 Jg^{-1} at 40–48°C. They also demonstrate thermal cycling ability and thermal stability.

1. Introduction

Despite the significant advances recorded in recent years, cancer is still one of the leading causes of human death worldwide [1]. Hyperthermia is an effective treatment method of general or regional warming of body tissue to kill tumor cells in the cancerous area [2] and [3]. Magnetite (Fe3O4) nanoparticles (MNPs) have been intensively investigated for biomedical applications due to their magnetic properties and dimensions similar to the biostructures. Recently, magnetic hyperthermia has also been arising as a promising approach for treatment of cancer [1] and [2]. In this method, when magnetic nanoparticles, locally injected through cancerous tissues, are subjected to an appropriate alternating magnetic field (AC), they generate heat due to the rotation of the nanomagnets themselves or the reversal of the direction of the magnetization vector μ of each single domain [2]. Cells are susceptible to heat damage because heat energy indiscriminately disrupts cellular pathways, especially through its effects on protein structure and function. For repeated applications of magnetic hyperthermia, it is highly preferred to keep the temperature constant at about 41–46°C while preventing the leakage of MNPs, to minimize secondary effects on surrounding healthy tissues. It would be therefore highly advantageous if a material could be developed that would allow the repeated heating of the MNPs [3]. To maintain the biocompatibility of magnetic nanoparticles in human body and to prevent their sedimentation within the tissues, they should be suspended in suitable fluids and coated by biocompatible polymers [4].

Previous studies have suggested developing polymeric formulations to coat or to encapsulate MNPs prior to their hyperthermia application [4], [5], [6] and [7]. Combining MNPs and polymeric matrices not only can lead to significant enhancement in the surface characteristics of the particles, but also polymers may serve as carrier for particles, as coating materials and encapsulating shells, as direct targeting and stabilizing agents, or as linker to couple functional biomolecules [8], [9], [10], [11], [12], [13] and [14]. Recently, Huang *et al.* [15] described the incorporation of MNPs into electrospun polystyrene fibres for developing novel mediators for magnetic hyperthermia as an anti-cancer strategy. They suggested that the use of electrospun fibres for the encapsulation of MNPs has the additional advantage of fibres displaying a very high surface area over volume ratio, which may enhance their interactions with the surrounding tissues and the possible use of coated fibres for cell binding applications. Owing to the very small dimensions of MNPs, electrospinning of nanofibers presents itself as a simplistic method to decorate or incorporate MNPs into polymeric nanofibers to add functionality. Up to now, numerous MNPs including magnetite have been successfully encapsulated into polymeric nanofibers where the principle of electrospinning remains similar. Most of these previous studies focused primarily on the production and characterization of magnetic fibres [16], [17], [18] and [19]. However, studies taking advantage of the potential of local heat-generating capability of magnetic electrospun nanofiber composites upon application of alternating magnetic field for hyperthermia treatment are limited [15] and [20]. One study reports on a magnetic nanofiber that encapsulated MNPs being used for cancer therapy, which was delivered by surgical or endoscopic methods precisely to the tumour site. Another study explains the fabrication of electrospun polyurethane (PU) nanofibers decorated with superparamagnetic Fe_3O_4 NPs and their potential use as heat-generating substrates for magnetic hyperthermia application [21]

Phase change materials (PCMs), which absorb and release a great amount of heat repeatedly during phase conversions between solid-liquid phases over a certain temperature range, have been receiving great attention in the last decades particularly to develop efficient thermal energy storage systems. A wide variety of PCMs are available with different heat storage capacities and phase change temperature intervals. Among them, fatty acids derived from the fats and oils, have recently attracted attention as potential bio-based and renewable PCM candidates in consequence of their attractive characteristics, such as high heat storage densities, low vapour pressure in the liquid phase, chemical and thermal stability, non-toxicity, biocompatibility and commercial availability at a relatively low cost [22]. In the last decade, a few research groups performed studies on the preparation of magnetic nanoparticle-PCM mixtures for testing their magnetic hyperthermia properties. Pradhan *et al.* prepared lauric acid-coated, superparamagnetic, nanoparticle-based magnetic fluids of different ferrites suitable for hyperthermia application [23]. Zhao *et al.* studied magnetic and inductive heating properties of Fe_3O_4/polyethylene glycol composite nanoparticles with core–shell structure [24].

The aim of this study is to develop novel nanowebs provided with magnetic and thermal buffering properties by encapsulating MNPs and Lauric Acid (LA) in poly(methyl ethyl acrylate) (PMEA) matrix via uniaxial electrospinning. Structural and thermal characterizations of the electrospun nanowebs were then carried out throughout the Fourier Transform Infrared (FTIR) spectroscopy, Differential Scanning Calorimetry (DSC), Thermogravimetry (TG) analyses. Magnetic properties of the NWs were characterized as a function of temperature and of applied magnetic field.

2. Experimental

2.1. Materials

The oleic acid functionalized Fe_3O_4 magnetic nanoparticles (MG3) and non-functionalized Fe_3O_4 magnetic nanoparticles (MG6) were synthesized by M D Carvalho, L P Ferreira and M M Cruz through the project UID/MULTI/04046/2013 and UID/MULTI/00612/2013 [Portuguese FCT foundation].

Lauric Acid [$CH_3(CH_2)_{10}COOH$] (LA), chloroform and methyl alcohol as solvents were all supplied from Sigma-Aldrich Chemicals (USA). For encapsulating LA and MNPs mixtures by uniaxial electrospinning, the copolymer of methacrylic acid and ethyl acrylate (PMEA) with the methacrylic acid groups of 46.0–50.6%, glass transition temperature of 96 ± 5°C and the molecular weight of 320,000 gmole^{-1}, was kindly supplied from Evonik Company (USA) under the trade name of Eudragit L100-55® in powdered form.

2.2. Methods

The nanowebs of PMEA, LA and MNPs were manufactured using uniaxial apparatus of the electrospinning device with 15 cm collector distance (Yflow Co.; Spain). The whole procedure was as follows. PMEA and LA were dissolved in chloroform-methanol (7 mL: 3 mL) solvent mixture. The final concentration of PMEA and of LA were calculated as 14%. Then, different amounts of MNPs with respect to the mass of electrospinning mixtures (either MG3 or MG6) were added to those solutions while stirring by a shaker for 5 minutes. The resultant MNP percentages were 1.25%, 1.75% and 2.50%. From then on, each mixture, instantly drawn into the pump injector, was electrospun under the processing parameters given in Table I. As spun nanofibers were collected on an aluminium sheet of 15 cm x 15 cm.

Table 1 Composition of the NW Samples and their electrospinning conditions

Sample	PMEA (g)	LA[a] (g)	MNP[b] (%)	η^c (Pas)	Pump rate (mLh^{-1})	Injector Voltage (kV)	Collector Voltage (kV)	Time (h)
PMEA-LA	2.5	2.5	-	0.84	0.040	+5.0	-5.0	1.0
PMEA-LA-MG3-1.25	2.5	2.5	1.25	0.87	0.030	+10.0	-10.0	1.0
PMEA-LA-MG3-1.75	2.5	2.5	1.75	0.95	0.030	+10.0	-10.0	1.0
PMEA-LA-MG3-2.50	2.5	2.5	2.50	1.02	0.030	+10.0	-10.0	1.0
PMEA-LA-MG6-1.25	2.5	2.5	1.25	0.89	0.040	+10.0	-10.0	1.0
PMEA-LA-MG6-1.75	2.5	2.5	1.75	1.06	0.030	+5.0	-5.0	1.0
PMEA-LA-MG6-2.50	2.5	2.5	2.50	1.11	0.030	+4.0	-4.0	1.0

[a] LA= Lauric Acid, [b]MNP%0 w.r.to the mass of electrospinning mixture, [c]η=dynamic viscosity of the electrospinning mixture

The Fourier transform infrared (FTIR) transmission spectra of the electrospun nanowebs were recorded between 4000 and 650 cm^{-1} at a resolution of 4 cm^{-1} using a Perkin Elmer Spectrum 100 FTIR spectrometer equipped with a universal attenuated total reflection (ATR) accessory. The thermal properties of the materials and nanoweb (NW) samples were investigated under nitrogen atmosphere (20 mLmin^{-1}) using a Perkin Elmer DSC 4000 differential scanning calorimeter (DSC). Temperature scans were run on samples that had been placed in a closed pan and that weighed 10–15 mg for ten successive heating and cooling cycles. All samples were brought to thermal equilibrium at the starting temperature. Then, the samples were heated to the final temperature at a heating rate of 10 °Cmin^{-1}. Finally, the samples were cooled to the starting temperature at a cooling rate of 10 °Cmin^{-1}.

Thermogravimetric (TG) analyses of the samples were performed from 20 to 650 °C at a heating rate of 10 °Cmin^{-1} under a dry nitrogen atmosphere purged at a rate 20 mLmin^{-1} by a SEIKO EXSTAR 6200 Model TG/DTA instrument. Magnetic properties of the NWs were characterized using a QD-MPMS SQUID magnetometer as a function of temperature, between 10 and 380 K, and of applied magnetic field up to 5.5 T. The temperature dependence was studied in an applied magnetic field of 2 mT after cooling from 300 K in zero magnetic field (zero field cooled–ZFC) and after cooling under the measurement field (field cooled–FC).

3. Results and Discussion

The FTIR transmission spectra of the electrospun PMEA-LA NWs, including different percentages of MG3 and MG6, as well as that of PMEA-LA are shown in Figure 1 (a) and (b). All distinctive bands of the PMEA and Lauric Acid are observed in the FTIR spectra of the NW samples. Asymmetric and symmetrical stretching vibrations of -CH_3 and -CH_2 groups of LA are observed at 2954, 2917 and 2849 cm^{-1}. The stretching bands for the ester bonds of the methacrylic acid and ethyl acrylate groups of the matrix PMEA, namely O-C = O (-COO) and the O-C and C = O, as well as those of Lauric Acid are observed at 1731, 1697 and 1159 cm^{-1}. The scissoring deformation band of long chain -CH_2 groups is observed at 1470 cm^{-1}. Consequently, the distinctive IR bands of PMEA-LA-MG3 and PMEA-LA-MG6 nanowebs and the IR bands of PMEA-LA NW sample are all overlapped, suggesting that the addition of MNPs (MG3 or MG6) does not have any adverse effect on the electrospinnability and physicochemical properties of PMEA-LA mixtures.

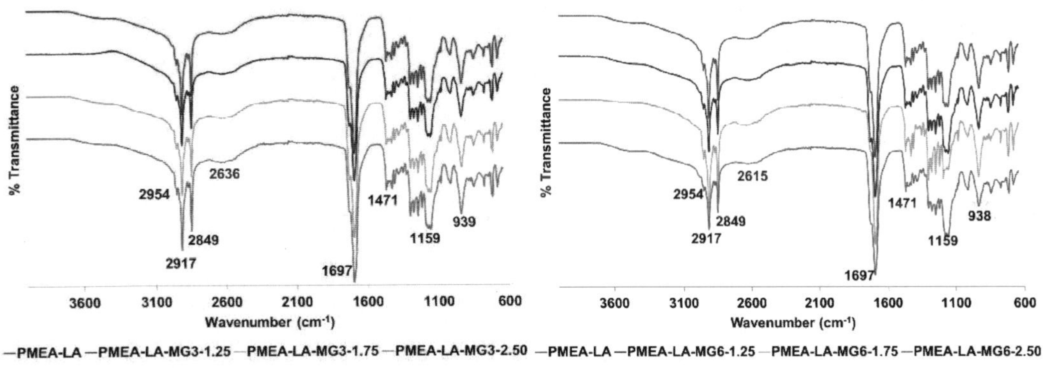

Figure 1. FTIR transmission spectra of (a) PMEA-LA and MG3 incorporated NWs of PMEA-LA-MG3; (b) PMEA-LA and MG6 incorporated NWs of PMEA-LA-MG6.

Table 2 summarizes the phase transition temperatures and enthalpies of all the NWs. Figure 2 illustrates DSC results for the PMEA-LA-MG3 NW samples obtained from the 10th heating-cooling cycles. The DSC results are consistent with the FTIR results. As seen from Table 2 and Figure 2, the onset, peak and end temperatures of the phase transitions of the PMEA-LA-MG3 and PMEA-LA-MG6 nanowebs, incorporated with increasing percentages of MG3 and MG6, mainly overlap with those of PMEA-LA.

The heat absorption capacities of the magnetite-containing nanowebs are measured as 70–82 Jg^{-1} between 42–48 °C. In the cooling cycles, the crystallizations slightly shift to the lower temperature interval of 41–36 °C, the heat release capacities are about 65–81 J.g^{-1}. As expected, there is a slight decrease in Lauric Acid encapsulation efficiency of PMEA-LA-MG3 and PMEA-LA-MG6 NWs with respect to PMEA-LA NW depending on the amount of MNP incorporated into the structure. Whatever happens, the phase change enthalpies of PMEA-LA-MG3 and PMEA-LA-MG6 samples are all remarkable regarding the related concentrations used in electrospinning as the evidence of successful

IOP Conf. Series: Materials Science and Engineering **460** (2018) 012025 doi:10.1088/1757-899X/460/1/012025

encapsulation of LA in the presence of MNPs, MG3 and MG6.

Table 2. Phase transition characteristics of LA and the NWs obtained from the DSC analyses (10°Cmin^{-1}).

Sample	10th Heating				10th Cooling			
	Phase transition (°C)			ΔH	Phase transition (°C)			ΔH
	T_{onset}	T_{peak}	T_{end}	(Jg^{-1})	T_{onset}	T_{peak}	T_{end}	(Jg^{-1})
LA	44	49	53	185	41	37	34	181
PMEA-LA	44	46	48	82	41	40	34	79
PMEA-LA-MG3-1.25	43	45	47	75	41	40	36	72
PMEA-LA-MG3-1.75	43	45	48	71	41	40	36	70
PMEA-LA-MG3-2.50	43	45	48	74	41	40	36	71
PMEA-LA-MG6-1.25	43	46	48	82	40	39	36	81
PMEA-LA-MG6-1.75	42	44	46	70	40	39	37	65
PMEA-LA-MG6-2.50	44	46	47	74	42	41	26	68

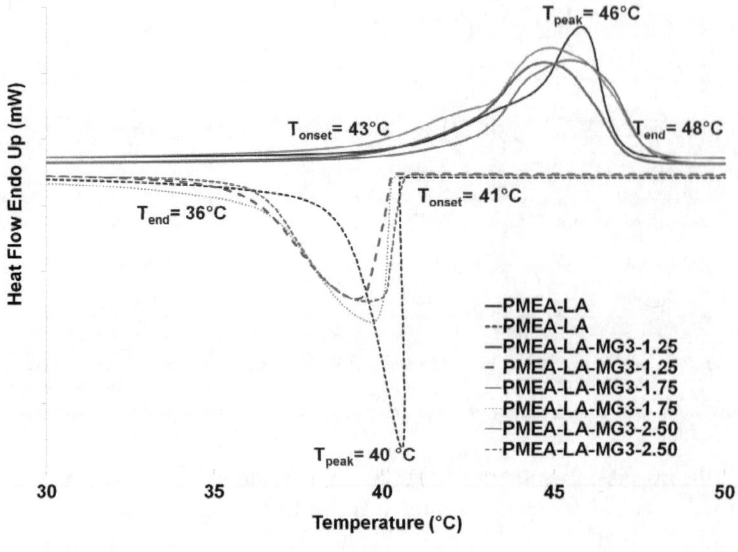

Figure 2. DSC curves of PMEA-LA and MG3 incorporated PMEA-LA-MG3 NWs for 10th heating-cooling cycle at 10 °Cmin^{-1}.

The thermograms of the NW samples are shown in Figure 3 and Figure 4. The PMEA-LA-MG3 and PMEA-LA-MG6 samples exhibit a two-step decomposition, starting at 175–179 °C and ending at 409–426 °C. The residues of the NWs at 650 °C are related to the non-volatile hard segments of the polymer chains and magnetite nanoparticles incorporated to their structures. These results confirmed that the PMEA-LA-MG3 and PMEA-LA-MG6 NWs are thermally stable.

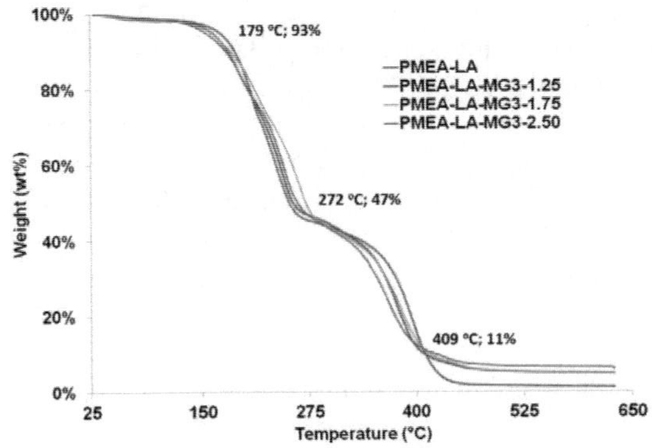

Figure 3. TG curves of PMEA-LA and MG3 incorporated NWs of PMEA-LA-MG3.

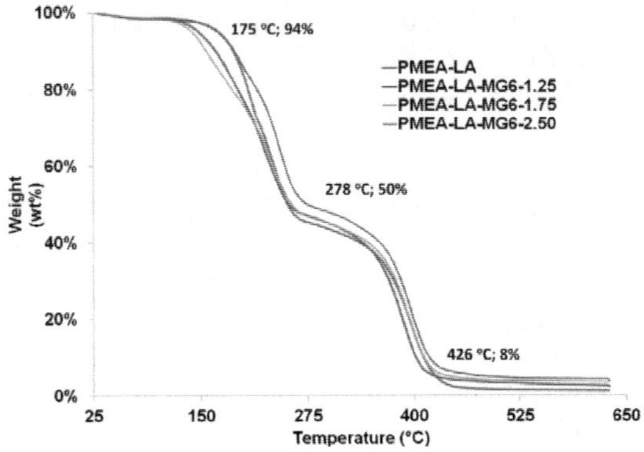

Figure 4. TG curves of PMEA-LA and MG6 incorporated NWs of PMEA-LA-MG6.

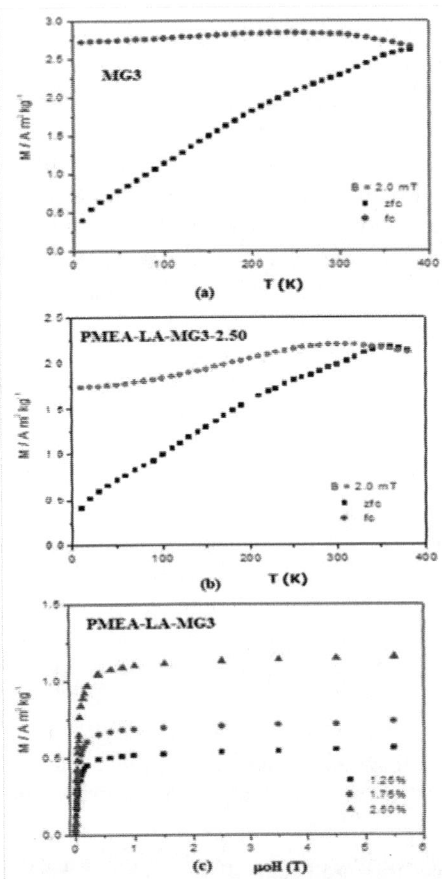

Figure 5. (a) Magnetization of oleic acid covered MG3 as a function of temperature; (b) Magnetization of PMEA-LA-MG3-2.50 as a function of temperature; (c) Magnetization of PMEA-LA-MG3 samples as a function of applied magnetic field.

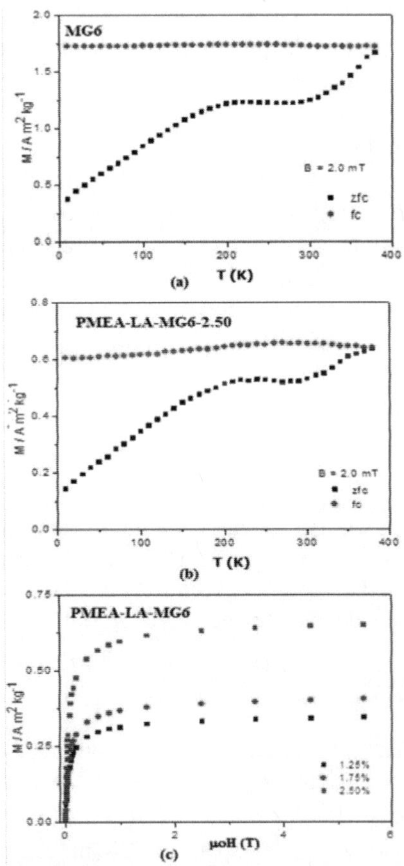

Figure 6. (a) Magnetization of non-functionalized MG6 as a function of temperature; (b) Magnetization of PMEA-LA-MG6-2.50 as a function of temperature; (c) Magnetization of PMEA-LA-MG6 samples as a function of applied magnetic field.

Figure 5 and Figure 6 display the magnetizations of MG3, MG6 and of PMEA-LA-MG3-2.50 and PMEA-LA-MG6-2.50 as a function of temperature and as a function of applied magnetic field. The magnetization results show that the nanoweb produced without magnetic nanoparticles, namely PMEA-LA, is diamagnetic with a magnetic susceptibility of -7.0×10^{-9} m^3kg^{-1} and that all the nanowebs became magnetic when fabricated with integrated MNP. The comparison of ZFC/FC curves for the MNPs and for the NWs with MNP shows that the production process does not affect the nanoparticles magnetic behavior. For two types of MNPs, namely MG3 and MG6, the ratio of obtained spontaneous magnetization is consistent with the ratio of the nominal mass concentration. For the oleic acid functionalized magnetic nanoparticles, MG3, the magnetization for the corresponding nanowebs is higher by a factor of two for all concentrations than the ones with non-functionalized magnetic nanoparticles MG6, indicating a more effective integration of the functionalized nanoparticles in the nanoweb.

4. Conclusion

We developed novel nanowebs provided with magnetic and thermal buffering properties by encapsulating oleic acid functionalized MNPs or non-functionalized MNPs and Lauric Acid in PMEA matrix via uniaxial electrospinning. There are numerous benefits of the NWs produced in this study. These are as follows: Lauric Acid, used as the PCM, and PMEA, used as the polymer matrix, are bio-based and biocompatible compounds that do not mainly pose human health hazards. The encapsulation of MNPs and LA in PMEA matrix prevents the leakage of MNPs and LA, which in turn minimizes their secondary effects on surrounding healthy tissues in prospective magnetic hyperthermia applications. They exhibit solid-liquid phase transitions between 41–48 °C as required in magnetic hyperthermia. The nanowebs have high enthalpies of fusion and crystallization, thermal cycling ability and thermal stability as well as the considerable magnetization capacity. Consequently, the as-spun nanowebs of MNPs, oleic acid covered MG3 or uncovered MG6, and LA in PMEA matrix, along with many other applications in biomedicine, are highly advantageous in their potential applications in the field of magnetic hyperthermia.

Acknowledgement

This work was carried out with the financial support of The Scientific & Technical Research Council of Turkey through project TUBITAK 213M281 and of Portuguese FCT Foundation through projects UID/MULTI/04046/2013 and UID/MULTI/00612/2013. The authors would like to acknowledge networking support by the COST Action TD1402 RADIOMAG.

References

[1] Jordan A, R Scholz, P Wust, H Fahling, J Krause, W Wlodarczyk, B Sander, T Vogl, R Felix. 1997 Effects of magnetic fluid hyperthermia MFH on C3H mammary carcinoma in vivo. *Int J Hyperthermia*, 136, pp 587−605

[2] Gupta A K, M Gupta. 2005 Synthesis and surface engineering of iron oxide nanoparticles for biomedical applications. *Biomaterials*, 2618, pp 3995−4021

[3] Mamiya H, B Jeyadevan. 2011 Optimal design of nanomagnets for targeted hyperthermia. *J Magn Magn Mater*, 32310, pp 1417−1422

[4] Ruiz A, G Salas, M Calero, Y Hernández, A Villanueva, F Herranz, S Veintemillas−Verdaguer, E Martínez, D F Barber, M P Morales. 2013 Short−chain PEG molecules strongly bound to magnetic nanoparticle for MRI long circulating agents. *Acta Biomater*, 9, pp 6421−6430

[5] Li Y S, J S Church, A L Woodhead, F Moussa. 2010 Preparation and characterization of silica coated iron oxide magnetic nano–particles. *Spectrochim Acta*, Part A 765, pp 484–489

[6] Chung T H, C C Hsieh, J K Hsiao, S C Hsu, M Yao, D M Huang. 2016 Dextran–coated iron oxide nanoparticles turn protumor mesenchymal stem cells MSCs into antitumor MSCs. *RSC Advances*, 6, pp 45553–45561

[7] Masur S, B Zingsem, T Marzi, R Meckenstock, M Farle. 2016 Characterization of the oleic acid/iron oxide nanoparticle interface by magnetic resonance. *J Magn Magn Mater*, 415, pp 8–12

[8] Liao Z Y, H J Wang, R C Lv, P Q Zhao, X Z Sun, S Wang, W Su, R Niu, and J Chang. 2011 Polymeric liposomes–coated superparamagnetic iron oxide nanoparticles as contrast agent for targeted magnetic resonance imaging of cancer cells. *Langmuir*, 27 6, pp 3100–3105

[9] Tassa C, S Y Shaw, R Weissleder. 2011 Dextran–coated iron oxide nanoparticles: a versatile platform for targeted molecular imaging, molecular diagnostics, and therapy. *Acc Chem Res*, 44 10, pp 842–852

[10] Yang X Q , Y H Chen, R X Yuan, G H Chen, E Blanco, J M Gao, *et al.* 2008 Folate–encoded and Fe_3O_4–loaded polymeric micelles for dual targeting of cancer cells. *Polymer*, 49 16, pp 3477–3485

[11] Chen D Y , X W Xia, H W Gu, Q F Xu, J F Ge, Y G Li, *et al.* 2011 pH–responsive polymeric carrier encapsulated magnetic nanoparticles for cancer targeted imaging and delivery. *J Mater Chem*, 21 34, pp 12682–12690

[12] Zhang J L , R S Srivastava, R D K Misra. 2007 Core–shell magnetite nanoparticles surface encapsulated with smart stimuli–responsive polymer: synthesis, characterization, and LCST of viable drug–targeting delivery system. *Langmuir*, 23 11, pp 6342–6351

[13] Shukoor M I, F Natalio, V Ksenofontov, M N Tahir, M Eberhardt, P Theato, *et al.* 2007 Double–stranded RNA polyinosinic–potycytidylic acid immobilized onto y–Fe_2O_3 nanoparticles by using a multifunctional polymeric linker. Small, 3 8, pp 1374–1378

[14] Boyer C, M R Whittaker, V Bulmus, J Q Liu, T P Davis. 2010 The design and utility of polymer–stabilized iron–oxide nanoparticles for nanomedicine applications. *NPG Asia Mater*, 2 1, pp 23–30

[15] Huang C B, S J Soenen, J Rejman, J Trekker, C X Liu, L Lagae, *et al.* 2012 Magnetic electrospun fibers for cancer therapy. *Adv Funct Mater*, 22 12, pp 2479–2486

[16] Chen I H, C C Wang, C Y Chen. 2010 Fabrication and characterization of magnetic cobalt ferrite/polyacrylonitrile and cobalt ferrite/carbon nanofibers by electrospinning. *Carbon*, 48 3, pp 604–611

[17] Song T, Y Z Zhang, T J Zhou. 2006 Fabrication of magnetic composite nanofibers of polyepsilon–caprolactone with FePt nanoparticles by coaxial electrospinning. *J Magn Magn Mater*, 303 2, pp 286–289

[18] Zhu J H, S Y Wei, D Rutman, N Haldolaarachchige, D R Young, Z H Guo. 2011 Magnetic polyacrylonitrile–Fe@FeO nanocomposite fibers–electrospinning, stabilization and carbonization. *Polymer*, 52 13,pp 2947–2955

[19] Mincheva R, O Stoilova, H Penchev, T Ruskov, I Spirov, N Manolova, et al. 2008 Synthesis of polymer–stabilized magnetic nanoparticles and fabrication of nanocomposite fibers thereof using electrospinning. *Eur Polym J*, 44 3,pp 615–627

[20] Lin T C, F H Lin, J C Lin. 2012 In vitro feasibility study of the use of a magnetic electrospun chitosan nanofiber composite for hyperthermia treatment of tumor cells. *Acta Biomater*, 8 7, pp 2704–2711

[21] Amarjargal A, L D Tijing, C H Park, I T Im, C S Kim. 2013 Controlled assembly of superparamagnetic iron oxide nanoparticles on electrospun PU nanofibrous membrane: a novel heat–generating substrate for magnetic hyperthermia application. *Eur Polym J*, 49, pp 3796–3805

[22] Onder E and Sarier N. 2015 Chapter 2: Thermal Regulation Finishes For Textiles, in Functional Finishes For Textiles: Improving Comfort, Performance And Protection, Edited By R Paul, *Woodhead Publishing Series in Textiles* No. 156, pp 17−98, ISBN 0 85709 839 X ISBN−13: 978 0 85709 839 9 print

[23] Pradhan P, J Giri, G Samanta, H D Sarma, K P Mishra, J Bellare, R Banerjee, and D Bahadur. 2007 Comparative evaluation of heating ability and biocompatibility of different ferrite-based magnetic fluids for hyperthermia application. *J Biomed Mater Res*, 81B, pp 12−22

[24] Zhao D L, P Teng, Y Xu, Q S Xia, J T Tang. 2010 Magnetic and inductive heating properties of Fe3O4/polyethylene glycol composite nanoparticles with core–shell structure. *J Alloy Comp*, 5022, pp 392−395

New Possibilities of Raw Cotton Pre-treatment before reactive dyeing

A G Demir[1], F R Oliveira[2], T Gulumser[1] and A P Souto[3]

[1]*Department of Textile Engineering, Ege University, İzmir, Turkey*
[2]*Department of Textile Engineering Universidade Federal de Santa Catarina, Blumenau, Brasil*
[3]*2C2T- Centro de Ciência e Tecnologia Têxtil, Minho University, Guimarães, Portugal*

Corresponding author: souto@det.uminho.pt

Abstract Cotton fiber is one of the most used materials among other natural fibers in the textile and clothing industry. On the other hand, environmental pollutants are formed during the processing of cotton industry especially in conventional wet processes such as bleaching, scouring for raw cotton. In order to make alteration to those processes enzymes can be also used in some of them such as: bleaching and scouring. Due to the fact that conventional textile finishing processes with intensive energy, water and chemical consumption are not environmentally friendly; it is necessary to develop more "green" and economic production methods. Plasma technology and enzymes are promising techniques to support eco-friendly production in textile industry. These eco-friendly methods can also be combined to produce high quality materials and to reduce production cost. The aim of this study is to investigate the effect of plasma treatment combined with enzymatic bioscouring system over raw cotton fabric.
Several techniques of characterization were used to study the effects caused by the interaction between plasma discharge/enzymes and cotton fabric, such as: static and dynamic contact angle, Scanning Electron Microscopy (SEM), Fourier Transform Infrared Spectrometer (FTIR) and reflectance spectrophotometer. All analyses performed in this study showed that DBD plasma discharge combined with enzymes, when applied on cotton fiber, produces significant modifications on the surface of this substrate, improving hydrophilicity and whiteness without altering its intrinsic properties. Thus, proving the effectiveness of the synergy of two important technologies to textile industry.

***Keywords*—Plasma, enzyme, reactive dye, raw cotton, uniformity**

I. INTRODUCTION

In the couple last decades, it has become important to save inputs in a process to create a product. Textile industry is one of the locomotive sectors for many countries, which is using certain amounts of water, energy, chemical and raw materials. In this case it is important to use ecofriendly and energy saving methods for production [1], [2].

In finishing processes of textile materials, there are several modification methods to make some changes on fibers. It is possible to make these modifications in three ways; chemical, biochemical and physicochemical methods [3], [4]. Conventional wet finishing processes are in the group of chemical modifications. In chemical modifications, hydrogen bridges or covalent bonds can be formed between certain groups of fibers. With chemical process it is possible to improve or modify the wettability, adhesion and staining properties of the material's surface [5-7]. However, during the chemical alteration processes, the need for chemical amount and the requirement of the energy is much higher than in the other modifications [8].

Physicochemical methods include applications with ionized gases such as plasma, corona discharge, heat treatment, UV and ozone.

Plasma is defined as a partially ionized gas composed of electrons (0-10 eV), ions (10-30 eV), photons, atoms, molecules and is named as the fourth state of matter [9-11]. Free electrons, photons and ion clouds begin to be formed, and some atoms continue to remain neutral and the mixture of atoms, ions and electrons form the plasma [12-14]. Figure 1 illustrates the main plasma constituents.

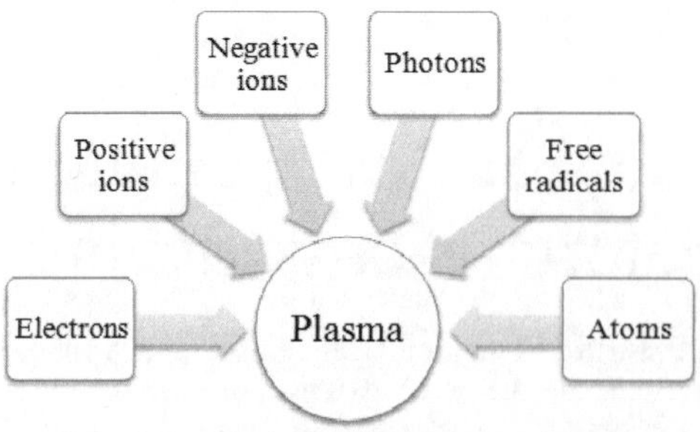

Fig1. Plasma constituents

Plasma technology is an effective modification method, to improve the surface properties of materials by taking advantage of the electrons, ions and radical components formed as a result of electrical discharges [13], [15], [16].

Plasma treatment of textile materials has become one of the high growing technology topics and appealing method due to its advantages [11]. Plasma creates effective modifications on the material surface by minimizing (or no use) the load of waste, usage of water and chemical to compare with conventional wet finishing methods [17], [18].

In this method, surface modification and formation of some functional groups can be achieved. Unlike other methods, it modifies the surface of the substrate only a few micrometers thick without any change in the bulk properties of the material [19], [20].

Dielectric barrier discharge (DBD) plasma applied in atmospheric conditions is a widely used process in order to modify the surface properties of different textile polymers [21]. DBD plasma has low energy consumption, low cost, quiet operating conditions and ease of use can be pointed as advantages. The system of DBD plasma has two electrodes (at least one of them coated with insulating dielectric material), which are placed parallel to each other [22]. In many studies DBD plasma technology when applied to textile processing has revealed to be an efficient technique to modify the surface properties of different types of fibers, such as cotton [23], [3], [24], polyamide 6,6 [25], PA/PES [26], wool/cashmere [27], PA [18], [28], PET/CO [29], wool [30], linen [31], silk [32], aramid [33], PET [34].

Biochemical modification is one of the improved methods for eliminating the disadvantages of chemical modifications. Studies have been carried out on fiber modifications using a number of microorganisms for instance fungi, bacteria and enzymes [8].

Biotechnological studies have led to use enzymes in textile industry. Enzymes are high molecular proteins with catalytic activity. They are working on specific substances that they have interest in. Processes with enzymes have advantages due to their safer and environmentally friendly working conditions. Moreover, enzymes did not produce effluents at the end of the processes.

Some of conventional wet processes such as bleaching, scouring for raw cotton, involve the use of chemicals that may cause pollution problems. In order to make alterations to those processes enzymes can be used in some of them like bioscouring. Bioscouring provides enzymatic removal of the natural hydrophobic substances present in the fibre/fabric, which is an ecofriendly method of textile production [35], [36]. Silva et al., studied on enzymatic pool in bioscouring with cellulase, lipase and pectinase of cotton fabric. They reported in this work, bioscouring with three enzyme combinations has led to best results in terms of degree of whiteness, pectin removal and hydrophilicity [40]. There are many studies on the use of several types of enzymes for cleaner production, such as laccase [37], [38], lipase and pectinase [39] applied on cotton fibers.

Because it is an environmentally friendly and gentle process, plasma discharge treatment is being studied in several applications together with enzymes.

Zhang and Wang applied DBD plasma on wool fabrics and enzymatic treatment with protease and transglutaminase to modify wool fabrics cleaner and eco-friendly [36]. Radetic et al. applied a RF plasma discharge on flax followed by a treatment with cellulose enzyme. After the treatment, the fabrics were dyed at low temperature with direct and acid dyes. The results of the combined plasma/enzyme treatment showed a decrease of the dye bath exhaustion attributed to plasma-facilitate enzymatic degradation of the amorphous areas of the hemp fibres [41]. Wong et al. obtained similar results after treating flax fibres in low-temperature plasma using oxygen and argon gases [42]. Pectinase enzyme treatment subsequent to atmospheric air or argon plasma surface modification was applied in flax fabric by Karaca et al. Results clearly proved that the efficiency of pectinase in improving flax water absorbency can be significantly enhanced by plasma pre-treatment [43].

Cotton fiber includes cellulosic and non-cellulosic substances. These non-cellulosic components such as waxes, pectin, lignin, protein and others give hydrophobic characteristic of cotton nature. They need to be removed from the fiber before dyeing and finishing processes. Table I. shows the chemical composition of the raw cotton fibre.

TABLE I
THE CHEMICAL COMPOSITION OF RAW COTTON

Percentages in the fiber composition	Substance
% 88-96	Cellulose
% 1-1.9	Protein
% 0.4-1.2	Waxes
% 0.4-1.2	Pectin
% 0.7-1.6	Others

Conventional methods consume huge amounts of chemicals, water and energy. Moreover, they cause environmental problems due to their heavy effluents. In this case it is important to use environmentally friendly methods in every step of textile finishing processes as much as we can. Thus, the aim of this study is to investigate the effect of DBD plasma treatment combined with enzymes bioscouring system before reactive dyeing of cotton fabric.

II. MATERIAL AND METHODS

A. *Materials*

A commercial 100% raw cotton fabric with a taffeta structure, a mass per unit area of 162 g m^{-2} and a texture of 19 threads/cm in warp and 14 threads/cm in weft direction was used in this study.

Bioprep 3000L containing pectinase and lipase were kindly supplied by Aquitex. Procion® (Dystar) reactive dyes were used for dyeing processes.

Conventional bleaching was made with sodium hydroxide and hydrogen peroxide and they were of analytical grade.

B. *Plasma treatment*

Plasma treatment was carried out with a semi-industrial prototype (Softal/University of Minho). The working conditions were: velocity 4 m/min, power 1 kW and 5 passages per each side. Working plan of the experiment is presented in Fig. 2.

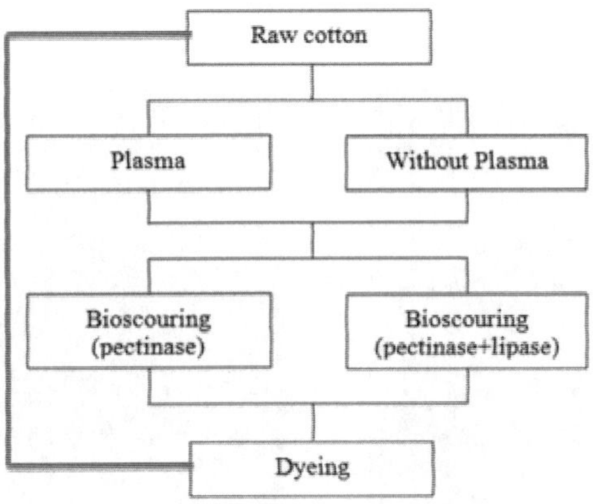

Fig. 2 Scheme of work plan used in this study.

C. *Scouring of cotton fabrics*

After plasma treatment bioscouring process was performed with pectinase or lipase+pectinase combination. Scouring with pectinase process conditions were pH 6-7, temperature 50°C, time interval between 30-60 min, bath ratio of 1:20 with appropriate amount of enzyme.

Scouring with lipase was carried out with pH 5, temperature 50°C, time 60 min, bath ratio of 1:30 and appropriate amount of lipase enzyme was used. The enzymes inactivation process was performed with hot distilled water washing.

Conventional bleaching was performed using sodium hydroxide and hydrogen peroxide with a standard recipe

IOP Conf. Series: Materials Science and Engineering **460** (2018) 012026 doi:10.1088/1757-899X/460/1/012026

at boiling temperature.

D. Dyeing of cotton fabrics

The dyeing method was carried out with 5 g of cotton fabric in a laboratorial Ahiba Spectradye-Datacolor machine equipped with infra-red heating with a liquor ratio of 1:20 with dye concentration of 1% (o.w.f) Procion®.

For dyeing process, the program was started at 25°C and raised at a rate of 2°C min^{-1} up to 80°C, and kept at this temperature for 60 minutes. The chemicals and auxiliary agents were used according to the producers' recommendations. After dyeing washing steps were applied.

E. Scanning electron microscopy (SEM)analysis

Morphological analysis of cotton fabrics was carried out in an ultra-high resolution field emission gun SEM instrument (NOVA 200 Nano SEM, FEI Co. Oregon, USA). Previously, the samples were coated with a thin layer of gold palladium.

F. Contact angle measurement

To determine hydrophilicity of the samples before and after the treatments processes, contact angle measurement was carried out in Dataphysics instrument (Filderstadt, Germany) using OCA20 software with a video system, for the capturing of images in static and dynamic modes.

G. Fourier-transform infrared spectroscopy (FTIR) measurements

A Shimadzu Iraffinity-1S spectrometer (Japan) was used to determine changes in the main characteristic groups of raw cotton surface. The spectra were collected over the range 400-4000 cm^{-1} at room temperature.

H. Whiteness and color strength

The whiteness degree, colorimetric coordinates and color strength of the bleached and dyed fabrics were measured using Datacolor Spectraflash SF600 Plus CT spectrophotometer at standard illuminant D65 and observer 10° combination.

III. Results and Discussions

The effects of the plasma and enzymes bioscouring treatments in the wettability, whiteness degree and dyeing properties of raw cotton fabrics were evaluated with different characterization techniques.

A. Surface evaluation

Plasma and enzyme treated fiber surface become cleaner with a smooth effect. The reason of the smoothness can be attributed to the etching effect of plasma treatment. Fig. 3 shows the cotton surface morphology by SEM images under different treatment conditions.

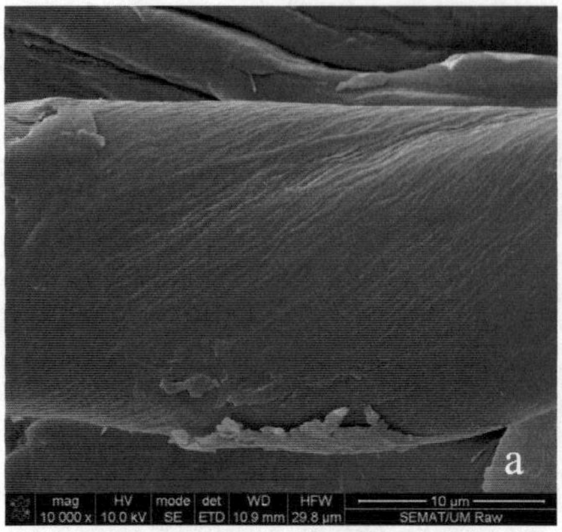

Fig. 3a SEM images of raw cotton

Fig. 3b SEM images of conventionally bleached cotton

Fig. 3c SEM images of raw cotton + plasma

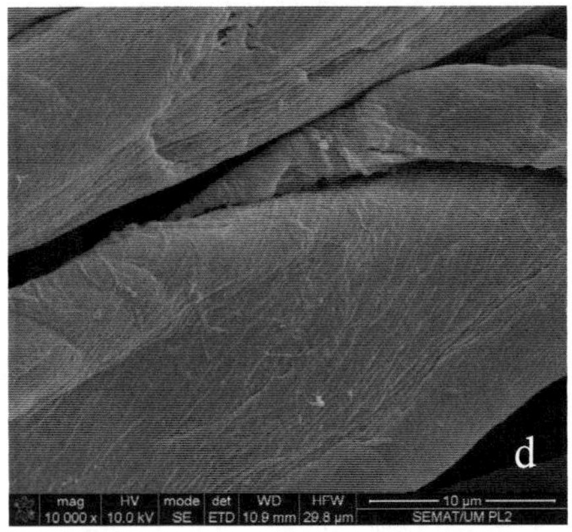

Fig. 3d SEM images of raw cotton + plasma + pectinase + lipase

SEM images of cotton fabrics indicate that with plasma and enzyme treatments there are considerable surface changes over fibers. Conventional bleaching of raw cotton uses harsh chemicals and high temperature that cause the damages of the cotton surface (Fig. 3b). Fig. 3c shows the images of cotton fiber treated with plasma.
The surface of the cotton fabric is smoother with the combined treatment of plasma application and enzyme (Fig. 3d). With the use of plasma and enzymes for treatment of raw cotton, conditions are much milder and do not create any damages of the surface on cotton fabrics.

B. Hydrophilicity

Hydrophobic nature of raw cotton fabric is not able to absorb water easily due to non-cellulosic substances. Due to these compounds, raw cotton fabric has a highly hydrophobic character. Plasma and enzymatic treatment create changes on the fiber surface improving the hydrophilicity as can be observed in the Table 2.

TABLE 2
CONTACT ANGLE RESULTS

Fabric	Result (°)
Raw cotton	161,3
Raw cotton + Plasma	107,4
Raw cotton + Pectinase + Lipase	137,5
Raw cotton + Plasma + Pectinase + Lipase	103,6

The results obtained by dynamic contact angle measurement show higher hydrophilicity with plasma and enzyme combination treated fabrics.

Figure 4 shows the wetting behavior of the treated and untreated cotton fabrics. Especially, after the application with plasma and enzyme combinations wetting properties of raw cotton were significantly improved, as we can observe in Figure 4. The hydrophilicity of the cotton fabrics was measure in terms of water contact angle. When a water drop was placed onto the fabric surface (with plasma and enzymatic treatment combination), it is instantly absorbed. Thus, fabrics treated with plasma and then followed by enzymatic incubations have significantly improved the wetting time, absorption rates and spreading speed results.

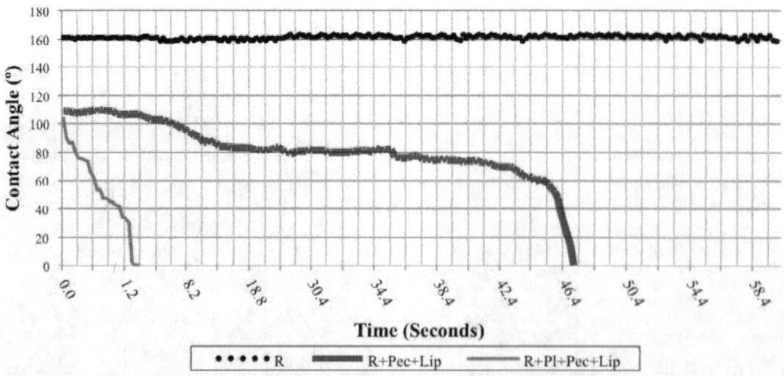

Fig. 4 Dynamic contact angle measurements of Raw cotton (R), Raw cotton + Pec + Lip (R+Pec+Lip), Raw cotton+ Plasma + Pectinase + Lipase (R+Pl+Pec+Lip).

C. Fourier-transform infrared spectroscopy (FTIR) measurements

Surface chemistry of cotton fabrics was analyzed with FTIR. The spectra obtained gave similar results among the samples treated and untreated because the bulk compositions of the fabrics are similar (Fig. 5). However, the results confer that the pectin and wax impurities were removed from the cotton surface in both the raw pectinase-lipase and plasma pectinase lipase scoured fabrics. The characteristic peak around 1730 cm^{-1} which represents the impurities such as waxes and pectin, were present in the raw cotton but almost disappeared in the bioscoured samples. The purpose of the scouring process was to remove hydrophobic substances (wax components, long chain alcohols, fatty acids) in the fiber to produce high wettability. It proved that the enzymes had successfully removed all the pectin and wax impurities from cotton thus increasing hydrophilicity as previously observed.

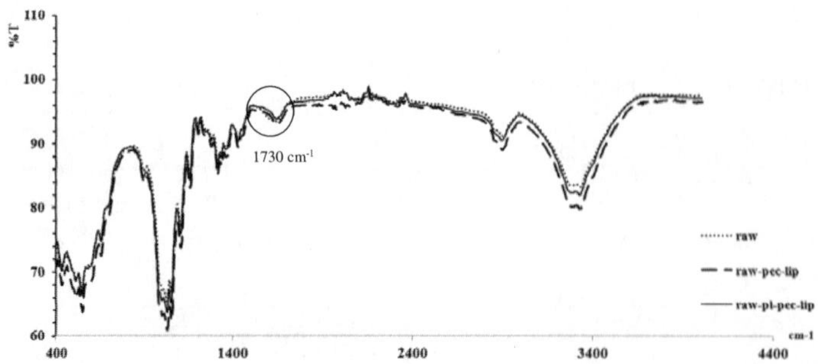

Fig. 5 FTIR measurements of samples

D. Color strength and whiteness degree

Conventional wet treatments of cotton fabrics aim to remove the hydrophobic substances such as natural non-cellulosic particles and starch, from the fiber. These impurities cause serious dyeing problems. For this reason, it is important that these impurities should be removed before the dyeing process. Application with plasma, wetting properties of raw cotton were significantly improved especially treatments when combined with enzymes. Before the dyeing processes the color measurements of the samples were measured (Table 3). The results showed higher whiteness degree and lower yellow (coordinate b) for the samples treated with enzyme and plasma when compared with raw cotton fabric. The combined application of enzymes + plasma and just enzyme treatment did not show any significant difference.

TABLE 3
COLOR MEASUREMENTS BEFORE DYEING PROCESSES

Samples	L*	a*	b*	C*	h	Whiteness Deg. (Berger)
R	84,42	1,99	13,14	13,29	81,38	10,9
R+Pec+Lip	84,69	1,82	11,48	11,62	80,98	17,4
R+Pl+Pec+Lip	84,63	1,97	11,43	11,60	80,20	17,3

R: Raw cotton, R+Pec+Lip: Raw cotton + Pectinase + Lipase, R+Pl+Pec+Lip: Raw cotton + Plasma + Pectinase + Lipase

Table 4 shows the color coordinates and color strength of the dyed cotton samples treated with plasma + enzyme and untreated raw cotton fabrics (Table 4).

TABLE 4
COLOR MEASUREMENTS AFTER DYEING PROCESSES

	Procion	L*	a*	b*	C*	h	K/S
BLUE	R+ Pec +Lip	46,11	-7,56	-16,54	18,18	245,44	64,70
BLUE	R+Pl+Pec+Lip	45,47	-7,44	-16,74	18,32	246,03	71,56
RED	R+Pec+Lip	49,30	46,98	-7,39	47,56	351,06	62,08
RED	R+Pl+Pec+Lip	49,66	47,97	-7,56	48,57	351,05	64,37
YELLOW	R+Pec+Lip	69,11	17,37	46,51	49,64	69,52	54,41
YELLOW	R+PL+Pec+Lip	67,66	17,36	45,07	48,30	68,93	54,28

The results of the combined plasma/enzyme treatments showed similar dye bath exhaustion and the equilibrium uptake of a reactive dye remained.

IV. CONCLUSION

The aim of this study was to investigate a green and alternative process instead of wet conventional treatments of cotton fabrics. As implied before, conventional wet treatments cause environmental problems. In order to improve treatment methods of the cotton fabrics, this study applied atmospheric plasma and enzymatic treatments combinations on the raw cotton fabrics. Application of the pectinase and lipase removed non-cellulosic impurities in order to create scouring effect on the cotton fiber.

Results clearly proved that the efficiency of pectinase in improving cotton water absorbency can be significantly enhanced by plasma pre-treatment.

No significant difference was observed in the dyeing properties of the samples with and without treatment.

Operating under controlled conditions, plasma/enzymatic treatments of textiles are particularly intriguing for research on different types of fibers. However, enzymatic treatment and plasma parameters must be strictly controlled in order to avoid unacceptable weight loss and reduction of strength. Despite that, the synergetic combination of plasma and enzyme technologies reveals without any doubt a great potential for textile applications.

The overall results showed that eco-friendly pre-treatment methods are acceptable when compared with conventional ones.

ACKNOWLEDGMENT

This work is supported by FEDER funds through Operational Programme for Competitiveness Factors – COMPETE and National Funds through FCT – Foundation for Science and Technology within the scope of the projects POCI-01-0145-FEDER-007136 and UID/CTM/00264.

REFERENCES

[1] V. Nehra, et al., "Atmospheric non-thermal plasma sources," *Int. J. Eng.*, vol.2, no.1, pp. 53-68, 2008.
[2] N. Carneiro, et al., "Preparation of cotton materials using corona discharge," *Journal of Natural Fibers*, vol. 2, no. 4, pp. 53-65, 2006.
[3] H. A. Karahan et al., "Effects of atmospheric plasma treatment on the dyeability of cotton fabrics by acid dyes", *Color. Technol.*, vol.124, pp. 106-110, 2008.
[4] M. J. John, R. D. Anandjiwala, "Surface modification and preparation techniques for textile materials," in *Surface modification of textiles*, no. 97, Q. Wei, Ed. Woodhead Publishing Lim., 2009, pp. 1-25.
[5] J. George, et al, "A review on interface modification and characterization of natural fiber reinforced plastic composites", *Polymer Engineering &Science*, vol. 41, no. 9, pp. 1471-1485, 2001.
[6] K. Mojsov, "Application of enzymes in the textile industry: A rewiev," in *II. International Congress "Engineering Ecology and Materials in the Processing Industry"*, Jahorina, 2011, pp. 230-239.
[7] A. Patino, et al., "Surface and bulk cotton fibre modifications: plasma and cationization. Influence on dyeing with reactive dye," *Cellulose*, vol. 18, pp. 1073-1083, 2011
[8] N. Lu, et al., "Surface modification of natural fibers for reinforcement in polymeric composites," *Surface Modification of Biopolymers*, 1st ed., vol. 1, 2015, pp. 224-237.
[9] W. B. Thompson, "An introduction to plasma physics," Addison-Wesley Series in Advanced Physics, Addison-Wesley Publishing Inc. 1962.

[10] N. Vandencateele, F. Reiners, "Plasma-modified polymer surfaces: characterization using XPS," *Journal of Electron Spectroscopy and Related Phenomena*, pp. 394-408, 2010.
[11] A. Zille, F. R., Oliveira, A. P. Souto," Plasma treatment in textile Industry," *Plasma Processes and Polymers,* vol. 12, no.2, pp. 98-13, 2015.
[12] W. G. Graham, "The physics and chemistry of plasmas for processing textiles and other materials," in *Plasma technologies for textiles*, R. Shishoo, Ed. Woodhead Publishing Lim., 2007, pp. 3-24.
[13] S. Korpayev, "Modification of cotton nonwoven fabric surface with graft polymerization by using plasma technique and As(V) adsorption studies," Ankara, MSc Thesis, 104 p., 2016.
[14] S. Mukhopadhyay, R. Fangueiro, "Physical modification of natural fibers and thermoplastic films for composites, a review", *Journal of Thermoplastic Composite Materials*, vol. 22, pp. 135-162, 2009.
[15] C. E. Cornelius, "Novel applications of atmospheric pressure plasma on textile materials," Raleigh, North Carolina, Phd. Thesis, 380 p., 2009.
[16] C. Hao, et al, "Saturated resin ectopic regeneration by non-thermal dielectric barrier discharge plasma," *Catalysts*, vol. 7, pp. 1-15, 2017.
[17] N. Seventekin, 2008; "Atmosferik Plazma Tekniğinin Sentetik Tekstil Materyallerine Uygulanabilirliğinin Araştırılması", 104 p., 2008.
[18] A. P. Souto et al., "Influence of DBD plasma modification in the dyeing process of polyamide", *Journal of Textiles and Engineer*, vol. 19 (85), pp 20-25, 2012.
[19] T. A. Otitoju, et al., "Superhydrophilic (superwetting) surfaces: A review on fabrication and application," *Journal of Industrial and Engineering Chemistry*, vol. 47, pp. 19-40, 2017.
[20] H. U. Poll et al., "Penetration of plasma effects into textile structures," *Surface and Coating Technology*, pp. 489-493, 2001.
[21] M. Nerurkar, M. Joshi, R. Adivarekar, "Bioscouring of cotton using lipase from marine bacteria Bacillus sonorensis," *Applied Biochemistry and Biotechnology*, vol 175, pp. 253-265, 2015.
[22] U. Kogelschatz, et al., "Non equilibrium air plasma at atmospheric pressure," IOP Publishing, 2004.
[23] R. Molina et al., "Hydrophobic coatings on cotton obtained by in situ plasma polymerization of a fluorinated monomer in ethanol solutions", *ACS Appl. Mater. Interfaces*, vol. 9, pp. 5513-5521, 2017.
[24] M. Zhang, "Antimicrobial cotton textiles with robust superhydrophobicity via plasma for oily water separation", *Applied Surface Science*, vol. 419, pp. 16-23, 2017.
[25] F. R. Oliveira, A. Zille, F. R., A. P. Souto, "Dyeing mechanism and optimization of polyamide 6,6 functionalized with double barrier discharge (DBD) plasma in air", *Applied Surface Science*, vol. 293, pp. 177-186, 2014.
[26] E. Mattsson, "Plasma treated water and oil repellent textiles", Sweden, Master thesis, 2013.
[27] S. Zanini, "Characterization of atmospheric pressure plasma treated wool/cashmere textiles: Treatment in nitrogen", *Applied Surface Science*, vol. 427, pp. 90-96, 2018.
[28] N. Yaman et al., "Improving physical properties of polyamide fibers by using atmospheric plasma treatments", *Tekstil ve Konfeksiyon*, vol.2, pp. 102-105, 2012.
[29] C. Cornelius et al., "Desizing of PVA sized pet/cotton fabrics with atmospheric pressure plasma", *Cellulose*, vol. 25, pp. 869-881, 2018.
[30] G. Sahan, "A research about the utilization of chitosan biopolymer in various forms with wool fibers", Izmir, MSc Thesis, 121 p., 2014.
[31] C.S. Ren et al., "Graft co-polymerization of acrylic acid onto the linen surface induced by DBD in air", *Surface & Coatings Technology*, vol. 201, pp. 2867-2870, 2006.
[32] K. V. Rani, "Optimization and surface modification of silk fabric using DBD air plasma for improving wicking properties", *The Journal of the Textile Institute*, vol. 109, no. 3, pp. 368-375, 2018.
[33] A. Demir et al., "An ecological approach for the surface modification of aramid fibers", *Tekstil ve Konfeksiyon*, vol.26, no. 3, pp. 256-261, 2016.
[34] S. Liu et al., "Research on the effect of dielectric barrier discharge (DBD) plasma remote treatment on drag force of polyethylene terephthalate (PET) yarns in air flow with different humidities", *Plasma Chem. Plasma Process*, vol. 37 pp. 1573-1586, 2017.
[35] S. Kalantzi et al., 2010, Fibers&Textiles in Eastern Europe 2010, 18, No.5 (82), 86-92.
[36] R. Zhang, A. Wang, "Modification of wool by air plasma and enzymes as a cleaner and environmentally friendly process", *Journal of Cleaner Production*, vol. 87, pp. 961-965, 2015.
[37] S. Kim et al., "Laccase-mediated grafting of polyphenols onto cationized cotton fibers to impart UV protection and antioxidant activities", *J. APPL. POLYM. SCI.*, vol. 135, 2018.
[38] H. A. Eren et al., "Pamuğun Ağartılmasında lakkaz, ozon ve hidrojen peroksitin combine kullanımı", *Tekstil ve Konfeksiyon*, vol.4, pp. 299-303, 2009.
[39] S. Kalantzi et al., "Improved properties of cotton fabrics treated with lipase and its combination with pectinase", *Fibres & Textiles*, vol. 18, No. 5 (82), pp. 86-92, 2010.
[40] L. Silva et al., "Study and application of an enzymatic pool in bioscouring of cotton knit fabric", *The Can. J. Chem. Eng.*, vol. 95, pp. 1253-1260, 2017.
[41] M. Radetic et al., "The influence of low-temperature plasma and enzymatic treatment on hemp fabric dyeability," *Fibres & Textiles*, vol. 15, no. 4, pp.93-96, 2007.
[42] K. K. Wong et al., "Effect of plasma and subsequent enzymatic treatment of linen fabrics," *Journal of the Society of Dyers and Colourists*, vol. 116, pp. 208-214, 2000.
[43] B. Karaca et al., "Effects of atmospheric plasma pretreatments on pectinase efficiency in bioscouring of linen fabrics," *Plasma Chem. Plasma Process*, vol. 31, pp. 623-633, 2011.

IOP Conf. Series: Materials Science and Engineering **460** (2018) 012027 doi:10.1088/1757-899X/460/1/012027

Development of Carbon Nanofiber Yarns by Electrospinning

A DEMIR[1], B ACIKABAK[2], Z AHAN[3]

[1] *Istanbul Technical University, Textile Engineering Department, Istanbul, TURKEY*
[2] *INOVENSO Ltd. Istanbul, Istanbul, TURKEY*
[3] *Istanbul Technical University, Nanoscience and Nanoengineering Department, Istanbul, TURKEY*

E-mail: ademir@itu.edu.tr

Abstract. In recent years the commercial importance of carbon nanofiber has increased due to its high mechanical properties, electrical and thermal conductivity and lightness. Therefore, carbon nanofibers are candidate to replace many commercial products. In general carbon nanofibers are produced from precursor fiber. Polyacrylonitrile (PAN) is usually used as precursor fiber because of comparatively high carbon yield and a thermally stable and highly oriented molecular structure. Large number of scientific publications have been demonstrated with the help of progress in nanofiber production methods, such electrospinning, centrifugal-spinning, air blowing from melt or solution, and optimization of process parameters of those methods. In this study, PAN nanoyarn is first obtained by the electrospnning method which is very convenient process for producing thin fiber. Stabilization is then applied by exposure to air at 230°C to provide thermally stable structure so that the resulting PAN nanofibers could be thermally durable during the carbonization process at high temperature. Finally, the process for the carbonization of the stabilized sample is operated in a argon gas atmosphere at 600°C. The produced material is characterized by SEM to observe its morphology.

1. Introduction

Production of nanofibers from polymer materials has been a very important part of the nanotechnology, which is the most important area of scientific and technological activities in the last 10 years. Large number of scientific publications has been published with the help of progress in nanofibre production methods, such as electrospinning, centrifugal-spinning, air blowing from melt or solution, and optimization of process parameters of those methods.

Nanofibers use of many areas such as composite applications, filtration membranes, biomedical materials and tissue engineering because of high area and good strength properties provided by unit weight. Despite all this, in the electrospinning method, randomly oriented nanofibers are produced which limit the application area of the final product. Thus many researchers have recently started to work on nanofiber yarn production. Because the nanofiber yarns have higher mechanical strength than the randomly positioned nanofiber webs, in order to produce nanofiber yarns, researchers have improved many novel electrospinning techniques. One of these is the double electrode disc twisting device which was developed by Ali et al. [1]. In this method, an earthed funnel collector, two opposite charged nozzles and a take up roller were used and the nanofibers formed from the opposite charged nozzles became nanofiber yarn in the middle. In recent studies, it seems that double needle disc twisting yarn device is commonly used for preparing twisted nanofiber yarns. This device was modified using multiple-needle by He et al. [2]. In this work, there was just a difference that was the number of the needles. He et al., used two opposite charged infusion tubes which have four needles in each tube. It resulted in increased nanofiber production therefore more nanofiber yarn formation. On the other hand, in recent years, activated carbon fibers have been widely used in textiles, air filters, supercapacitors, wound dressings and insulation materials due to their high porosity, high adsorption ability, excellent electrochemical properties and flexibility, therefore carbon nanofiber yarns have become the focus of researchers. Carbon nanofiber are obtained by carbonization after the stabilization processes. In generally, carbonization is carried out in argon or nitrogen atmosphere at high temperature and stabilization takes place in air atmosphere at high temperature. Moon and Farris studied on the electrospun PAN carbon fiber yarns [3], they formed good aligned nanofiber yarn via electrospinning method and the produced twisted PAN nanofiber yarn is stabilized in air at a specific high temperature and then carbonized at high temperature in a nitrogen atmosphere. The mechanical strength of carbonized and untreated PAN nanofiber yarns was evaluated and as expected the carbon nanofiber yarn had high mechanical strength because of the conversion to a stable and strong cross-linked structure from a ladder structure. Another study on this subject was made by Chawla et al. [4]. They examined the conductivity of nanofiber yarns which was affected the level of twisting. In this work, the PAN nanofiber yarns obtained via electrospinning technique and they were then stabilized at 250°C in air in an oven for 1 hour. After stabilization, the nanofiber yarns were carbonized in a tube furnace at 1100°C for 1 hour in a nitrogen atmosphere. According to the results of this study, the level of twisting enhanced the molecular orientation and increased the alignment of nanofiber yarns. However the further increment of this level gave rise to fiber breakage and conductivity drop. It is possible to increase the properties of the fiber by carbonization, for example the conductivity of the material. Works on the conductivity properties of carbon nanofibers have increased with reasons such as carbonization of the material becoming similar to the graphitic structure, reduction of the diameter and increase of the strength. The carbonization temperature affects the properties of the ultimate product. Panapoy M. et al. [5] showed that increasing carbonization temperature increases the conductivity of the material. By carbonization, the nitrogen groups disappear so that the structure of the fiber is similar to the graphite structure with high carbon content. The increase in the graphite structure due to the high temperature causes the resonance vibration effect of the π bond in the graphite structure and it leads to the electrons in the carbon layer to move. As a result, the electrical conductivity increases.

Thanks to the superior properties of the carbon nanofiber yarns in each application area, the study for the use of carbon nanofiber yarns has increased day by day. In this context, in this work we aimed to produce the PAN nanofiber yarn via electrospinning method and to improve the properties of PAN nanofiber yarn by carbonization in argon atmosphere at high temperature.

2. Experimental

2.1. Materials

The polymer used was polyacrylonitrile (PAN) and the solvent used was N,N-dimethylformamide (DMF). The %8 electrospinning solution was prepared by dissolving the required amounts of polymers in the solvents at the required temperature.

2.2. Electrospinning of nanofiber yarn

Nanofiber yarn electrospinning system consists of three needle nozzles placed on a circular path 120° apart from each other, a purpose-designed cylindrical collector, a high-voltage power supply and yarn winding system (Figure 1 represents a schematic view of the system). The collector was encircled by three spinnerets, and winding system was placed underneath the rotating collector.

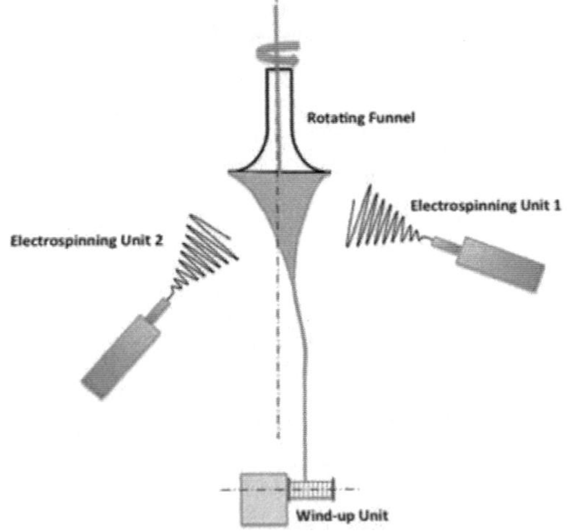

Figure 1. Schematic view of nanofiber yarn production system

All three spinnerets were connected to 20 kV positive high power supply. Electrospun nanofibers were deposited onto the rotating cylinder collector and formed a nanofiber web that covers the end of the collector. Nanofiber yarn was drawn by the help of a conventional guide yarn which was loosely attached to the collector.

2.3. Stabilization and carbonization of nanofiber yarn

The produced PAN nanofiber yarn was wrapped on a glass rod (Figure 2) then it was stabilized at 230°C under air with a heating rate of 1°C/min before carbonization to provide thermal stability. The stabilized nanofiber yarn (SNFY) was then carbonized. The sample was carbonized by three stage heating in argon gas to avoid degradation of the sample with rapid temperature rise. In the first step, the sample was heated to 50°C (heating rate: 1°C/min), after that the temperature reached to 300°C at a heating rate of 0.5°C/min. In the last step the temperature came up to 600°C with a heating rate of 3°C/min and it was left at this temperature for 30 min.

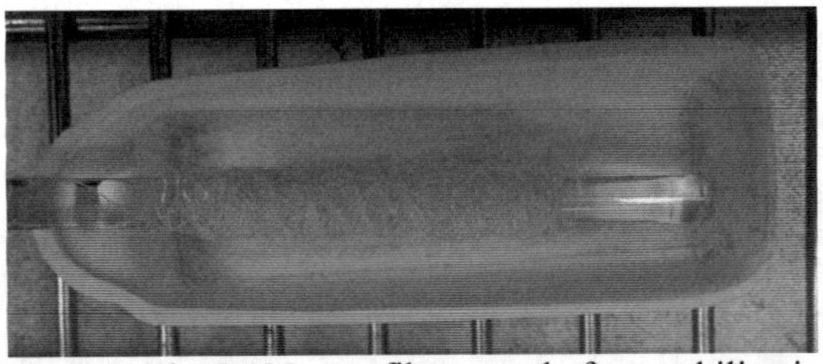

Figure 2. The PAN nanofiber yarn before stabilization

2.4. Characterization

The PAN nanofiber yarn, SNFY and carbonized nanofiber yarn (CNFY) were examined by SEM analysis. The morphology and diameter of these samples were observed from the SEM images.

3. Results

3.1. Nanofiber yarn

Nanofiber yarn was electrospun continuously by using the Inovenso NanoYarn machine as shown in the Figure 3, from 8% PAN solution. Optimized results were obtained by 1,5 ml/h feeding rate, 20 kV voltage and 350 rpm rotation speed of the collector. Nanofibers were electrospun from positively charged three spinnerets and deposited onto the grounded rotating cylindrical collector. Nanofibers were mainly deposited onto the mouth end of the collector due to the structure of the collector. A conventional yarn was used a guide to draw nanofiber yarn from collector to the winding system. Some of the deposited nanofibers were joined to the cone shaped web formation and some of them were also deposited at the inner edges of the collector. The reason could be electrostatic interaction between positively charged nanofibers. They are in tendency of repelling each other and depositing on grounded collector.

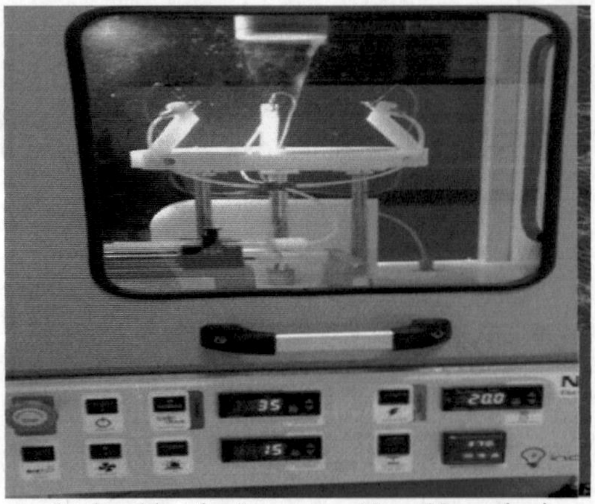

Figure 3. Nanofiber yarn obtained by the Inovenso NanoYarn device using 8% PAN solution

3.2. Stabilized and carbonized nanofiber yarns

As it can be seen in Figure 4, during the stabilization the macromolecular chain of PAN turns into ladder type structure due to the chemical reactions such as oxidation, dehydrogenation and cyclization and a result of C≡N bonds convert to C=N and finally to C-N bonds [6]. Therewithal the colour of the sample turns dark brown and the weight of sample and fiber diameter decrease due to the removing of CN and CH_2 groups from the sample. The temperature effects the reaction rates of stabilization process therefore the temperature should be optimum level. If the temperature is low, the reaction takes place with slow kinetic or it may not become true. On the other hand the high temperature leads to degradation of the sample and loss of molecular orientation. In the Figure 5, the SNFY and CNFY are shown.

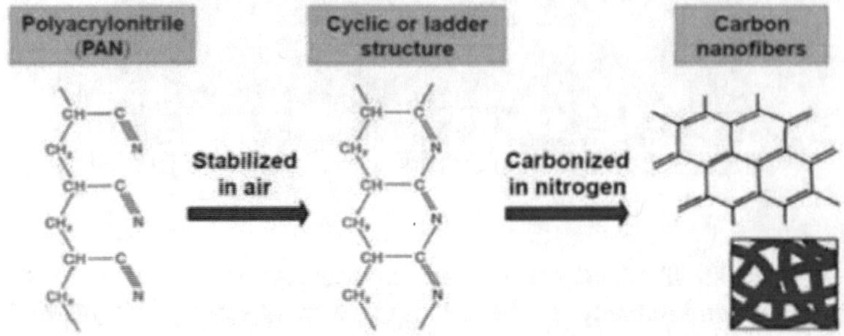

Figure 4. Steps of preparation of carbon nanofibers

During the carbonization process some non-carbon elements such as hydrogen, nitrogen, oxygen and sulfur are eliminated from fibers. As a result shrinkage occurs in the fibers and the diameter

IOP Conf. Series: Materials Science and Engineering **460** (2018) 012027 doi:10.1088/1757-899X/460/1/012027

decreases. Carbonization process causes the nitrile groups to disappear and that indicates cyclization of the structure like graphite fiber [7].

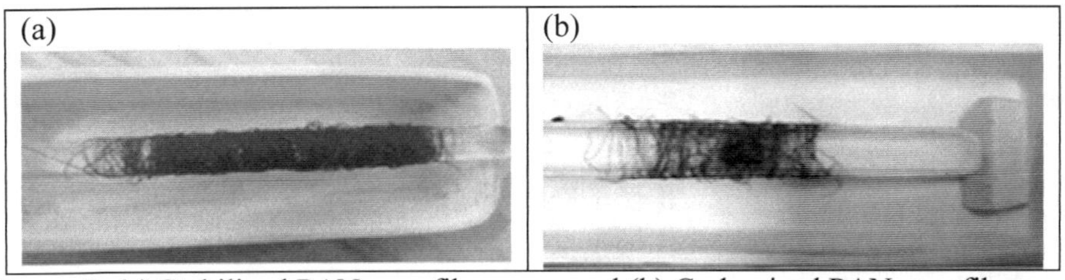

Figure 5. (a) Stabilized PAN nanofiber yarn and (b) Carbonized PAN nanofiber yarn

As seen in Figure 6, continuity of yarn was maintained after the carbonization and the diameter of fibers also decreased. The range of the diameter of the untreated PAN nanofiber yarn was between 210 nm and 340 nm while CNFY had a diameter between 160 nm and 300 nm. Carbonization temperature is important parameter for ultimate material properties. If they are low, the removal of nitrogen groups can not be achieved and carbon nanofiber yarn can not be produced in the desired yield. On the other hand it is possible to perform a complete burning operation in the case that these parameters are high.

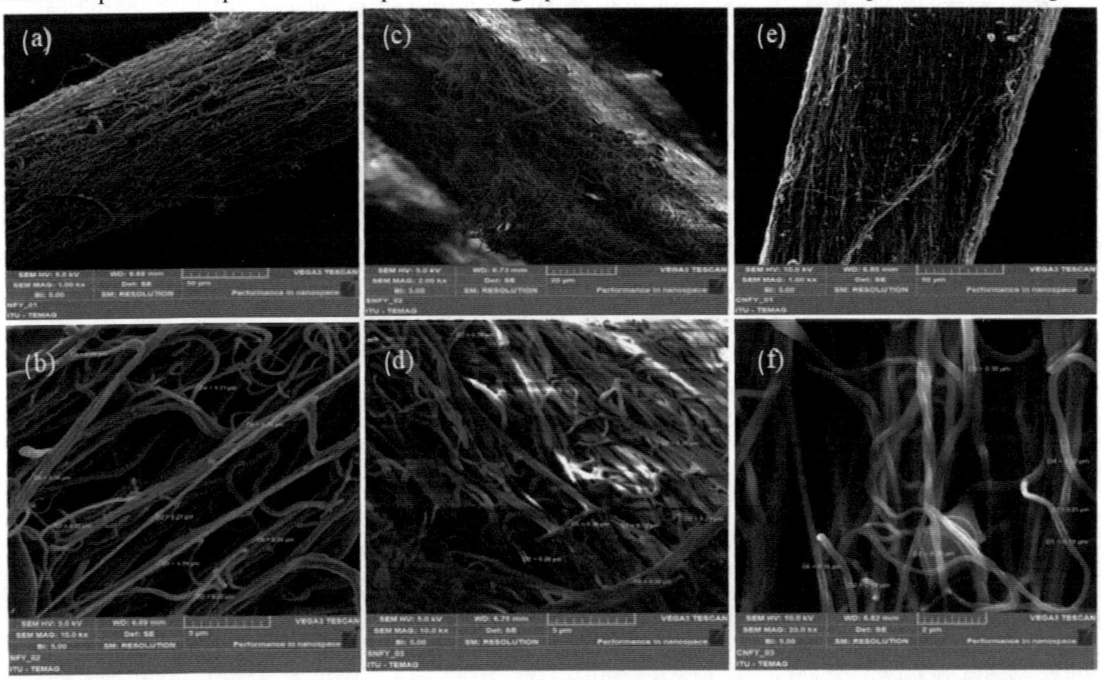

Figure 6. SEM images of (a) PAN nanofiber yarn with 1.00kx (b) PAN nanofiber yarn with 10.0kx (c) stabilized PAN nanofiber yarn with 1.00kx (d) stabilized PAN nanofiber yarn with 10.0kx (e) carbonized PAN nanofiber yarn with 1.00kx and (f) carbonized PAN nanofiber yarn with 20.0kx

4. Conclusion

The experimental work has successfully shown that PAN based carbon nanofiber yarn could be achieved. It has been discussed that PAN/DMF solutions prepared by low molecular weight PAN could not be carbonised due to the lack of required intramolecular forces. However, as the molecular weight of the PAN exceeds 150 000 g/mol, both stabilization as well as the carbonization stages of the conversion could be easily managed.

Having achieved carbon nanofiber yarns by utilising the electrospinning-stabilization-carbonization process chain in a multi step operation, it is now our intention to design and develop an on-line process to achieve all these three steps sequentially or if possible simultaneously.

Acknowledments
This work has been carried out by the Research and Development grant by TUBITAK and INOVENSO with the R&D Project Number 0937.STZ.2015.

References

[1] Ali U, Zhou Y, Wang X and Lin T 2012 Direct electrospinning of highly twisted, continuous nanofiber yarns *The journal of the textile institue* **103** 80-88
[2] He J, Qi K, Zhou Y and Cui S 2014 Multiple conjugate electrospinning method for the preparation of continuous polyacrylonitrile nanofiber yarn, *Journal of applied polymer science* **131** 40137
[3] Moon S C and Farris R J 2009 Strong electrospun nanometer-diameter polyarcylonitrile carbon fiber yarns *Carbon* **47** 2829-2839
[4] Chawla S, Naraghi M and Davoudi A 2013 Effect of twist and porosity on the electrical conductivity of carbon nanofiber yarns *Nanotechnology* **24** 255708
[5] Panapoy M, Dankeaw A and Ksapabutr B 2008 Electrical conductivity of PAN-based carbon nanofibers prepared by elektrospinning method, *Thammasat international journal of science and technology* **13** special edition 11-17
[6] Esrafilzadeh D, Morshed M and Tavanai H 2009 An investigation on the stabilization of special polyarcrylonitrile nanofibers as carbon or activated carbon nanofiber precursor *Synthetic metals* **159** 267-272.
[7] Su C I, Huang Y X, Wong J W, Lu C H and Wang C M 2012 PAN-based carbon nanofiber adsorbents prepared using elektrospinning *Fibers and polymers* **13** 436-442

Silkworm excrement used for dyeing textiles

N T Vila[1]*, M G Silva[1], A L Musialak[1], M Fernandes[2], A P Souto[2] and A J S Ferreira[3]

[1]*Departamento de Engenharia Têxtil, Universidade Estadual de Maringá, Goioerê, Paraná Brasil*

[2]*2C2T-Centro de Ciência e Tecnologia Têxtil, Universidade do Minho, Guimarães, Portugal*
[3]*Departamento de Engenharias, Universidade Federal de Santa Catarina, Blumenau, Santa Catarina, Brasil*

Corresponding author: ntvila@uem.br

Abstract The use of natural dyes for textile fibers has become an alternative to solve issues as environmental pollution and health risks caused by synthetic dyes. With this purpose, the use of silkworm excrement to dye textile fibers was targeted in silk (SK) and polyamide 6.6 (PA) weave fabric. This extract was characterized by spectroscopy to evaluate the color intensity and the color strength (K / S) and, finally, measured color fastness to wash, to friction and perspiration. The SK fiber as a better dye yield with silkworm excrement. Although the PA fiber shows also reasonable dye yield. With the results presented, it is possible to conclude that the SK and PA fabrics can be easily dyed with the natural dye extracted from the excrement of the silkworm, obtaining yellowish-brown colors.

Keywords—**Silkworm Excrement, Silk, Polyamide 6.6, Natural Dye**

I. Introduction

Wiyh the recent environmental policies and the introduction of more stringent legislation, the use of natural dyes in fiber dyeing tends to become a viable alternative, since this class of dyes show desirable characteristics such as: they do not cause allergies, they are not toxic and are eco-friendly.

The natural dyes are mostly derived from plant material and can be simultaneously renewable and biodegradable products. This range of dyes allows to dye a large part of natural and even some synthetic fibers. The dyes, such as those obtained from safflower flowers, onion peel and pomegranate peel are by-products of agricultural activity. The curcuma, safflower, marigold, and indigo are obtained from fruits, flowers and tree seeds that regenerate at least annually without compromising the potential for sustainability [1–3].

In this case, the present research intends to study the potential of using silkworm excrement as a natural dye alternative for dyeing fibers, especially silk (SK) and polyamide 6.6 (PA). It is a non-toxic and low-cost material, naturally generated in the SK production. Silkworm excrement is the result of feeding entirely based on mulberry leaves (*Morus*). When the silkworm ingests mulberry leaves, about 60% are excreted without being digested [4]. Therefore, this excreta are probably composed of leaves of mulberry and several constituents biotransformed by microbes or enzymes in the intestine of the beetle [5]. Constituents isolated from mulberry leaves include flavonoids, such as rutin, quercetin, quercitrin, isoquercitrin, steroids and triterphoids, substances present in various materials used as natural dyes [4], [6].

Thus, it is used widely, from healing iron-deficient research to its use in removing pesticides in activated carbon form [7], [8]. In dyeing context, the extract was used for dyeing cotton and wool. Wool fiber have better acceptance to this dye [9].

II. Procedures

A. *Materials and methods*

Commercial pre-washed pure PA (warp 52 yarns cm^{-1}, weft 32 yarns cm^{-1}, 110 GSM) and pure SK woven fabrics (warp 60 yarns cm^{-1}, weft 50 yarns cm^{-1}, 50 GSM) were employed in this study.

B. *Preparation of dye extract from silkworm excrement*

Dye extract from the silkworm excrement was obtained by aqueous extraction, scaling 10 g of silkworm excrement and mixing with 1 L of distilled water. The sample was then heated with stirring at 90 ° C in a thermostatic bath for 60 minutes. After heating, the contents were cooled to room temperature and filtered. Prior to dyeing, the pH of the extract was adjusted from 9.6 to 3.5 with acetic acid.

C. *Pre-mordanting process*

Potassium alum and tannic acid were used as mordants. The pre-mordanting procedure consisted of fabrics immersed in a mordant solution of 5g L^{-1} each at 90ºC for 45 min. The mordanting process was also conducted in the dyeing machine Kimak AT1-SW. After treatment, samples were dried at room temperature.

D. *Dyeing*

The samples were dyed for 60 minutes at 40 rpm on the Kimak AT1-SW equipment, employing a bath ratio of 1: 100 and a temperature of 90 ºC. Following, the dyed samples were rinsed with distilled water and dried at ambient conditions.

III. CHARACTERIZATION

A. *Color parameters*

The color strength of the dyed fabrics expressed as K/S value was calculated from the Kubelka–Munk [10] equation (1) as given below:

$$\frac{K}{S} = \frac{(1-R)^2}{2R} \qquad (1)$$

where K is the absorbance coefficient, S is the scattering coefficient, and R is the reflectance ratio measured at the maximum absorbance of dye using a Datacolor 550 reflectance spectrophotometer. Measurements were taken at five different positions on the fabric surface and averaged.

The color intensity (I) of the samples was also determined, using equation (2), as described below [10].

$$I = \sum_{\lambda=400nm}^{\lambda=700nm} \frac{K}{S}(\lambda) \times \Delta\lambda \qquad \Delta\lambda = 10 \qquad (2)$$

where **λ** represents the wavelength.

B. *Washing fastness test*

For the washing fastness tests, the dyed fabrics were treated in a Kimak AT1-SW, for 30 min at 40ºC in accordance with ISO Standard 105-C06 A1S-2010. A solution of ECE detergent type B was prepared with a concentration of 4 g L^{-1} and volume of 150 mL. When the cycle was finished, the samples were washed with distilled water dried under ambient conditions.

TABLE I
COLOR INTENSITY (I), CIELAB COLOR COORDINATES AND COLOR REPRESENTATION OBTAINED WITH THE MORDANTS

Fibre	Mordant	Intensity (I)	CIELAB Color Coordinates			Color
			L*	a*	b*	
Silk	Tannic acid	1080,33	43,20	8,52	21,14	
	Non-mordanted	933,15	49,20	6,59	25,75	
	Potassium alum	777,67	51,91	6,30	25,85	
PA 6.6	Tannic acid	515,30	54,84	6,27	20,01	
	Potassium alum	479,93	59,77	5,66	27,47	
	Non-mordanted	477,30	59,71	5,85	27,35	

C. *Rubbing fastness test*

Rubbing fastness tests were performed to analyse the resistance of the natural dyed fabrics to physical action, in accordance with ISO Standard 105-X12:2016. Each sample was abraded against cotton abrasive fabric for ten cycles in a dry and wet state.

D. Perspiration fastness test

For perspiration fastness tests each sample was tested in acid and alkaline form, in accordance with ISO Standard 105-E04:2013.

After the fastness tests, all staining and color change measurements were carried using a Datacolor 550 spectrophotometer. Results related to color strength (K/S) and wash fastness of non-mordanted and pre-mordanted fabrics were compared.

IV. Results and Discussion

The SK and PA samples dyed with silkworm excrement showed different color for each mordant used, however, all the samples were verified to be yellowish brown.

The K/S values as a function of the wavelength were plotted for the dyed SK without mordant, SK mordanted with tannic acid and for SK mordanted with potassium alum, shown in figure 1. The color intensity and the respective coordinates for the samples are presented in table I.

The results show a better color intensity for SK treated with tannic acid, as well as a higher K/S, restating the dye uptake. Following, the SK non-mordanted showed a relevant K/S curve when compared to the SK treated with potassium alum.

A better yield was noticed for the PA pre-mordanted with tannic acid, as it can be seen in the figure 2 and the table I. The pre-mordanted PA with potassium alum did not show any significant influence on the PA that did not go through the pre-mordanting process.

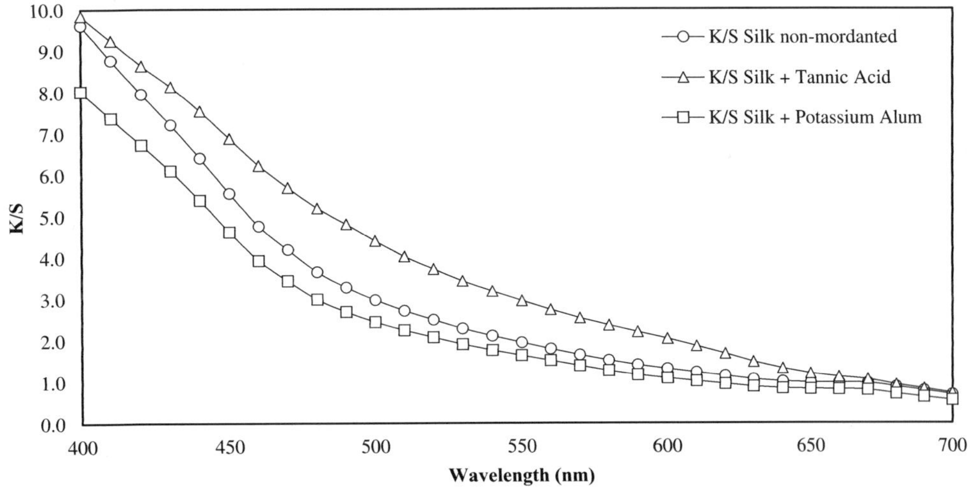

Fig. 1. Color strength of silk with different mordants.

The analysis of the samples dyed in the different conditions tested also allows to verify that the obtained shades have changed as a function of the mordant used before the dyeing, a situation that is confirmed by the existence of crossings in the curves of K/S of the dyed samples, as evidenced into variations in the respective chromatic coordinates, as shown in figure 2.

From the analysis of the color intensity (I) values of SK and PA samples dyed with silkworm excrement (table I), it is possible to analyse that the dye uptake was higher in SK for all conditions tested. This behaviour is possibly related to the greater number of functional groups present in the SK fibers, available to bond with the dye molecules as it is possible to see in Wei [11].

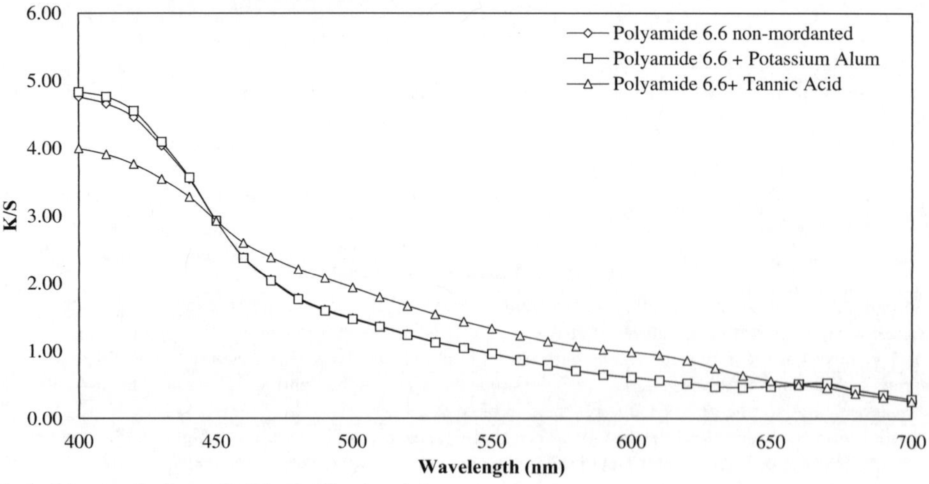

Fig. 2. Color strength of polyamide 6.6 with different mordants.

By analysing the chromatic coordinates of dyed SK and PA samples, the pre-mordanting process with tannic acid, followed by dyeing, produced darker shades (lower luminosity L *), more reddish (a* coordinate increase), and less yellowing (decrease of the b * coordinate) in relation to the other samples. The SK samples pre-mordanted with potassium alum, however, presented lighter, less reddish and more yellowish than the samples without pre-treatment with mordant. In the case of PA dyeing, pre-treatment with alum did not significantly influence L * luminosity, providing less reddish and more yellowish tones compared to the sample without mordant pre-treatment.

The results for washing fastness, as well as the results for rubbing fastness are presented in table II for SK and PA samples.

TABLE II
WASHING AND RUBBING FASTNESS TEST RESULTS

Sample	Washing Fastness							Rubbing Fastness	
	Color Change	*Staining on adjacent fibers*						Dry	Wet
		WO	PAC	PES	PA	CO	AC		
SK	4-5	5	5	5	4-5	4-5	5	3	3
SK/Tannic	4-5	4-5	5	5	4-5	5	5	4-5	4
SK/Alum	4-5	4-5	5	5	4-5	5	5	4	4
PA	3-4	5	5	5	5	5	5	5	5
PA/Tannic	3-4	5	5	5	5	5	5	5	5
PA/Alum	3-4	5	5	5	5	5	5	5	4-5

SK – Silk non-mordanted, **SK/Tannic** – Silk pre-mordanted with tannic acid, **SK/Alum** - Silk pre-mordanted with Potassium alum, **PA** – Polyamide non-mordanted, **PA/Tannic** – Polyamide pre-mordanted with tannic acid, **PA/Alum** - Polyamide pre-mordanted with Potassium alum, **AC**- Acetate, **CO**- Cotton, **PA**- Polyamide, **PES**- Polyester, **PAC**- Acrylic, **WO**- Wool, **CC**- Color change.

The washing and rubbing fastness index were classified as good (3) to excellent (5). Comparing the SK samples to the PA samples for washing fastness, it is explicit that the PA had a greater color change, due to a worse fastness property for this dyeing.

The results or rubbing fastness for SK indicated a benefit when using the mordants. However, for the PA, the addition of the pre-mordanting with Potassium Alum presented worse fastness to rubbing than without mordant.

In table III, the results for the perspiration fastness tests are presented.

TABLE III
PERSPIRATION FASTNESS RESULTS

Sample	**Perspiration Fastness**													
	Acid							**Alkaline**						
	Color change	*Staining on adjacent fibers*						*Color change*	*Staining on adjacent fibers*					
		WO	PAC	PES	PA	CO	AC		WO	PAC	PES	PA	CO	AC
SK	4-5	4-5	4-5	4-5	3-4	4	4-5	4	4	4-5	4-5	4	4-5	4-5
SK/Tannic	4-5	4-5	4-5	4-5	4	4-5	4-5	4-5	4-5	4-5	4-5	4	4-5	4-5
SK/Alum	4	4-5	4-5	4-5	4	4-5	4-5	4	4-5	5	4-5	4	4-5	4-5
PA	3	4-5	5	5	4-5	5	5	3-4	5	5	5	4-5	4-5	5
PA/Tannic	2-3	4-5	5	5	4-5	5	5	3	4-5	5	5	4-5	5	4-5
PA/Alum	3-4	4-5	5	5	4-5	5	5	3	4-5	5	5	4-5	4-5	5

It is possible to verify that the silk samples presented good perspiration fastness indexes, both regarding the color change and the staining of the samples, from grade 3-4 to 5. In the case of polyamide samples, the fastness indexes were slightly lower, ranging from 2-3 to 5.

The use of the mordants generally did not significantly improve the perspiration fastness indexes of the samples, and it has even worsened in some cases.

The PA samples, either the PA non-mordanted, the PA+Tannic and the PA+Alum showed a more relevant change in color, stating a worse fastness compared to SK. For the multifiber fabric tested in acidic and alkaline solutions, there were no relevant improvement using mordants. Nevertheless, it can be observed a decrease in fastness in alkaline solution to PA using PA/Alum and PA/Tannic, what invalidates the use of mordants when in an alkaline solution.

V. CONCLUSION

According to the results presented, it is possible to conclude that the SK and PA fabrics can be easily dyed with the natural dye extracted from the excrement of the silkworm, obtaining yellowish-brown colors. Nevertheless, more work will be done, namely with fluorescence microscope technique to confirm the degree of the dye penetration in PA fiber.

For both fibers, the best dye yield was obtained using tannic acid as a mordant. Potassium alum, however, negatively influenced the color intensity of the SK samples and showed no significant influence on the dye yield of the PA samples. The colors obtained were more intense for the SK samples, due to the greater dye uptake as a consequence of the greater number of groups functioning in the chemical structure of the fiber. For the wash fastness tests, it does not influence the addition of mordants. In the friction tests, tannic acid showed a positive influence. In this way, we can conclude that for both fibers (SK and PA), the pre-mordanting process with tannic acid was the best regarding dye uptake, but none of the pre-mordanting processes are significantly effective for the fastness properties.

Acknowledgment

The authors gratefully acknowledge the funding by Fundação Araucária, FA, Brasil, under grants Proc. 11040/2016 DCI/CCE - Universidade Estadual de Maringá (UEM) and thank the CENTRO DE TECNOLOGIA of UEM for his support.

This work is supported by FEDER funds through Operational Program for Competitiveness Factors – COMPETE and National Funds through FCT – Foundation for Science and Technology within the scope of the projects POCI-01-0145-FEDER-007136 and UID/CTM/00264.

References

[1] B. Chengaiah, K. M. Rao, K. M. Kumar, M. Alagusundaram, and C. M. Chetty, "Medicinal importance of natural dyes a review," Int. J. PharmTech Res., vol. 2, no. 1, pp. 144–154, 2010.

[2] N. A. Ibrahim, A. R. El-Gamal, M. Gouda, and F. Mahrous, "A new approach for natural dyeing and functional finishing of cotton cellulose," Carbohydr. Polym., vol. 82, no. 4, pp. 1205–1211, 2010.

[3] P. S. Vankar, R. Shanker, and S. Wijayapala, "Dyeing of cotton, wool and silk with extract of Allium cepa," Pigment Resin Technol., vol. 38, no. 4, pp. 242–247, 2009.

[4] B.-H. Sohn et al., "Isolation and identification of lipids from the silkworm (Bombyx mori) droppings," J. Korean Soc. Appl. Biol. Chem., vol. 52, no. 4, pp. 336–341, 2009.

[5] J.-H. Park et al., "Isolation of megastigmane sesquiterpenes from the silkworm (Bombyx mori L.) droppings and their promotion activity on HO-1 and SIRT1," Arch. Pharm. Res., vol. 34, no. 4, pp. 533–542, 2011.

[6] A. K. Samanta and A. Konar, "Dyeing of textiles with natural dyes," in Natural dyes, InTech, 2011.

[7] S. G. Mohammad and S. M. Ahmed, "Preparation of environmentally friendly activated carbon for removal of pesticide from aqueous media," Int. J. Ind. Chem., vol. 8, no. 2, pp. 121–132, 2017.

[8] X. Huang, J. Wu, Q. Li, S. Gu, Z. Chen, and K. Wang, "Silkworm feces extract improves iron deficiency anemia via suppressing hepcidin expression and promoting iron-regulatory proteins expression," RSC Adv., vol. 7, no. 79, pp. 50378–50388, 2017.

[9] M. G. da Silva, A. B. da Silva, F. Giacomini, V. C. Arzani, F. L. Ferreira, and T. Ruggeri, "Tingimento de têxteis com o corante natural extraído dos excrementos do bicho-da-seda," 4o CONTEXMOD, vol. 4, no. 1, 2016.

[10] M. Štěpánková, J. Wiener, and K. Rusinová, "Decolourization of vat dyes on cotton fabric with infrared laser light," Cellulose, vol. 18, no. 2, pp. 469–478, 2011.

[11] B. Wei, Q. Y. Chen, G. Chen, R. C. Tang, and J. Zhang, "Adsorption properties of lac dyes on wool, silk, and nylon," J. Chem., vol. 2013, pp. 1–7, 2013.

Compressibility of weft knitted reinforcement fabrics from glass yarn

K S Gunes[1], M E Ince[1] and H I Icoglu[2]

[1]Textile Engineering Department, Gaziantep University, Turkey
[2]Metallurgical and Materials Engineering Department, Gaziantep University, Turkey
E-mail: eince@gantep.edu.tr; meince@ncsu.edu

Abstract

Compression, and relaxation of a reinforcement fabric during composite manufacturing dictates mold design, resin flow, and fiber volume fraction of resultant composite; and single layer fabric architecture, and numbers of layers in a fabric stack change the compressibility character of overall reinforcement material. This study dealt with the effect of knit architecture and number of layers on dry compression and relaxation responses of weft knitted fabrics from glass yarn. Fabric architectures of 1x1, 2x2, English, and Fisherman ribs were selected. The thicknesses of single-, double-, and triple-fabric layers from same knit architectures were measured under compression and recovery pressures ranging from 2 to 200 kPa. Areal densities of single layer fabrics were measured. 2x2 and fisherman rib fabric architectures exhibited higher thickness, areal density and fiber volume fraction than 1x1 and English rib fabric architectures due to their nested and compact structures. Due to nesting between adjacent layers; multi-layer fabrics demonstrated higher fiber content than single layer fabrics. The difference between the compression and recovery thicknesses proved the existence of dissipated energy. Compression curves – the relationship between fiber volume fraction and compression pressure - classified by number of layers were fitted by square equation with fairly high determination of coefficients (R^2).

Keywords: fabric compressibility, fabric relaxation, glass yarn, reinforcement fabric, weft knitted fabric.

1. Introduction

Having knowledge about single and multi-layer compressibility of a reinforcement fabric is crucial to design a stable mold, to model resin flow, and to estimate fiber content of final composite product. This research is about the effect of knit pattern and number of layers on compression and recovery responses of weft knitted fabrics from glass yarn, and the literature review within this context is given below.

Pearce and Summerscales [1] monitored compressibility and repetitive relaxation responses of single and multiple (2, 3, 4, and 5) layers plain woven fabric from glass tow at maximum compression pressure levels of 100, 200, and 300 kPa. Specimen surface area exposed to compression was also varied as 50x50, 71x71, and 100x100 mm^2. The relationship between fiber volume fraction and compression pressure for initial loading was drawn on log-log scale and modeled by power-law equation. Single layer fabrics exhibited higher fiber volume fraction and lower exponent in power-law equation than multiple layer fabric stacks. Thickness - measured at maximum compression and normalized by number of layer - increased with number of layers, which indicated the lack of nesting between adjacent layers. Strain energy stored in compressed fabric (stack) dissipated during relaxation that reduced the pressure on the plates. Repetitive relaxation and reloading to maximum compression pressure improved fiber volume fraction. No lateral spreading of fabric was observed during compression test that verified the absence of areal density change.

Robitaille and Gauvin [2] reviewed published theoretical and experimental data on compression and relaxation behavior of commercial random mats and woven reinforcement fabrics from glass fiber. They listed parameters of compression and relaxation processes as; fabric architecture of reinforcement material, number of fabric layers, stacking sequence, presence of a lubricant, compression velocity, maximum compression pressure, number of compression cycles. The researchers underlined that due to complex internal structures of textile fabrics; theoretical compressibility studies failed to model the behavior textile reinforcements accurately. Compression curves obtained from experimental works were modeled by Equation 1 (power-law). The researchers assumed no in-plane global deformation during compression and relaxation that was parallel with the results of Pearce and Summerscales's study [1].

$$\boldsymbol{v_f = A * P^B} \tag{1}$$

Where $\boldsymbol{v_f}$ is fiber volume fraction, $\boldsymbol{A}$ **is** fiber volume fraction for a compaction pressure P equals to 1 Pa, $\boldsymbol{P}$ is compaction pressure (Pa), $\boldsymbol{B}$ is compaction stiffening index ($B < 1$).

The main findings of Robitaille and Gauvin's study [2] for compression curves of woven fabrics and random mats from glass fiber were;

- While number of fabric layers increased rigidity of the stack (*M*), and initial fiber volume fraction (*A*); it lowered stiffening index (*B*), and representative fiber volume fraction (Rv_f).

- While number of compression cycles increased rigidity of the stack (*M*), initial fiber volume fraction (*A*), and representative fiber volume fraction; it lowered stiffening index (*B*)

- While compression speed slightly reduced rigidity of the stack (*M*), and initial fiber volume fraction (A); it slightly increased stiffening index (*B*), and increased representative fiber volume fraction (Rv_f).

Similar to the findings of Pearce and Summerscales [1]; Robitaille and Gauvin [2] emphasized the importance of repetitive relaxation and reloading to maximum compression pressure to enhance rigidity and fiber volume fraction of overall fabric stack.

Luo and Verpoest [3] investigated the compressibility of a new sandwich fabric, Multimat, and its components, a mat and a weft knit with a woven fabric as comparison. The relaxations of the fabrics were tested. The fabric compression tests were done on varying number of fabric layers, by taking the relaxation behavior of the fabrics into account. The compressibility data are compared with the actual clamping pressure and a simple model was established. They mentioned that the relaxation of the knit was negligible in terms of volumetric dissipation energy and fabric stiffness; the fabric compression curves gave an idea of the practical range of fiber volume fraction in resin transfer molding, and the multi-layered knit preforms were easier to compress than the single-layered ones, whereas the number of layers did not influence the compressibility of other fabrics.

Weft knitting is a rapid, low-cost and versatile fabric manufacturing technique. Complex internal structure of weft knitted fabrics - created by interlocked stitches - make them unique among reinforcements family. They are stretchable, and shapeable that enables the direct production of intricate 2D or 3D near-net-shape preforms. Due to three dimensional interlacement of stitches within a layer and allowable nesting between porous layers; knitted fabrics exhibit better impact resistance than other textile forms as they are converted into layered composites. However, low load carrying capability arising from loose structure, and low fiber content of weft-knitted fabrics are disadvantageous as they are considered for in-plane reinforcement [4-7]. This disadvantageous character of weft knitted fabrics renders their compressibility and relaxation responses critical to improve their in-plane fiber volume fraction. To the best knowledge of the authors of this study; no special attention was devoted to characterize compression and recovery response of weft knitted fabrics from glass yarns. Focuses in previous studies were directed to mat and woven fabric materials. This study revealed the effect of two important material parameters -knit pattern, and number of fabric layers – on compression and recovery response of weft knitted fabrics.

2. Methods and methods

Three-ply 133 tex, E-glass multi filament yarn was consumed to manufacture fabrics on Brother KH-864 manual, flat weft knitting machine with a fineness of 5 gauges. Table 1 indicates the experimental plan of our study. Two input variables; knit architecture, and number of fabric layers were considered; in total 12 (4x3) different samples were considered. Figure 1 indicates technical notations (needle diagrams), three dimensional drawings, actual knit-fabric architectures both on the machine under tension in wale direction, and on the table in relaxed equilibrium state.

Table 1. Experimental Plan

Variables:	**Knit architecture**	**Number of fabric layers**
Levels:	1x1 Rib	1
	2x2 Rib	2
	English Rib	3
	Fisherman Rib	

Name	Technical notations	3D drawings	On the machine	On the table
1x1 rib	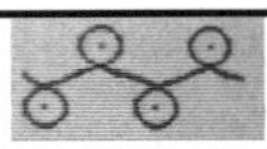			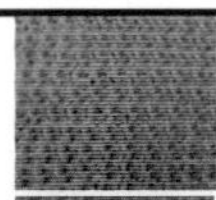
2x2 rib	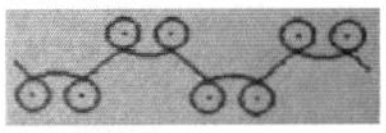	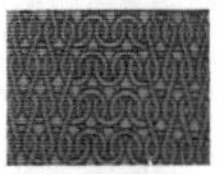		
English rib	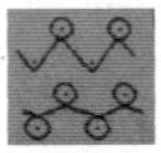	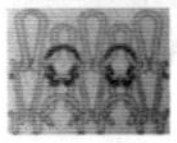	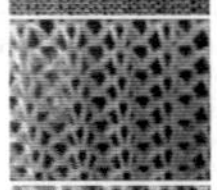	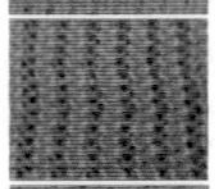
Fisherman rib	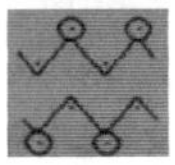	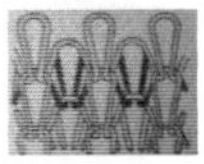		

Figure 1. Technical notations, 3D drawings [8] [9], actual fabric images on the machine under tension, and on the table in relaxed state of the knit architectures

Knitted fabric swatches were laid on anvil of digital thickness gauge with a pressure foot diameter of 21.15 mm. The pressure foot that has a compression pressure of 2 kPa (due to its weight) was lowered on the fabric. After 30 seconds waiting time, once the display of the gauge stabilized, the thickness was recorded as compression thickness at 2 kPa. Then additional weights were successively added on pressure foot to achieve compression pressures of 5, 10, 20, 50, 100, 150, 200 kPa, respectively. After addition of each weight, 30 seconds were waited to record the thickness shown on the stable display of the thickness gauge. Once compression cycle was finalized at 200 kPa compression pressure, the weights on the pressure foot were gradually removed, and recovery thicknesses starting at 150 kPa pressures were recorded until recovery pressure of 2 kPa was achieved. This procedure was repeated five times from different spots of a fabric swatch for all knit architecture and number of fabric layer combinations. In total, 900 (12x5x15) thickness measurements were performed; 480 of them at compression pressures, and 420 of them at recovery pressures. For each measurement cycle; one compression and recovery curve couple was drawn; in total, 60 different compression and recovery curve couples were drawn. The areal densities of 5x5 cm^2 single layer pieces cut from fabric swatch were measured on the precision balance according to ASTM [10]. Six areal density measurements were completed for each different knit pattern.

3. Results

3.1. The Relationship between Number of Layers and Normalized Thickness

To reveal the presence of nesting between adjacent layers, we normalized all thickness measurements by their matching number of layers, and drew the graph of normalized thickness versus number of layers (Figure 2). Normalized thicknesses grouped by number of layers did not show statistically significant difference at significance level of 0.05 (Table 2) that indicated the lack of nesting between adjacent knitted fabric layers. Wale-on-wale and course-on-course placement of fabric layers impeded nesting between adjacent layers. However, the decrease in thickness as number of layer was increased showed a tendency in nesting that could have been revealed at compression pressures higher than 200 kPa. Thus, the difference between normalized thicknesses of single and two layers fabric stack was found as statistically significant.

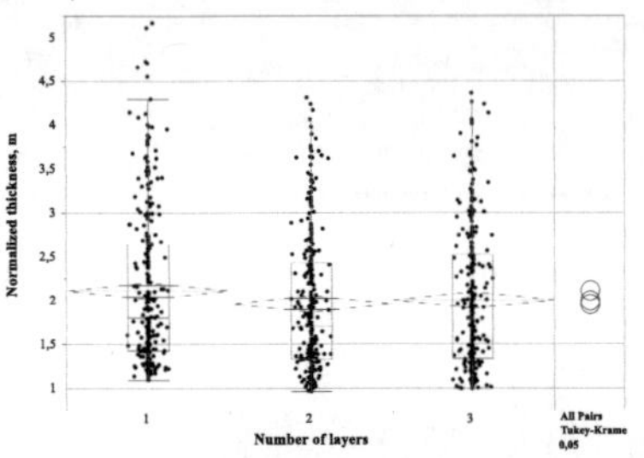

Table 2. Number of layers versus normalized thickness report

Number of layers		mean mm	sd mm	LL mm	UL mm
1	A	2,11	0,88	2,01	2,21
2	A	1,96	0,80	1,87	2,05
3	A	2,00	0,80	1,91	2,10

Note: levels not connected by same capital letter are significantly different. **sd**: standard deviation, **LL**: lower limit, **UL**: upper limit. Limits are based on 95% confidence level.

Figure 2. The relationship between number of layers and normalized thickness
Note: The top and bottom of each green diamond represent the 95% confidence interval for each layer level. Comparison circles (given on the right column) representing means those are significantly different either do not intersect, or intersect slightly. The height of red box - a quantitative indication of variation - is called interquartile range of mean.

3.2. The Effect of Compression and Recovery Pressures on Thickness

Comparison of thicknesses classified by compression and recovery pressures showed that recovery pressure thicknesses were significantly lower than compression thicknesses (Figure 3, Table 3). Compressed knitted fabrics could not recover to their initial thicknesses once the pressure was removed, which is advantage in the case of composite production. Because in many composite production technique, dry fabric is initially compressed before resin infusion, and later compression pressure on the fabric is offset by resin during infusion. Lack of complete recovery from compression lowers the final thickness of composite that enhance fiber volume fraction and mechanical properties of resultant composite. Lack of complete recovery also is an indication of dissipation of energy that was transferred to fabric during compression. Dissipated energy lowers the load on platens of two-sided mold that enhance the life of the mold.

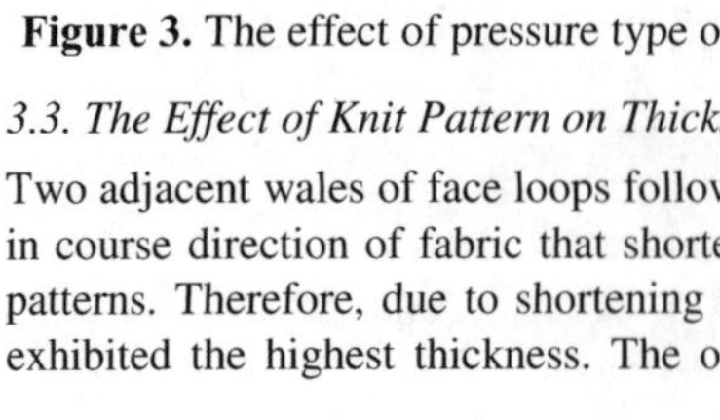

Table 3. The effect of pressure type on thickness

Pressure type	mean, mm	sd, mm	LL, mm	UL, mm	p-value
compression	4,33	2,61	4,10	4,56	<0,0001
recovery	3,66	1,99	3,47	3,85	

Figure 3. The effect of pressure type on thickness

3.3. The Effect of Knit Pattern on Thickness, Areal Density, and Fiber Volume Fraction

Two adjacent wales of face loops followed repeatedly by two adjacent wales of back loops enhanced internal tension in course direction of fabric that shortened the knit pattern of 2x2 ribs in width direction more than the other knit patterns. Therefore, due to shortening in width direction more than the other knit architectures; 2x2 rib structures exhibited the highest thickness. The only statistically significant thickness difference was observed between knit patterns of 1x1 and 2x2 ribs (Figure 4 and Table 4). Similar trends were observed for the effects of knit patterns on areal density and fiber volume fraction (Figures 5 and 6, Tables 5 and 6, respectively).

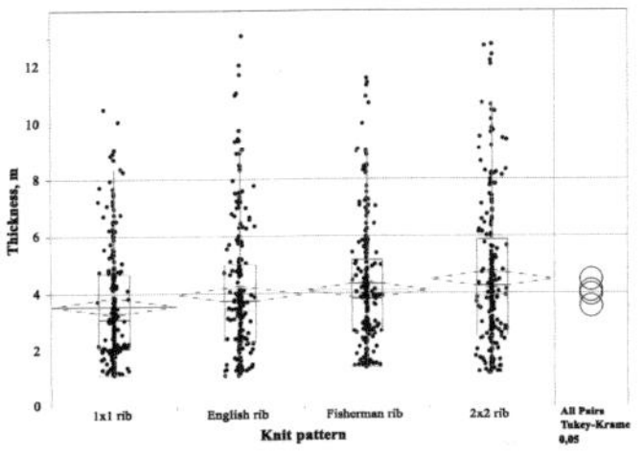

Figure. 4. The effect of knit pattern on thickness

Table 4. The effect of knit pattern on thickness

Knit pattern			mean, mm	sd, mm	LL, mm	UL, mm
2x2 rib	A		4,48	2,66	4,13	4,82
Fisherman rib	A	B	4,09	2,23	3,80	4,38
English rib	A	B	3,95	2,42	3,63	4,27
1x1 rib		B	3,54	2,01	3,28	3,80

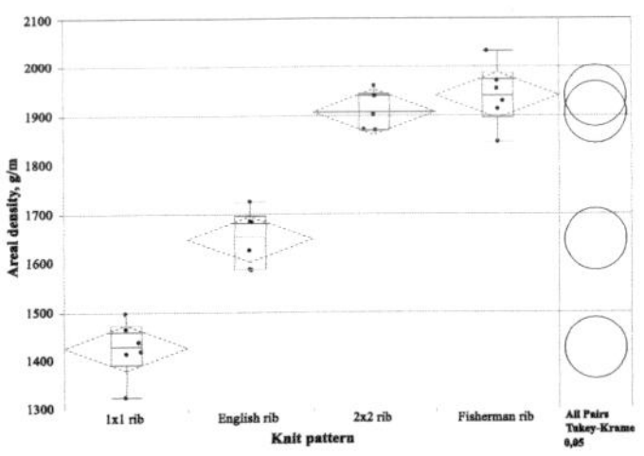

Figure 5. The effect of knit pattern on areal density

Table 5. The effect of knit pattern on areal density

Knit pattern				mean g/m^2	sd g/m^2	LL g/m^2	UL g/m^2
Fisherman rib	A			1941,67	62,41	1876,2	2007,2
2x2 rib	A			1907,73	36,01	1869,9	1945,5
English rib		B		1648,80	57,90	1588,0	1709,6
1x1 rib			C	1426,27	58,12	1365,3	1487,3

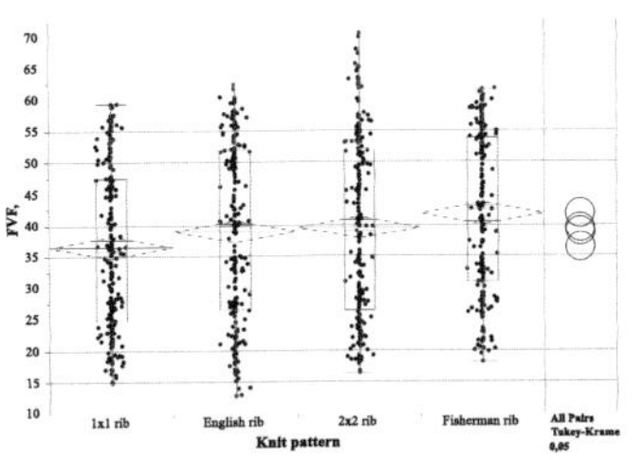

Figure 6. The effect of knit pattern on fiber volume fraction

Table 6. The effect of knit pattern on fiber volume fraction

Knit pattern				mean %	sd %	LL %	UL %
Fisherman rib	A			41,79	12,66	40,13	43,46
2x2 rib	A	B		39,54	14,37	37,65	41,43
English rib	A	B		38,95	14,11	37,10	40,80
1x1 rib			C	36,42	12,75	34,75	38,10

3.4. The Effect of Number of Layers on Fiber Volume Fraction

Number of layers increased fiber volume fraction of knitted fabrics. Nesting between adjacent layers in multiple fabric stacks lowered thickness and enhanced fiber content of knitted fabrics (Figure 7, Table 7).

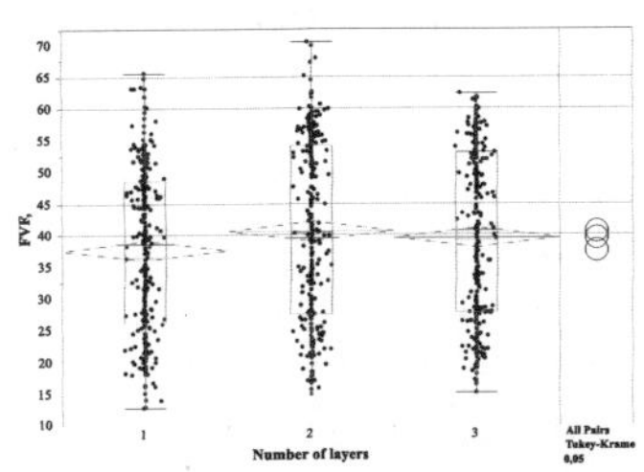

Table 7. The effect of number of layers on fiber volume fraction

Number of layers			mean %	sd %	LL %	UL %
2	A		40,52	14,15	38,91	42,12
3	A	B	39,48	13,60	37,94	41,03
1		B	37,52	12,92	36,06	38,99

Figure 7. The effect of number of layers on fiber volume fraction

Figure 8 shows the compression curves of knitted fabrics with different number of layers. Fiber volume fraction improvement was observed again as the number of layer was increased. The compression curves were fitted by quadratic equation with fairly high coefficient of determinations, R^2 (Table 8).

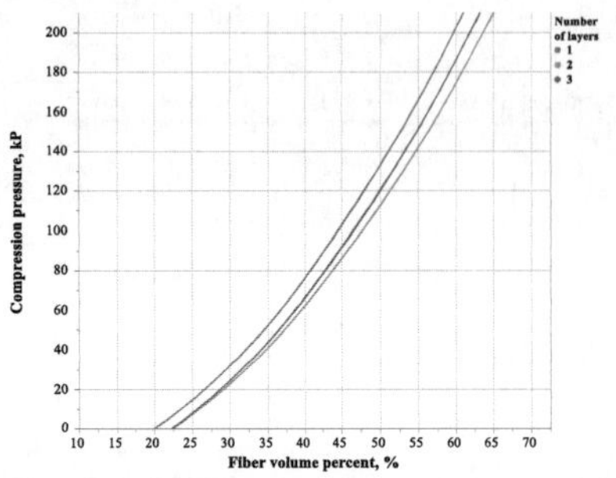

Figure 8. Compression curves grouped by number of layers

Table 8. Parameters of compression curves

Number of layers	R^2	Equation
1	0,82	$y = -24,73 + 0,063x^2$
2	0,86	$y = -27,57 + 0,056x^2$
3	0,90	$y = -29,35 + 0,060x^2$

4. Conclusion

This study revealed the effect of fabric architecture and number of fabric layers on compressibility and recovery from compression response of knitted fabrics from glass yarn. The onset of nesting between adjacent layers of multilayer fabric stack was indicated by number of layers versus normalized thickness graph. Lower recovery thicknesses than compression thicknesses showed the presence of dissipated energy during recovery that enhanced fiber volume fraction of the fabric stack and life cycle of the two-sided mold. Knit architecture emerged as a significant parameter to manipulate thickness, areal density and corresponding fiber content of the fabric. Due to internal course-wise tension that shortened the width of the fabric, once it was removed from the machine; 2x2 rib architectures demonstrated higher thickness, areal density and fiber volume fraction than other knit architectures. Fiber volume fraction increased by number of fabric layers. For future studies individual compression and recovery curves would be drawn and the effect of knit architecture and number of layers on dissipated energy would be revealed.

References

[1] N. Pears, J. Summerscales, "The compressibility of a reinforcement fabric", *Composites Manufacturing.*, vol. 6, no. 1, pp. 15-21, 1995.
[2] F. Robitaille, R. Gauvin, "Compaction of Textile Reinforcements for Composite Manufacturing. I: Review of Experimental Results", *Polymer Composites*, vol. 19, no. 2, pp. 198-216, 1998.
[3] Y. Luo, I. Verpoest, "Compressibility and Relaxation of a New Sandwich Textile Preform for Liquid Composite Molding", *Polymer Composites*, vol. 20, no. 2, pp. 179-191, 1999.
[4] G. Pamuk, F. Çeken, "Manufacturing of Weft-Knitted Fabric Reinforced Composite Materials: A Review", *Materials and Manufacturing*, vol. 23, no. 7, pp. 635-640, 2008.
[5] B. Gommers, I. Verpoest, P. Van Houtte, "Analysis of knitted fabric reinforced composites: Part I. Fiber orientation distribution", *Composites Part A: Applied Science and Manufacturing*, vol. 29, no. 12, pp. 1579-1588.
[6] L. Ciobanu, "Development of 3D Knitted Fabrics for Advanced Composite Materials", in *Advances in Composite Materials – Ecodesign and Analysis,* B. Attaf, Ed. InTech, 2011, pp. 161-192.
[7] S. D. Pandita, D. Falconet, I. Verpoest, "Impact properties of weft knitted fabric reinforced composites", *Composite Science and Technology*, vol. 62, no. 7-8, pp. 1113-1123, 2002.
[8] A. B. Marmaralı, *Atkı Örmeciliğine Giriş*. İzmir, Turkey: E.Ü. Tekstil ve Konfeksiyon Araştırma – Uygulama Merkezi, 2004, pp. 102-103.
[9] A. B. Marmaralı, S. D. Kretzschmar, *Örme Terimleri ve Tanımlamaları*. İzmir, Turkey: E.Ü. Tekstil ve Konfeksiyon Araştırma – Uygulama Merkezi, 2004, pp. 110-111.
[10] ASTM D3776, Standard Test Methods for Mass per Unit Area (Weight) of Fabric, 2009.

Surface activation of textile fibers by plasma DBD for dyeing with teak leaf

A J S Ferreira[1], M B Goede[1], N T Vila[2], M G Silva[2], S Fiori[2], T A Nascimento[2], M Fernandes[3] and A P Souto[3]

[1]*Departamento de Engenharias, Universidade Federal de Santa Catarina, Blumenau, Santa Catarina, Brasil*
[2]*Departamento de Engenharia Têxtil, Universidade Estadual de Maringá, Goioerê, Paraná, Brasil*
[3]*2C2T-Centro de Ciência e Tecnologia Têxtil, Universidade do Minho, Guimarães, Portugal*

Corresponding author: alexandre.ferreira@ufsc.br

Abstract The use of natural dyes is a field of great interest to solve the problems related to environment pollution and health risks. The objective of this research was to study the plasma discharge of double dielectric barrier (DDB) to modify the fabrics surfaces of polyamide 6.6 (PA66), silk (SK) and cotton (CO) dyed with leaf extract of teak (*Tectona grandis*). It was also aimed the improvement in the penetration and fixation properties to the natural dyes even without using metal mordants, which are toxic and usually used in this type of dyeing. The dye extract obtained from the teak leaves was chemically characterized by using UHPLC/QTOF-MS. The color parameters were calculated by means of spectroscopy to evaluate the color intensity (K/S) and its fastness was evaluated for washing, friction and perspiration.

***Keywords*— Plasma DBD, *Tectona grandis*, Silk, Polyamide 6.6, Natural Dye, UHPLC/QTOF-MS.**

I. Introduction

THE extracts of organic dyes from animal, plants and fungus oriented to several industries are currently the focus of attention for several research groups. Unlike its alternative synthetic, the natural dye revealed a product with a reduced environmental impact. The natural dyes may be based on tannin, indigo, anthraquinone, naphthoquinone, benzoquinone, flavonoids, carotenoids and, with various applications in food, cosmetics, pharmaceuticals and textile industry. The natural dyes are emerging as novels agents to produce textiles with antimicrobial, insect repellent, fluorescent and UV protection properties [1-3]. The variety of natural dyes as well as the color palette is quite large. *Tectona grandis* is an Asian native plant known also as Teak. This plant can be also found in Brazil. Teak wood is widely used by the naval industry due to its durability, moreover which has beautiful texture and resistance to fungal and termite damage and, its dried leaves are used as natural dye. Phytochemical investigations on teak has revealed the presence of different classes of compounds, such as: tannin, coumarins and amino acids mostly [4]. Anthraquinone based coloring compounds can be found in leaves of Teak [5].

Dielectric Barrier Discharge (DBD) plasma is one of the most efficient atmospheric plasma. The DBD has aroused the interest of the textile industries due to their scalability for large systems [6-8]. Natural dyes are applied to substrates that have been subjected to plasma treatment [9] in a few studies. The low temperature plasma is characterized by being a dry process, eco-friendly, simple processing by changing the surfaces without modifying the intrinsic properties of the material [10]. In particular, in textile substrates, atmospheric plasma is particularly recommended due to its non-thermal characteristic, appearing as an advantage when limited to heat-sensitive polymer fibers [11]. It has already been showed that the creation of functional groups on the surface of the substrates, namely hydroxyl and carboxyl groups ensure effective covalent bonds, whose, are indispensable in order to create a covalent attachment of the natural precursors to the substrate and thus ensuring the color fastness [12,13].

II. Experimental Section

A. Materials

Commercial pre-washed pure PA66 woven fabric (warp 52 yarns cm^{-1}, weft 32 yarns cm^{-1}, 110 GSM), pure SK woven fabrics (warp 60 yarns cm^{-1}, weft 50 yarns cm^{-1}, 50 GSM) and CO bleached woven fabric (warp 58 yarns cm^{-1}, weft 22 yarns cm^{-1}, 160 GSM) were employed in this study.

B. Preparation of dye extract from teak

Dye extract from the teak was obtained by aqueous extraction, weighing 10 g of teak and mixing with 1 L of distilled water. The sample was then heated with stirring at 90°C in a thermostatic bath for 60 minutes. After heating, the contents were cooled to room temperature and filtered. Prior to dyeing, the pH of the extract was adjusted to 3.0 with acetic acid for SK, 5.5 to PA66 and 10.5 to CO.

C. UHPLC/QTOF-MS analysis

Extracts were analysed by ultra high-performance liquid chromatography (Shimadzu, Nexera X2, Japan) coupled to high resolution mass spectrometry (Impact II, Bruker Daltonics Corporation, Germany) equipped with an electrospray ionization source. The capillary voltage was operated in negative ionization mode, set at 3000 V and with an end plate offset potential of −500 V. The dry gas parameters were set to 8 L min^{-1} at 180°C with a nebulization gas pressure of 4 bar. Data were collected from *m/z* 50-1300 with an acquisition rate of 5 spectrums per second, and the ions of interest were selected by auto MS/MS scan fragmentation. Chromatographic separation was performed with a gradient mixture of solvents A (H_2O 0.1% formic acid v/v) and B (acetonitrile 0.1% formic acid v/v) using a C18 column (Waters, ACQUITY UPLC®). The gradient system was structured as follows: 5% of B during 0-2 min, 50% of B through 2-10 min, 95% of B, through 10-18 min and maintained at 95% of B 18-20 min, 5% of B, during 20-22 min and maintained at 5% of B, during 22-25 min. Identification of the compounds was performed based on similarity of the high-resolution fragmentation spectra and exact mass information of the compounds, with literature spectra and public databases.

D. Plasma treatment

The DBD plasma treatment was conducted in a semi-industrial prototype machine (Softal Electronics GmbH/University of Minho) working at room temperature and atmospheric pressure, using a system of metal electrode coated with ceramic and counter electrodes coated with silicon with 50 cm effective width, gap distance fixed at 3 mm and producing the discharge at high voltage 10 kV and low frequency 40 kHz. The machine was operated at 1 W of power, velocity of 4 m min^{-1}. Plasmatic dosage was defined by the equation 1:

$$Dosage = \frac{N \times P}{v \times l} \quad (1)$$

Where, N=number of passages, P=power (W), v = velocity (m min^{-1}), and l = width of treatment (0.5 m).

E. Dyeing

The samples were dyed for 60 minutes at 40 rpm on the Ahiba Texomat Turbo Color equipment, employing a bath ratio of 1: 100 and a temperature of 90 ºC. After dyeing the samples were rinsed with distilled water and dried at ambient conditions.

III. CHARACTERIZATION

A. Contact Angle

Contact angle measurement was carried out in a Dataphysics instrument (Filderstadt, Germany) using OCA20 software (Germany) with a video system for the caption of images in static and dynamic modes. A drop of 5μL of distilled water was placed on the composite surface with a microliter syringe and observed with a special CCD camera. At least five measurements at different places were taken for each fabric. The camera takes an image every 0.04 s.

B. Color parameters

The color strength of the dyed fabrics expressed as K/S value was calculated from the Kubelka–Munk equation (2) as given below [14]:

$$\frac{K}{S} = \frac{(1-R)^2}{2R} \quad (2)$$

where K is the absorbance coefficient, S is the scattering coefficient, and R is the reflectance ratio measured at the maximum absorbance of dye using a Datacolor Spectraflash SF 600 Plus CT spectrophotometer. Measurements were taken at five different positions on the fabric surface and averaged.

The color intensity (I) of the samples was also determined, using equation (3), as described below [14].

$$I = \sum_{\lambda=400nm}^{\lambda=700nm} \frac{K}{S}(\lambda) \times \Delta\lambda \qquad \Delta\lambda = 10 \quad (3)$$

where **λ** represents the wavelength.

C. Washing fastness test

For the washing fastness tests, the dyed fabrics were treated in a Linitest C1-20 Original Hanau, Germany, for 30 min at 40ºC in accordance with ISO Standard 105-C06 A1S. The solution of ECE detergent was replaced by a non-ionic detergent with a concentration of 4 g L^{-1} and volume of 150 mL. This replacement is justified in natural dye field research by a several number of research groups. They use this substitution due a color change when the ECE detergent is used. With this particular situation the measurement of color parameter was impracticable [15-17]. When the cycle was finished, the samples were washed with distilled water dried under ambient conditions.

D. Rubbing fastness test

Rubbing fastness tests were performed to analyse the resistance of the natural dyed fabrics to physical action, in accordance with ISO Standard 105-X12:2016. Each sample was abraded against cotton abrasive fabric for ten cycles in a dry and wet state.

E. Perspiration fastness test

For the perspiration fastness tests each sample was tested in acid and alkaline form, in accordance with ISO Standard 105-E04:2013. After the fastness tests, all staining and color change measurements were carried using a Datacolor Spectraflash SF 600 Plus CT spectrophotometer.

IV. Results and Discussion

A. Chemical characterization of the extract by UHPLC/QTOF-MS

Figure 1 shows the base peak ion chromatogram of the teak leaf extract obtained in the negative ionization mode.

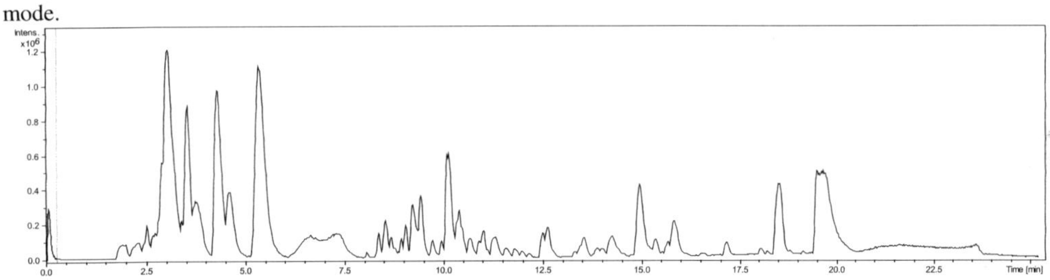

Fig. 1. Chromatogram of teak leaves extract, negative ionization mode

The mass spectra of teak leaf extract presented spectral match with ten characteristic ions of phenolic compounds. Their identities were confirmed by comparing mass accuracy and fragmentation behaviours with data available in the literature. The mass error for the molecular ions of all the compounds identified was within ± 8 ppm.

A list of compounds identified in the teak extract using UHPLC/QTOF-MS, retention time, experimental $[M-H]^-$, ion fragmentation profile and the literature consulted are described in Table 1.

TABLE I.
CHEMICAL COMPOSITION OF TEAK LEAVES AND CORRESPONDING FRAGMENTATION PROFILE

No.	T_R (min)	Compound name	Formula	$[M-H]^-$ (m/z)	$[M-H]^-$ experimental (m/z)	Error (ppm)	Fragment ions (m/z)	Reference
1	2.97	Quinic acid	$C_7H_{12}O_6$	1,910,556	1910552	1.90	127.0391, 111.0449, 93.0336, 85.0285	(Abdel-Hameed et al. 2013)
2	7.84	Mesaconic acid	$C_5H_6O_4$	1,290,188	1290181	5.30	85.0283, 66.3217	HMDB ID: HMDB0000749
3	9.18	Chlorogenic acid	$C_{16}H_{18}O_9$	3,530,873	3530856	4.69	209.0286, 191.0549, 147.0291, 85.0288	(Santos et al. 2011)
4	9.90	Rutin	$C_{27}H_{30}O_{16}$	6,091,456	6091420	5.84	300.0254, 178.9973	(Wu et al. 2017)
5	9.93	Caffeic acid	$C_9H_8O_4$	1,790,344	1790337	4.10	135.0443, 107.0490	(Kivrak et al. 2017)
6	10.29	quercetin-3-glucoside	$C_{21}H_{20}O_{12}$	4,630,877	4630866	2.27	415.1589, 300.0257, 271.0217, 255.0304, 191.0557, 149.0445, 119.0344, 99.0081, 89.0234	(Abdel-Hameed et al. 2013)
7	10.30	Kaempferol 7-O-glucoside	$C_{21}H_{20}O_{11}$	4,470,927	4470901	5.90	327.0459, 285.0376, 269.1388, 225.1491, 209.1163	(Sánchez-Rabaneda et al. 2003)
8	11.26	Coumaric acid	$C_9H_8O_3$	1,630,395	1630387	5.02	119.0491	(Koley et al. 2017)
9	12.52	Luteolin	$C_{15}H_{10}O_6$	2,850,399	2850377	7.76	272.0303, 257.0440, 243.0659, 223.0331, 213.0555, 199.0379, 175.0373, 151.0020, 133.0282, 107.0119, 91.0371	(Santos et al. 2011)
10	12.65	Quercetin	$C_{15}H_{10}O_7$	3,010,348	3010328	6.74	273.0350, 255.0251, 189.0158, 178.9991, 161.0234, 151.0017, 121.0271, 107.0118	(Wang et al. 2015)

B. Plasma Treatment

The DBD plasma treatment was optimized for 5 passages on PA66, SK and CO at the power of 1 kW and velocity of 4 m/min. The Table 2 presents de results for contact angle in the silk fabric.

The plasma treatment has enabled an increase in hydrophilicity of fabric, which can be understood by the decrease of the static contact angle.

TABLE II.
STATIC CONTACT ANGLE IN THE SILK FABRIC

	Untreated	**1500 W min m^{-2}**	**3000 W min m^{-2}**
Static contact angle	132.53° 	11.13° 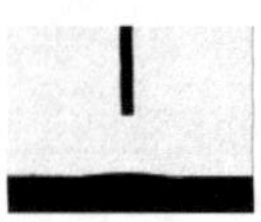	6.06° 

C. Color strength

It was possible to observe changes in the dyeing behaviour of silk and polyamide fibers after being submitted to plasma
treatment and dyed with the natural dyes. Silk was the fiber that presented a higher increase in dye yield after the plasma treatment as can be observed in the graph of K/S in Figure 2.

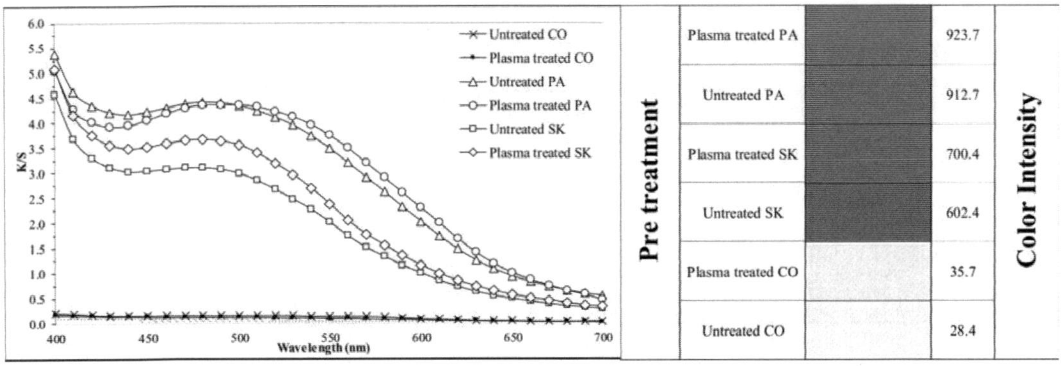

Fig. 2. Color strength of teak leaves and color intensity.

Indeed, plasma treatment may have potentiated the silk and polyamide dye sensitivity due to changes in chemical and possibility structural effect (increased chemical groups containing nitrogen and oxygen and creating microroughness) on the surface of that fibers.

On the other hand, the cotton fiber did not absorb the dye, which may indicate that there is no affinity between the dye and the fiber under the conditions tested. Thus, the strength tests were not performed on the cotton samples.

D. Washing and rubbing fastness

The results of the washing and rubbing fastness tests of the polyamide and silk samples dyed with teak leaves extracts are shown in Table 3. The wash fastness indices can generally be considered to be reasonable at very good, with the exception of the silk samples, which presented color change of grade 1-2 for sample without treatment and grade 2 for the sample treated with plasma. The colors change were lower with samples treated with plasma Furthermore, plasma treatment also showed no influence on color transfer indexes for adjacent fibers.

TABLE III.
WASHING AND RUBBING FASTNESS TEST RESULTS

Sample	*Color change*	Washing Fastness						Rubbing Fastness	
		Staining on adjacent fibers						Dry	Wet
		WO	PAC	PES	PA	CO	AC		
Polyamide Untreated	4	4	5	5	3-4	4-5	4-5	4-5	4-5
Polyamide Plasma Treated	5	4	5	5	3-4	4-5	4-5	5	5
Silk Untreated	1-2	3-4	4-5	4-5	3-4	4	5	5	4-5
Silk Plasma Treated	2	3-4	4-5	4-5	3-4	4	5	5	5

E. Perspiration Fastness (acid and alkaline)

The acid and alkaline indexes may, in an overview, be considered reasonable to very good, with the exception of the untreated silk sample, which showed a color change specifically of grade 2-3 in the alkaline perspiration test, as shown in Table 4.

With polyamide fiber, plasma did not contribute to the improvement of perspiration indexes, and even influenced negatively in some cases. With silk, plasma treatment improved color change indexes for both acid and alkaline perspiration, however, it had no influence on transfer indexes, except for wool fiber in the acid and acetate fiber for alkaline perspiration, in which the indices were worse for treated samples.

Acid perspiration, in general, was better than alkaline perspiration for all samples tested. This behaviour is possibly related to the degradation suffered by the natural dye in basic conditions [18].

TABLE IV.
PERSPIRATION FASTNESS TEST RESULTS

Sample	Perspiration Fastness													
	Acid							Alkaline						
	Color change	*Staining on adjacent fibers*						*Color change*	*Staining on adjacent fibers*					
		WO	PAC	PES	PA	CO	AC		WO	PAC	PES	PA	CO	AC
Polyamide Untreated	4	4	4-5	4-5	4-5	5	4	4	3-4	4-5	4-5	3	4	4
Polyamide Plasma Treated	4	4	4-5	4-5	3-4	4-5	4	3-4	3-4	4-5	4-5	3	4	4
Silk Untreated	3-4	4-5	5	5	4-5	4-5	4	2-3	3	4	4	3	3-4	4-5
Silk Plasma Treated	4-5	3-4	5	5	4-5	4-5	4	3	3	4	4	3	3-4	4

V. CONCLUSION

The results show that PA66 and SK fabrics are possible to dye with teak extract, being obtained the reddish brown and brown colors for PA66 and SK, respectively.

The chemical characterization showed that the aqueous extract of the teak leaves consists of several types of compounds from the flavonoid class. Plasma treatment promoted increased dye yield for silk and polyamide fibers. This was also detected washing and friction fastness indexes were improved with plasma treatment.

ACKNOWLEDGMENT

The authors gratefully acknowledge the funding by Fundação Araucária, FA, Brasil, under grants Proc. 11040/2016 DCI/CCE - Universidade Estadual de Maringá (UEM) and thank the CENTRO DE TECNOLOGIA of UEM for his support.

This work was supported by FEDER funds through Operational Programme for Competitiveness Factors – COMPETE and National Funds through FCT – Foundation for Science and Technology within the scope of the projects POCI-01-0145-FEDER-007136 and UID/CTM/00264.

REFERENCES

[1] M. Shahid, I. Shahid ul, and F. Mohammad, "Recent advancements in natural dye applications: a review," Journal of Cleaner Production, vol. 53, no. 0. pp. 310–331, 2013.
[2] F. Mohammad, "Natural colorants in the presence of anchors so-called mordants as promising coloring and antimicrobial agents for textile materials," ACS Sustain. Chem. Eng., vol. 3, no. 10, pp. 2361–2375, 2015.
[3] M. Shahid and F. Mohammad, "Green chemistry approaches to develop antimicrobial textiles based on sustainable biopolymers□ A review," Ind. Eng. Chem. Res., vol. 52, no. 15, pp. 5245–5260, 2013.
[4] R. B. Nidavani and A. M. Mahalakshmi, "Teak (tectona grandis linn.): A renowned timber plant with potential medicinal values," Inter J Pharm Pharm Sci, vol. 6, no. 1, pp. 48–54, 2014.
[5] A. K. Prusty, T. Das, A. Nayak, and N. B. Das, "Colourimetric analysis and antimicrobial study of natural dyes and dyed silk," J. Clean. Prod., vol. 18, no. 16–17, pp. 1750–1756, Nov. 2010.
[6] G. Borcia, C. A. Anderson, and N. M. D. Brown, "Surface treatment of natural and synthetic textiles using a dielectric barrier discharge," Surf. Coatings Technol., vol. 201, no. 6, pp. 3074–3081, 2006.
[7] H. Zeng, C. Gao, Y. Wang, P. C. P. Watts, H. Kong, X. Cui, and D. Yan, "In situ polymerization approach to multiwalled carbon nanotubes-reinforced nylon 1010 composites: Mechanical properties and crystallization behavior," Polymer (Guildf)., vol. 47, no. 1, pp. 113–122, 2006.
[8] U. Kogelschatz, "Dielectric-Barrier Discharges: Their History, Discharge Physics, and Industrial Applications," Plasma Chem. Plasma Process., vol. 23, no. 1, pp. 1–46, 2003.
[9] A. Zille, F. R. Oliveira, and A. P. Souto, "Plasma treatment in textile industry," Plasma Process. Polym., vol. 12, no. 2, pp. 98–131, 2015.
[10] F. S. Denes and S. Manolache, "Macromolecular plasma-chemistry: an emerging field of polymer science," Prog. Polym. Sci., vol. 29, no. 8, pp. 815–885, 2004.
[11] R. Morent, N. De Geyter, J. Verschuren, K. De Clerck, P. Kiekens, and C. Leys, "Non-thermal plasma treatment of textiles," Surf. Coatings Technol., vol. 202, no. 14, pp. 3427–3449, 2008.
[12] F. R. Oliveira, A. Zille, and A. P. Souto, "Dyeing mechanism and optimization of polyamide 6,6 functionalized with double barrier discharge (DBD) plasma in air," Appl. Surf. Sci., vol. 293, pp. 177–186, 2014.
[13] F. R. Oliveira, A. P. Souto, N. Carneiro, and J. H. O. Nascimento, "Surface modification on polyamide 6.6 with double barrier discharge (DBD) plasma to optimise dyeing process by direct dyes," in Materials Science Forum, 2010, vol. 636, pp. 846–852.
[14] M. Štěpánková, J. Wiener, and K. Rusinová, "Decolourization of vat dyes on cotton fabric with infrared laser light," Cellulose, vol. 18, no. 2, pp. 469–478, 2011.
[15] D. Jothi, "Extraction of natural dyes from African marigold flower (Tagetes ereectal) for textile coloration," Autex Res. J., vol. 8, no. 2, pp. 49–53, 2008.
[16] R. Räisänen, P. Nousiainen, and P. H. Hynninen, "Emodin and dermocybin natural anthraquinones as mordant dyes for wool and polyamide," Text. Res. J., vol. 71, no. 11, pp. 1016–1022, 2001.
[17] M. Montazer, M. Parvinzadeh, and A. Kiumarsi, "Colorimetric properties of wool dyed with natural dyes after treatment with ammonia Coloration Technology," Color. Technol., vol. 120, pp. 161–166, 2004.
[18] V. R. Giri Dev, J. Venugopal, S. Sudha, G. Deepika, and S. Ramakrishna, "Dyeing and antimicrobial characteristics of chitosan treated wool fabrics with henna dye," Carbohydr. Polym., vol. 75, no. 4, pp. 646–650, Feb. 2009.

Thermal strain in motor sport: effect of protective garments

Ada Ferri[1], Giulia Beghini[1], Francesca Dotti[1] and Angelo Bardascino[2]

[1]*Department of Applied Science and Technology, Politecnico di Torino, Corso Duca degli Abruzzi 24, 10129 Torino (Italy)*

[2] *Sparco S.p.A., Via Leinì 324, 10088 Volpiano (TO), Italy*

Corresponding author: ada.ferri@polito.it

Abstract The objective of this paper is the assessment of heat shield offered by a racing suit in extreme environmental conditions. The method was based on driving simulations in a climatic chamber, during which thermal strain was monitored by measuring a number of physiological and psychological parameters. It was found that a conventional racing suit offered some protection from heat flow but a greater level of protection could be obtained by a reflective outfit. During the test evaporative cooling was negligible because of poor breathability of the outfit and the seat. In conclusion, it was proven that a conventional racing suit offers protection from extreme environmental conditions but it does not guarantees thermal comfort. Use of reflective elements in the racing suit could be beneficial for thermal protection.

***Keywords*—motor sport, racing, thermal comfort.**

I. INTRODUCTION

In motorsports athletes can drive at speeds higher than 300 km/h for 3–4 h of competition in a cockpit that can reach temperatures of about 50°C [1,2]. Extreme heat in motorsports can pose safety concern due to extreme pressure on the cardiovascular system with large blood flow displaced to the skin to dissipate heat. Fatigue caused by extended hours of driving a racing car has considerable influences on driver alertness, performance and safety.

The physiological stress faced by professional racing drivers during competition is not adequately investigated but some studies have evidenced that competitive racing requires a significant degree of physical effort, similar to elite athletes competing in other sporting activities [3].

One way to assess and train drivers is to use drive simulators [4], which are mainly used to monitor cognitive aspects of driving [5] but should also pose a physical effort in the virtual environment comparable to real world.

Racing suits have to comply with international standards, such as UNI ISO 9151, which predict conductive properties of the suit when invested by the high radiative heat flow typical of a fire but no standard exists to predict thermo-physiological comfort condition of a driver during motor racing. In this work, racing sessions were simulated by using a virtual simulator in extreme hot conditions. The same driving session was carried out with three different outfits (a racing suits, a tropical suit and a reflective outfit). The aim of the work was to assess the ability of the racing suits to dissipate body heat and protect the driver from the extreme environmental conditions encountered during the competition.

II. MATERIALS AND METHODS

A racing suits approved by FIA (Federation International de Automobile) was tested by three testers in extreme environmental conditions inside a climatic chamber with air temperature 48,3±0.15°C and air relative humidity 48.8±2.2% and compared with a tropical outfit and a reflective outfit.

Each tester carried out three identical racing simulation sessions with the three outfits during three different days at the same time to avoid the effect of circadian rhythm.

Each outfit configuration included:

- *tropical outfit*: 100% PES T-shirt and shorts, ankle-high socks and sneakers;
- *racing suit outfit*: long-sleeve and long-leg Nomex underwear, ankle-high socks, balaclava, racing suit, racing gloves and helmet;

- *reflective outfit*: the same as the racing suit outfit with a isothermal blanket on top.

Figure 1 shows the driving simulation run in the climatic chamber for one tester wearing the racing suit or the reflective outfit.

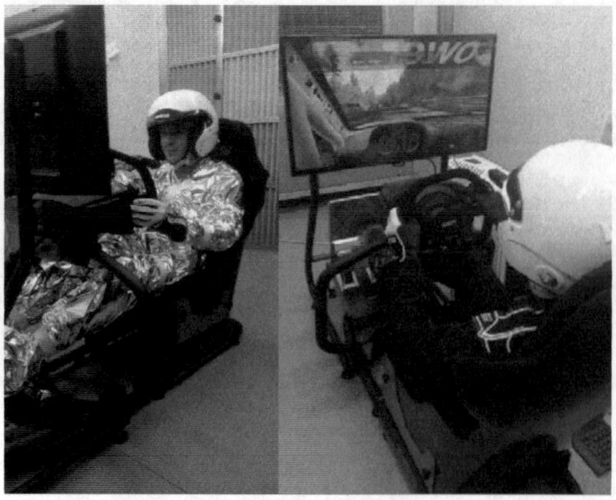

Fig. 1 Driving simulation in the climatic chamber with reflective and racing suit outfit.

The test is made of three phases:

- 10-min acclimatization inside the climatic chamber at rest with light air velocity (2 km/h)
- 30-min active driving simulation without forced ventilation
- 5-min recovery outside the climatic chamber

BMW M235i Racing car and Nürburgring circuit were selected on the simulator.

During the test, the following parameters were recorded: average skin temperature according to ISO 9886:2004 and average skin humidity, microclimate and interlayer temperature and humidity, amount of sweat produced during the driving simulation, amount of sweat still on the garment at the end of the test, heart rate. At the end of the test two thermografic images of the front and back of the suits were taken by sung the thermocamera Nec G100ex. Finally, subjective assessments of the thermal perception were collected via a questionnaire.

Temperature and humidity were measured by the wireless I-Button Maxim Integrated DS1923, with resolution of 0.0625°C and 0.04%. The sensors were placed on the skin through plasters as shown in Figure 2, on the skin districts prescribed by ISO9886:2004 to calculate average skin temperature according to the following equation:

$$\mathbf{T_{sk}} = T_{forehead}*0{,}07 + T_{scapula\ dx}*0{,}175 + T_{chest\ sx}*0{,}175 + T_{deltoid\ dx}*0{,}07 + T_{elbow\ sx}*0{,}07 + T_{hand\ sx}*0{,}05 + T_{tigh\ dx}*0{,}19 + T_{calf\ sx}*0{,}20 \quad (1)$$

The same multi-linear formula was adapted to the calculation of average skin humidity.

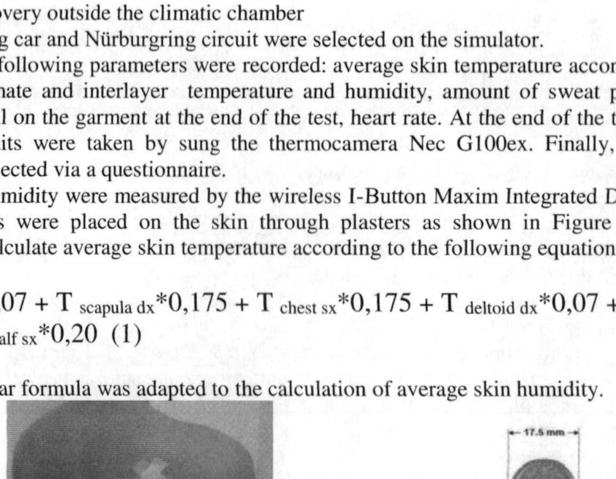

Fig. 2 Temperature and humidity sensors Maxim Integrate DS1923 placed on the skin.

III. RESULTS

Average skin temperature during driving simulation is shown in Figure 3. It can be observed that average skin temperature increased steeply during the acclimatization phase from the physiological level (32-33°C) to uncomfortably high level (over

38°C). During active driving, skin temperature stayed almost steady and it was slightly higher in the tropical outfit than in the racing and reflective suit.

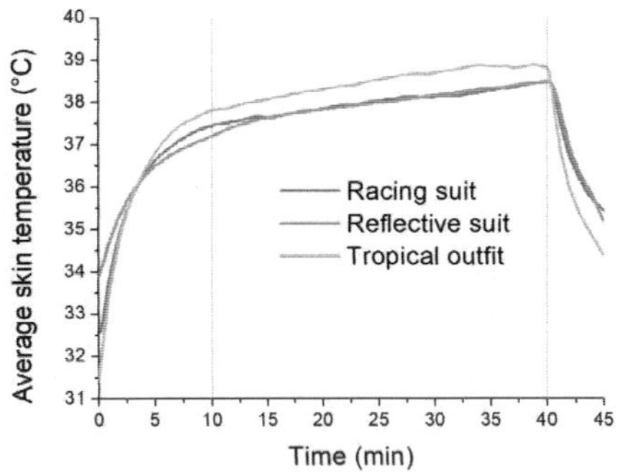

Fig. 3 Average skin temperature during the test (average value over three testers).

To appreciate the difference among the outfits, average skin temperature was shown in Figure 4 in terms of increase with respect to the initial value. It can be noticed that driving in a tropical outfit in extremely hot environment produced the greatest skin temperature increase while wearing a reflective outfit limited its increase to the minimum value. This observation confirmed that the racing suit had a protective role against heat flow from the environment and showed that incorporation of reflective components in the outfit could improve heat protection.

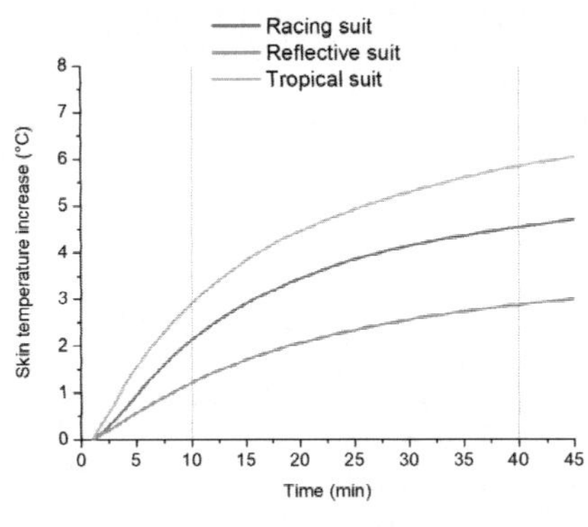

Fig. 4 Average skin temperature increase during the test (average value over three testers).

Microclimate humidity, given in Figure 5, confirmed that heat strain was more severe in case of the tropical suit, with humidity almost reaching saturation within the first two minutes of acclimatization due to abundant sweating. The results of Figure 5 also supported the hypothesis that heat reflection would be beneficial when racing in extreme conditions, with the lowest level of microclimate humidity measured in the reflective suit. .

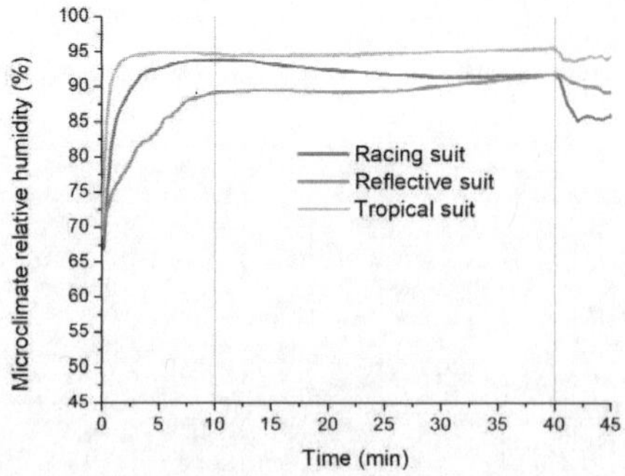

Fig. 5 Microclimate humidity (average value over three testers).

The amount of sweat produced during the racing simulation session (Table 1) was relevant, but not significant differences could be appreciated on the basis of the outfits.

TABLE I
AMOUNT OF PRODUCED SWEAT

TESTER	TROPICAL SUIT	RACING SUIT	REFLECTIVE SUIT
#1	655	675	640
#2	415	395	400
#3	1160	1065	1005

Strong individual differences among the testers in terms of amount of produced sweat can be observed but for all testers it was found that evaporative cooling was poor as most sweat was found on the outfit at the end of the test, especially with the racing and reflective suits. This is due to poor breathability of the racing outfits, made of three or four overlapped layers (underwear, 2-layer suit and optional reflective layer). Moreover, the driver was seated on an enveloping racing seat. Even though the percentage of residual sweat on the outfit was lower for the tropical suit, this result was not a signal of better thermal comfort but it was simply due to the smaller surface covered by the suit, much lower in case of the tropical outfit. Sweat, rather than evaporating from the skin, was most likely adsorbed by the seat cover during driving simulation.

TABLE II
PERCENTAGE OF RESIDUAL SWEAT IN THE OUTFIT

TESTER	TROPICAL SUIT	RACING SUIT	REFLECTIVE SUIT
#1	49%	71%	73%
#2	24%	66%	66%
#3	46%	78%	87%

Skin temperature variation of the different body districts during the test is shown in Figure 6 when wearing the conventional racing suit.

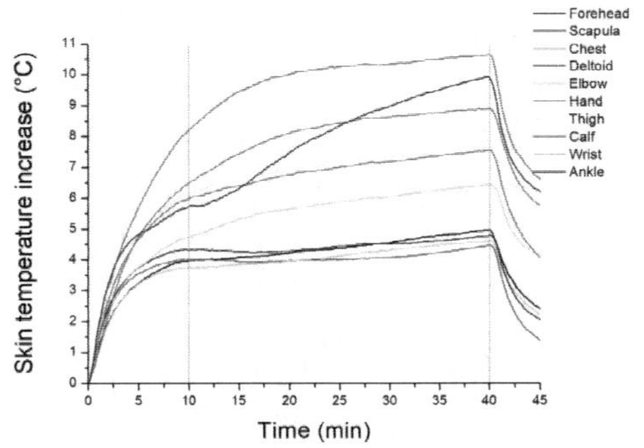

Fig. 6 Temperature increase of different skin districts.

Skin temperature of different districts at the beginning of the test and at the end in shown in Table 3. It can be observed that skin temperature reached an homogeneous high level at the end of the test, while it was much more variable at the beginning of the test, with extremities colder than head and torso, as expected. It is worth observing that body districts can be classified in three levels in term of temperature increase: 1) head and torso, whose temperature raised of 4-5°C; 2) peripheral districts, such as elbow, thigh and calf, with temperature rise around 6-7°C and 3) extremities, with temperature rise around 9-10°C.
The greatest increase was observed for the hand because it was covered by a single-layer glove, while most areas were covered by at least two layers. Hand and forehead were the only skin districts with temperature above 39°C.

TABLE III
SKIN TEMPERATURE IN DIFFERENT SKIN DISTRICTS

	Initial temperature (°C)	Final temperature (°C)	Temperature increase (°C)
FOREHEAD	34.1	39.1	5.0
SCAPULA	33.9	38.4	4.5
CHEST	33.9	38.5	4.6
DELTOID	33.6	38.4	4.8
ELBOW	32.1	38.5	6.4
THIGH	30.9	38.5	7.6
CALF	30.9	38.5	7.6
HAND	28.3	39.0	10.7
WRIST	29.9	38.8	8.9
ANKLE	28.4	38.3	9.9

Heart rate during the test is shown in Figure 7 for Tester #3 as example. It increased progressively during driving, although not reaching values as high as expected during racing. In the literature [6,7], heart rate could be as high as 190 bpm because of G-acceleration and high pressure on the sympatetic nervous system. These two aspects are missing or attenuated during driving simulation sessions.

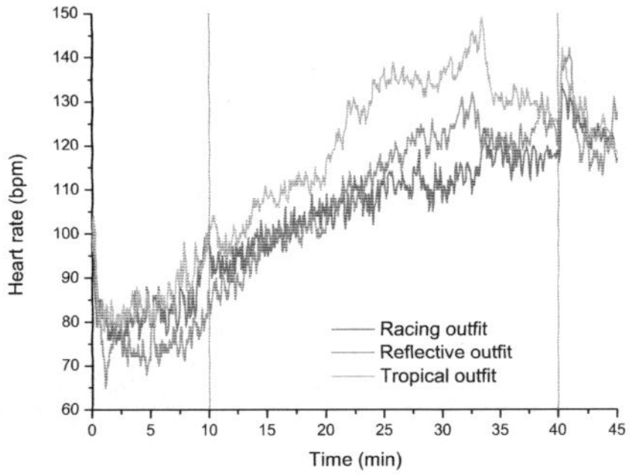

Fig. 7 Heart rate during driving simulation.

Finally, thermal perception after acclimatization and at the end of the test highlighted that the worst thermal and humidity perceptions (*very hot* and *dripping sweat*) were expressed when wearing the tropical outfit, as it did not shield heat flow from the environment, which was hotter than the human body inside the climatic chamber.
Front and back thermal images of the conventional racing suits at the end of the test showed that sewing are areas where heat shield was compromised. Small cold areas in blue represents the sweat soaked fabric, with lower temperature due to sweat evaporation.

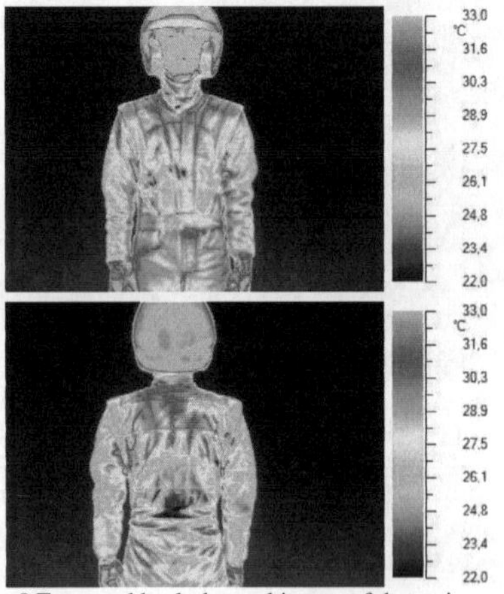

Fig. 8 Front and back thermal image of the racing outfit.

IV. Conclusion

Thermal strain during motor racing was simulated inside a climatic chamber. Three testers carried out a controlled driving simulation activity with three different outfits: a conventional racing suit, a reflective suit and a tropical suit. It was found that thermal strain can be partially attenuated by using a reflective outfit. The effort of driving in extreme conditions was demonstrated by huge increases of skin temperature, especially for extremities due to vasodilatation. It was found that evaporative cooling was very poor and most sweat stayed in the outfit or was absorbed by the seat cover.

Acknowledgments

The authors greatly acknowledge Regione Piemonte for the financial support of this work, within the regional project Sirtaki (POR FESR 2014-2020).

References

[1] Falkner, F., Isometric exercise and racing driving, in *Lancet* vol. 2, pp.1368–1369, 1972.
[2] Walker,S. M.,Ackland, T.R., Dawson,B., 2001. The combined effect of heat and carbon monoxide on the performance of motor sport athletes. Comp. Biochem. Physiol. A Mol. Integr.Physiol.128,709–718.
[3] Jacobs, P.L., Olvey, S.E., Johnson, B.M., Cohn, K.A. Physiological responses to high-speed, open-wheel racecar driving (2002) Medicine and Science in Sports and Exercise, 34 (12), pp. 2085-2090.
[4] Owen N.. King H., Lamb M., Literature review of race driver fatigue measurement in endurance motorsport, Procedia Engineering, 112 (2015) 344 – 348
[5] Carlson L.A, Ferguson D.P, Kenefick R.P., Physiological strain of stock car drivers during competitive racing, Journal of Thermal Biology, 44 (2014) 20-26.
[6] Baroody, N., J. Thomason, and E. O'Bryan. The heart of a 500-mile race. Am. Fam. Physician 8:184–189, 1973.

Multifunctionalization of cotton with onion skin extract

M G Silva[1]*, M A S D Barros[2], R T R Almeida[3], E J Pilau[3], E Pinto[4], A J S Ferreira[5], N T Vila[1], G Soares[6] and J G Santos[6]

[1]Department of Textile Engineering, State University of Maringá, Maringá, Brazil
[2]Department of Chemical Engineering, State University of Maringá, Maringá, Brazil
[3]Department of Chemical, State University of Maringá, Maringá, Brazil
[4]Faculty of Pharmacy, University of Porto, Porto, Portugal
[5]Department of Engineering, Federal University of Santa Catarina, Blumenau, Brazil
[6]2C2T-Center for Textile Science and Technology, University of Minho, Guimarães, Portugal

Corresponding author: mgsilva4@uem.br

Abstract The present study aimed to develop a dyeing process of cotton fabrics with natural onion skin extracts in order to obtain dyed fabrics with antimicrobial and anti-UV properties. The extract was characterized by LC-ESI-MS/MS and the antimicrobial activity of the extract was also evaluated. The dyeing process was optimized considering the influence of the temperature, pH and electrolytes. The effect of cotton pre-treatment with chitosan in the dye uptake was also evaluated. The dyed fabrics were characterized according to antimicrobial activity, anti-UV protection and fastness properties.

***Keywords*— Cotton, natural dyes, onion skin, UV protection, antimicrobial, chitosan**

I. Introduction

IN recent years, the textile industry has been severely criticized for their contribution to environmental pollution. Therefore, ecological and economic restrictions have been established, including the prohibition of certain synthetic dyes and toxic finishing agents. All these factors led to a greater motivation for the use of natural products, improving sustainability of dyeing and finishing textile processes [1,2]. Moreover, it is possible to extract natural dyes from byproducts such as food, wood and agricultural waste at lower costs. Additionally, these extracts have other interesting properties that allow the functionalization of textiles with UV protection [3] and antimicrobial activity [4]. Among the natural extracts, those obtained from onion skin, mainly composed by quercetin, seem to be interesting for textile applications [5]. The skin, as well as the non-edible apical dry parts of the onion bulbs, are removed prior to processing, constituting an abundant, cheap and readily available agricultural byproduct. Such residues have been used for textile dyeing, since quercetin is a natural colorant that provides several brown hues to natural fibers [5].
In present investigation, the main objective was to develop a dyeing process of cotton fabrics with natural onion skin extracts in order to obtain dyed fabrics with antimicrobial and anti-UV properties.

II. Experimental Section

A. *Materials and methods*

Bleached cotton fabrics (225 gm^{-2}) were used in this study. The onion skin extract was obtained in aqueous medium. The mass spectrometry analysis was performed using a Waters pump model 1525µ (Milford, USA) using a Waters Symmetry® C18 (4.6 x 75 mm x 3.6 µm) column, along with Quattro micro API Waters (Beverly, EUA) triple-quadrupole mass spectrometer with electrospray as ionization source. The optimization of the dyeing process was performed in an Ahiba Texomat Turbo Color. The dyed samples were evaluated through their color strength (K/S) and color intensity (I) using a Datacolor 550 spectrophotometer. Washing fastness tests were performed according to ISO 105-C06:2010 standard test in a Linitest. The rubbing fastness tests were carried out according to ISO 105-X12:2016 standard method using a Crockmeter. The UPF was measured using a SDL-M350 spectrophotometer, according to the standard procedure AS/NZS 4399:1996. The antimicrobial activity of fabrics was performed according to a non-standard qualitative method as described by Magalhães et al. [6].

III. Results and Discussion

A. *Optimization of the extraction process*

The UV-Vis spectra of onion skins extracts obtained at different extraction temperatures are shown in Figure 1. It was observed that the spectra of onion skin extract change with the temperature, as it can be seen in Figure 1a).

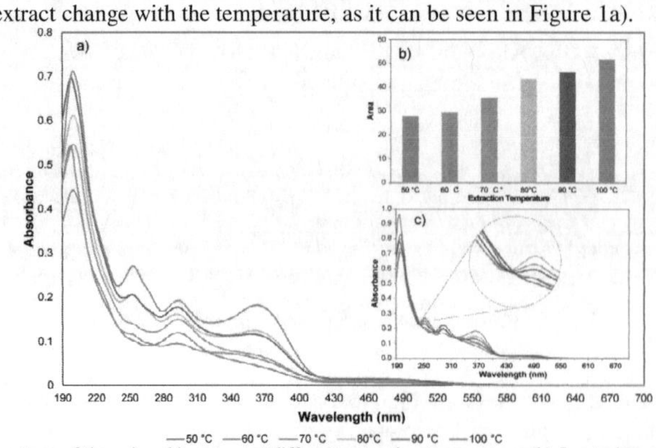

Figure 1. (a) UV-Vis absorption spectrum of the onion skin extract at different extraction temperatures (b) Spectral area of the solutions obtained at different temperatures and (c) Normalized spectra of extracts of onion skin obtained at different temperatures.

For all extraction temperatures, the maximum absorption peak was found at 199 nm. However, at temperatures under 80 °C, the spectra present lower definition for peaks above 199 nm. This possibly indicates that the extraction may not complete under this temperature. Also, the extracts of onion skin showed an intense brown coloration at pH 4.

Figure 1b) shows the areas of the spectra of solutions obtained at different extraction temperatures. It was observed that increasing temperature favored a higher absorbance of the onion skin extract. Therefore, at higher temperatures there is a higher concentration of extracted compounds and, consequently, a greater efficiency of the extraction process. The normalized spectra of the different solutions (Figure 1c) present three isosbestic points. This means that the qualitative composition of the extract remains unchanged regardless of the extraction temperature used. The samples are characterized by the presence of the same compounds but in different proportions [7].

It is also verified that the onion skin extract is able to absorb radiation in the UVA (315-400 nm) and UVB (280-315 nm) regions. Thus, it can be expected that their absorption by textile fabrics can increase the protection level conferred to harmful UV radiations.

B. *Chemical characterization of the extract*

In the mass spectra of onion skin extract were identified five characteristic ions of phenolic compounds. Their identities were confirmed through the analysis of ion fragments of the molecules and the data was compared to those provided in the literature.

The compounds identified in the onion skin extract using HPLC-MS/MS is described in Table 1, in addition to its retention time, observed mass $[M\text{-}H]^-$, ion fragmentation profile and the consulted literature. It has been found that among the five compounds identified in the extract, four are quercetins, in free or glycosylated form. In fact, quercetin is the most abundant flavonoid found in many onion species [8], and it can be used as a natural dye that offers different shades of brown in textile materials produced with natural fibers [5].

TABLE I.
CHEMICAL COMPOSITION OF ONION SKIN EXTRACT AND CORRESPONDING FRAGMENTATION PROFILE. IDENTIFICATION ACCOMPLISHED BY MS/MS SPECTRA COMPARISON WITH LITERATURE

Number	T_R (min)	Compound name	$[M-H]^-$ (m/z)	MS/MS	Reference
1	7.08	Protocatechuic acid	153	109	[13]
2	7.56	Quercetin diglycoside	625	463, 301	[14]
3	7.93	Quercetin diglycoside	625	463, 301	[14]
4	8.66	Quercetin-hexoside	463	301, 300, 271, 254, 179, 151	[15]
5	11.07	Quercetin	301	273, 257, 229, 179, 151, 107	[16]

C. Antimicrobial properties of the extract

Antimicrobial activity against Gram-positive and Gram-negative bacteria and *C. albicans* was evaluated using the plate microdilution method. The results showed that the *S. aureus* was relatively sensitive to onion extract, whereas *E. coli* was very resistant until the maximum concentration tested. The fungus showed complete resistance [9].

D. Dyeing process

By analyzing the results presented in Fig. 2 it is possible to observe that the increase of the dyeing temperature favored the exhaustion of the dye extract in the fiber. This can be seen by the color intensity of samples. The best results were obtained at 100 °C. This behavior can be assigned to the higher kinetic energy of the dye molecules and consequently, to their increasing migration ability at higher temperatures [10-11], which favors the multiplication of the number of colored molecules that interact with the active sites on the surface of the fabrics [12].

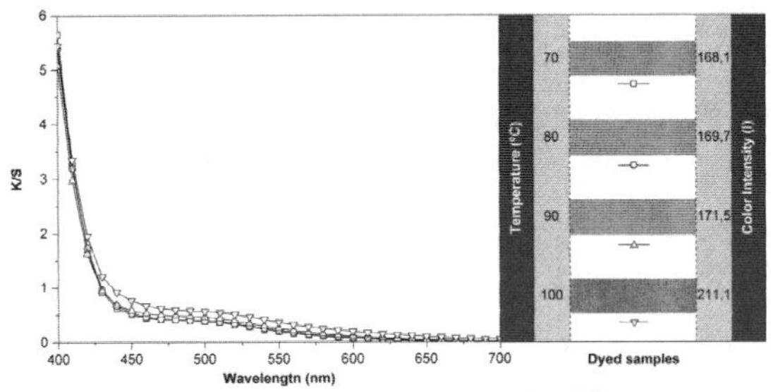

Figure 2. Dyeing of cotton with onion skin extract - K/S curves, Color intensity and samples dyed at different temperatures tested.

The study of the influence of pH on the dyeing of cotton with the onion skin extract shows that the color intensity decreases with the pH of the dyeing solution (Figure 3). The best results were obtained at pH 4.0.

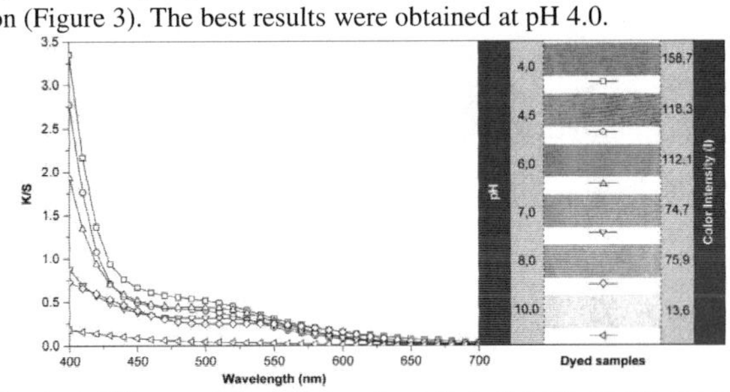

Figure 3. Dyeing of cotton with onion skin extract - K/S curves, Color intensity and samples dyed at different pH values tested.

The effect of the dyebath pH on the dyeing process yield is related with the ionic interactions between ions in solution and chemical groups on surface of cotton [4]. Some of the hydroxymethyl (-CH_2OH) groups presented in the cotton fibers are naturally oxidized to carboxylic groups during growth and subsequent processes. Consequently, cellulosic fibers generally exhibit a slightly negative charge when immersed in aqueous dyeing solutions. If the pH of the solution is increased above 8, some of the hydroxyl groups present on the hydroxymethyl side chains may also ionize, increasing significantly the negative charge. This negative electric potential at the surface of the cellulose fibers contributes to repel anions at higher pH values. At pH below 4, the carboxyl and hydroxyl groups along the polymer chain of cellulosic fibers are poorly ionized, which considerably reduces the negative surface electrical potential of the fiber. As a result, it reduces the repulsive action on the dyebath anions. In this context, the dye molecules can easily reach the surface of the cotton fibers and interact with them throughout the formation of hydrogen bonds, along with some ion-dipole interactions. Therefore, the negative charge on the adsorbent surface of the cellulose clearly creates an adverse situation for the adsorption of the onion skin extract at alkaline pH. For this reason, the best results were obtained under acidic conditions [17].

The addition of salt to the dyebath increased the color intensity of dyed cotton fabrics (Figure 4). This behavior can be attributed to the neutralization of the negative surface charge by the sodium ions in the dyeing solution, which favors the absorption of colored compounds [11].

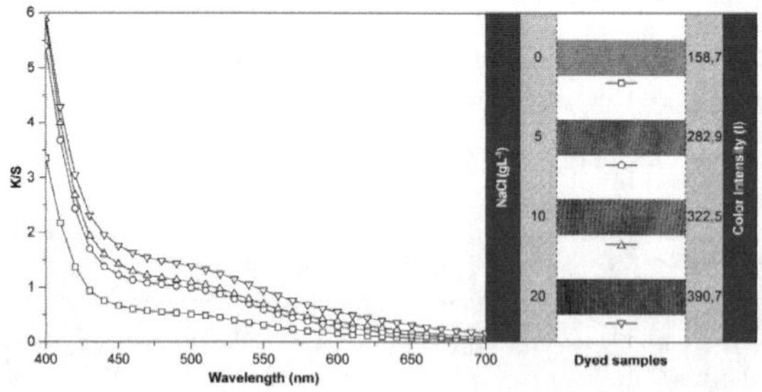

Figure 4. Dyeing of cotton with onion skin extract - K/S curves, Color intensity and samples dyed with different concentrations of NaCl.

Pre-treatment of cotton with an aqueous solution of chitosan (1.5% w/v) showed a significant increase in the color strength of dyed samples (Figure 5). As can be seen, the color intensity of the pre-treated chitosan sample is considerably higher than that of the bleached samples dyed with and without NaCl. Therefore, it can be concluded that chitosan pre-treatment increases the dye uptake due to the interaction between the cationized surface and colored natural compounds.

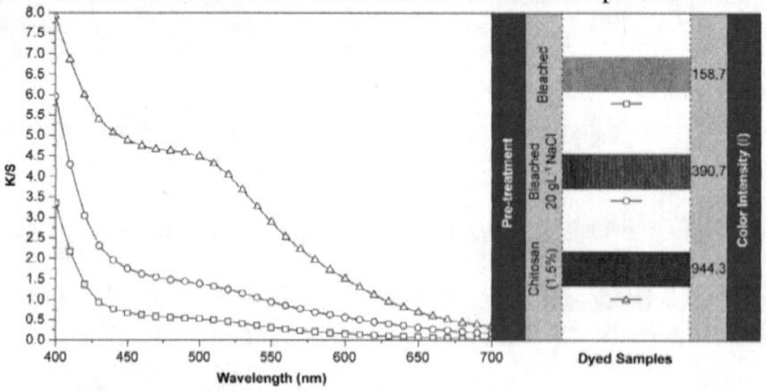

Figure 5. Dyeing of cotton with onion skin extract - K/S curves, Color intensity and samples with and without pre-treatment with chitosan dyed in the presence and absence of NaCl.

E. UV protection factor of dyed materials

The UPF results of dyed cotton fabrics with onion skin extract are shown in Figure 6.

In general, samples dyed without pre-treatment with chitosan showed a statistically significant increase of the average UPF index when compared to the bleached samples (A1).

However, the differences between the UPF mean values of the samples dyed with extract concentrations higher than 30 gL^{-1} (A4 to A6) are not statistically significant according to the Tukey test at a significance level $\alpha = 0.05$. Based on these results, dyed samples with extract concentration of 10 gL^{-1} have a UPF index that can be classified as very good, while dyed samples with extract concentrations equal to or higher than 20 gL^{-1} can be classified as having excellent protection against UV radiation.

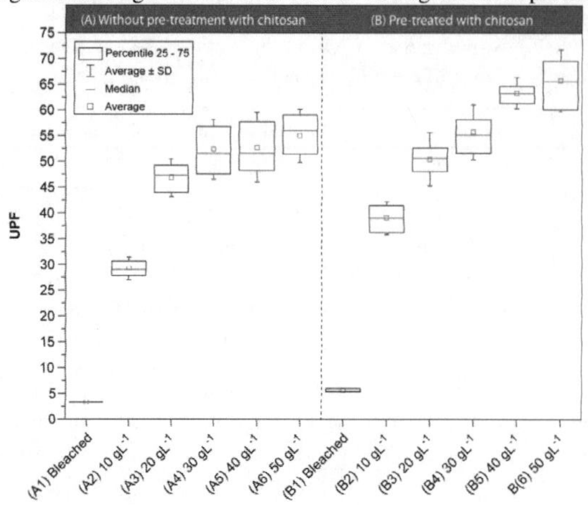

Figure 6. Boxplot from de UPF values of samples with and without previous treatment with chitosan, after dyeing with onion skin extract at different concentrations

Similarly, samples dyed after the pre-treatment with chitosan showed a statistically significant increase of their UPF index average when compared to the bleached samples pre-treated with chitosan (B1).

It was also observed that the UPF index undergoes a statistically significant increase when the concentration of the extract used in the dyeing increases to 40 gL^{-1}. Thus, the materials pre-treated with chitosan and subsequently dyed had UPF indexes that ranged from very good, for samples dyed with 10 gL^{-1} extract (B2), to excellent in the case of samples dyed with extracts of concentration equal or higher than 20 gL^{-1} (B3 to B6).

F. Evaluation of antimicrobial activity of dyed materials

Table 2 presents the results of the antimicrobial evaluation of cotton samples.

TABLE II.
ANTIMICROBIAL ACTIVITY OF COTTON FABRICS DYED WITH ONION SKIN EXTRACT

Sample	Microorganisms		
	E. coli	*S. aureus*	*C. albicans*
Dyed (30 gL^{-1})	(+)	(-)	(+)
Dyed (50 gL^{-1})	(+)	(-)	(+)
Pre-treated with chitosan (1.5%) and dyed (30 gL^{-1})	(+)	(-)	(+)
Pre-treated with chitosan (1.5%) and dyed (50 gL^{-1})	(+)	(-)	(+)
Pre-treated with chitosan (1.5%)	(+)	(-)	(+)
Bleached	(+)	(+)	(+)

(+): Growth of the microorganism in the textile; (-): Absence of growth of the microorganism in the textile

In samples subjected to sterility control no growth of microorganisms was observed, confirming their sterility. On the other hand, in the samples used as negative control there was growth, as expected, for all the microorganisms.

All dyed samples showed activity only against *S. aureus.* whereas, against *E. coli* and *C. albicans*, no sample showed inhibitory effect.

Comparing the bioactivity results of the pre-treated chitosan samples with those without pre-treatment, it is possible to suppose that the application of chitosan provided higher diffusion of antimicrobials, since the zone of inhibition formed was significantly larger.

G. Evaluation of washing and rubbing fastness properties of dyed materials

The washing and rubbing fastness properties were evaluated. The washing fastness of bleached dyed fabrics was 2-3. However, the pre-treatment with chitosan promoted an improvement in the washing fastness level, especially regarding to the color change of the dyed substrate. In turn, the rubbing fastness obtained was good, ranging from 3-4 to 5 and it was slightly improved in the samples pre-treated with chitosan.

IV. CONCLUSIONS

The results showed that it is possible to dye cotton fabrics with natural extracts of onion skins, obtaining acceptable levels of fastness and antimicrobial and anti-UV properties. The best color intensity was obtained by adding 20 gL^{-1} of NaCl to dyebath and carrying out the dyeing at 100 °C and pH 4.0. However, the chitosan cotton pre-treatment proved to be more effective, since it allowed increased dye uptake compared to electrolyte use. In this context, chitosan could successfully replace the application of large amounts of salt in the dyeing process, making them more sustainable.

The pre-treatment with chitosan increases UPF values in all cases, due to the increase of dye exhaustion.

The onion skin extract showed antimicrobial activity only against *S. aureus*.

The fabrics pre-treated with chitosan showed better washing and rubbing fastness indexes.

To sum up, the developed dyeing and finishing process allows us to obtain sustainable dyed and multifunctional cotton materials as they are produced with low environmental impact and using only natural based products.

ACKNOWLEDGMENT

This research was partially supported by the Strategic Funding UID/Multi/04423/2013 through national funds provided by FCT – Foundation for Science and Technology and European Regional Development Fund (ERDF), in the framework of the program PT2020.

This work was also financed by FEDER funds through the Competitiveness Factors Operational Program- COMPETE and by national funds through FCT- Foundation for Science and Technology within the scope of project POCI-01-0145-FEDER-007136

References

[1] M. Shahid and F. Mohammad, "Recent advancements in natural dye applications: a review," Elsevier Ltd, 2013.
[2] T. Bechtold, R. Mussak, A. Mahmud-Ali, E. Ganglberger, and S. Geissler, "Extraction of natural dyes for textile dyeing from coloured plant wastes released from the food and beverage industry," J. Sci. Food Agric., vol. 86, no. 2, pp. 233–242, Jan. 2006.
[3] R. Mongkholrattanasit, J. Kryštůfek, and J. Wiener, "Dyeing, Fastness , and UV Protection Properties of Silk and Wool Fabrics Dyed with Eucalyptus Leaf Extract by the Exhaustion Process," Fibres Text. East. Eur., vol. 19, no. 3, pp. 94–99, 2011.
[4] N. F. Ali, R. S. R. El-mohamedy, and S. Rajput, "Improvement of antimicrobial activity for onion natural dyed fabrics through chitosan pretreatment," J. Appl. Sci. Res., vol. 9, no. 8, pp. 4993–5002, 2013.
[5] P. S. Vankar, R. Shanker, and S. Wijayapala, "Dyeing of cotton, wool and silk with extract of Allium cepa," Pigment Resin Technol., vol. 38, no. 4, pp. 242–247, Jul. 2009.
[6] A. P. A. Magalhães, "Atividade antimicrobiana em têxteis" Universidade do Porto, 2015.
[7] O. Thomas and V. Cerda, "From spectra to qualitative and quantitative results," in UV-visible spectrophotometry of water and wastewater, O. Thomas and C. Burgess, Eds. Oxford: Elsevier Ltd., 2007, pp. 21–45.
[8] K. S. Yoo, E. J. Lee, and B. S. Patil, "Quantification of Quercetin Glycosides in 6 Onion Cultivars and Comparisons of Hydrolysis-HPLC and Spectrophotometric Methods in Measuring Total Quercetin Concentrations," J. Food Sci., vol. 75, pp. 160–165, 2010.
[9] J. Santas, M. P. Almajano, and R. Carbó, "Antimicrobial and antioxidant activity of crude onion (*Allium cepa L.*) extracts," Int. J. Food Sci. Technol., vol. 45, no. 2, pp. 403–409, 2010.
[10] S. Ali, T. Hussain, and R. Nawaz, "Optimization of alkaline extraction of natural dye from Henna leaves and its dyeing on cotton by exhaust method," J. Clean. Prod., vol. 17, no. 1, pp. 61–66, Jan. 2009.
[11] N. Nasirizadeh, H. Dehghanizadeh, M. E. Yazdanshenas, M. R. Moghadam, and A. Karimi, "Optimization of wool dyeing with rutin as natural dye by central composite design method," Ind. Crops Prod., vol. 40, pp. 361–366, Nov. 2012.
[12] M. Chairat, S. Rattanaphani, J. B. Bremner, and V. Rattanaphani, "Adsorption kinetic study of lac dyeing on cotton," Dye. Pigment., vol. 76, no. 2, pp. 435–439, Mar. 2008.
[13] F. Sánchez-Rabaneda, O. Jáuregui, I. Casals, C. Andrés-Lacueva, M. Izquierdo-Pulido, and R. M. Lamuela-Raventós, "Liquid chromatographic/electrospray ionization tandem mass spectrometric study of the phenolic composition of cocoa (Theobroma cacao)," J. Mass Spectrom., vol. 38, no. 1, pp. 35–42, 2003.
[14] J. Lee and A. E. Mitchell, "Quercetin and isorhamnetin glycosides in onion (Allium cepa L.): Varietal comparison, physical distribution, coproduct evaluation, and long-term storage stability," J. Agric. Food Chem., vol. 59, no. 3, pp. 857–863, 2011.
[15] E. S. S. Abdel-Hameed, S. A. Bazaid, and M. S. Salman, "Characterization of the phytochemical constituents of taif rose and its antioxidant and anticancer activities," Biomed Res. Int., vol. 2013, 2013.
[16] M. Wang, A. Cao, C. Ouyang, Y. Li, and Y. Wei, "Rapid screening and identification of non-target flavonoid components in invasive weeds by LC/MS-IT-TOF," *Anal. Methods*, vol. 7, no. 24, pp. 10207–10216, 2015.
[17] S. Rattanaphani, M. Chairat, J. B. Bremner, and V. Rattanaphani, "An adsorption and thermodynamic study of lac dyeing on cotton pretreated with chitosan," Dye. Pigment., vol. 72, no. 1, pp. 88–96, Jan. 2007.

Effect of polyamide ratio on the flammability of wool/PA blended aircraft seat fabrics

M E Üreyen[1,2], M Torsun[3], Ü.Y. Yilmaz[1,4] and A S Koparal[4]

[1] *Anadolu University, Civil Aviation Research Center, İki Eylül Campus Eskişehir 26555, Turkey*
[2] *Anadolu University, Faculty of Architecture and Design, Fashion and Textile Design Department, Yunusemre Campus, Eskişehir, 26473, Turkey*
[3] *Kadifeteks Mensucat San. A.Ş., İkitelli OSB Mah. Eski Turgut Özal Cad. No:40 Başakşehir/İstanbul, Turkey*
[4] *Anadolu University, Faculty of Engineering, Environmental Engineering Department, İki Eylül Campus Eskişehir 26555, Turkey*

Corresponding author: meureyen@anadolu.edu.tr

Abstract Wool and polyamide blended fabrics are widely preferred for aircraft seats. Due to the requirements of the flammability regulations of airline authorities, aircraft fabrics have to pass vertical flammability and smoke density tests. In this work, polyamide/wool blended yarns produced in five different ratios. Yarn samples were knitted on a circular knitting machine. Fabric samples were dyed in overflow dyeing machine. Following the washing of the fabric samples, well known low smoke Zirpro flame retardant treatment based on a zirconium salts were applied. The effect of polyamide content on the flammability properties, the thermal behavior and heat release properties of the wool/polyamide blended knitted fabric samples were investigated. For this aim, cone calorimetry, thermogravimetric analyses, smoke density and vertical burn tests were performed. Results showed that polyamide content decreased the flame retardancy of the wool fabrics.

***Keywords*—aircraft seat, flame retardant, polyamide, thermal behavior, wool**

I. Introduction

Textiles used in aircraft interiors including carpeting, curtains and seat coverings must meet flammability regulations throughout the world. Wool is known as the most flame retardant fiber among natural fibers because of its protein structure and chemical composition. Wool contains high level of moisture, nitrogen and Sulphur. The high Sulphur (3-4 wt%) and nitrogen content (15-16 wt%) present in both chain and side groups provide low flammability to wool fiber. Wool also contains about 15wt% moisture under standard atmospheric conditions (20 °C and 65% RH) [1]. The natural flame resistant properties of the wool (LOI value is approx.. 25%) are associated with these properties [2].

Polyamide (PA) fibers offer high strength and abrasion resistance to wool fabrics. On the other end, small percentage of PA fibers may improve the processing performance and durability of the final products. PA is also preferred for the blending with wool when the fabric is to be dyed because PA can be dyed with wool in the same bath. Therefore, aircraft seat fabrics are generally produced by blending with wool and PA fibers. On the other aliphatic polyamide fibers such as nylon 6 and 6.6 are less flame retardant than wool. Their LOI values are about 21.

Many flammability treatments for wool and wool blends did not satisfy the requirements of the flammability regulations of airline authorities. Therefore, a new flame retardant treatment process (Zirpro) were developed by the former International Wool Secretariat, based on the reaction of zirconium and titanium salts with wool [3]. Durable flame retardant finishing of wool fabrics based on Zirpro process is dominated by finishing agents includes the potassium hexafluorozirconate (K_2ZrF_6). Zirpro treatments are mainly based on the exhaustion of negative charged zirconium or titanium salts, under acid conditions, onto positively charged wool. Finishing agent should be applied under acidic (pH≤3) conditions. It was claimed that the application of only about 3% flame retardant to the fiber causes a negligible effect on properties such as handle. Zirpro application generally increases the smoke generation. Therefore generally, zirconium acetate solution is also applied to the fabrics after zirconium or titanium salt application for low smoke emission [4].

In this work, the effect of polyamide content on the flammability properties and the thermal behavior of the wool-PA blended knitted fabric samples were investigated. For this aim, cone calorimetry, thermogravimetric analyses, smoke density and vertical burn tests were used.

II. Methods and Procedures

Potassium hexafluorozirconate (Aflammit ZR) and Zirconium acetate solution (Aflammit ZAL) are obtained from THOR Specialities Ltd. Company. Citric acid (monohydrate) and formic acid were supplied from Merck, Germany. Polyamide 6.6 (3.3 dtex, 85-90 mm) and wool fibers (23 micron, 68 mm) in top form were blended on gilling machine in five different blend ratio (100%wool, %95-5 wool-PA, 90-10% wool-PA, 85-15%wool-PA and 80-20% wool-PA). Nm 36/2 yarns were produced on ring spinning and doubling/twisting machines. Yarn samples were knitted on a circular knitting machine (Ipekcioglu, Turkey).

Fabric samples were dyed in overflow dyeing machine (Mathis AG, Switzerland). Following the washing of the fabric samples, low smoke Zirpro treatment was applied with 3% Aflammit ZR and 10% Aflammit ZAL for 45 min at 70ºC (liquor ratio was 1:20) in the same machine. Citric acid and formic acid were added to the solutions in order to maintain a pH 2 to 2.5 during the exhaustion process and decrease the smoke production. The treated samples were rinsed with deionized water and dried at room temperature.

Flame retardant flexible polyurethane foams, which are suitable for aircraft seating (100x100x30 mm^3), were covered by the fabric samples according to ASTM E1474-14 test standard. Fabric/foam combinations were tested by cone calorimetry test device (Fire Testing Technology, UK) under 35 kW/m^2 heat flux. Samples were tested without a retainer frame. Time to ignition (TTI), peak heat release rate (HRR), total heat release (THR), smoke production rate (SPR), specific extinction area (SEA), mass loss rate (MLR) and maximum average rate of heat emission (MARHE) values were measured and compared. Thermal analyses were carried out by SDT Q600 (TA Instruments, USA) at heating rates 10ºC/min under N_2. Vertical flammability tests were performed by FAA multi-purpose small scale flammability test chamber (Govmark, FAA MP-1-X, USA) according to FAR Part 25 App. F Part 1(ii). Burn length, after flame time and drippings flame time after exposing to bottom of the fabric sample (305x75 mm) for 12 s were calculated. According to FAA, the average burn length must not exceed 203 mm (8 in), and the average after flame time must not exceed 15 s. Smoke density tests were performed according to the FAR 25.853 (d), (App. F to part 25, part V). Specific optical smoke density value (D_s) which is a dimensionless measure of the amount of smoke produced per unit area when a material is exposed to both flaming and radiant heat sources was measured. The maximum value of D_s that occurs during the first 4min of the test, ^4D_m should be less than 200.

All samples were conditioned at 20 °C, 50% RH for at least 24 hours before testing.

III. Results and Discussion

Fabric weight test results before and after FR treatment are shown in Table I. According to test results fabric weight of the samples increased about 40% percent after dyeing and flame retardant treatment.

TABLE I.
FABRIC WEIGHT TEST RESULTS OF THE SAMPLES.

	Wool (%)	Fabric weight (g/m^2)	
		Before application	After application
1	100	227.5	336.5
2	95	236.6	330.9
3	90	243.6	334.8
4	85	255.1	344.9
5	80	249.3	348.1

Thermogravimetric (TG) and derivative thermogravimetric (DTG) curves of the wool, polyamide fibers and dyed fabric samples before flame retardant finishing are shown in Fig. 1. The TGA data are listed in Table II where $T_{10\%}$ and $T_{50\%}$ are defined as the temperatures corresponding to weight loss of 10% and 50%.

Fig. 1 (a) shows that the thermal degradation of the untreated fabric samples includes two mass loss steps. In first step changes of the samples occur in its physical properties and weight loss were observed from 40 to 125 °C which was attributed to the desorption of water physically adsorbed by wool fibers. The second step is the main pyrolysis step occurred from 200 to 550 °C which was attributed a number of pyrolysis reactions take place in wool and polyamide fibers. The hydrogen-bond peptide helical structure ruptures and the ordered part of the wool protein undergo a solid-to-liquid phase change at 230 to 240 °C; also, the destruction of cystine disulphide bonds occurs and a number of volatiles including hydrogen sulfide and sulphur dioxide are released between 250 and 295 °C. Char forming reactions with dehydration and loss of other volatiles occur above 250 °C [5, 6]. The DTG curve of the polyamide fiber showed peak at 441 °C corresponding to thermal decomposition of polyamide. Major weight loss of the polyamide fibers are observed in the 360-500 °C region. Polyamide fibers lost about 94% of its weight at 600 °C.

Thermogravimetric test results indicated that the decomposition of wool and polyamide interact in the 350-500 °C region. $T_{10\%}$ and $T_{50\%}$ values are increased while the polyamide content increased because of the very low water content of the polyamide fibers and loss of H_2O, NH_3 and loss of H_2S, SO_2 and other small molecular weight volatiles from wool in 200-300 °C region [6]. Whereas the polyamide content increase, char residues of the samples decreased at 600 °C and 700 °C because of the thermal degradation of polyamide before 500 °C and enhanced char formation of wool above 450 °C. As the polyamide ratio increased in the fabric content, a second peak appeared prominently in the DTG curves at 410 °C.

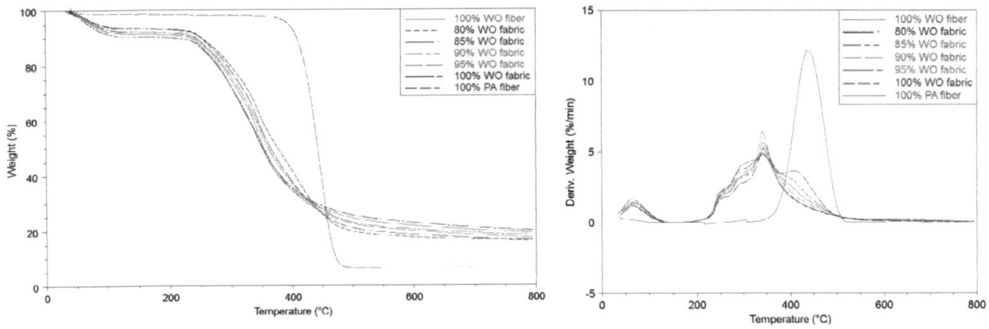

(a) (b)

Fig. 1. TG (a) and DTG (b) curves of the wool, polyamide fibers and untreated samples.

TABLE II.
TGA DATA OF THE WOOL, POLYAMIDE AND UNTREATED SAMPLES UNDER NITROGEN

Sample Code	$T_{10\%}$(°C)	$T_{50\%}$(°C)	Char Residues at 600 °C	Char Residues at 700 °C	pDTG (°C)
1	197.7	351.7	22.8	21.1	345.3
2	239.4	353.3	20.9	19.4	345.0
3	242.7	360.3	20.5	19.1	347.9
4	249.9	364.3	18.5	17.3	343.5
5	251.9	375.8	17.6	16.8	343.6
WO fiber	233.2	350.3	21.8	20.1	342.4
PA fiber	408.7	442.1	6.2	6.1	440.9

The thermogravimetric (TG) curves and derivative thermogravimetric (DTG) curves of the treated fabric samples are shown in Fig. 2. The TGA data are listed in Table III. Similar reaction peaks were observed when compared with the untreated fabrics. $T_{10\%}$ and $T_{50\%}$ temperatures of the treated fabrics are slightly higher than untreated fabrics.

DTG curves showed that second step happen below the thermal degradation of untreated fabric samples. It means that flame retardant treatment reacts on condensed phase [7], catalyzes the thermal decomposition of wool and accelerates the formation of volatiles from the samples. Second peak were also observed at treated fabrics about 380 °C.

When the samples compared, samples having higher polyamide fiber ratio exhibit lower weight loss at low temperatures and higher weight loss at high temperatures. This shows that the addition of polyamide changes the decomposition behavior of wool and decrease the thermal stability of the fabric. Char residues of the samples decrease while the polyamide content increase. On the other hand, amount of residue of the all samples were increased about 5% at 600 and 700 °C. It showed that the flame retardant finishing contribute the formation of more thermally stable carbonaceous char.

IOP Conf. Series: Materials Science and Engineering **460** (2018) 012033 doi:10.1088/1757-899X/460/1/012033

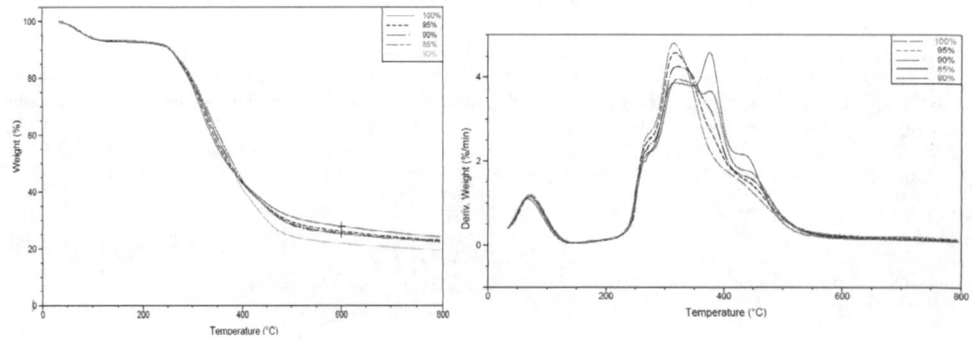

(a) (b)

Fig. 2. TG (a) and DTG (b) curves of the treated samples.

TABLE III.
TGA DATA OF THE TREATED SAMPLES.

Sample Code	$T_{10\%}$(°C)	$T_{50\%}$(°C)	Char Residues at 600 °C	Char Residues at 700 °C	pDTG (°C)
1	255.0	368.8	27.9	25.8	314.8
2	257.8	371.7	26.1	24.3	316.8
3	257.8	372.3	26.0	23.9	321.7
4	257.9	380.9	25.3	23.7	380.9
5	266.3	387.9	22.9	21.6	382.4

Cone calorimetry test results of the untreated and treated fabric samples are shown in Table IV. As reported by Price et.al [8] flexible polyurethane foams are easily ignited during the cone calorimeter test, even if they were treated by flame retardant. Therefore, very low TTI values were obtained.

HRR graphs of the samples are given in Figure 3. The peak heat release rate (HRR) of the treated samples decreased when compared with untreated samples under the same conditions. Flame retardancy of the fabrics decreased when the polyamide content increased. The lowest value was obtained by 100% WO fabrics. Specific extinction area values and SPR values were increased slightly after treatment (except 100% WO fabrics). It shows that samples continued to release additional smoke due to smoldering. MLR values of the 100%, 95% and 90% WO samples decreased whereas others remained the same. It is well known that wool fabric forms carbonaceous char while thermoplastic polyamide is melt during burning. These results showed that char was cracked and flaming increased. As a result, HRR and THR (MJ/m^2) values of samples having higher polyamide content higher than the other samples. MARHE values of the treated fabrics reduced when compared with untreated samples.

TABLE IV.
CONE CALORIMETRY TEST RESULTS OF THE UNTREATED AND TREATED SAMPLES.

	TTI (s)		HRR (peak) (kW/m^2)		THR(0-300s) (MJ/m^2)		SPR(av) (m^2/kg)		SEA (300s) (m^2/kg)		MLR(300s) (g/s)		MARHE (kW/m^2)	
Sample Code	BT*	AT*	BT*	AT*	BT*	AT*	BT*	AT*	BT*	AT*	BT*	AT*	BT*	AT*
1	12	13	320.65	174.94	14.44	11.20	0.0021	0.0014	188.73	172.18	0.0222	0.0206	158.13	100.20
2	11	13	347.79	238.84	13.93	11.13	0.0039	0.0027	136.94	261.68	0.0212	0.0200	158.81	119.14
3	12	14	349.49	243.89	15.21	14.76	0.0030	0.0032	237.14	277.14	0.0232	0.0213	166.96	133.28
4	10	13	343.22	264.00	14.71	14.02	0.0034	0.0043	290.15	308.60	0.0216	0.0216	170.19	145.07
5	14	14	340.36	303.91	16.30	13.69	0.0011	0.0028	166.99	258.34	0.0231	0.0231	179.23	153.99

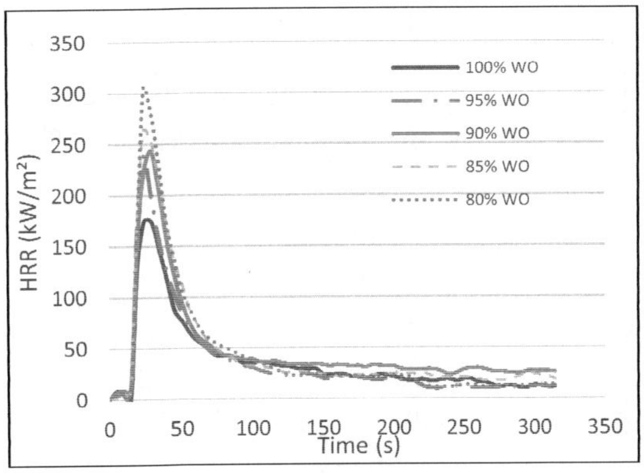

Fig. 3. HRR graphs of the treated fabric samples.

Ds values of the treated samples after smoke density test are given in Figure 4. As pointed out maximum ^4D_m value of the sample should be less than 200. Measured ^4D_m values of all treated samples were ranged between 31 to 59.

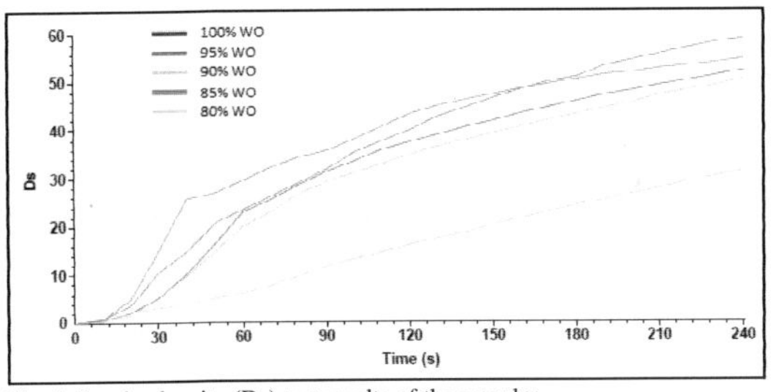

Fig. 4. Smoke density (Ds) test results of the samples.

Images of the samples after vertical flammability test are shown in Fig. 5. After flame time and burn length results are given in Table V. It is observed that an increase in polyamide content leads to an increase in after flame time and burn length. The results showed that the flame retardant application increased the flame retardancy of the samples. It can be concluded that %85 WO and %80 WO samples completely burnt and cannot pass the test. Therefore, more flame retardant is required because polyamide fibers are thermoplastic and less resistance to flame.

TABLE V.
VERTICAL FLAMMABILITY TEST RESULTS OF THE SAMPLES.

Sample Code	Flame Time (sec)	Burn length (mm)	Drip Flame Time (sec)	Pass/Fail
1	5.2	116	0	Pass
2	6.9	143	0	Pass
3	9.2	154	0	Pass
4	33.0	ND	0	Fail
5	39.9	ND	0	Fail

Fig. 5. Vertical flammability test results of the samples

IV. Conclusion

In this work wool/polyamide blended yarns were produced in five different ratios. Yarns were knitted, dyed and flame retarded by the Zirpro process. Thermogravimetric analysis showed that flame retardant chemical act in a condensed phase. Char residue of the samples containing polyamide fibers decreased. Zirpro process reduced the peak heat release rate of the wool and wool/polyamide blended fabrics. Cone calorimetry test results showed that fabrics having higher wool content form a stable char layer over the foam. All flammability tests confirmed that polyamide decreased the flame retardancy of the wool fabrics. Smoke density test results showed that the applied Zirpro process reduced the smoke production rate significantly.

Acknowledgment

This study was supported by Anadolu University Scientific Research Projects Commission under grant no. 1701F026 and TUBITAK-1511 program under grant no. 1150520.

References

[1] [1] D. Price and A.R. Horrocks, *"Combustion processes of textile fibres,"* in Handbook of Fire Resistant Textiles, F.S. Kilinc, Ed; Woodhead Publishing, 2013, pp. 3-25.

[2] M. Forouharshad, M. Montazer, M. B. Moghadam, and O. Saligheh, "Flame retardant wool using zirconium oxychloride in various acidic media optimized by RSM," *Thermochimica acta,* vol. 516, no. 1-2, pp. 29-34, 2011.

[3] M. Forouharshad, M. Montazer, M. B. Moghadam, and O. Saligheh, "Preparation of flame retardant wool using zirconium acetate optimized by CCD," *Thermochimica Acta*, vol. 520, no. 1, pp. 134-138, Oct. 2011.

[4] Shen J. 2009 Advances in wool technology Ed. N A G Johnson and I M Russell (Cambridge: Woodhead Publishing) p 347

[5] X.-W. Cheng, J.-P. Guan, G. Chen, X.-H. Yang, and R.-C. Tang, "Adsorption and flame retardant properties of bio-based phytic acid on wool fabric," *Polymers*, vol. 8, no. 4, p. 122, 2016.

[6] A. R. Horrocks and P. J. Davies, "Char formation in flame-retarded wool fibres. Part 1. Effect of intumescent on thermogravimetric behaviour," *Fire and materials*, vol. 24, no. 3, pp. 151-157, 2000.

[7] P. G. Gordon, D. T. W. McMahon, and L. J. Stephens, "Investigations into the Mechanisms of Flame Retardation on Wool," *Textile Research Journal*, vol. 47, no. 11, pp. 699-711, 1977.

[8] D. Price, Y. Liu, T. R. Hull, G. J. Milnes, B. K. Kandola, and A. R. Horrocks, "Burning behaviour of fabric/polyurethane foam combinations in the cone calorimeter," *Polymer International*, vol. 49, no. 10, pp. 1153-1157, 2000.

IOP Conf. Series: Materials Science and Engineering **460** (2018) 012034 doi:10.1088/1757-899X/460/1/012034

Effect of Compression Bandages on Muscle's Behavior

Abdelhamid R. R. Aboalasaad [1], Brigita S. Kolčavová [1], and Gözde G. Berk [2]

[1] *Technical University of Liberec, Faculty of Textiles, Department of Technologies and Structures, Studentská 1402/2 461 17 Liberec 1, Czechia*

[2] *Istanbul Technical University, Faculty of Textile Technologies and Design, Fashion Design Department, Inonu Cad. No.65 Gümüşsuyu Beyoglu, Istanbul, Turkey*

Corresponding author: eabdo6@gmail.com

Abstract: This study aims to explore the relationship between compression bandage (CB) and muscles' performance. Hand wrist, mid-calf, and ankle muscles are subjected to electrical voltage test with and without wearing CB. Electromyography (EMG) analysis is a substantial component often used for the assessment of muscles activity. Flexor Carpi (FC), Medial Gastrocnemius (MG), and Soleus (SO) muscles are selected to represent wrist, mid-calf, and ankle muscles respectively. The standardized activities protocol used to test FC muscles are (flexion-extension and squeezing a soft roll), while the activities for MG and SO muscles are (flexion-extension and while walking). The obtained data are analyzed using Mega-win and Mat-lab software. Wearing CB was associated with significantly lower muscle activation and higher median frequency for MG and SO muscles during different actions. These results suggest that using CB can improve muscles function, which might enhance walking performance and reduce muscles fatigue.

Key words—**Compression bandages, Electromyography, Mega-win software, Muscle performance.**

1. Introduction

1.1. Compression bandages

Compression bandages (CBs) consist of elastic textile that exert pressure onto muscles. Bandages can provide the required optimal environment by controlling moisture and compression levels for wound healing. These medical elastic textile structures can improve athletic performance and reduce sports injury, which exerts compression and pressure onto muscles to relieve muscle stiffness and fatigue during sports [1-6]. In clinical practice, CBs are applied in the form of overlapping layers which results in multiple layers of fabric that overlay a particular point of the surface of the limb [7]. CB which applied with spiral 50% overlap technique will overlay the leg with two layers of bandage, CB applied with 33% overlap will result in three layers of bandage and CB applied with the Fig.-of-eight technique with 50% overlap will result in four layers bandaging [7, 8]. Wei-Chun Hsu et al selected eight healthy male recreational runners to perform 40 min. treadmill running trials, one with compression garment (CG), and the other with control garment made of normal cloth. The rating of perceived exertion (RPE) and the surface EMG test of 5 lower extremity muscles including gluteus maximus (GM), rectus femoris (RF), semitendinosus (ST), tibialis anterior (TA), and gastrocnemius (GAS) were measured during the running trial. Participants wearing CG had lower muscle activation in GAS, ST, and RF muscles, despite no additional benefit to lactate clearance or RPE [9].

1.2. Detection and analysis of muscles activation

Electromyography is the subject which deals with the detection, analysis and utilization of electrical signals emanating from skeletal muscles. The electric signal produced during muscle activation, known as the myoelectric signal, is produced from small electrical currents generated by the exchange of ions across the muscle membranes and detected with the help of electrodes [10]. Many researchers have evaluated the effect of sports compression apparel using EMG, applied testing methods and mathematical modeling. However, there is a limited scientific work which has explored that wearing compression garments has a positive influence on muscle activation during running [11]. So that it was necessary in current study to discuss the enhancement of Medial Gastrocnemius (MG) and Soleus (S) muscles activation while wearing CB using eMotion wireless EMG system moreover describe muscle's performance of Flexor Carpi (FC) at different hand wrist actions.

2. Experimental work

A. Experimental Samples

Viscose/Polyamide bandages (VI-PA) and two types of Cotton CBs (as shown in "Fig. 1") were used for hand and leg muscles testing respectively. The three bandages structures are plain weave. Yarn counts and densities are different depends on the construction and technology of the final product.

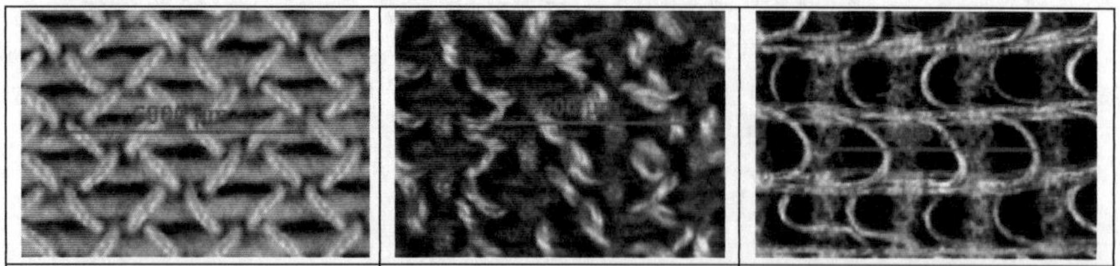

A) Bleached Cotton bandage	B) Cotton/Polyamide/ Polyurethane (CO-PA-PU)	C) Viscose/Polyamide bandage (VI-PA)
Warp set: 8 ends/cm	**Warp set:** 11 ends/cm	**Warp set:** 12 ends/cm
Weft set: 15 picks/cm	**Weft set:** 18 picks/cm	**Weft set:** 14 picks/cm
Warp count: Cotton, 20x2 tex, 2x1200 turns/m, SS/Z, ZZ/S	**Warp count:** Cotton, 10x2 tex / Polyamide, 7.8 tex / Polyurethane, 42.5 tex	**Warp count:** Viscose,16.5 tex, open end (OE), Polyamide – 7.8 tex
Weft count: Cotton, 75 tex, OE.	**Weft count:** Cotton, 36.9 tex	**Weft count:** Viscose, 16.5 tex.

Fig. 1. Experimental Compression Bandages description.

B. Testing Methods

Viscose/Polyamide bandages were used to test the FC muscle voltage during different wrist actions (flexion-extension, squeezing a soft roll) with and without wearing CB (as shown in "Fig. 2"). VI-PA compression is adjusted and standardized to medium compression ranges 22±2 mmHg (through 70% bandage extension and 50% overlap). Bleached Cotton and CO-PA-PU bandages were used to test MG and SO muscles behavior during the standardized protocol actions (flexion-extension and while walking). The lower leg bandage pressure was adjusted to compression ranges 30±2 mmHg (by 100% bandage extension and 50% overlap) [12]. All these tests were carried out on 3 healthy men (age ranges 29- 37 years) using eMotion wireless EMG system at different metronome beats (20, 30, and 40 beats/min). Surface electrodes were pasted on the mentioned muscles of the human skin as shown in "Fig. 2", three trials for each activity [13]. In order to investigate the change of muscles activity, root mean square (RMS) is processed by exporting the filtered signals with band-pass filtering between 20–500 Hz using Matlab software.

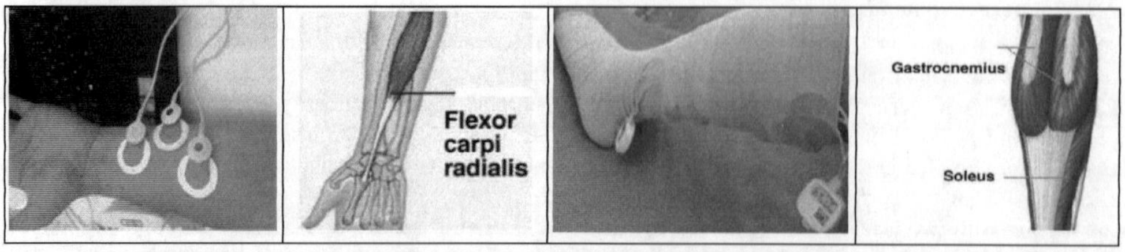

Fig. 2. Emotion EMG system for Flexor Carpi, Medial Gastrocnemius and Soleus muscles [14], [15]

3. Results and Discussion

"Fig. 3" illustrate the FC muscle voltage with and without wearing the VI-PA CB during the standardized action (flexion-extension), average muscle voltage was 85 and 93.33 μV respectively (i.e. wearing CB decreases muscle activation by a percent 8.92 % (as listed in Table I).

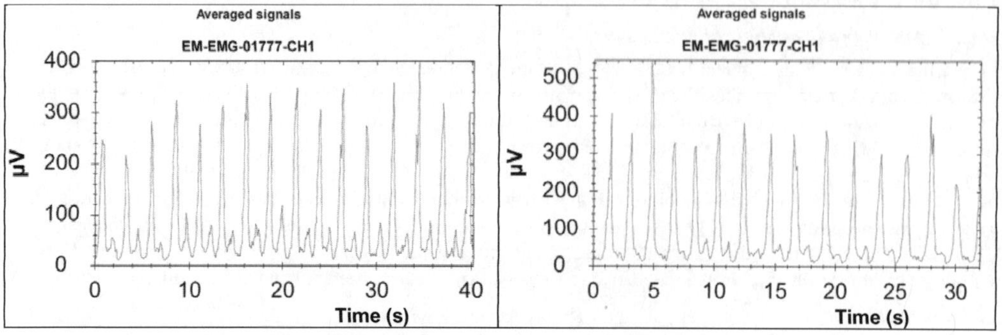

Fig. 3. Flexor Carpi muscle voltage with and without CB, (flexion-extension) action.

"Fig. 4" shows the FC muscle voltage with and without wearing the VI-PA CB during the action (squeezing a soft roll), the average EMG measured signal was 90.67 and 97.67 μV respectively (i.e. wearing CB enables lower muscle activation by a percent 7.17 %) as illustrated in (Table I). The obtained results in "Figs. 3 & 4" confirm that wearing CB enhances the Flexor Carpi Radials muscle performance while performing the standardized activities. The frequency of flexion and extension are similar and therefore, the effect of CB focuses on a reduced muscle oscillation and improves muscle function and efficiency.

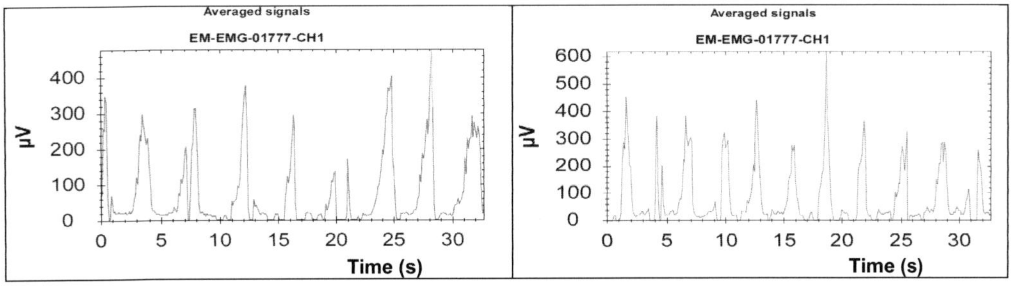

Fig. 4. Flexor Carpi muscle voltage with and without CB, (squeezing a soft roll) action.

3.1. EMG Test of Medial Gastrocnemius Muscle

Surface EMG signals were obtained from the MG and SO muscles by pre-amplified bipolar surface electrodes [16]. "Figs. 5 & 6" show MG muscle performance with and without wearing the CO-PA-PU CB during the standardized action (flexion-extension) and using the bleached Cotton bandage for walking action. Wearing CB enables a significant decrease in MG muscle activity during flexion-extension action by a percent 26.67% and 4.65% while walking (see Tables III and II respectively). This decrease may be due to the increase in the mean muscle fascicle length and the reduction in the mean muscle thickness and mean pennation angle [17]. Researchers have also claimed that muscle force being exerted for a limb's motion and stability may be wasted on muscle flexion-extension, while compression garment may prevent muscle vibrations during sports activities which can enhance athletic performance [11].

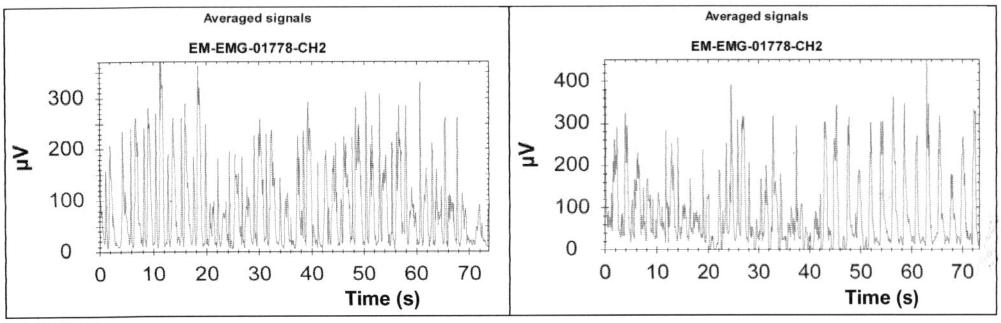

Fig. 5. MG muscle voltage with and without wearing CO-PA-PU bandage during (flexion-extension) action.

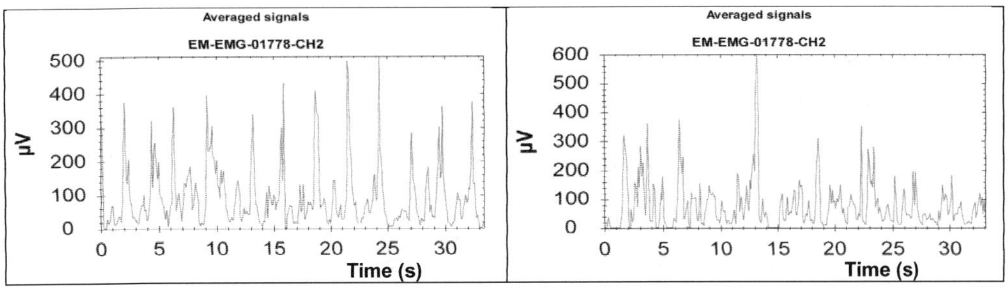

Fig. 6. MG muscle voltage with and without wearing bleached cotton bandage while walking action.

3.2. EMG Test of Soleus Muscle

"Figs. 7 & 8" show SO muscle behavior with and without wearing the CO-PA-PU and bleached Cotton CBs during the activities (flexion-extension and walking) respectively at same speed (using metronome beats 20, 30, and 40 beats/min). Wearing CB decreases SO muscle activity during flexion-extension action by a percent 21.88% and 34.13% while walking as summarized in (Tables III and II) respectively. These significant reductions in SO muscle activation clarify the enhancement of ankle muscle behavior wearing CB, because SO muscle is the main factor of controlling the walking performance while MG muscle is more effective for flexion-extension action.

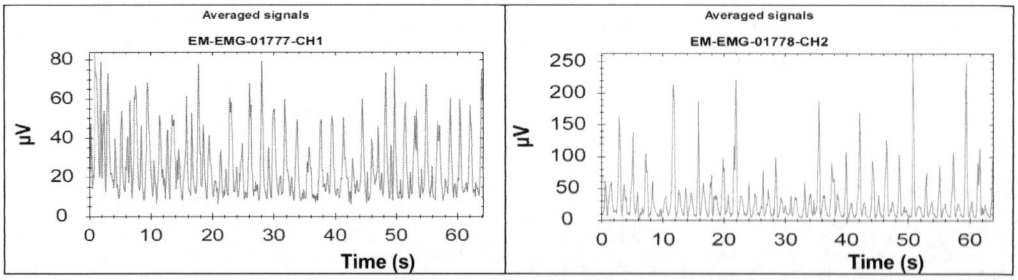

Fig. 7. Soleus muscle voltage with and without wearing CO-PA-PU bandage during (flexion-extension) action.

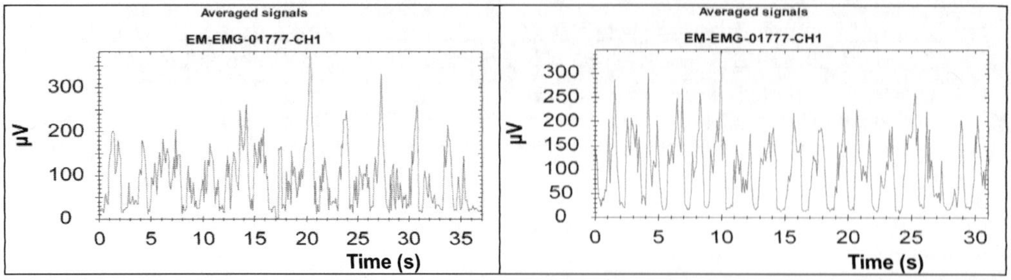

Fig. 8. Soleus muscle voltage with and without wearing bleached cotton bandage while walking action.

3.3. EMG Mean Voltage for FC, MG, and SO Muscles

Emotion EMG system and Mega-win analysis were used to investigate the relationship between three types of CBs and muscles activation. EMG mean voltage for Flexor Carpi muscle (FC) during standardized activities (squeezing a soft roll and flexion-extension) are illustrated as shown in (Table I). Average FC muscle voltages with and without wearing the VI-PA CB were 90.67 and 97.67 μV respectively (i.e. wearing CB decreases muscle activation by a percent 7.17 % during squeezing a soft roll action). Moreover using CB decreases muscle activation by a percent 8.92 % during flexion-extension action.

TABLE I

EMG MEAN VOLTAGE OF THE HAND SIGNALS FOR FLEXOR CARPI MUSCLE

Case	Metronome beats (BPM)	FC, squeezing soft roll action		FC, flexion-extension action	
		Mean voltage (μV)	Standard deviation	Mean voltage (μV)	Standard deviation
With bandage	20	75	101	78	85
	30	81	108	84	86
	40	116	124	93	103
	Mean	90.67	111	85	91.33
Without bandage	20	76	95	81	93
	30	98	121	85	101
	40	119	128	114	134
	Mean	97.67	114.67	93.33	109.33

EMG mean voltage for MG and SO muscles using the C-p-p and bleached cotton compression bandages during the standardized activities (flexion-extension and walking) at the same speed (using metronome beats 20, 30, and 40 beats/min) can be summarized and compared as listed below in (Tables II and III). Wearing bleached cotton CB while walking was associated with a decrease in average MG and SO muscles activation by a percent 2.91 and 18.18% respectively, whereas using C-p-p CB decreases MG and SO muscles activation by a percent 4.65 and 34.13% respectively. While wearing C-p-p CB during flexion-extension action decreases MG and SO muscles activation by a percent 26.67 and 21.88% respectively.

TABLE II

EMG MEAN VOLTAGE FOR LEG MUSCLES WHILE WALKING (BLEACHED COTTON BANDAGE)

Case	Metronome beats (BPM)	Medial Gastrocnemius		Soleus Muscle	
		Mean voltage (μV)	Standard deviation	Mean voltage (μV)	Standard deviation
With bandage	20	76	69	74	59
	30	80	82	88	67
	40	110	114	117	102
	Mean	88.67	88.33	93	76
Without bandage	20	77	93	96	68
	30	94	106	111	98
	40	103	136	134	102
	Mean	91.33	111.67	113.67	89.33

TABLE III

EMG MEAN VOLTAGE FOR LEG MUSCLES WHILE WALKING (BLEACHED COTTON BANDAGE)

Activity	Case	Medial Gastrocnemius		Soleus Muscle	
		Mean voltage (μV)	Standard deviation	Mean voltage (μV)	Standard deviation
Flexion – extension	with bandage	22	27	25	16
	without	30	36	32	60
While walking	with bandage	82	79	83	87
	without	86	86	126	112

3.4. Data Analysis Using Matlab Software

The RMS value for each muscle's activation is calculated using Matlab software to clarify and compare the differences between different standardized actions for hand and leg muscles as shown in (Tables IV to VI). RMS values of the FC muscle using the VI-PA bandage are illustrated in Table IV. Wearing VI-PA bandage was associated with lower muscle activation by a percent of 8.13 % for FC muscle during the standardized activity (squeezing a soft roll) and 7.10% during (flexion-extension action).

TABLE IV

RMS VALUES FOR FLEXOR CARPI MUSCLE SIGNALS

Case	Metronome beats (BPM)	FC, squeezing a soft roll RMS value	FC, flexion-extension RMS value
With bandage	20	136.98	132.36
	30	161.78	147.70
	40	180.43	164.72
	Mean	159.73	148.26
Without bandage	20	132.10	127.86
	30	189.11	155.44
	40	200.40	195.46
	Mean	173.87	159.59

Table V concludes the RMS values for the human leg muscles' signals using the bleached cotton bandage while walking. Wearing CB was associated with a reduction of muscle activation as assured by lower RMS values for MG and SO muscles as average value of the obtained results at metronome beats 20, 30, and 40 beats/min. Table VI summarizes the RMS values for the leg muscles signals using the CO-PA-PU bandage. Wearing C-p-p bandage decreases muscle activation as confirmed by lower RMS values by percentages of 21 and 13.42 % for MG and SO muscles respectively during flexion-extension action, and a percent reduction of RMS values as 6.03 and 22.31 % for MG and SO muscles while walking action. The enhancement of muscles' performance wearing CB may be due to a small increase in intramuscular pressure and in conjunction with the proposed reduction in muscle vibration [18].

TABLE V

RMS VALUES FOR MG AND SO MUSCLES' SIGNALS

Case	Metronome beats (BPM)	RMS value	
		Medial Gastrocnemius	Soleus muscle
With bandage	20	147.85	145.81
	30	165.65	151.47
	40	183.88	237.15
	Mean	165.79	178.14
Without bandage	20	138.4	167.53
	30	181.33	205.74
	40	191.13	190.94
	Mean	170.29	188.07

TABLE VI

RMS VALUES FOR MG AND SO MUSCLES' SIGNALS WEARING (C-P-P BANDAGE)

Activity	Case	RMS value	
		Medial Gastrocnemius	Soleus Muscle
Flexion – extension	with bandage	31.26	35.68
	without	39.57	41.21
While walking	with bandage	112.74	145.97
	without	119.97	187.88

CONCLUSION

Using VI-PA CB reduced the average FC muscle activation by a percent 7.17 and 8.92 % during the standardized actions (squeezing a soft roll and flexion-extension) respectively. Wearing bleached Cotton CB enabled lower muscle activation and higher median frequency for MG and SO muscles by a percent of 4.65 and 34.13 % during walking action. Using CO-PA-PU CB was associated with significantly reduction of MG and SO muscles activation by a percent 26.67 and 21.88 % during flexion-extension action. The obtained RMS values using Matlab software confirmed that wearing CB improved the performance of FC, MG, and SO muscles and could enhance muscles' fatigue.

ACKNOWLEDGMENT

I wish to acknowledge assistance and encouragement of my colleagues (Sinem Kahveci and Shahrukh Shahbaz); specially during testing of my samples. This work is supported under Student Grant Scheme (SGS 21249) by Technical University of Liberec, Czech Republic.

REFERENCES

[1] W. Ping, L. Lei, Y. Mao-de, et al., "Effects of compression garments on lower limb muscle activation via electromyography analysis during running", Journal 01 Donghua University, vol. 32, no.1, pp. 48-52, 2015
[2] N. Miyamoto, K. Hirata, N. Mitsukawa, et al., "Effect of pressure intensity of graduated elastic compression stocking on muscle fatigue following calf-raise exercise", Journal of Electromyography and Kinesiology, vol. 21, pp. 249–254, 2011
[3] K. Goto, S. Mizuno, A. Mori, "Efficacy of wearing compression garments during post-exercise period after two repeated bouts of strenuous exercise: a randomized crossover design in healthy, active males. Sports Medicine, 2017.
[4] Martorelli S., Martorelli A., Pereira M., et al.: Graduated compression sleeves: effects on metabolic removal and neuromuscular performance. Journal of Strength and Conditioning Research 2015, vol. 29, 5.
[5] Jariyapunya N., Musilová B., Geršak J., et al.: The Influence of stretch fabric mechanical properties on clothing pressure. Vlákna a textil (Fibres and Textiles) 2017, Vol. 24, Issue 2, pp. 43-48.
[6] Fu W, Liu Y, Fang Y. (2013). Research advancements in humanoid compression garments in sports. International Journal of Advanced Robotic Systems. Volume 10(66), pp. 1-6. Available: www.intechopen.com
[7] Khaburi J., Dehghani-Sanij A.: The Effect of multi-layer bandage on the interface pressure applied by CBs. International Journal of Mechanical, Aerospace, Industrial, Mechatronic and Manufacturing Engineering 2011, 5, pp.1169-1174.
[8] Rimaud D., Convert R., Calmels P.: In vivo measurement of CB interface pressures: The first study. Annals of Physical and Rehabilitation Medicine 2014. Doi: 10.1016/j.rehab.2014.06.005
[9] Hsu W., Tseng L., Chen F., et al.: Effects of compression garments on surface EMG and physiological responses during and after distance running. Journal of Sport and Health Science 2017. Doi:10.1016/j.jshs.2017.01.001.
[10] Jamal M. (Intech 2012). Signal acquisition using surface emg and circuit design considerations for robotic prosthesis [Online]. Available: http://dx.doi.org/10.5772/52556.
[11] P. Wang, J. McLaren, K. F. Leong, et al., "A Pilot study: evaluations of compression garment performance via muscle activation test". Procedia Engineering 2013, 60, pp. 361-366.
[12] A. Aboalasaad, K. Brigita, "Analysis and prediction of CBs tension," in Conf. CEC, 2017. International Ph.D. Students Day, pp. 6-9.
[13] C. Gatti, D. L. Case, J. Langenderfer. (2008 February). Evaluation of three methods for determining EMG-muscle force parameter estimates for the shoulder muscles. Clin Biomech (Bristol, Avon). Volume 23(2), pp. 166–174.
[14] Gvs14. (Sep. 2012). Anterior Forearm Muscles [Online]. Available: https://anatomystudybuddy.wordpress.com/2012/09/20/flexor-carpiradialis/
[15] Luqman. (May 02, 2017). Pearson Benjamin Cummings [Online]. Available: https://humananatomyly.com/gastrocnemius-muscle/
[16] Sell K., Ghigiarelli J., Kitsos J., et al.: Electromyographic analysis of abdominal and lower back muscle activation during abdominal exercises using an objective biofeedback device. JEP online 2011, 14(5), pp. 54-65.
[17] James M. W., Meghan J. and Ana I. N.: The Effect of external compression on the mechanics of muscle contraction. Journal of Applied Biomechanics 2013 vol. 29, pp. 360-364
[18] R. Duffield, J. Cannon, M. King, "The effects of compression garments on recovery of muscle performance following high-intensity sprint and plyometric exercise", Journal of Science and Medicine in Sport, pp. 136-140, 2010.

IOP Conf. Series: Materials Science and Engineering **460** (2018) 012035 doi:10.1088/1757-899X/460/1/012035

Optical Attenuation of Linear Composites Containing SEPOF

J Militky[1], D Kremenakova[1] and J Saskova[1]

[1]Technical University of Liberec, Faculty of Textile Engineering, Department of Material Engineering, Studentska 1402/2, 461 17 Liberec 1, Czech Republic here

jiri.militky@tul.cz

Abstract. Start Optical fibers are one category of specialty fibers originally developed for light transmission (fiber optic) and information (optical cables). The fiber consists of a core surrounded by a cladding. Standard polymer optical fiber (POF) is transferring light across its axis by the mechanism of total internal reflection. In the side emitting plastic optical fibers (SEPOF) the light leaks out from their surface. The conditions for obtaining side illumination can be derived from POF geometrical model composed of two materials with different refractive indexes. The main aim of this contribution is description of SEPOF optical properties and their efficient embedding into fibrous structures for creation of line illumination hybrid structure useful for design purposes and creation of safety textile structures with active visibility in shadows. The special device for measurement of light intensity on surface and cross section at various distances from light source is described. Light intensity of textile structures is compared with light intensity of fibers.

1. Introduction

Standard polymer optical fibre (POF) is a dielectric waveguide transferring light or infrared radiation across its axis by the mechanism of total internal reflection on the interface of two materials with different refractive indices [1]. In the side emitting plastic optical fibres (SEPOF) the light leaks out from their surface. Side emission occurs if the light incidence angle is smaller than critical angle. This effect can be obtained by the increasing of cladding refractive index, decreasing of core refractive index or by the change of incident light angle. It is also possible to use multiple micro-bending of core or cladding; additives causing reflection or fluorescence into core/cladding or to create geometric asymmetry in the core/cladding system.

Comprehensive information about properties and applications of POF in the textile field are presented in the book [1]. There are various commercial types of patented SEPOF including methods of their preparation [2, 3, 11].The SEPOF can be used for creation of optically active textile structures providing opportunities to highlight people and objects without the need for external light exposure. Due to the transmission loss, the intensity of radiation emitted in any direction decays exponentially along the straight fibre axis with increasing distance from the light source [4, 5]. An optical fibre integrated in weaving patterns can be described by a sequence of locally bended and straight sections. For this case the transmission loss is based on the distance between threads, thickness of threads and weaves. Local bends should be suppressed in the cases of using SEPOF for achieving active visibility at higher distances from light source. It is therefore necessary to use special embedding of SEPOF into textile structures e.g. in the form of tubes wrapped by textile yarns.

The main aim of this contribution is description of SEPOF optical properties and their efficient embedding into fibrous structures for creation of line illumination hybrid structure useful for design purposes and safety textile structures with active visibility in shadows. For preparation of textile structures containing SEPOF the special weaving technology is used [6]. The special device for measurement of light intensity on surface and cross section at various distances from light source is described. Light intensity of textile structures is compared with light intensity of fibres.

2. Basic of fiber Optics

Standard i.e. *end emitting optical fiber* is a special form of a relatively long distance (hundreds meters to some kilometers) optical wave guide and consists typically of a cylindrical core and a cladding (see. fig. 1), both made from highly transparent materials. The core refractive index n_1 should be slightly higher than that of the cladding n_2 to avoid loss of illumination power due to side emission. Depending on ultimate applications, on the surface there is additional protective layer called as jacket.

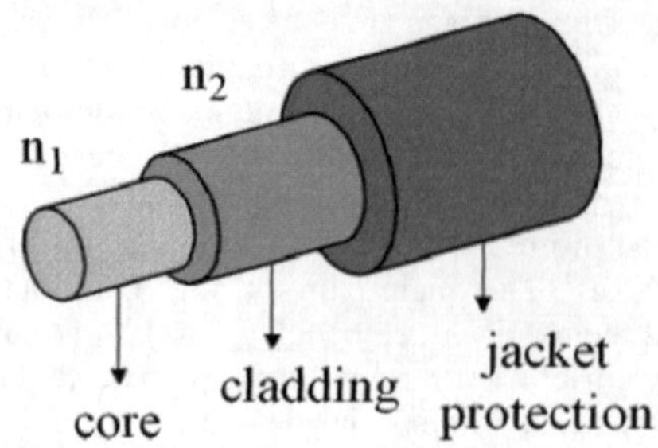

Figure 1. Typical structure of optical fiber.

For purposes of line lighting in short distances (few meters) the so called *side emitting optical fibers* are used. The simple way how to obtain side emission is to change magnitude of refractive indices i.e. using slightly smaller core refractive index n_1 than that of the cladding n_2. Still there are a limited number of companies only producing side emitting optical fibers based on this simple idea. Second possibility is to use the end emitting optical fiber with modification enabling side emission. There are in fact possibilities to promote scattering of core by adding particles or changing super molecular structure; creating imperfections (grooves) in cladding; using structures with sharp bends (as in woven fabrics). In fact all these modification are a little bit against aims of producers to suppress attenuation i.e. side emission. The side emitting optical fibers will be probably in future produced according to their main goal, i.e. side emission, with more even and approximately constant decay of illumination power along their working length. The properties and functionality of optical fibers are dependent on the refractive indices of core and cladding.

When light passes from a medium with a lower refractive index to a medium with a higher index, it is bent towards the normal plane between media. Otherwise light bends in the opposite direction (away from the normal plane, see fig. 2).

If the angle of incidence in this case is greater than the critical angle N_c all the light energy is reflected back into the medium with a higher refractive index. This phenomenon known as total internal reflection is used to guide the light in optical fibers (fig. 2).

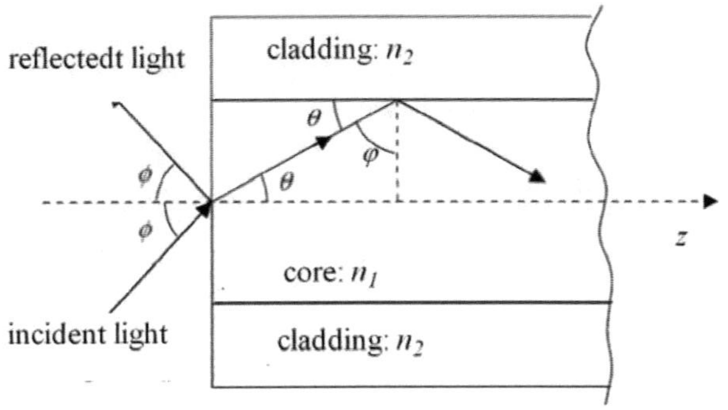

Figure 2. Internal reflection in optical fibers.

In the standard end emitting optical fibers, refractive index of core n_1 is always higher than refractive index of cladding n_2, i.e. $n_1 > n_2$. Over a certain angle N_c the total reflection then takes place at the boundary between the core and the cladding. If light enters under lower angle, the partial refraction (side emission) occurs and only part of the total light energy is reflected back to core. Fig. 2 shows that total internal reflection (generally electromagnetic wave) occurs in the optical fibers at the boundary between the core and the sheath. The maximum angle φ_m (see fig. 2) when there is total internal reflection of light is equal to

$$\varphi_m = \arcsin\left(\frac{n_2}{n_1}\right) \tag{1}$$

According to Snell's law, the upper critical angle $\phi = N_c$ at input into the optical fiber is equal to [12].

$$N_c = \arcsin\left(\frac{\sqrt{n_1^2 - n_2^2}}{n_0}\right) \tag{2}$$

Here n_o is refractive index of surroundings (for air $n_o = 1$). The acceptance angle ($2N_c$) corresponds to the vertex angle of the largest cone of rays that can enter the core of the fiber.

Assuming that there is a total internal reflection (i.e. input angle ϕ is below the critical angle N_c) the whole electromagnetic radiation is captured in core and propagates along the fiber axis. If the angle of incidence on the optical fiber is lower, the part of the light is refracted and there appears so-called leakage or partial side emission. Side emission occurs if the input angle of light incidence ϕ is higher than critical angle N_c. This effect can be obtained by the increase of n_2 or decrease of n_1 or by the change of input angle of incident light ϕ.

The sine of the maximum incident angle N_c is defined as the *numerical aperture* $NA = n_o sin(N_C)$. The angle N_c is often denoted to as the acceptance angle, and twice the acceptance angle is denoted as the aperture angle. The NA is defined by relation

$$NA = \left(n_1^2 - n_2^2\right)^{1/2} \tag{3}$$

The NA is a very important parameter of optical fibers, since it indicates their capacity for accepting and guiding light. The size of the NA is solely dependent on the difference in the refractive indices of

the core and cladding material. For standard PMMA fiber is n_1 = 1.49 and n_2= 1.40; thus NA = 0.50 and N_c = 30°.
Compared with other fiber types [8], POF has the largest numerical aperture and the largest core diameter. This is one of the most important advantages of POF, since the connection technology that can be used for POF fibers is more economical to apply than that used for glass fibers.
Due to the transmission loss, the power of radiation P [W] emitted in any direction decays exponentially along the fiber axis with increasing distance from the light source of the fiber as observed by Zajkowski [13, 14], while the percentage of light emitted per unit length is uniform over the entire fiber length.
The simple exponential model for prediction of this attenuation is proposed. Illumination power $P(L)$ for straight optical fiber is decreasing with increasing distance from source L according to relation [15]

$$P(L) = P(0)\, 10^{-\alpha L} \tag{4}$$

where $P(0)$ is illumination intensity of source and α is attenuation coefficient of optical fiber. Coefficient α describes attenuation dependent on fiber length

3. Measurement of Optical Attenuation

Illumination efficiency characterized by loss of light intensity generally depends on: light wavelength, fiber type, fiber structure, impurities, accompanying substances, outer geometric shape and mainly on distance from the source. For POF, the goal is to suppress attenuation because it is source of information loss. For SEPOF is side emission wanted and can be supported. The standard approach to characterize attenuation, dispersion and bandwidth of POF is described i.e. in book [1].
Special measurement and evaluation system was constructed for evaluation of LIHS attenuation in straight state and creation of the attenuation profile P(L) i.e. dependence of illumination intensity on distance from light source L till maximum length in the range of 1-20 meters [15]. Main principle is measurement of illumination intensity by using of integrating cylinder. The light emitted by the LIHS falls on the inner surface of the cylinder, which has a high reflectivity and is opaque, thus incident rays are scattered randomly in all directions (into the interior of the cylinder). For sufficiently big dimensions of this cylinder the random light scattering extends substantially to the statistically uniform illumination of its internal surface. Irradiation of inner surface E(A) [W m^{-2}] depends on the light power (illumination intensity) P(L) [W m^{-2}] escaping from measured surface area A of the LIHS to the inner surface of the cylinder at distance L from light source.

$$E(A) = \eta\, P(L) \left(\frac{\pi d^2}{4} l \right)^{-1} \tag{5}$$

where η [-] is efficiency coefficient, l is length and d is inner diameter of integrating cylinder. The sensor of light power has active area S_s. Then the illumination intensity $P(L)$ is calculated from measured radiant flux Φ_V [W] by using of equation

$$P(L) = \frac{\pi d^2}{4} \frac{\Phi_V}{\eta S_S} l \tag{6}$$

This principle was used for construction of on-line computer controlled measurement and evaluation system for creation of the attenuation profile (fig. 3). System is composed from radiant flux light sensor (THORLABS PM 1000 SB), step driver, control unit, measuring channel, input/ output rolls and illumination unit based on LED.

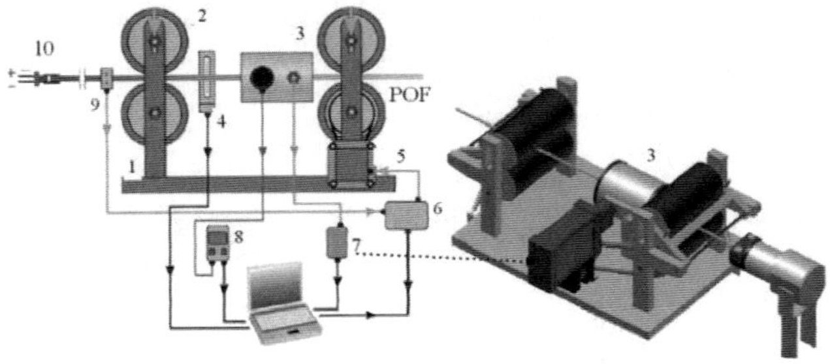

1 frame
2 rubber roller
3 optical integration cylinder
4 raster camera
9 end of fiber detector
5 stepping motor
6 controller
7 spectrophotometer
8 power measuring unit
10 illumination unit

Figure 3. Device for measurement of SEPOF light intensity in straight state [10].

Illumination unit is composed from light emitting diode (LED) connected with power source. Electric power supply is 3W. At 10% conversion efficiency the light output power is of 300 mW. Moreover, about 30% radiates outward from the fiber, so that the radiant flux at the input is 100 mW.
The LIHS is located between the feed rollers, which guide them to the integrating cylinder. The tow rollers are driven by step motor. The actual measurement is performed in predefined step lengths. Step lengths are processed by a step motor that drives the two rollers. The measurement process is controlled by a computer program created in MATLAB.
The computer unit is used also for calculation of attenuation profile including statistical analysis. Due to relative high variability of results, 20 attenuation profiles are usually measured. Robust estimator of mean illumination intensity and corresponding 95% confidence intervals are then calculated by robust Horn procedure suitable for small sample sizes [16].
It was found, that the model defined by eqn. (4) is not sufficient for mean intensity profile experimental course approximation. It was found that at short distances from light source, illumination intensity is strongly decreasing especially for POF with higher diameter (higher than 1 mm). For better expression of P(L) behavior, the so called LLF2 model was proposed. LLF2 is in fact linear spline i.e. linear piece-wise function composed from two different straight sections [16]. This model is based on the assumption that in short distances from light source there are some nonuniformities in side emission due to accommodation to aperture and critical angle. In second phase the illumination intensity is slowly decreasing with distance L from light source (system is accommodated). Local slopes of LLF2 are in fact sensitivity coefficients a_1, a_2. Corrected illumination intensity on the fiber input is $P_{cor}(0)$. LLF2 model is described by equation (linear regression spline with one knot).

$$LLF2 = P_{cor}(0) + a_1 L + a_2 (L - L_c)_+ \qquad (7)$$

where function $(x)_+ = 0$ if x is negative and if x is positive, function $(x)_+ = x$. L_c is distance of transition between first and second phase. By using modified linear regression, [16] parameters of LLF2 for experimental mean attenuation profiles can be calculated.
As attenuation profiles characteristics of POF, the sensitivity coefficients a_1, a_2, corrected illumination intensity on the fiber input $P_{cor}(0)$ and distance of transition between first and second phase L_c can be used. Based on comprehensive tests, it was found that the main factor influencing the intensity of the side emission of SEPOF is their diameter. For straight SEPOF, 20-40% of illumination intensity is lost

at short distances from light source. The intensity of radiation is here falling sharply. After a certain length L_c the illumination intensity decreases slowly.
As example the mean attenuation profile of SEPOF Grace-standard having diameter 0.25 mm was evaluated.
Robust estimator of mean illumination intensity (points in fig. 4) and corresponding 95% confidence intervals (lines in fig. 39) were calculated.

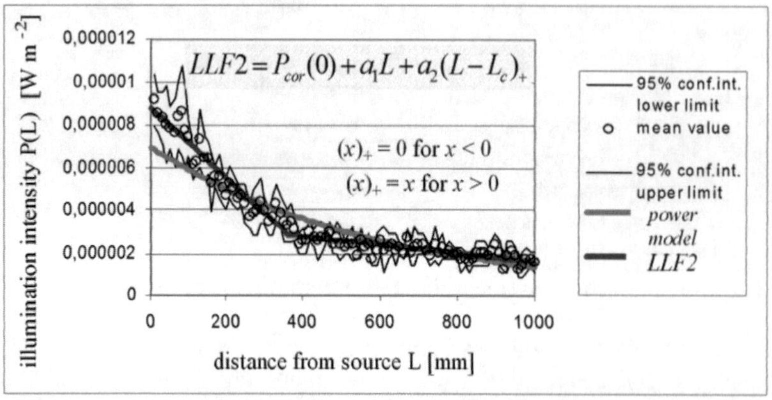

Figure 4. Mean attenuation profile for SEPOF Grace-standard with added line of $P(L)$ from eqn. (4) and line corresponding to *LLF2* - eqn. (7).

The simple power model (red curve) and LLF model (blue lines) are shown as well. It is visible that the LLF2 model is much better and allows to describe whole mean attenuation profile.

4. Line Illumination Textile Structure

Direct application of polymeric optical fiber - POF for embedding into textiles or creation of line lighting systems is not so easy. POFs are created from polymeric materials which are sensitive to surface damage, their mechanical properties are not as good as properties of synthetic fibers, durability in standard conditions of use are limited, they are degraded under UV radiation, sensitive to moisture, and their comfort properties including hand (touch) and drape are generally bad.
For side emitting optical fiber - SEPOF it is possible to have sufficient side illumination in straight state. One solution how to avoid the problems with POF and SEPOF is to use so-called line illumination hybrid structure - LIHS composed from SEPOF in core and cover (textile tube) from textile material. LIHS, have the following benefits:

- Enhancing illumination intensity by selection of textile layer composition and surface dopants.
- Protection of LIHS against weather including UV and combination of temperature/moisture.
- Enabling LIHS standard maintenance during use as washing.
- Suppression of mechanical damage as abrasion and sensitivity to repeated multiaxial deformation.
- Simplification of attachment of LIHS into textiles by standard techniques as sewing.

Each structure containing LIHS needs to have lighting system and power supply. Components of side illumination system containing LIHS are shown in fig. 5.

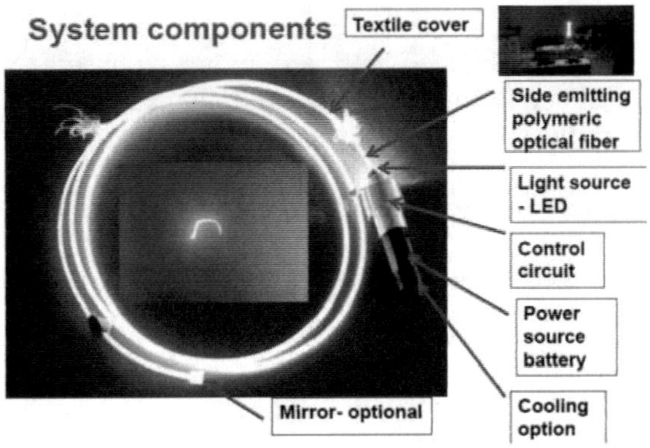

Figure 5. System components for side illumination.

The SEPOF end connected with light energy source is prepared by cutting with heated wire and then by polishing with diamond powder. Illumination system with light emitting diode (LED) is used as light source (illumination intensity of source is 43.9 Wm^{-2}).

For quantitative comparison of SEPOF and LIHS side illumination the attenuation profiles P(L) were created. Illumination intensity as function of distance from source for SEPOF and the same SEPOF embedded into textile tube are shown in fig. 6.

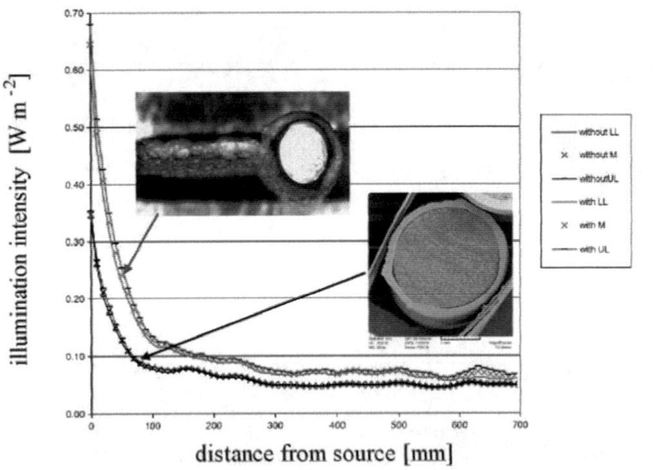

Figure 6. Side illumination intensity of SEPOF and LIHS.

It is visible that LIHS (SEPOF in textile cover) is producing higher side emission intensity in comparison with SEPOF only.

5. Discussion

LIHS can be used for creation of safety textiles, fashion purposes and line lighting. Safety textiles containing LIHS have these main advantages:

- The object shape is highlighted directly which significantly reduce the risk of incorrect interpretation of their size.
- Side emission is not dependent on the conditions of external exposure (typically vehicle headlights), these active lighting systems operate well in darkness.
- There is no glare disturbing e.g. road users.

IOP Conf. Series: Materials Science and Engineering **460** (2018) 012035 doi:10.1088/1757-899X/460/1/012035

Due to full integration into the fabric SEPOF are protected against environmental influences (moisture, UV radiation). It is simple to use reflective or fluorescence colors for dyeing or coating outer textile layer enhancing diffusion scattering.
The simple removal of power/ light source enable treatment of textiles according to standard procedures. Based on the practical testing the following care rules were specified:

- gentle machine wash, moderate mechanical treatment, rinses and spin drying,
- do not bleach with chlorine,
- do not iron,
- can be cleaned with perchlorethylene,
- do not tumble dry.

The examples of application of systems containing LIHS for clothing and accessories are shown in fig. 7, fig. 8 and fig. 9.

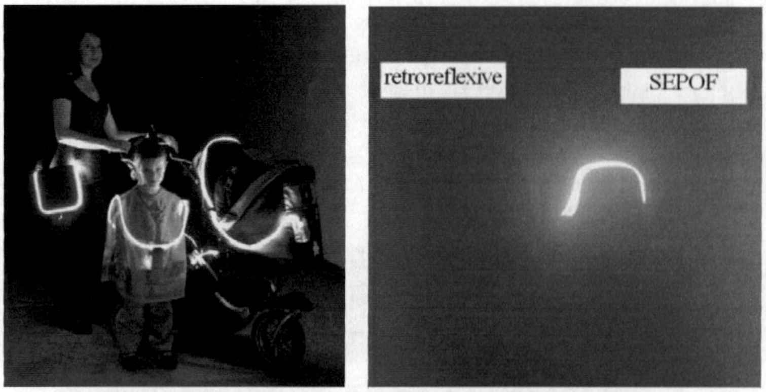

Figure 7. Systems using LIHS for safety purposes.

Figure 8. Systems using LIHS for fashion.

Figure 9. Systems using LIHS for accessories.

Line lighting systems based on LIHS have the following advantages:

- can be used for line illumination of complicated ways and significantly reduce the local variation of light intensity,
- low cost of lighting on the level of illumination standards,- not local over lighting,
- intensity of light is controlled by voltage,
- simple connections to permanent or portable electricity sources
- simple branching of line lighting
- maintenance according to standard procedures

Line lighting systems based on LIHS can be used for:

- setting of limits (parking barriers, end of carpets, stairs, etc.).
- emergency lighting (hospitals, hotel, dormitories, supermarkets)
- lighting in corridors, lifts, edge visualization, etc.

The example of line lighting are in fig. 10.

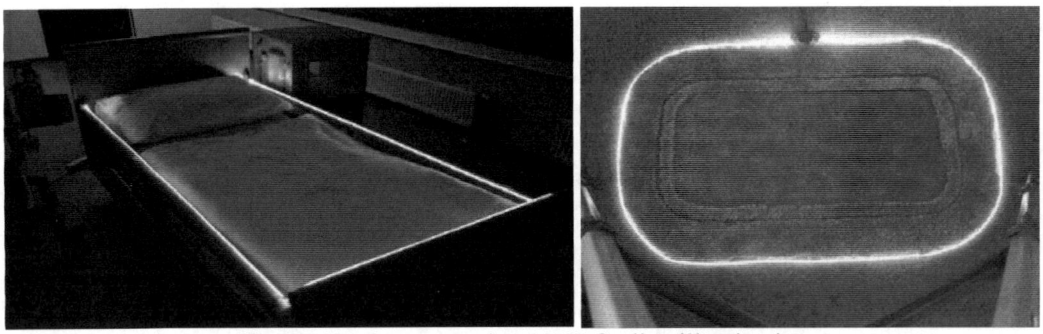

Figure 10. Systems using LIHS for line illumination.

6. Conclusion

Plastic optical fibers applied in illumination and decorations are as follows [86]:

• End emitting optical fiber POF are mainly using the lighting effect of the end face of fibers to achieve decorative lighting.

• Side emitting optical fiber SEPOFs are easy to be used to set off the outline of patterns of objects and produce a variety of artistic design.

• Flash point optical fiber are used to make special effect. These fibers can be made into fiber-optic curtains, fiber-optic waterfall and other fiber-optic craft products.

Recently, the POF and SEPOF technology has been developing quickly, the applications has been recognized in many branches. The LIHS are still in the first stage of industrial realization and it will be necessary to optimize their composition and functionality for various targeted applications including surface effects, doping and use of special dye stuffs. The system of creation of illumination intensity profiles (see fig. 3) will be simple tool for evaluation of new effects and optimization of LIHS based systems. Probably the most important element for efficient low cost active illumination system creation is now the cost of control unit with battery.

References

[1] Bunge Ch A et al eds 2017 *Polymer Optical Fibres Fibre Types, Materials, Fabrication, Characterisation and Applications*, Elsevier Amsterdam
[2] Zarian J R et al 1999 *Side lighting optical conduit.* US Patent 5987199
[3] Harlin A et al 2003 *AUTEX Res. J.*, **3**, 1-8
[4] Endruweit A et al 2006 *8th International Conference on Textile Composites (TEXCOMP-8)*, Nottingham, UK, 16-18 Oct.
[5] CaiI, Bo, Xiao-li, J., Zhang, Ch.: *Journal of Wuhan University of Technology – Mater. Sci. Ed.*, Vol.**18**, No.4, Dec 2003.
[6] Křemenáková, D., Lédl, V., Militký, J., Bůbelová, B., Meryová, B.: Utility model č.24997 *Active emitting safety means*. Industrial property office, entered 4th March 2013.
[7] Hecht E.: *Optic*, Pearson, Boston 2017
[8] Daum W. et al.: *POF - Polymer Optical Fibers for Data Communication*, Springer-Verlag Berlin 2002
[9] Koike Y 2015 *Fundamentals of Plastic Optical Fibers*, Wiley-VCH Verlag Weinheim
[10] Křemenáková D et al 2012 *Journal of Fiber Bioengineering & Informatics*, **5**(4) p. 423-431
[11] Zubia J and Arrue J 2001 *Opt Fiber Technol.* **7**, 101-140,
[12] Thévenaz L ed 2011 *Advanced fiber optics concepts and technology*, EPFL Press Lausanne
[13] Zajkowski M 2002 *Proc SPIE* **5125**,322–7
[14] Zajkowski M (2005) *Proc SPIE* **5775**, 440–5
[15] Křemenáková D et al 2012 *Int. Symp. TBIS*, Shinshu University, Ueda Japan August
[16] Meloun M and Militky J 2012 *Statistical Data Analysis*, Woodhead, New Delhi
[17] Mao Q 2010 *Communication technology and standards, Expert forum*, **9** China

Acknowledgments

This work was supported by project TH01021163 Efficient Line Lighting Systems, Program: TH – EPSILON and project FV10356, Program: FV – TRIO of Technology Agency of the Czech Republic

Processing of Biopolymers-based, Efficiently Integrated Bilayer Membranes for Use in Guided Tissue Regeneration Procedures

G. Selestina, V. Tomaž, and K. Vanja

Institute of Engineering Materials and Design, Faculty of Mechanical Engineering, University of Maribor, Smetanova uluca 17, 2000 Maribor, Slovenia

Corresponding author: selestina.gorgieva@um.si

Abstract

Periodontitis is a chronic inflammatory disease affecting the tooth supporting structures - the alveolar bone in particular, to extent of eventual tooth loss. Among others, the guided tissue regeneration (GTR) procedure including placement of occlusive barrier membrane is very promising approach. Herein we develop chitosan (CHT) / gelatin (GEL) bilayer membranes *via* successive solvent– and freeze-casting procedures and genipin (GEN)-mediated cross-linking chemistry. By utilizing the auto-fluorescence signal from GEN cross-linking products (i.e. the secondary CHT (GEL) amines and GEN esters), the Confocal Fluorescent Microscopy (CFM) identifies the chemical as well as physical integration between layers´ interfaces, supported also by Scanning Electron Microscopy (SEM) data. The presence of non- and highly μ-porous and pore-interconnecting regions is demonstrated within cross-sections of GEL-prevalent membranes in contrast to the sheet-like organization in membrane with equal components presence. The constant processing conditions onto variable compositions did not significantly affect the pore size distributions (in 1-230 μm range), while pore wall thickness increases up to 220 μm with GEL increase, which also improves the yield stress at compression from 10 to 19 kPa and elastic modulus from 26 to 34 kPa. The further applied rapid mineralization resulted in deposition of non-regular to spherical minerals, containing nonstoichiometric carbonated apatite with Ca/P ration in 1.7-2 range, which demonstrates formation of osseointegrative interface. The intensive, composition-dependent swelling (up to 580%), as well as 67% to 100% weigh loss in 4 weeks *in vitro* degradation experiment imply on bilayer membranes GTR relevance.

Keywords—auto fluorescence, chitosan, gelatin, guided tissue regeneration, membrane

I. Introduction

The Periodontitis is a chronic inflammatory disorder that affects ~750M people worldwide, being the fourth highest cost disease with up to 10% consumption of healthcare resources. As chronic inflammatory disease affecting the tooth supporting structures, it may destroy the alveolar bone to extent of eventual tooth loss. Further, often-complicated management of toot loss is positioning of dental bridge or implant, the latter being often compromised by further bone loss around the tooth, i.e. by peri-implantitis. Current therapeutic strategies are placement of autographs, allografts, bioglass, β-tricalcium phosphate, hydrogels, or s.c. flap surgery in conjunction with biomaterials combinations, the use of biological factors [1], etc.

The concept of barrier membrane placement over the bone defects, which allows specific cells to repopulate the isolated spaces present the guided tissue regeneration (GTR). The cell-occlusive, space making, tissue integrative, clinically manageable and biocompatible are attributes to "ideal" membrane, for which the large diversity of (non)biodegradable, and bioactive materials have been used [2, 3]. Chronologically of the GTR membranes starts with the non-bioresorbable types (polytetrafluoroethylene, titanium, bacterial cellulose, cellulose acetate, silicon, rubber), continues with the bioresorbable (poly(lactic acid), poly(glycolyc acid), poly (ε-caprolactone), collagen type I, porcine skin, bacterially-derived polymers, hydroxyapatite/polyamide mixture, amniotic membrane, polyethersulfone, calcium phosphate, β- tricalcium phosphate, latex), and up-grade to bioactive types containing antibiotics (metronidazole benzoate, tetracycline hydrochloride) and growth factors (fibroblast growth factor /FGF-2/, platelet-derived growth factor /PDGF/, insulin-like growth factor /IGF-I/, bone morphogenetic protein /BMP-2/, transforming growth factor /TGF-b/). [4]–[6]

Despite diversity, their limited and random success have been reported. Among reports is also obvious lack of attention to μ-structural features as highly relevant for materials intended to interface with- and integrate to- structurally diverse tissues, such in case of periodontal tissue.

Within presented study, we attempt to develop bilayer membranes with μ-structural diversity by merging two straightforward manufacturing processes, the solvent-casting and temperature-controlled freeze- casting, and by using mixture of biocompatible polymers being well elaborated in biomedical applications, i.e. the chitosan (CHT) and gelatin (GEL). Upgraded concept in μ-structuring follow-up by means of CFM and SEM microscopies is also presented, as well as comprehensive spectroscopic evaluation of physic-chemical events, with particular attention on rapid mineralization process outcome onto the membranes.

II. METHODS AND PROCEDURES

2% w/v of CHT solution in acetic acid and 10% w/v of GEL solution in 0.1M MES (pH 5.5) buffer were prepared as stock solution, future used in membranes casting. Membranes' library was obtained by using particular volume of each of the biopolymer solutions in following CHT/GEL weight ratios: 1/1 (A), 1/5 (B), 1/10 (C) and 1/20 (D). The 5% w/v genipin (GEN) solution was added to each of the solutions being further casted onto Teflon Petry dish and oven dry for 48h at 40 °C. The same biopolymer mixtures were subsequently applied onto the solvent-casted membranes and freeze-dried at constant temperature (-20 °C). The rapid mineralization procedure was applied on final bilayer structures, according to reported procedure [7] being modified in term of duration and final pH . Attenuated Total Reflectance- Fourier Transform Infrared (ATR-FTIR), X-ray Diffraction (XRD) and Energy Dispersive X-ray (EDX) spectroscopy, the SEM and CFM microscopy, swelling and degradation profiles (in PBS at 37 °C) as well mechanical testing at compression in dry and wet state were assessed for all CHT/GEL combination membranes and related to processing factors.

III. RESULTS AND DISCUSSION

Physical linking between both, CHT and GEL components is assessed form ATR-FTIR data before and after GEN cross-linking and compared to both controls with typical spectral band positions (as presented within Table 1).

Table 1. ATR-FTIR band positions for chitosan (CHT) and gelatin (GEL).

Chitosan (CHT)		**Gelatin (GEL)**	
Wavenumber (cm^{-1})	*peak assignment*	*Wavenumber (cm^{-1})*	*peak assignment*
3600-3000	OH	3600-3000	OH
3237	$-NH_2$ str.	3292	N-H str. (amide A)
2921	$-CH_2$ str. in pyranose ring	3070	N-H str. (amide B)
1635	-C=O (amide I)	1630	C=O str. (amide I)
1539	$-NH_2$ bend. (amide II)	1534	N-H bend.; C-N str. (amide II)
1403	OH, CH in the ring	1330	N-H bend.; C-N str. (amide III)
1350-1200	-C-N (amide III)	1404	C-H str.
1230	-C-O; C-H	1175	C-O str.
1152	-C-O-C	1036	C-O str.
1063, 1018	-C-N; –C-O str.		

The H-bonding as well as ionic interactions (due to the presence of fixed charges within CHT and GEL biopolymers) are physical interaction forces, which are however not sufficient to keep the integrated and stable membrane within application scenario. The complementary use of GEN for chemical linking is straightforward approach, being applied in mixing phase while cross-linking occurs within extended period.

The Fig.1. Identifies the cross-linking related conformational changes within CHT and GEL because of structural rearrangement during formation of new bonds. According to the literature[7] two-way reaction mechanisms are possible. First reaction is characterized by the increase of C-N band at ~ 1078 cm^{-1} on expense of C-O band at ~ 1026 cm^{-1}, combined with the decrease in intensity of protonated amine at ~ 1576 cm^{-1}, and should endows with opening of dihydropyran ring in GEN molecules, the formation of nitrogen-iridoid, the aromatic intermediates, as well as the formation of highly conjugated heterocyclic GEN-CHT (GEL) derivatives. The increase of ration between area of absorption bands at 1640-1650 cm^{-1} and 1570-1580 cm^{-1} for ν (C=O) and ν (N-H), respectively, in cross-linked (A, B, C and D) relative to non-cross-linked (Ac, Bc, Cc and Dc) membranes can be attributed to formation of secondary amino groups within CHT and/or GEL, and ester groups of GEN, being identified as second, slower reaction under which single bi-functional cross-link between CHT and/or GEL molecule are formed. The resented graph inset (Fig. 1, below) imply on strong correlation between the mentioned ratio and composition, i.e. on CHT promotion of second reaction products.

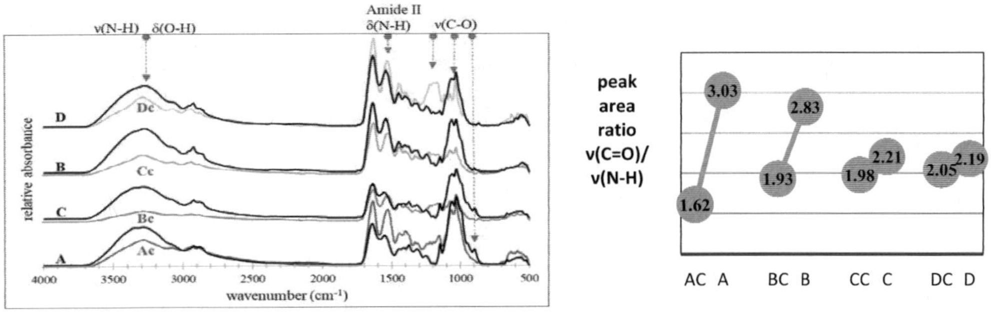

Fig. 1. FTIR spectral data of CHT/GEL membranes with different (1/1-A, 1/5-B, 1/10-C and 1/20-D) w/w ration before and after cross-linking with GEN (left), with extracted data for peak area ration between carbonyl /ν (C=O)/ and amino- related band /ν (N-H)/ at ~ 1630 cm^{-1} and ~ 1540 cm^{-1} (right). The subscript c holds for control.

By utilizing the auto-fluorescence signal from GEN cross-linking products (i.e. the secondary CHT (GEL) amines and GEN esters), the CFM micrographs identify the chemical inter-linking as well as physical integration between layers' interfaces. The presence of non- and highly μ-porous and pore-interconnecting regions is demonstrated (Fig. 2.) within cross-sections of membranes with (by weight) prevalent GEL contribution in contrast to the sheet-like organization in membrane with equal components presence as presented in Fig. 2.

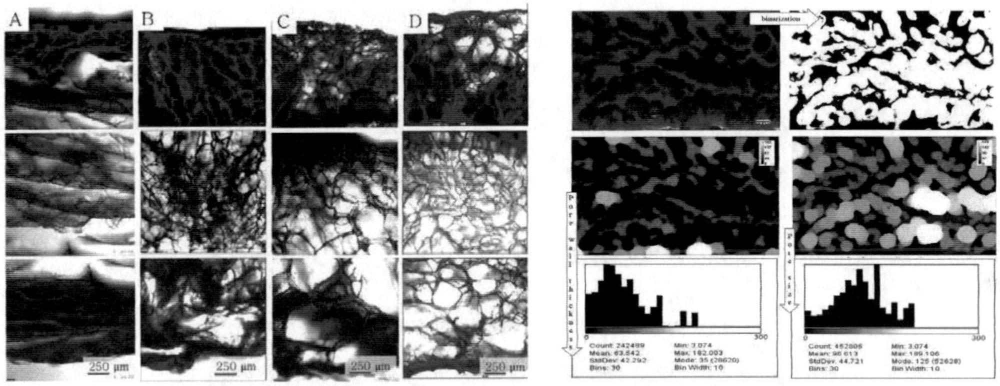

Fig.2. CFM images from different composition (A-C) membranes (left) with respective image analysis of pore size/wall thickness distributions (right).

The constant processing conditions onto variable compositions did not significantly affect the pore size distributions (in 1-230 μm range), while pore wall thickness increases up to 220 μm with GEL increase.

In parallel to CFM, the SEM microscopy was also undertaken, and the bottom-up inspection of SEM micrographs (Fig. 3) reveal microstructurally distinct regions also varying among different membranes.

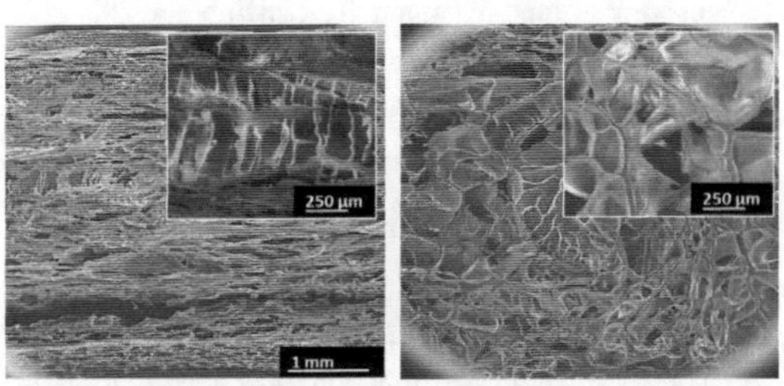

Fig. 3. SEM images from different composition (A-left and C-right) membranes in dry state.

Bottom-most section reveal the presence of dense skin (visible on membranes B and D, not presented) resulting from lyophilization process of solvent-casted layer when sharp increase in the polymer density at the film/vapor interface occurs, creating a polymer density gradient in layer thickness direction. In respect to GTR application, this layer is expected to interface with soft tissue, providing barrier function against fibroblasts penetration. Since the solvent casting was not completed to zero water content by freezing, the remaining porosity was induced also in a solvent-casted layer as seen in membranes A and C (Fig. 3) which is expected to maintain a good nutrition flow and metabolic exchange for cell proliferation and tissue growth in potential TE applications. Moreover, the pore wall thickness, as well as pore size were relatively lower in membrane A comparing to the others. This membrane significantly deviate between samples library due to specific, sheet like, anisotropic orientation parallel to the mould bottom, being assigned to typical CHT molecules arrangement. Such aligned structure combined with porous CHT was previously demonstrated to support the functional regeneration of periodontium, being demonstrated with extensive formation of mature collagen fibers perpendicular against the root surface and formation of tooth supporting mineralized tissue. On the other hand, the increment of GEL fraction increases the membrane density and at the same time reduce the structure heterogenicity, most prominently in GEL-rich membrane D.

The rapid mineralization process was successfully applied onto membranes surface. The random presence of spherical minerals (in Am and Dm) (Fig. 4) or their irregularly shaped clusters (in Bm and Cm) are visible, in higher quantity within Cm and Dm membranes. The respective EDX data (SEM images insets) reveal slight differences in Ca/P ration among and within different regions of membranes, all in range 1.7-2, which, when supplemented with FTIR (where carbonate peak position was assigned to substitution of phosphate ion when B-type HAp is formed) and XRD findings (the broadening of 2θ at 32° due to substitution) indicates on presence on nonstoichiometric carbonated apatite minerals. The Ca/P ratio within apatite mineral phase of different tooth- or tooth surrounding tissues are summarized in literature as 2.3–2.4 for enamel, 1.63– 2.01 and 1.48–1.55 for bone and pathologic bone, respectively, 1.3–1.97 for cementum, 2.1 for dentin, 1.9 for hypo mineralized dentin, and 0.9–1.7 for dental calculus. This suggest that compositions of minerals obtained by this rapid process demonstrate is chemically suitable for alveolar bone regeneration. The XRD and FTIR data (not presented), suggest on the same findings.

Fig.4. SEM micrographs with EDX (insets) of mineralization events in cross-sections of CHT/GEL membranes with different (1/1-Am, 1/5-Bm, 1/10-Cm and 1/20-Dm) w/w ration, cross-linked with 0.5 wt% GEN, after 12h incubation in 10xSBF media. The subscript m stand for mineralized membranes.

The other (GTR) relevant bulk property is mechanical stability, which is affecting the formation of tissues from proper cells phenotype and site-specific properties. On macroscopic level, GTR-intended materials must bear loads to provide stability to the tissue and to allow proper handling by surgeon handling during implantation, while microscopically, the mechanical input to cells is important clue impacting the cell growth and differentiation. Fig.5. describes the stress-strain behavior of dry (a) and fully wet (b), GEN cross-linked CHT/GEL membranes, being typical for porous biomaterials.

In both states, two major regions can be identified, first, the region of elasticity (before yield point), characterized by linear increase in stress with strain and second, the deformation (post yield point) region, where rapid increase in stress imply on pores irreversible collapse under the compression forces.

The very initial hyperbolic region which may be also identified, is excluded from these division as we assumed it as region for machine and sample adaption (including the air bubbles escape from hydrated material, as well as sample levelling to get in touch with both instrument plates between which compression occur), after which the real mechanical response curve is obtained..

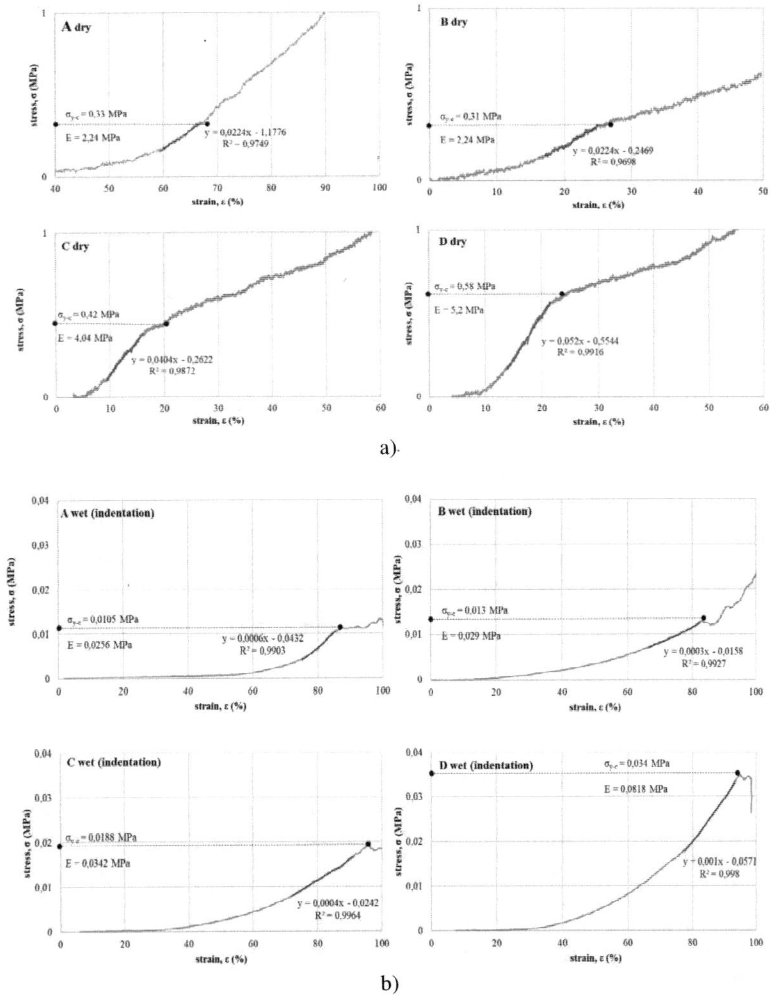

Fig. 5. Stress-strain curve from compression testing of dry (a) and wet composites (b) with different composition (denominated as A, B, C, D).

The GEL component increase (from A to D) obviously improves the yield stress at compression from 10 to19 kPa and elastic modulus from 26 to 34 kPa, both in wet state, as more relevant for envisioned application. On the other hand, the two log higher values are measured for parameters in dry state, which however indicates its mechanical stability at manipulation (in dry). The fast and high (up to 580%), composition dependent swelling, as well as 67% to 100% weight loss in 4 weeks *in vitro* degradation experiment point on membranes' relevance in GTR.

The material swelling profile as well as physiological stability, even bulk-related properties are important aspects in biomedical applications, since defect side (e.g. periodontal pocket) has limited space for membrane positioning and filler effect is desirable since provide multiple cues for efficient regeneration. Data (not presented) demonstrate fast weigh increase in first 5 min and equilibrium reached in less than 1 hour, irrespective of the composition, with increment of the swelling % at equilibrium (Figure 8b) in order from D (580%) to A (180%), i.e. from low to high

CHT content. Very convenient regression coefficient ($R^2 = 0.999$) obtained by curve fitting with second order polynomial function was obtained for CHT/GEL (w/w) ratio vs. swelling (%) at equilibrium correlation, implying on straightforward regulation of swelling profile by simple composition change, i.e. ~3 times reduction of swelling % was observed by 20 times reduction of CHT mass in respect to the GEL component. Obtained data for in vitro degradation test, reveal the reverse trend (from A to D) than swelling data after 4 incubation weeks, where GEL-rich membrane (D) was 100% degraded, while membrane with (by weight) equal components contribution (A) degrade to 2/3rds of ist weight, due to CHT insolubility in physiological media. It worth mentioned that irrespective of composition, the porous part was firstly degraded while dense layer maintain its shape up to 4 weeks in case of A and B membranes.

IV. CONCLUSION

Solvent casting, followed by T/time controlled, unidirectional freeze-casting process, is demonstrated to be feasible method for constructing physiologically and mechanically stable, well integrated, and μ-structurally diverse, 3D membranes from CHT and GEL, having proved biocompatibility. Important outcome herein is utilization of auto fluorescence signal of genipin-CHT/GEL cross-linking product by CFM for demonstration of physical (by identified μ-roughness) and chemical (by developed fluorescence signal) integration between layers. Further functional (biological) testing of developed composites (i.e. penetration test, viability and proliferation, as well as antimicrobial action) as GTR membrane for potential periodontal applications are envisioned.

REFERENCES

[1] Plessas A. Nonsurgical periodontal treatment: review of the evidence. 13(1):71–80, 2014
[2] Zhang Y, Zhang X, Shi B, and Miron R, Membranes for guided tissue and bone regeneration, 2013, *Ann. Oral Maxillofac. Surg.*, vol. **1**, no. 1, pp. 1–10.
[3] Dimitriou R, Mataliotakis G I, Calori M, and Giannoudis P V, The role of barrier membranes for guided bone regeneration and restoration of large bone defects: current experimental and clinical evidence, 2012, *BMC Med.*, vol. **10**, no. 1, p. 81.
[3] Y. Zhang, X. Zhang, B. Shi, and R. Miron, "Membranes for guided tissue and bone regeneration," *Ann. Oral Maxillofac. Surg.*, vol. 1, no. 1, pp. 1–10, 2013.
[4] G. Rasperini, "Surgical approaches based on biological objectives: GTR vs GBR techniques."*Stem Cell Biology and Tissue Engineering in Dental Sciences*, vol. 5. Academic Press, 2014.
[5] R. Dimitriou, G. I. Mataliotakis, G. Calori, and P. V Giannoudis, "The role of barrier membranes for guided bone regeneration and restoration of large bone defects: current experimental and clinical evidence," *BMC Med.*, vol. 10, no. 1, p. 81, 2012.
[6] Tas A C and Bhaduri S B, Rapid coating of Ti6Al4V at room temperature with a calcium phosphate solution similar to 10× simulated body fluid, 2004, *J. Mater. Res.*, vol. **19**, no. 9, pp. 2742–2749.
[7] L. Cui, J. Jia, Y. Guo, Y. Liu, and P. Zhu, "Preparation and characterization of IPN hydrogels composed of chitosan and gelatin cross-linked by genipin, *Carbohydr. Polym.*, vol. 99, pp. 31–38, Jan. 2014.

ACKNOWLEDGMENT

The authors acknowledge the project (ID Z7-7169) was financially supported by the Slovenian Research Agency. The authors are grateful to Dr. Andraž Stožer (Institute of Physiology, Faculty of Medicine, Maribor, Slovenia) for the allowed access to CFM unit.

Design and Production of Tailored Rear Trunk Covering for Acoustic Applications

H. Baskan[1], B. Isık[1], E. Erarslan[1], P. Ejder[1], S. Sezer[2] and H. Karakas[1]

[1]*Istanbul Technical University, Textile Technologies and Design Faculty, Textile Engineering Department, Inonu Street 65, 34437, Gumussuyu, Beyoglu, Istanbul, Turkey*
[2]*HP Pelzer Pimsa, TOSB, 1st Street, 14th Avenue, No:4, 41420, Sekerpınar, Cayırova, Kocaeli*

Corresponding author: baskan@itu.edu.tr

Abstract. Using recycled nonwovens as sound absorption materials are of great interest because of the increased nonwoven waste. In this study, needle-punched recycled polyester nonwoven was coated with polyethylene film, vinyl-based recycled polymer film and blends of these two materials in order to obtain new sound absorptive materials. Abrasion, tensile and acoustic features of the produced samples were measured and compared with the commercial PP-Fusing sample. It was seen that use of VBR polymer with polyethylene enhanced sound absorption properties of samples as well as their tensile strength and abrasion resistance performance. It is suggested to use needle-punched recycled polyester 'PET- 50% VBR- 50 %PE' sample for automobile rear trunk where acoustic insulation is required.

***Keywords*— Nonwoven, recycled polyester, recycled vinyl based polymer, rear trunk, sound absorption**

I. INTRODUCTION

ALTERNATION of pressure propagating through an elastic medium like air is called sound [1]. Undesirable and unpleasant sound is defined as noise. Being exposed to noise causes considerable problems like sleep disturbance, hearing loss, decrease in productivity and daily performance, increase in stress and blood pressure etc. [2]. Especially, noise is a big problem in closed space areas such as in automobiles [3]. Therefore, sound absorptive materials have come into prominence. Sound absorption is a process that converts the kinetic energy of the sound to heat energy when the sound strikes the sound absorptive material, which can be wall, cell, fiber, nonwoven, fibrous glass, mineral wools, felt and foams [1, 4]. In automotive industry, textile materials in the form of woven, knitted and nonwoven are used as sound absorptive materials [5].

Acoustic textiles can be bulky, high-loft, behaving as a rigid, porous sound absorber or light-weight, compact woven and nonwoven textiles that behave as porous screen [5]. Since their low production costs, small specific gravity and high total surface, nonwoven textiles are more preferred over woven and knitted types [2]. It was stated that noise absorption coefficients of nonwovens were comparable to rockwool and glass fiber in the high frequency range [3].

Lee and Joo produced thermally bonded nonwoven polyester fabrics with different fiber linear densities and investigated the sound absorption properties of produced samples. It was mentioned that nonwoven fabrics generated by using finer fibers showed higher noise insulation coefficients and absorption properties [3]. Tascan and Vaughn investigated the effect of fiber linear density and shape on the acoustical performance of needle-punched nonwoven fabrics. It was summarized that needle-punched nonwoven fabrics with finer fiber linear densities were superior than nonwoven fabrics with coarser fiber deniers in any fiber shape [6]. Needle-punched nonwoven with jute and blends of jute can be regarded as sound absorptive material. Sengupta tried to reduce sound by using polyester, polypropylene and jute / polypropylene needle-punched nonwovens. It was observed that the best sound reduction was achieved by jute/ polypropylene nonwovens when they were blended in equal amount [7].

On the other hand, using recycled nonwovens as sound absorption materials are of great interest because of the increased nonwoven waste. Using waste textile materials gives the chance for reducing environmental pollution, using the sources more effective and producing a new material for many applications [2, 8]. Factories generally burns or buries the selvages of nonwovens, which causes environmental destruction. For this reason, Kalebek examined the effects of the mass per unit area and thickness on the sound absorption performance of recycled needle-punched polyester nonwoven selvages. It was noticed that increasing the fabric mass per unit area and thickness also increases the sound insulation of the produced fabric [2]. Tiuc et al. tried to improve sound absorption properties of rigid polyurethane closed-cell foam by adding textile waste with changing amounts and measured the sound absorption coefficients of produced samples by impedance tube. According to the test results, it was observed that produced sample comprising 40 % textile waste and 60% rigid polyurethane foam had higher sound absorption coefficient than pure rigid polyurethane sample [8]. Moreover, Arenas mentioned that by addition of elastomer to needled recycled fibers and spunbonded polyurethane can be considered as a substrate for a foamed rubber material in case of trunk covering.

In this study, a new type of automotive rear trunk covering was designed and produced by coating recycled vinyl-based polymer onto needle-punched polyester nonwoven. It was aimed to measure sound absorption properties of produced recycled polyester-recycled vinyl-based polymer rear trunk covering and compare the results with the available commercial types. In addition to sound absorption coefficient test, abrasion test and tensile test were also applied to the samples. According to the test results, it can be suggested to use the sample 'PET- 50% VBR- 50 %PE' as a sound absorptive material for rear-trunk covering between 1500-2500 Hz frequencies.

II. MATERIALS

Needle punched recycled polyester (PET) nonwoven fabrics were used as substrate. Vinyl based recycled (VBR) polymer was used as coating film. Polyethlyene (PE), polypropylene (PP) and fusing were used as coating film in order to compare the sound absorption coefficient test results with the produced 'PET-VBR' sample.

III. METHOD

Recycled polyester nonwoven fabrics were produced by needle-punching process. Polymer films were produced by extrusion process and obtained films were simultaneously attached to recycled polyester nonwoven fabric in the presence of pressure and hot cylinders in extruder, separately.

Three samples consisting different amounts of VBR polymer, 'PET- 100 % VBR', 'PET- 50 %VBR- 50 % PE', 'PET-100 % PE' were produced. On the other hand, PP-Fusing commercial sample was supplied in order to compare the acoustic and abrasion performance of the produced samples with the commercial one. Fig. 1 shows the produced needle-punched recycled polyester samples containing different amounts of VRB polymer.

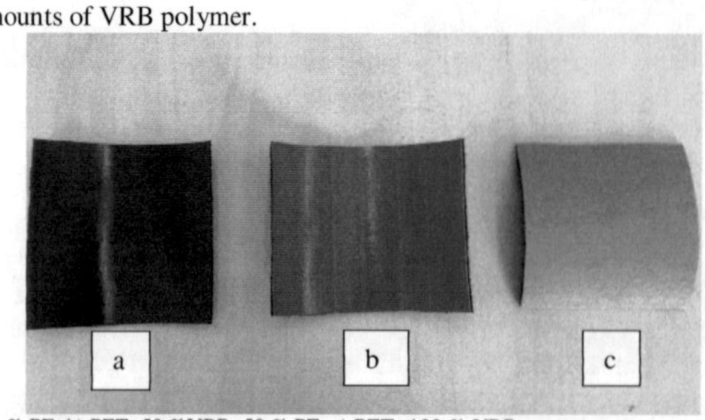

Figure 1: Produced samples a) PET-100 % PE, b) PET- 50 %VBR- 50 % PE, c) PET- 100 % VBR

For determining the sound absorption coefficient of the samples, impedance tube method was conducted according to the ASTM E1050 standard using a tube, a microphone and a digital frequency analyzer. A sound source is mounted at one end of the impedance tube and the material sample is placed at the other end [4]. Fig. 2 demonstrates impedance tube used for analyzing sound absorption coefficients.

Figure 2: Impedance tube used for analyzing sound absorption coefficients

In addition to acoustic tests, tensile and abrasion properties of the samples were also measured according to the standards EN ISO 1421:2000 and ISO 5470-2: 2003, respectively. Fig. 3 displays the tensile strength test and Fig. 4 shows abrasion test devices.

Figure 3: Martindale abrasion tester

Figure 4: Tensile tester

IV. RESULTS

Sound absorption coefficient test results were conducted between 16 and 6300 Hz frequencies. When all frequency ranges were considered, it can be said that PET-100 % VBR sample had the lowest sound absorption coefficient. However, when it had the highest sound absorption coefficient between 1500-2500 Hz frequencies. The sample composed of PP-Fusing had the highest sound absorption coefficient value in the range of 4000-4500 Hz. PET-100 % PE sample had average value in all frequency ranges. Fig. 5 shows the sound absorption coefficients of polyester samples consisting 100 % PE, 100 % VBR, 50 % PE-50 % VBR and the sample PP-Fusing versus frequency.

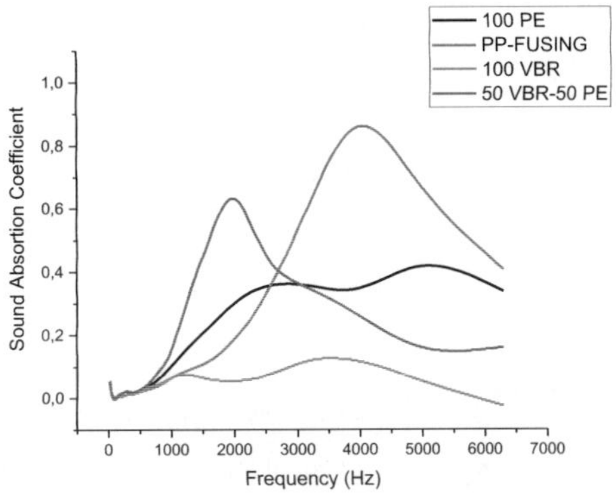

Figure 5: Sound absorption coefficients of polyester samples consisting of 100 % PE, 100 % VBR, 50 % PE-50 % VBR and the sample PP-Fusing versus frequency (Hz).

Tenstile and elongation test results of the polyester samples consisting different amounts of VBR polymer are summarized in Table I.

TABLE I
TENSILE AND ELONGATION TEST RESULTS OF THE PRODUCED POLYESTER SAMPLES CONSISTING OF DIFFERENT AMOUNTS OF VBR POLYMER.

Sample	Tensile Strength (N) Warp Direction	Tensile Strength (N) Weft Direction	Elongation (%) Warp Direction	Elongation (%) Weft direction
PET-100 %PE	742	720	105	114
PET-50 % PE-50 % VBR	622	695	107	118
PET-100 % VBR	599	568	159	134

According to the results given in Table I, it can be seen that addition of VBR polymer into the PET-PE structure decreased the tensile strength in both warp and weft direction. However, a slight increase was observed in elongation for both warp and weft direction. Further increase in the amount of VBR also allowed elongation to increase but tensile strength to decrease.

It was seen from the abrasion test results obtained according to the standart ISO 5470-2:2003, no abrasion occured although 10,000 cycle abrasion application. Fig. 6 shows the samples after the abrasion testing.

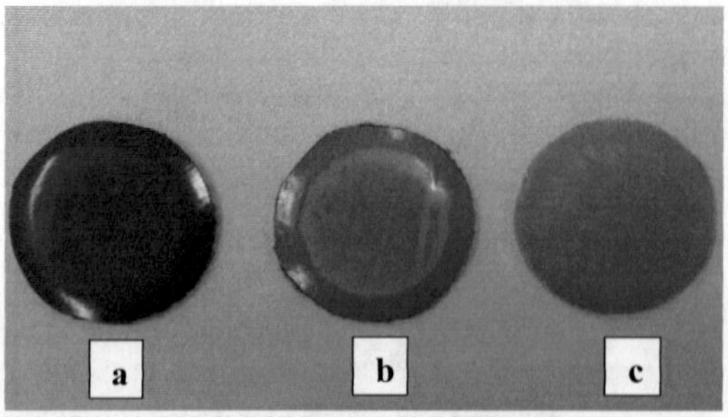

Figure 6: Polyester samples consisting a) 100 % PE, b) 50 % PE-50 % VBR and c) 100 % VBR

V. CONCLUSION

In this study, recycled needle-punched polyester nonwovens were coated with 100 % PE film, 50 %-50 % PE-VBR film and 100 % VBR film in order to generate novel structures with satisfactory sound absorption features. Acoustic features of samples were tested as well as abrasion and tensile properties. It was observed that using VBR polymer with polyethylene enhanced sound absorption properties of samples. In addition, it was seen that the produced samples had good abrasion, tensile and elongation properties when compared with the commercial types. As a conclusion, it can be suggested that using the sample containing 'PET-50% VBR- 50 %PE' can be an appropriate candidate for automobile rear trunk where acoustic insulation is desired.

ACKNOWLEDGEMENT

The authors would like to express their thanks to HP Pelzer Pimsa Automotive Textiles, especially İlhan Demirbaş for his contribution to the study.

REFERENCES

[1] V.K. Midha, MD. V. Chavcan, ''Nonwoven Sound Absorption Materials'', *International Journal of Textile and Fashion Technology*, vol. 2, no.2, pp. 45-55, June 2012.
[2] N. Avcioglu Kalebek, ''Sound Absorbing Polyester Recycled Nonwovens for the Automotive Industry'', *Fibres & Textiles*, vol. 24, no. 1(115), pp. 107-113, 2016.
[3] Y. Lee, C. Joo, ''Sound Absorption Properties of Recycled Polyester Fibrous Assembly Absorbers'', vol. 3, no.2, pp.1 -7, June 2003.
[4] R.S. Kumar, S. Sundaresan, ''Acoustic Textiles-Sound Absorption'', *Textile Technology*, pp. 1-10.
[5] J.P.Arenas, *Acoustic Textiles*. Singapore: Springer, 2016, ch. 7.
[6] M. Tascan, E.A. Vaughn, '' Effects of total Surface Area and Fabric Density on the Acoustical Behaviour of Needle-Punched Nonwoven Fabrics'', *Textile Research Journal*, vol. 78, no.4, pp. 289-296, 2008.
[7] S. Sengupta, ''Sound Reduction by Needle-Punched Nonwoven Fabrics'', *Indian Journal of Fibre and Textile Research*, vol. 35, no. 3, pp. 237-242, September 2010.
[8] A.E. Tiuc, H. Vermeşan, T. Gabor, O. Vasile, ''Improved Sound Absorption Properties of Polyurethane Foam Mixed With Textile Waste'', *Energy Procedia*,

vol. 85, pp. 559-565, January 2016.

IOP Conf. Series: Materials Science and Engineering **460** (2018) 012038 doi:10.1088/1757-899X/460/1/012038

Effects of Binder Solution on Color Fastness of Digital Printed Cotton Fabric

U K Sahin[1] and H Acikgoz Tufan[1]
[1] Istanbul Technical University, Textile Technologies and Design Faculty, Textile Engineering Department, İnönü Cad. No: 65, Beyoğlu/İstanbul, TURKEY

Email: acikgozh@itu.edu.tr

Abstract. Digital printing of textiles first appeared in early 90's, and its share in total printing market dramatically increased due to developments in the last five years and advantages of technique, but there is still much to achieve to gain a respectable place in the market. Pigments were most commonly used inks in digital printing and they have nearly no solubility in water or no affinity for any fiber, that results in a demand for binders to create a strong bond between pigment and fiber. In this study, 12 binder solution recipes were prepared with various amounts of KEOJET PGT® binder (0, 20, 25, 30%) and Catalizzatore PGT® catalyst (0, 9, 12, 15%), as well as distilled water. Then, binder solutions were applied onto 100% cotton twill fabrics before printing of Cyan (C), Magenta (M), Yellow (Y) and Black (K) colors via ink jet machine, PicassoTex. After a fixation process, color fastness to rubbing, light, washing and perspiration tests were conducted to printed fabrics. Color fastness to washing and perspiration test results were around 4/5, color fastness to light test results were 5 for all printed samples. However, color fastness to rubbing test results were observed between 3 and 4/5. When all of fastness results were taken into consideration, best results were obtained at the recipe which has 25% binder and 12% catalyst.

1. Introduction

Digital printing is based on delivering a specific design or media onto the fabric by using computerized machinery [1]. Ink jet printing technology appeared as an alternative to the conventional printing technology. It achieved to be an alternative in such ways including quick delivery, short run production, and photographic print. These features were much harder to apply in conventional printing machines [2]. There are many advantages of digital printing but there are some limitations, as well. The most important limitation is cost of printer heads. Heads are part of the machine, which are responsible for each ink droplet formation and application. In digital printing process, printer heads are moving back and forth along the face of the fabric in order to apply a tight line on the fabric with each pass. That is to say a fast-running production rate of the printer. However, most of the digital printing machines manufacturers reported that the production rate is approximately 150 meters per hour.

Bonding is achieved by attraction of colorant on fiber molecules, by using different colorant types, various bonding types can be achieved. Reactive dyeing application is done by making covalent bonds between colorant and the fiber, which is one of the strongest bond types. In order to decrease this strong bonding of electron sharing, an alkali may be used. In addition, there are ionic bonds, which can be explained as one or more electrons donating links with each other. Especially acid dyes use ionic bonds. Disperse dyes form hydrogen bonds, which are basically connecting atoms by a linking hydrogen atom.

Van der Waals forces are also available in disperse dyes, which are based on molecule attraction, and these are weaker bonds than the other type of bonds.

The most popular ink used in digital printing is pigments; however, they do not have water solubility or affinity for any fiber, and demand binders to be fixed with the fiber. Binders are typically polymers that form strong bonds with the fiber and the pigment. Thanks to these bonds, binders become materials called 'films', which are also insoluble in water and rubbing-resistant. The stability of the binder film defines color fastness and handle of the printed cloth [3-5].

Inkjet is the printing method which makes possible the delivery of liquid ink without any contact. PicassoTex is a textile inkjet printing machine produced by Optimum Digital Planet, which is based in Turkey. This machine is generally used for cotton-based fabrics in the industry.

In this study, the effects of concentration of chemicals used in binder solution on resulting color fastness of digital printed cotton fabrics are investigated.

2. Materials and method

2.1. Materials

100% cotton twill (56 ends/cm 30 picks/cm) made of ring yarns with count of Ne 46 were bleached at 98°C for one hour using 4g/L 37% Hydrogen peroxide (L.R.:1/10).

KEOJET PGT which is a special chemical that increases the performance in printing fabrics with digital pigment inks and increases the interest of the dyestuff is used as binder in this study. Keojet PGT is a white liquid which contains blend of resins in its chemical composition. Keojet PGT improves printing definition and color intensity without affecting the handle [6].

In this study, Catalizzatore PGT is used as a component of binder solution. The mission of catalyst is catalyzing the bonding of binder with the fabric. Catalizzatore PGT is a colorless liquid which is an organic acid solution containing <1% oxalic acid and 10-20% diethylenglycole.

PicassoTex is a textile inkjet printing machine produced by Optimum Digital Planet and it has eight printing heads. This machine is generally used for printing on polyester and cotton-based fabrics in the industry.

2.2. Method

Varying amounts of KEOJET PGT® binder (0, 20, 25, 30%) and Catalizzatore PGT® catalyst (0, 9, 12, 15%) were used to prepare binder solutions with distilled water. 12 binder recipes were calculated that have different amounts of binder, catalyst and distilled water. Other than binder recipes, all the parameters were identical, such as weave design, printing inks, fixation temperature and fixation time.

Each binder solution was padded on a fabric which has dimensions of 40 cm x 10 cm, followed by printing main colors (CMYK) using locally manufactured Picasso-Tex® digital printing machine equipped with Konica Minolta® 1024i printing heads, having a print resolution of 2880 dpi, and printing speed up to 420 m^2/h [7]. Fixation was done in a Thermo Scientific® Hereus model drying oven at 170°C for 4 minutes.

Color fastness to rubbing, light, washing and perspiration are conducted to each of the Cyan (C), Magenta (M), Yellow (Y) and Black (K) printed fabrics according to standard test methods, ISO 105-X12, ISO 105-B02, ISO 105- C06, ISO 105- E04, respectively.

3. Results and Discussion

Results showed that all color change and staining values for washing fastness tests are 4/5. Even some color was seen in washing bath but nearly there is no penetration to the adjacent fabric. Since the dye has very limited affinity to the adjacent fabric, both acidic and basic perspiration test results follow the same trend, resulting as 4/5. Light fastness results of all samples were 5, as the pigments used are highly stable against the influence of light.

As can be seen from Figure 1., 12% of catalyst is enough to reach highest rubbing fastness value regardless of the amount of binder among the levels studied. Moreover, 25% of binder gives better rubbing fastness score than both 20% and 30%. It is assumed that 20% of binder is not enough to form a stable film, and 30% causes excessive hard print which will cause more ink loss during crocking.

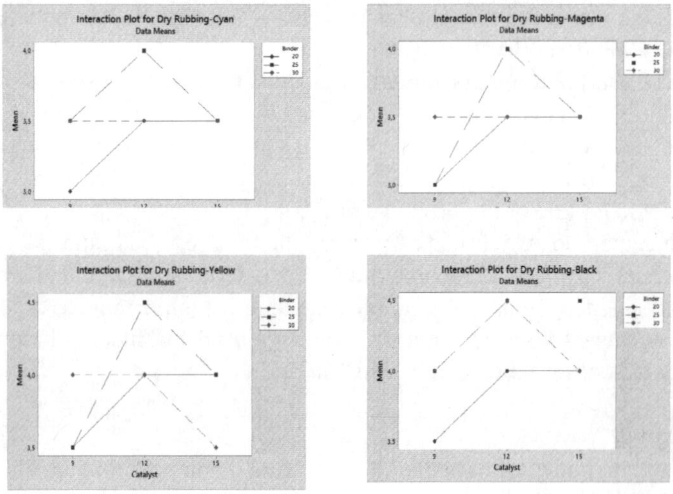

Figure 1. Interaction plots for dry rubbing

As can be seen from Figure 2, 12% of catalyst and 25% binder are enough to reach highest wet rubbing fastness values, and higher amount of binder and catalyst does not further contribute to wet rubbing fastness. It is apparent from both dry and wet rubbing fastness values that for cost-effective pigment printing with better quality, investigation of interaction between binder solution chemicals and pigment type must carefully be taken into consideration.

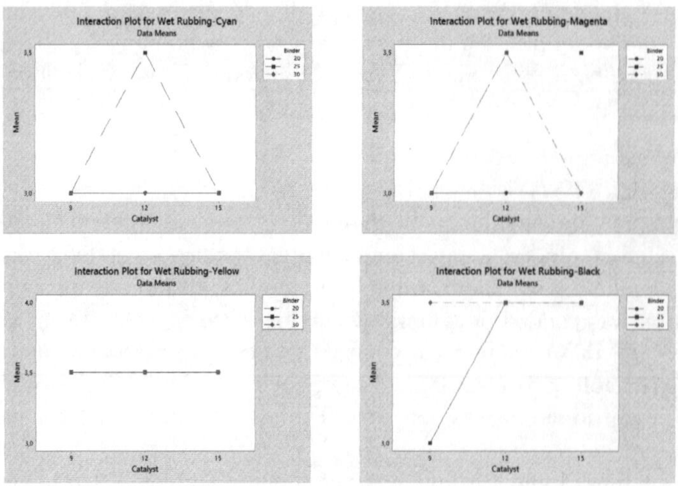

Figure 2. Interaction plots for wet rubbing

4. Conclusion

In this study, effects of binder solution in digital pigment printing of cotton fabrics were investigated. For this aim, different amounts of KEOJET PGT® binder (0, 20, 25, 30%) and Catalizzatore PGT® catalyst (0, 9, 12, 15%) were used to prepare 12 binder solutions with distilled water, and binders were applied onto 100% cotton fabrics before digital pigment printing. While investigating for the best recipe, color fastness to rubbing, light, washing and perspiration tests were conducted to each of the Cyan (C), Magenta (M), Yellow (Y) and Black (K) printed fabrics. Color fastness to washing, perspiration and light resulted from 4/5 to 5. Dry and wet rubbing fastness results had lower fastness degrees and showed differences ranging from 3 to 4/5. When all of fastness results were taken into consideration, best results were obtained at the recipe which has 25% binder and 12% catalyst. It is apparent from both dry and wet rubbing fastness values that for cost-effective pigment printing with better quality, investigation of interaction between binder solution chemicals and pigment ink type must carefully be taken into consideration prior to printing, and binder recipe must be carefully adjusted for every ink combination to be applied.

Acknowledgement
The authors would like to thank Mete AKSU and Ümit TOKGÖZ from Optimum Digital Planet for their support throughout the experimental study.

References
[1] Tyler D J 2005 *Textile Digital Printing Technologies, Textile Progress.* Woodhead Publishing **37-4** p 1-65
[2] Cie C 2015 *Ink Jet Textile Printing.* Woodhead Publishing ch.11
[3] Hees U, Frenche M, Kluge M, Provost J and Weiser J 2002 Developments in Textile Ink Jet Printing with Pigment Inks *Int. Conf. on Digital Printing Technologies* p 242-245
[4] Gupta S 2001 Inkjet printing - A Revolutionary Ecofriendly Technique for Textile Printing *Journal of Fiber & Textile Research* **26(1-2)** p 156-161
[5] Lin J W, Fague G R, Wong R R, Pigment ink jet ink compositions for high resolution printing USA Patent No: US6143807 A; 07 Nov 2000
[6] Keojet PGT, (2018, April 04). [Online]. Available: http://www.cplchimica.com/en/search.html
[7] PICASSO TEX, (2018, April 04). [Online]. Available: http://www.optimumdigital.com/tr/urunler/tekstil-baski-makineleri/picasso-tex/i-64

IOP Conf. Series: Materials Science and Engineering **460** (2018) 012039 doi:10.1088/1757-899X/460/1/012039

Design of a Zwitter-Ionic Nanocomposite for Dyestuff Removal

E Emekdar[1] and U K Şahin[1]
[1] Istanbul Technical University, Textile Technologies and Design Faculty, Textile Engineering Department, İnönü Cad. No: 65, Beyoğlu/İstanbul, TURKEY

Email: emekdar@itu.edu.tr

Abstract. Today, with the increasing awareness of environmental conservation, many scientists are directing their work on environmental improvements. In the textile industry, dyestuff load of wastewater increases, and improvements are made to reduce this load. As a contribution to these improvements, a nanocomposite is designed with clay, chitosan and its modifications, which are totally natural. As a preliminary study, the clay was modified and treated with dyestuff solutions to investigate the absorption properties of Methylene Blue with various clay types. According to the results obtained by UV visible spectrophotometer, Cloisite 30B, which is the clay with the best absorption property, has been determined as the reference material. With the lights of preliminary study, a nanocomposite was designed to adsorb dyestuffs from textile waste waters using modified clay and modified chitosan. With the anionic modification of chitosan, it can be converted into carboxymethylchitosan (CMCTS) and it can also have affinity to cationic structures. With this structure, colored composites are obtained while contributing to recycling and environment by providing adsorption of dyestuff. According to UV-vis and FT-IR results, CMCTS that is modified with Butanetetracarboxylic acid (BTCA) has better adsorption from textile wastewaters.

1. Introduction

Clay is accepted as the most used filling material thanks to its high visibility rate and its laminated morphology [1]. Economic value of clay is determined according to its mineral types and presence of non-clay material inside the clay. Clay minerals that have white, green, pink, grey and hues of brown colors can be found with different chemical compositions and at different contents, and pure clay mineral is rarely found in nature [2]. Due to their superior chemical and physical properties, easy accessibility in nature and having technologic properties such as plasticity, moisture adsorption, high strength, replacement with every kind of ions that are organic or inorganic, high adsorption amount and catalytic activity, currently more than 100 fields of production use clay as raw material [3].

Clay minerals are results of the alteration of the volcanic tuffs or hydration of the laminated silicates that have layer thickness of 1nm each and lateral diameter that is more than 100 nm.

Unless modification of clay surfaces is done, silicate layers cannot make good interaction with water repellent polymer matrices due to water absorbency of the layers. Many types of mechanisms can be used for modification of clay surfaces such as ion exchange using organic and inorganic cations, the attachment of organic and inorganic ions, getting into reaction with acids, burning, ultrasound, plasma etc. After modifications, clays are also used at removing organic pollutants from water, air and oil with adsorption, as rheological control factor and at dye and cosmetics fields [4].

Changeable cations in clay affect its physical structure. Organic and inorganic anions and cations that can penetrate into layers of clay are adsorbed by clay minerals and they are hold in changeable state. When they are treated with ions in aqueous solutions, ions in changeable state can be replaced with new ones. The sum of changeable cations equivalent mass number (meg) at 100 grams of mineral is known as the cation exchange capacity (CEC) and it is an important parameter for selecting the most suitable clay type to use [5]. The surface polarity of the clay and the polarity of the polymers pairs increase and the middle layers of the clays enlarge when the organic onium ions that are located at the surface of the middle layer take the place of the inorganic changeable cations. Thus, polymers, dyestuffs and such matters can penetrate more properly when ions that have long alkali chains and organo-modified silicates (organoclays) are used [6]. According to their different element groups and orientations many organoclay types with many different characteristics are commercially available for using in polymer composites. However, Cloisite 30B has better absorbency, higher CEC and bigger gap values thanks to its long tallow ammonium chain. Also, previous study of the authors gives detailed information about clays and proves that among the commercially available modified clays Cloisite 30B is better one to use in this study [7].

Clay based nanocomposites are a good example of nanocomposites which are developed with high absorbency properties. They are preferred in dye removal applications due to their features such as having high aspect ratio, layered structure and providing good properties even at low add-on percentages. This structure provides a better and more environmentally friendly refining with natural absorbent structure by reducing the amount of chemical amounts and number of refining processes. Depending on the needs of the area of use, custom-made modifications of this nanocomposite are manufactured [8].

Further modifications can be done with intercalation of different polymers to clays. This polymer can be selected according to its characteristics and variety of nanocomposite structure can be increased by modification of polymer. In this study chitosan is selected but it is modified for adsorption of Methylene Blue dyestuff which is cationic.

Chitosan is soluble form of chitin which has the second biggest supply at the world as a polymer after cellulose and it produced when approximately 50% of deacetylation is done on chitin. This amount at the definition depends on the origin of the polymer which is a polysaccharide. Solubility of the Chitosan is seen when it is turned into polyelectrolyte form at acidic conditions and to make this protonation of $-NH_2$ group of unit of D-glucosamine where the position of it is C-2 is done. It has many application areas including depollution and protein recovery as flocculent because of the specific property as being the only pseudonatural polymer which is cationic. Also the biggest reason for selecting Chitosan at different application areas with different forms from solution to fiber is its high solubility property [9].

Deacetylation of chitin is the main procedure of Chitosan production. With the usage of potassium/sodium hydroxide solutions or mixture of anhydrous hydrazine and hydrazine sulfate that produces an alkaline media, acetated part of chitin is hydrolyzed and removed. Resultant Chitosan is yield when chitin is treated for 4-5 hours with 40–45 % (w/v) NaOH solution and the N-deacetylation reaction occurs at 90–120 °C [10-11].

Structure of Chitosan is semi crystalline when it is solid. If the low molecular-weight chitin is deacetylated totally, Chitosan with single crystals can be yield [12]. Degree of acetylation (DA), pH, protonation acid, ionic concentrations, acetyl group locations of Chitosan, drying and isolating conditions are the parameters that determines the final solubility. When DA is around 50% at neutral pH Chitosan becomes soluble. At pH 7- 7.1 and room temperature solution of Chitosan is stable, however when temperature is about 40° C it turns into gel form. Chitosan has two hydrogen bonds at its chain structure as first one is between O5 and OH3 and second one is between N and OH6 while chitin has two hydrogen bonds as first one is between O5 and OH3 and second one is between O of COO and OH6 [12].

Chitosan has many different classes according to its characteristics with different modifications as O-and N-Carboxymethlchitosans, Chitosan 6-O-sulfate, N-methylene phosphonic chitosans, Trimethylchitosan ammonium, Carbohydrate branched chitosans, Chitosan-grafted copolymers,

Alkylated chitosans and Cyclodextrin-linked chitosans. In this study, chitosan is modified as O-and N-Carboxymethylchitosan to adsorb cationic dyestuff. The most known type of Chitosan is Carboxymethylchitosan (CMCTS) which has amphoteric characteristic that means changing ionic character according to pH of the medium. When sodium hydroxide is present in the medium Chitosan gives O and N-carboxymethylation reaction with the help of sodium monochloracetate and forms O and N-CMCTS. pH range of being soluble is enlarged with the help of carboxymethylation reaction as pH>7, however its charges are balanced during the reaction, and this balancing causes phase separation between pH 2.5 and 6.5. Furthermore, when there is a reductive agent Chitosan reacts with glyoxylic acid and forms N-CMCTS [13].

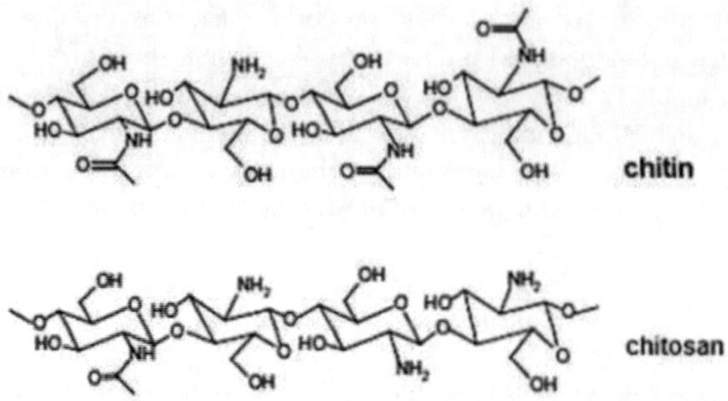

Figure 1. Chitin and Chitosan Structure [9].

Nowadays, there are many applications of Chitosan at nanotechnology with different modifications of it in composites for different purposes and properties at cosmetics, biomedical engineering, biotechnology, ophthalmology, pharmaceuticals, agriculture, textiles, food processing nutrition etc. while application of chitin remains less. Chitosan is the only pseudo-natural polymer that has cationic character thanks to this property is inimitable. Together with this property, film forming capability brings more application areas to Chitosan. Properties are enhanced with modifications and combined at nanocomposites for needed products. Fibers, sponges, hydrogels, films and biomedical products can be manufactured from Chitosan since processability of it easier than that of chitin while its stability is less because of being hydrophilic and pH sensitive [9].

Nanocomposites of nanolayered silicates with Chitosan are the most popular method to enhance the characteristics physically, thermally, mechanically and chemically. Especially, utilization of montmorillonite is very popular as nanolayered silicates and enhanced properties vary according to harmony between Chitosan and montmorillonite and production processes [10]. When Chitosan is intercalated to Na+ montmorillonite for a nanocomposite, film forming tendency decreases which is resulted with resistant and stable products [14].

Chitosan become a good material of many studies at the literature whether used alone or with a clay combination. Firstly, Chitosan was used alone in adsorption of anionic dyestuff and sulphonated azo dyestuff by coagulation of Chitosan [15,16]. In case of combination of Chitosan with clay, N, O-Carboxymethyl Chitosan/ Montmorillonite with 5:1 weight ratio is used for anionic dyestuff removal and it shows an increased the dye adsorption with decreasing of pH [17]. Secondly, bentonite filled with chitosan and acrylic copolymer is used to remove synthetic dyes from waters and examined for effects of co-monomers [18]. Also crosslinked Chitosan and Bentonite is used for cationic dyestuff and anionic azo dyestuff removal [19,20]. Chitosan and CTAB modified bentonite is used for removing of dyestuff while magnetic carboxymethyl chitosan-gpoly(acrylamide)/laponite RD with improved dye adsorption capacity is produced for use at acidic media [21,22].

With the light of all these information, the purpose of the project is to create a nanocomposite from nanoclays that is chemically activated, having more branched structure with intercalation of Chitosan and Carboxymethyl Chitosan (CMCTS) and removing dyestuff from waste water with its enhanced absorbency for textile industry.

2. Materials and method

2.1. Materials

In this study, alkyl ammonium based organoclay (Cloisite 30B) which is supplied by Southern Clay Company, is used. Chitosan with medium molecular weight is bought from Sigma Aldrich to produce nanocomposite with clay. Chloroacetic acid is an organochlorine compound which has a molecular formula of $ClCH_2CO_2H$. It has a molecular weight of 94.49702 g/mol and it is used as a modifier to modify Chitosan to produce CMCTS. 1,2,3,4-Butanetetracarboxylic acid is an organic compound which has a molecular formula of $C_8H_{10}O_8$. It has a molecular weight of 234.16 g/mol and it is used as a modifier to modify Chitosan to produce CMCTS. Sodium Hydroxide (NaOH), which has a molecular weight of 39.99 g/mol and ethanol (ethyl alcohol), which has a molecular formula of C_2H_5OH are used to modify Chitosan. At the experiment nanocomposites are used to adsorb Methylene Blue (C.I 52015) which is a standard basic dyestuff with a molecular formula of $C_{16}H_{18}N_3SCl$. It has a molecular weight of 319.85 g/mol. 60% (v/v) Perchloric acid ($HClO_4$) which has a density of 1,764 g/mL, is used to protonate Methylene Blue dyestuff solution. Dichloromethane is an organic compound which has a general formula of CH_2Cl_2. It is used as dissolver material of Methylene Blue dyestuff solution.

2.2. Method

2.2.1. Modification of Chitosan (Preparation of CMCTS). 5 g Chitosan is soaked into 50 mL NaOH solution (20 wt%) for 12 hours and mixture is filtered. Filtered Chitosan is soaked into 50mL ethyl alcohol at volumetric flask and stirred for 30 minutes. While stirring continues 4 g butanetetracarboxylic acid (BTCA) or chloroacetic acid is added and stirring is continued for 30 minutes more. Then mixture is cooled to 20°C and kept at the temperature for 1 hour. Finally, treated sample is filtered, dissolved in distilled water and pH value is set as 7.0 by adding acetic acid. The mixture is poured into ethyl alcohol for precipitation as foam gel and the white precipitate is taken into petri dishes by separating with pouring. It is washed three times by using ethanol/water mixture with ratio of 70%/30% (volume on volume) and one more time by using pure ethyl alcohol. Samples in petri dishes are placed into oven at 80°C for drying of the CMCTS. Finally, CMCTS is scraped off from the petri dishes and it is turned into powder form by mechanical milling. FT-IR analysis is done to CMCTS sample and resultant peaks are compared with those presented in the literature [17].

2.2.2. Combination of Modified Chitosan and Clay. Firstly, swelling of 1 g clay is done at 100 mL distilled water. Secondly, CMCTS solutions are prepared separately by dissolving CMCTS at distilled water which refers to 5:1 weight ratio at nanocomposite. After that, CMCTS solution is added to clay suspension slowly and stirred at 60°C for 6 hours. Then until the supernatant becomes pH 7.0 nanocomposites are washed by using distilled water and placed in oven at 60°C for 12 hours. Finally, CMCTS-clay nanocomposite is scraped off and turned into powder form by mechanical milling [17]. Chitosan (CTS)-clay combination is produced following the same procedure with the exception of preparing CTS solution with water instead of CMCTS.

2.2.3. Dyestuff Adsorption of Clay and its Combinations. Sivathasan J.'s study which comprises dissolving of 2 g Cloisite 15A at 200 mL dichloromethane (DCM) and adding 2mmol cationic dyestuff with 5 mL HCl solution with ratio of 70% is taken as reference to determine the solution ingredients [23]. Firstly, 200 mL of DCM is put into each flask. Secondly, combined clays are treated with the certain dyesuff solution to examine the adsorbed amount and comparisons are made. 2 g per each sample with weight ratios as Chitosan 5:1, CMCTS 5:1 that gives the best adsorption result is used. Then, 5.83 mL perchloric acid with ratio of 60% is added to each flask for protonation. Finally, 0.6397 g Methylene Blue (equivalent of 2 mmol dyestuff) is added to mixtures at each flask and each mixture stirred by magnetic stirrer at 60 rpm for 12 hours. After stirring they are let to precipitate for 24 hours.

3. Results and Discussion

With this final structure, a colored composite will be obtained while contributing to recycling and environment by providing adsorption of dyestuff. FT-IR for characterization of the nanocomposite and UV-vis spectrophotometer for dyestuff adsorption results are used. Methylene Blue dyestuff solution.

3.1. UV-Visible Spectrophotometer (UV-vis)

The formula below is used to calculate the removal rate of dyestuff from the solutions by using UV-vis results. At the formula, R is removal percentage, c_0 (mg L^{-1}) is the initial concentration and c_t (mg L^{-1}) is the final concentration of adsorbate by adsorbents [24].

$$R = \frac{c_0 - c_t}{c_0} \times 100\% \quad (1)$$

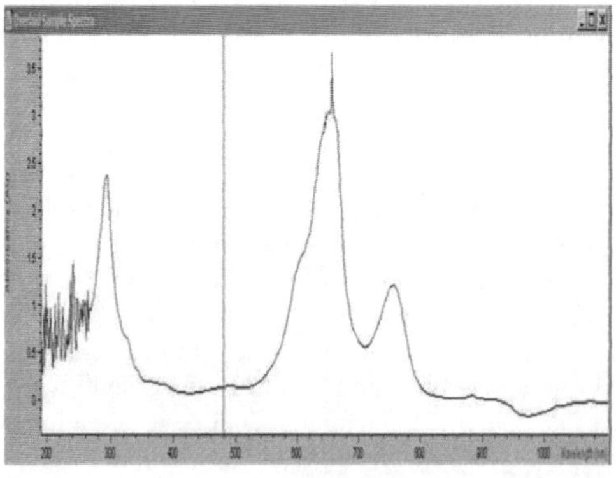

Figure 2. UV-Vis of Chitosan-Cloisite 30B Nanocomposite-Dyestuff Solution

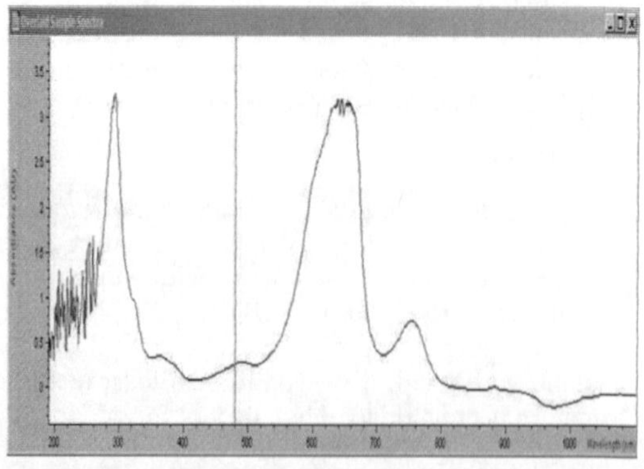

Figure 3. UV-Vis of Chloroacetic Acid-Chitosan-Cloisite 30B Nanocomposite -Dyestuff Solution

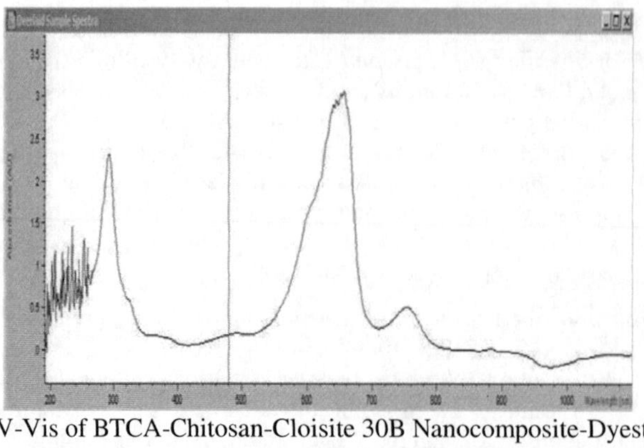

Figure 4. UV-Vis of BTCA-Chitosan-Cloisite 30B Nanocomposite-Dyestuff Solution

According to UV-vis results, Chitosan and clay composites decrease the amount of dyestuff in waste water but modified ones adsorb more dyestuff on themselves thanks to highly anionic characteristics. BTCA works more than Chloroacetic acid (CAA) in modification, thus BTCA- CMCTS has best removal capacity as %67.77. This phenomenon is probably due to higher number of available reactive side groups BTCA has when compared to those offered by CAA.

Table 1. Removal Percentage of Nanocomposites

Sample	Absorbance (AU)	Number of Peaks	Removal Percentage (%)
Dyestuff Solution	9.0	too many	-
Chitosan-Clay	3.6	4	60.00
CMCTS with Chloroacetic Acid-Clay	3.1	3	65.55
CMCTS with BTCA-clay	2.9	3	67.77

3.2. Fourier Transform Infrared Spectrophotometer (FT-IR)

As it can be seen from the FT-IR graph, Chitosan makes peaks at 1090 cm^{-1}, 3336 cm^{-1}, 1597 cm^{-1}, 879.88 cm^{-1} and at 1355 cm^{-1}. At 1090 cm^{-1}, bending of C-O bonds occur, at 3336 cm^{-1}, bending vibration of -OH and amine bonds occur, at 1597 cm^{-1}, NH2 bonding occurs and lastly at 1355 cm^{-1}, amide III bonds are observed. When these peaks are compared with those of Chitosan which is modified with Butanetetracarboxylic acid, at BTCA/Chitosan, the 1597 cm^{-1}, 879.88 cm^{-1} and 1355 cm^{-1} shifted to 1565 cm^{-1}, 865 cm^{-1} and 1399 cm^{-1}, respectively and two of the peaks which are 1090 cm^{-1}, 3336 cm^{-1} disappeared but two new peaks at 1216 cm^{-1} and 622 cm^{-1} are formed which can be the characteristics of BTCA.

After the modification of Chitosan, the BTCA/Chitosan polymer is combined with Cloisite 30B and its characterization showed the characteristics of both BTCA/Chitosan and Cloisite 30B. The formed nano composite gives peaks at 463.87 cm^{-1}, 527 cm^{-1}, 618 cm^{-1}, 665 cm^{-1}, 1034 cm^{-1}, 1212 cm^{-1}, 1399 cm^{-1} and 1561 cm^{-1}. The 618 cm^{-1}, 1212 cm^{-1}, 1399 cm^{-1} and 1561 cm^{-1} peaks are derived from the peaks of the BTCA modifier and Chitosan which are respectively 622 cm^{-1} and 1216 cm^{-1} from the modifier and 1355 cm^{-1} and 1597 cm^{-1} from Chitosan. The other peaks are derived from the structure of Cloisite 30B which were 515 cm^{-1}, 630 cm^{-1}, 689 cm^{-1} and 1125 cm^{-1}. With the peaks of 622 cm^{-1} from Chitosan and 630 cm^{-1} at the nanocomposite the peak at 618 cm^{-1} strengthened.

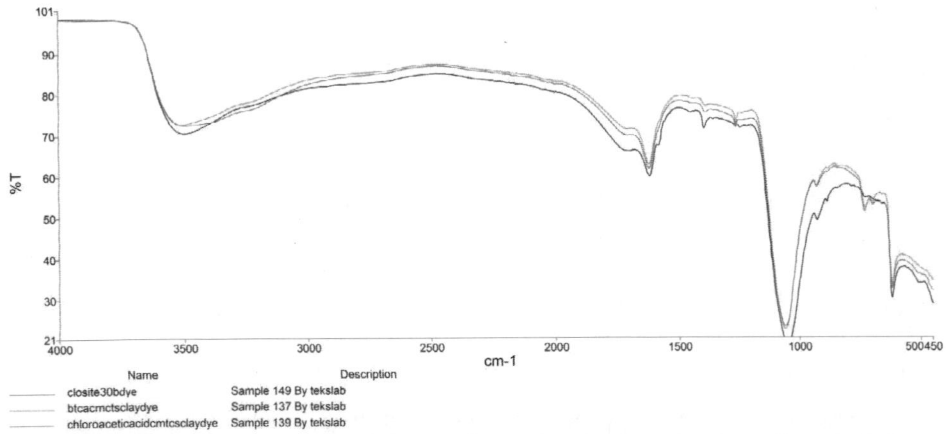

Figure 5. FT-IR result of clay and chitosan based composites after dyestuff removal stage

Lastly, the nanocomposite is treated with Methylene Blue dyestuff and the formed structure has the peaks of both the nanocomposite and the Methylene Blue dyestuff. The dyestuff treated nanocomposite has new peaks formed that are at 3509 cm^{-1}, 1625 cm^{-1} and at 1264 cm^{-1} and it can be the characteristic of the Methylene Blue dyestuff. Instead of these points the peaks of dyestuff-treated nanocomposite are parallel to those of un-dyed one but 1561 cm^{-1}, 1034 cm^{-1} and 618 cm^{-1} peaks are shifted to 1625 cm^{-1}, 1061 cm^{-1} and 619 cm^{-1}.

When combination of Clay-Chitosan modified with Chloroacetic acid graph is examined it is seen that 1050.3 cm^{-1} comes from clay component referring stretching of Si-O-Si bonding, 3336.4 cm^{-1} referring to bending vibration of -OH and amine bonds and 1549.5 cm^{-1} referring to NH_2 bonding comes from Chitosan. Peaks at 1640.6 cm^{-1}, 1406.8 cm^{-1} and 1260.2 cm^{-1} appeared at this sample and they can be coming from chloroacetic acid. In case of dyestuff-treated form of the sample, dyestuff peaks which are 3498.67 cm^{-1}, 1625.70 cm^{-1} and 1264.80 cm^{-1} appear while 1549.5 cm^{-1} and 1406.8 cm^{-1} disappear. Peaks of clay component are preserved and peak at 1060.06 cm^{-1} is strengthened.

Results achieved from both CMCTS appear to be similar but CMCTS modified with chloroacetic acid could not make the peak at 702.43 cm^{-1} which means there is less bonding therefore CMCTS modified with BTCA offered better results. FT-IR results show that BTCA- CMCTS makes higher number of bonding with dyestuff than other composites, therefore it refines more precisely.

4. Conclusion

Textile industry needs a refinement for wastewater that includes dyestuff, and easier and shorter refinement processes gain an importance in case of economy and time consumption. Clay combinations are examples of natural refining chemicals with enhancing characteristics. Within this study, clay combinations are prepared with Chitosan, and Chitosan that has cationic characteristic is modified anionicly to produce an attraction for adsorption of a standard cationic dyestuff, namely Methylene Blue. This modification is done by using two different chemicals, BTCA and Chloroacetic acid, and dyestuff adsorption potentials of such modified products are compared with each other, and with that of ordinary Chitosan. Results of yield nanocomposites show that anionic modification is maintained properly with both chemicals and thanks to carboxymethylation cationic dyestuff adsorption capacities are significantly increased. However, BTCA-CMCTS-clay combination is more precise option among them as a result of making more bonds with dyestuff. With regards to its unique structure, the composite has a potential to get different chemicals from a single refining bath thus to decrease refining process steps and chemicals that are used to refine until the quality of wastewater becomes within waste management standard limits.

Acknowledgement

This study is supported by ITU-BAP with Project number 40852. The authors would like to express their gratitude to Istanbul Technical University for its support. Results presented here are part of MSc dissertation prepared by Elçin Emekdar under the supervision of Umut Kıvanç Şahin.

References

[1] Lei, W., 2005, Preparation, Morphology and Thermal/Mechanical Properties of Epoxy-Nanoclay Composites, B. Sci, University of Science & Technology China, China.

[2] Grim, R.E.,1962, *Applied Clay Mineralogy*, McGraw-Hill, New York.

[3] Adams, J.M. and Evans, S., 1979, Determination of Cation Exchange Capacity (Layer Charge) Of Small Quantities of Clay Minerals by Nephelomenty, *Clays and Clay Minerals*, **27**, 137-9.

[4] Paiva L., Morales A. and Diaz F., 2008, Organoclays: Properties, preparation and applications, *Applied Clay Science*, **42**, 8-24.

[5] İpekoglu, B., Kurşun, İ., Bilge, Y. and Barut, A., 1997, Türkiye Bentonit Potansiyeline Genel Bir Bakış, 2. Endüstriyel Hammaddeler Sempozyumu, 51- 6, İzmir.

[6] Yılmazbahyah A., 2006, Maleik Anhidritle Graftlanmış Oligomerlerin ve İ-Pp/Silikat Nanokompozitlerin Tepkimeli Ekstrüzyon Yöntemiyle Sentezi ve Karakterizasyon, Doktora Tezi, Hacettepe Üniversitesi Fen Bilimleri Enstitüsü, 16-26, Ankara.

[7] Emekdar E., 2016, Designing a Nanocomposite that Absorbs dyestuff, BSc Thesis, Department of Textile Engineering, Istanbul Technical University, Istanbul, Turkey.

[8] Şen F., Palancıoğlu H. ve Aldaş K., 2010, Polimerik Nanokompozitler ve Kullanım Alanları, *Makine Teknolojileri Elektronik Dergisi*, **(7)1**, 111-8.

[9] Rinaudo.M., 2006, Chitin and chitosan: Properties and applications, *Scince Direct, - Prog. Polym. Sci.* **31**, 603–32
[10] Yılmaz.N.C., Yılmaz.O., Vasil.C., 2015, Eco-Friendly Chitosan-Based Nanocomposites: Chemistry and Applications
[11] Ahlafi, Hammou; et al. (2013). "Kinetics of N-Deacetylation of Chitin Extracted from Shrimp Shells Collected from Coastal Area of Morocco" (PDF). medjchem.com. *Mediterranean Journal of Chemistry*.
[12] Cartier N, Domard A, Chanzy H. Single crystals of chitosan. *Int J Biol Macromol* 1990; **12**:289–94.
[13] Le Dung P, Milas M, Rinaudo M, Desbrie`res J. Water soluble derivatives obtained by controlled chemical modification of chitosan. *Carbohydr Polym* 1994;**24**: 209–14.
[14] Zuber.M., Zia.K.M., Barikani.M., Chapter 3 Chitin and Chitosan Based Blends, *Composites and Nanocomposites*
[15] Cestari.A.R., Vieira.E.F.S., Mota.J.A., 2008, The removal of an anionic red dye from aqueous solutions using chitosan beads—The role of experimental factors on adsorption using a full factorial design, *Journal of Hazardous Materials* **160** 337–43
[16] Szygula.A., Guibal.E., Ruiz.M., Sastre.A.M., 2008, The Removal of Sulphonated Azo-dyes by Coagulation with Chitosan, *Colloids and Surfaces A: Physicochem.Eng.Aspects* **330** 219-26
[17] Wang L., Wang A., 2008, Adsorption Behaviours of Congo Red on the N,O-Carboxymethyl Chitosan/ Montmorillonite Nano Composite, *Chemical Engineering Journal* **143** 43-50
[18] Bhattacharyya.R., Ray.S.K., 2014, Micro- and Nano-sized Bentonite Filled Composite Superabsorbents of Chitosan and Acrylic Copolymer for Removal of Synthetic Dyes from Water, *Applied Clay Science* **101** 510-20
[19] Bulut.Y., Karaer.H., 2015, Adsorption of Methylene Blue from Aqueous Solution by Crosslinked Chitosan/Bentonite Composite, *Journal of Dispersion Science and Technology*, **36:1**, 61-7, DOI: 10.1080/01932691.2014.888004
[20] Liua. Q., Yangb.B., Zhanga.L., Huanga.R., 2015, Adsorption of an anionic azo dye by cross-linked chitosan/bentonite composite, *International Journal of Biological Macromolecules* **72** 1129–35
[21] Guo.J., Chen.S., Liu.L., Li.B., Yang.P., Zhang.L., Feng.Y., 2012, Adsorption of dye from wastewater using chitosan–CTAB modified bentonites, *Journal of Colloid and Interface Science* **382** 61–6
[22] Mahdavinia.G.R., Karami.S., 2015, Synthesis of magnetic carboxymethyl chitosan-gpoly(acrylamide)/laponite RD nanocomposites with enhanced dye adsorption capacity, *Polym. Bull.* **72**:2241–62 DOI 10.1007/s00289-015-1402-8
[23] Sivathasan J., 2007, Preparation of clay-dye pigment and its dispersion in polymers, PhD Thesis, RMIT University, Melbourne, Australia.
[24] C G Joseph and L Elilarasi, 2017, Removal of Methylene Blue Dye from Aqueous Solution Using a Newly Synthesized TiO2-SiO2 Photocatalyst in the Presence of Active Chlorine Species, *IOP Conf. Ser.: Mater. Sci. Eng.* 206 012090

Virtual prototyping as an evaluation method for functional clothing

R Boldt[1], **M Carvalho**[1]

[1]School of Engineering, Textile Engineering Department (University of Minho), Campus de Azurem, 4800-058 Guimaraes, PT

E-mail: rachelsagerb@gmail.com

Abstract. In the process of developing new products, validation through prototypes is of fundamental importance. This traditional task accelerates the process, allowing detection of errors, identification of changes in the styles, turning possible the observation of new potential solutions. CAD 3D software for textile-based products contributes to the aesthetic and constructive visualization of new products. This paper has as objective the evaluation of the contributions of the virtual prototyping systems in the aid of the ergonomic validation of functional garments, related to the compression and the tension produced on the body of the user. The research procedure of this study lies within the exploratory field of qualitative character. Through the experimental process, it was possible to identify the characteristics and differentials between the traditional and virtual procedure in the ergonomic validation of clothing. Finally, it is possible to conclude that the virtual simulation offers significant gains in the development process for new products. Nevertheless, this technique does not eliminate the need to produce the physical pilot garment as a final evaluation of the development process. Keywords. Anthropometrics, Ergonomics, Functional clothing, Virtual prototyping.

1. Introduction

The process of developing functional clothing has different methodological characteristics in relation to conventional garment project design. The importance of fashion runway trends and the elaboration of collections based on seasons, branding and aesthetics, essences for the fashion market, are less significant in the environment where specific needs must be met.

The functional garment project the premise of, understanding the context of its use, considering all aspects of the situation of its use, as well as possible, direct and indirect, interactions of the garment with the system where it will be inserted. [1]

During the development process of clothing, taking as a case study the functional clothing specific of workwear, we can highlight the importance of the freedom of movements and comfort to the users, to carry out their work tasks safely and efficiently. Thus, it is important to study the interaction between the user, the elements surrounding him and the context of his actions, with the goal of improving his well-being, achieving better performance. [2] The ergonomic factors are the ones with the greatest impact in the flat pattern design process, fundamental for the performance of the garments. Similarly, the physical constitution of the user and their postural dynamics, interfere in the garment design process and consequently in the performance of the garment. [3]

The verification of the functionality proposal and its validation through prototype development allows strategic gains in the development of the new product. It enables an acceleration of the process, allowing detection of errors, identification of changes in the styles, turning possible the observation of new potential solutions. According to Baxter [4], during the construction of a product prototype, the designer will be able to validate or not, the ideas proposed in the project and continue its development. In the clothing universe, the prototype, besides checking the style aesthetics, has the function of verifying its functional aspects, such as the adequate garment construction, fit, patterns, draping of the used materials, ability to dress, and the various aspects of comfort related to its use. [5]

The traditional prototyping process can be time-consuming and expensive for companies, especially considering the increasing need of the fashion industry to present regularly new products to market, each time with shorter development times. With the physical prototype, it is possible to identify possible defects in the pattern design process. Therefore, basic patterns must be updated and new prototypes made until their final approval. [6]

The use of CAD 3D systems as a prototyping tool is widely used in engineering projects, namely in mechanical engineering, and in some segments of product design development, such as furniture, electronics and others. [7]

An example is *SolidWorks* software. It enables the virtual prototyping from the simulation of design, strength and mechanical operation of the product during the design stage, in addition to other specific features. For virtual prototyping, the properties of the materials applied in the manufacturing stage are considered in the function of the shape of the product. Thus, making possible the visualization of the performance of the project, through a colour chart, showing the points of greater and lesser fragility of the project. Figure 1 represents an example of a resistance simulation in *SolidWorks*.

Fig. 1. SolidWorks interface and resistance simulation.

However, the use of CAD 3D in the textile and garment industry is recent, aiming mainly textile materials simulation and virtual prototyping in the clothing universe. There is also a high contribution potential in the development of fashion products, due to its flexibility and speed, resulting in important cost reductions in the process of products validation.

According to Sayem, Kennon, and Clarke [8], there are currently available different software programs responding to the demand of virtual prototyping in the fashion industry. Being the main ones: *Vstitcher*™ from the Israelis of *Browzwear*; *3D Runway* from *OptiTex International*; *Accumark 3D* from the North American *Gerber*; *TUKA3D* from *Tukatech*; *Modaris 3D* from the French *Lectra*; *Vidya* from the Germans of *Assyst*. In an updated review Pires [9] mentions the potential of other solutions with similar proposals, adding two solutions from *Clo Virtual Fashion*, *Clo 3D*, and *Marvelous Designer systems*.

The different companies that supply CAD 3D systems promise greater agility in the process of creation, validation, and production. In other words, "*CAD systems used in the textile industry are, in general, so-called specific systems, whose ultimate function is to replace the manual process*". [10] Analogously, they have similar interfaces, which combines the 2D and 3D universes into independent windows. Figures 2, 3 and 4 represent examples of some of the existing solutions in the market.

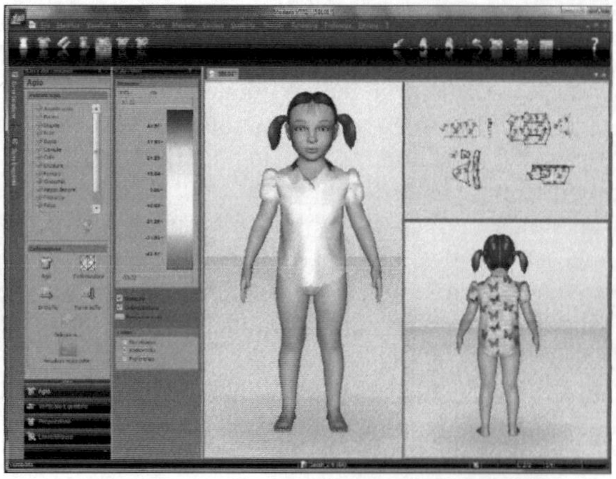

Fig. 2. Lectra - Modaris 3D interface.

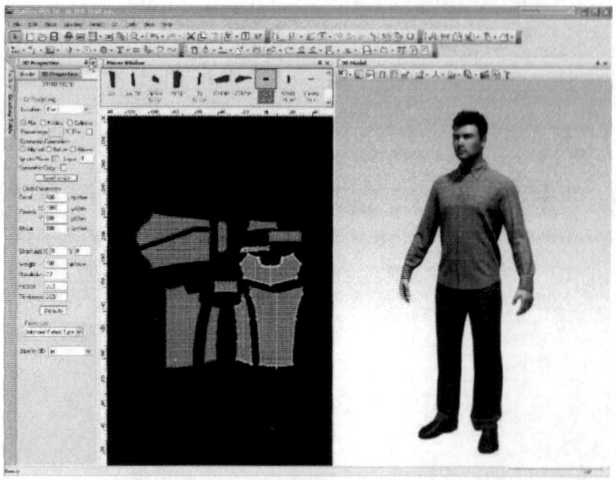

Fig. 3. OptiTex International - 3D Runway interface.

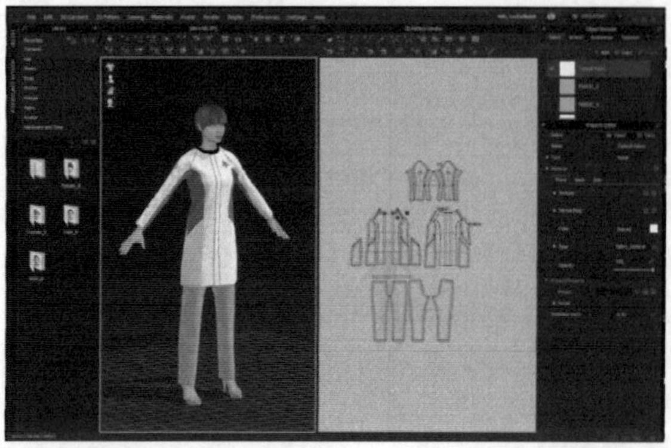

Fig. 4. Clo Virtual Fashion – Clo 3D interface.

Equally similar is the simulation feedback proposal. The software proposes similar assembly features and three-dimensionalization of the flat patterns; relative representation to the fit of the garment, as a function of the virtual proof body and effect of the trim of the fabric, related to the characteristics of the used fabric.

2. Methodology

The article presented here is classified as an exploratory research of qualitative approach, according to its study objectives. Its rationale is based on "providing greater familiarity with the problem, with a view to making it more explicit" [11]. The main objective of this study was the evaluation of 3D CAD system capacity as a methodological tool to analyze problems related to the ergonomic aspects in the design of functional work clothes.

For this, in the first moment of the research, the bibliographic survey was adopted as a technical procedure of data collection and analysis. Its objective was to contextualize and provide bases for the evaluation of problems such as those related to the development of functional workwear, the limits of the traditional methods of verification of clothing functionality and the technological availability of Cad 3D systems intended for the textile and of clothing.

In the second moment, experimental procedures were performed with the objective of producing situations that promoted the analysis of examples that stimulated the understanding of the research problem [12]. The results obtained were directly documented, mainly through images taken from the interface of the software and compared with the experience of evaluation of the traditional method of validation of clothing modelling.

As a methodology for the preparation of the experiment were developed design proposals for functional work clothing. These had the purpose of acting as pre-existing variables of the experiment [13], that is, previously established bases of comparison of results for different scenarios

Clo 3D system version 4.0 was used in order to identify the ergonomic analysis capability promoted through virtual prototypes. The software served as a constructive platform in 3D models and provided model behaviour data based on the shape of the flat patterns, applied fabric physics, and the test body.

The project simulation and evaluation procedure cover the work steps, from the prototype preparation to the final model, as shown in the diagram in Figure 5.

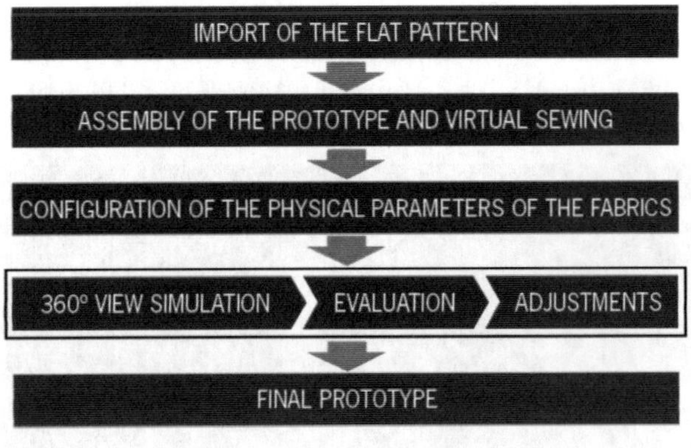

Fig. 5. Stages of work.

The steps characteristics will be exposed and discussed on the following topic.

3. Results and discussion

The first stage was characterized by the export of the flat patterns files, developed in the 2D CAD system, and then imported into the 3D CAD system. This way, the technical characteristics of the flat patterns were preserved and can be manipulated directly in the 3D software.

In the assembly stage of the prototype, it was possible to visualize dynamically the three-dimensionalization of the developed flat patterns. Through seam simulation, it was better understood the construction lines of the developed flat patterns, and, how to fit all parts into the final product. Figure 6 represents the prototype assembly process and flat patterns visualization.

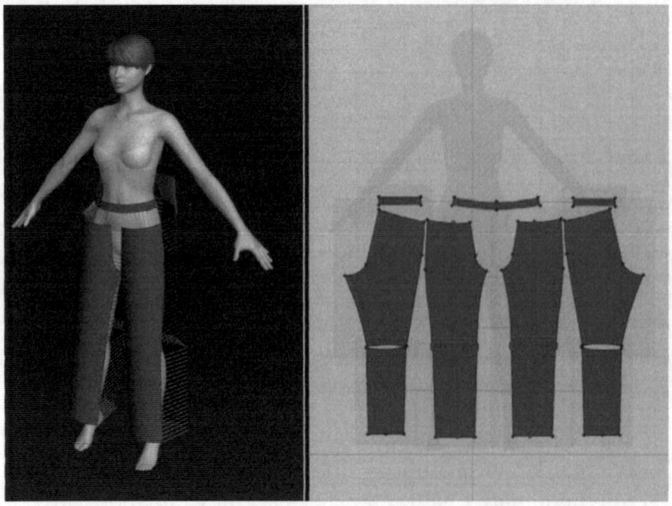

Fig. 6. Prototype assembly process and flat patterns visualization.

With the purpose to obtain greater simulation fidelity, it was developed the physical parameters configuration of the fabric in a virtual environment. For this, it was requested preliminary tests of the real fabrics sample. Checking material physical aspects of toughness, elasticity, hardness, draping, etc.

The materials physical characterization can be easily defined in laboratory tests. However, in the context of development in apparel companies, this step is seen as an obstacle. The CAD 3D software's offer a diverse range of pre-programmed materials. This facilitates the operation, but do not guarantee a simulation compatible with the actual result.

3.1. Simulation, Analysis, and adjustments

For the ergonomic evaluation of the project, images were extracted from the virtual environment for analysis. Three typologies of images were analysed, namely: simulation sketch; realistic; and tension map. Figure 7 represents the three typologies of images.

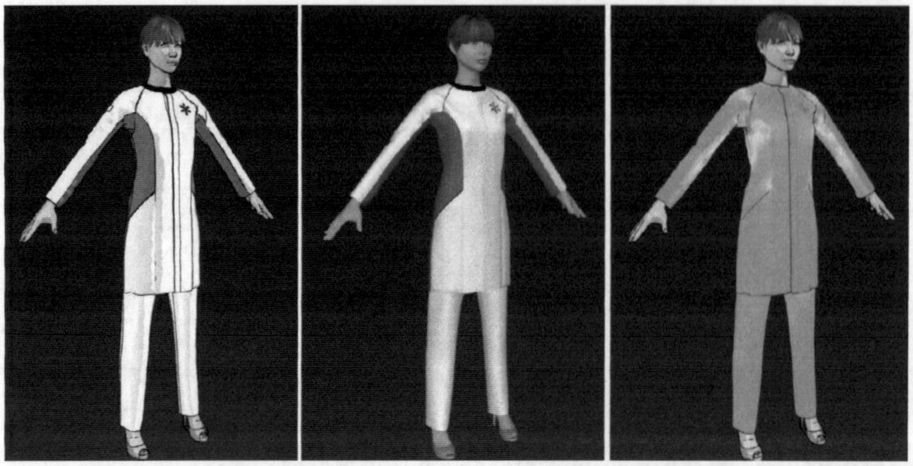

Fig. 7. Images extracted from the software: sketch simulation, realistic and tension map.

The typologies with sketch and realistic effect show the aspects of aesthetic order, namely the fall and volume of the fabrics, constructive details of the cut-outs and aesthetic elements such as colours, textures, etc. The simulation with effect sketch allows to evidence the constructive lines, seams, and folds, through the prominence of black lines.

The stress simulations are subdivided into three new types of representation: Stress Map, Strain Map and Fit Map. The simulations using a voltage graph show the aspects related to the ergonomics of the parts. They indicate, through colour diagrams, the areas of greater and lesser tension in the simulated fabric. The simulations differ from each other and complement the understanding of the behaviour of clothing during use.

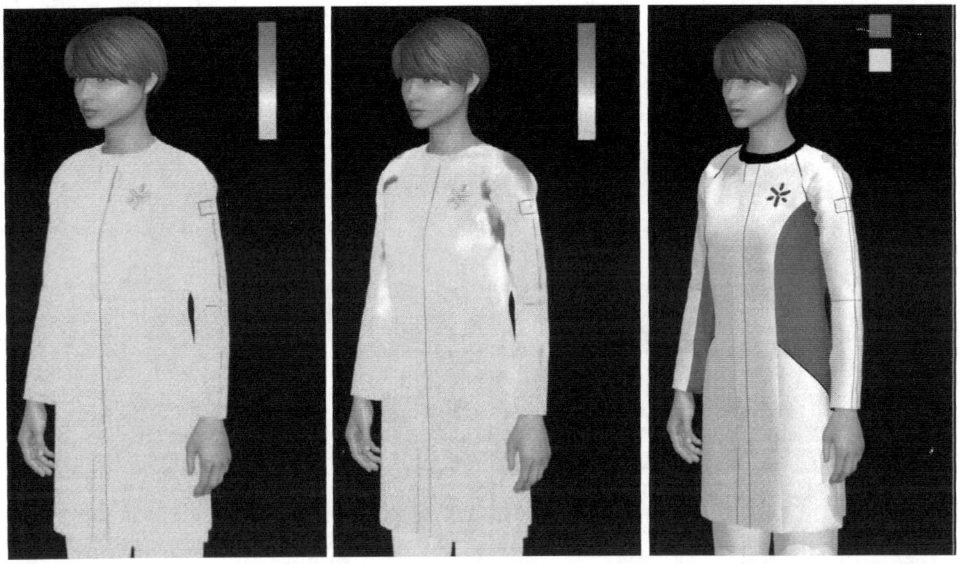

Fig. 8. Typologies of images extracted from the Clo 3D software: Stress Map, Strain Map and Fit Map.

The *Stress Map* shows the external stress causing garment distortion per area of the fabric, appears in the range of colour and numbers. Stress Map appears in total eight colours: With the red colour indicating the strongest stress (100kPa) while the blue colour indicates zero distortion (0.00kPa). Other numbers in between are expressed as the gradation of two colours.

The *Strain Map* represents the rate of distortion of clothing due to external stress. This stress is quantified in percentage form and is represented by a colour diagram. Being a total of eight colours: Red indicates 120% of the distortion rate while blue indicates 100% (no distortion). The numbers between them are expressed as two-colour gradation.

The *Fit Map* shows how tight clothes are on the user's body. It is shown in the colour diagram, where red indicates the impossibility of wearing; orange indicates as very tight area; yellow as a slightly tight area and white indicates that the area is loose.

From the observation of the simulation images with sketch effect, it was possible to identify some project aspects able to be improved: small adjustments in tweezers and heights; positioning of aesthetic details, such as the chest-height logo; and evaluation of the proposed aesthetic-ergonomic attributes.

On the other hand, the evaluation from the ergonomic perspective was based on the observation of the images of tension graphs with the test body in a work situation, in a static and dynamic position. It was possible to observe the clothing's behaviour and the tension generated in a simulation of use. Figure 9 represents the static and dynamic positions used.

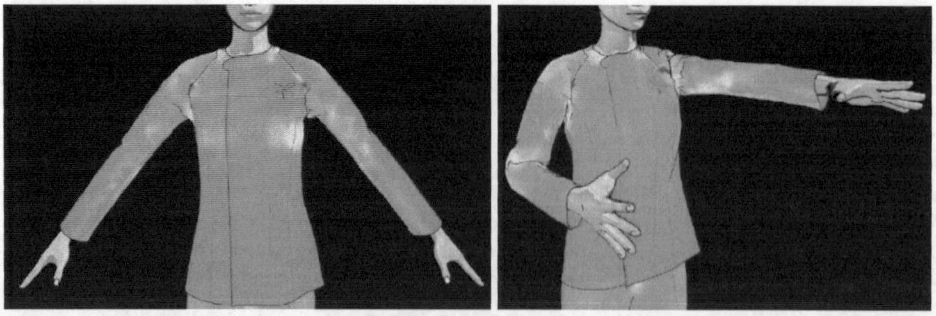

Fig. 9. Bodywork situation, in a static and dynamic position

In Figure 10 it is possible to observe the garment's behaviour in two different design proposals, one using a fold in the middle of the back panel and one a simple seam in the centre back, without fold or extension capacity. This simulation provides relevant data to compares both designs effectiveness, where it was possible to conclude that the use of the fold supported the user's movements.

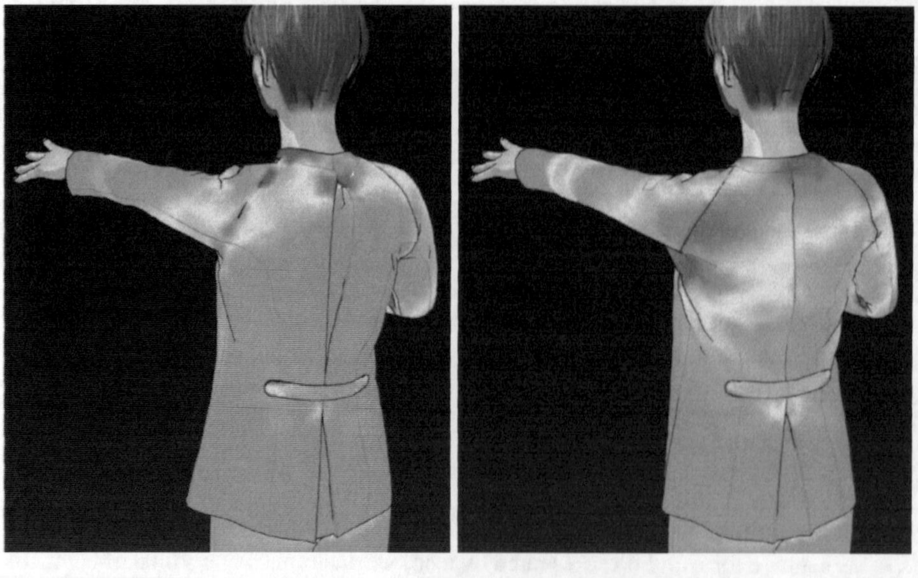

Fig. 10. Comparative evaluation of ergonomic attributes. Proposal with a fold in the centre back (left) and proposal with a regular seam in the centre back

The simulation, evaluation and adjustment process was repeated until the desired effect was reached, resulting in the final virtual prototype, without producing any physical prototype, wasting time and material.

4. Conclusion

The results obtained show relevant gains in relation to the product traditional development process, regarding the evaluation of the ergonomic effectiveness, the gain in flexibility of execution of small adjustments and the aesthetic variations in the product. The CAD 3D software used contributed to the evaluation of the various modifications and proposals for improvement, without the need to produce the final product, and, in a lower extent, use the software test models, since it was possible to evaluate the products on virtual manikins in conventional positions and in working positions.

The results were presented graphically. The compression forces are visualized by the representation of tension from the variation of colours between green, yellow and red, with the cold tones representing a lower tension and the hot tones representing the greater strain.

It is also possible to make a pre-evaluation of the materials draping in the body, since the virtual representation demonstrates the formation of roughness, volume, and deformation, according to the cut of the pattern and the physical characteristics of the simulated material. However, for a more effective simulation of the draping, it is necessary to adjust parameters of physical simulation of the fabrics having the actual fabric as a comparative.

Finally, it was possible to conclude that the virtual simulation offers significant gains in the development process. Nevertheless, this technique does not eliminate the need to produce the physical pilot garment as a final evaluation of the development process, largely increasing its success in its first development.

Acknowledgement
We would like to acknowledge 2C2T-Science Center for Textile Technology from University of Minho. This work is financed by FEDER funds through the Competitive Factors Operational Program (COMPETE) POCI-01-0145-FEDER-007136 and by national funds through FCT-Portuguese Foundation for Science and Technology, under the project UID/CTM/000264.

References

[1] Gupta D. 2011 Design and engineering of functional clothing. *Indian J Fibre Text Res* 36:**327–35**.
[2] Iida I. 2005 *Ergonomia: projeto e produção*. 2nd ed. (São Paulo: Edgard Blücher) 2005.
[3] Saltzman A. 2004 *El Cuerpo diseñado: sobre la forma en el proyecto de la vestimenta*. 1st ed. (Bueno Aires: Paidos) 2004.
[4] Baxter M. 2011 *Projeto de produto: guia prático para o design de novos produtos*. 3rd ed. (São Paulo: Blucher) 2011.
[5] Gragnato L. 2015 *Fundamentos do design de moda*. 1st ed. (Rio de Janeiro: Europa) 2015.
[6] Treptow DE. 2005 *Inventando moda: planejamento de coleção*. 3°. (Brusque: Doris Treptow) 2005.
[7] Junior JOB, Nunes LEN do P. 2011 Modelagem de objetos em sistemas Cad 3D com a utilização de geometrias paramétricas. *J Exact Sci* 17.
[8] Sayem ASM, Kennon R, Clarke N. 2010 3D CAD systems for the clothing industry. *Int J Fash Des Technol Educ* 3:**45–53**. doi:10.1080/17543261003689888.
[9] Pires GA. 2015 O Cad 3D Aplicado Na Validação De Protótipos Na Indústria Do Vestuário. 2015.
[10] Neves M. 2000 Desenho têxtil malhas. *TecMinho* 2:**129**.
[11] Gil AC. 2002 *Como elaborar Projetos de Pesquisa*. 4th ed. (São Paulo: Atlas) 2002. doi:10.1111/j.1438-8677.1994.tb00406.x.
[12] Selltiz, Wrightsman, Cook. 1987 *Métodos de pesquisa nas relações sociais*. 2nd ed. (São Paulo: E.P.U.) 1987.
[13] Freixo MJV. 2011 *Metodologia Científica : Fundamentos, métodos e técnicas*. 3rd ed. (Lisboa: Instituto Piaget) 2011.

Energy and Water Saving Finishing Method for Producing Durable Flame Retardant Cotton Fabric

Z. Omerogullari Basyigit[1], D. Kut[2]

[1]*Uşak University, Faculty of Engineering, Textile Engineering Department, Usak, Turkey*
[2]*Uludag University, Faculty of Engineering, Textile Engineering Department, Bursa, Turkey*

Corresponding author: zeynep.omerogullari@usak.edu.tr

Abstract

This study presents the improvement of durable flame retardant (FR) effect of 100 % cotton fabrics via chemical foaming system which provides energy and water saving significantly, compared to conventional methods. In order to carry out the experiments, dialkylphosphonocarboxylic acid amide based flame retardant agent and auxiliary chemicals were applied on cotton fabrics in a foam formed. Foam parameters such as blow ratios and wet-pick-up ratios were varied and flame retardant properties of foamed cotton fabrics were compared with cotton fabrics treated through pad-dry-cure process in order to determine the performance effect of cotton fabrics in flame retardancy via different finishing methods. After the finishing processes, flame retardant effect of all treated fabrics was examined with vertical burning test method. As durability of flame retardant applied cotton fabrics against washing process is one of the most important subjects that has been tried to be improved with different approaches for a long time, all treated fabrics were washed and dried 50 times, separately. After repeated washing and drying cycles, burning behavior of washed and FR treated fabrics were also tested vertically for determining flame retardancy effect. Beside the vertical burning test, tearing strength test, color spectrums, SEM and SEM-EDX analysis of fabrics were carried out. According to test results, even at the low wet-pick-up ratios, a significant increase was observed in flame retardant effect of cotton fabrics that were treated with FR agents via foam application. Moreover, all treated fabrics were durable against 50 washing and 50 drying processes.

Keywords: *Chemical foam system, cotton fabric, flame retardancy, durable effect, water-saving*

I. INTRODUCTION

Fires occurred from upholstered furniture, floor coverings, curtains, drapes or clothing worn by people can result in high rate of fatalities or serious burns in their bodies [1-6] that's why researches have been carried out for years in order to improve flame retardant effects of cellulosic based materials which have high flammability [3, 7-9].There are several studies performed with conventional methods on flame retardant finishes, however one of the biggest issue of conventional methods is high amount of water and energy consumption during the finishing processes. In order to save water and energy which is one of the most important subjects nowadays that has been studying on for our world's sustainability, some other new technologies have been improved. Beside conventional methods, there are some alternative finishing application methods that allow low pick-up ratios as foaming technology. The use of air to extend the volume of a chemical solution in the form of bubbles showed in Fig. 1 is known as a foaming process. Foaming liquor to a large volume allows volumetric control and hence ensures even distribution of the liquor on the fabric. Foaming technology has offered solutions to the basic problems encountered with the other low-add-on topical and expression systems [3, 10]. Thus, in this study, FR agents were applied to 100 % cotton fabrics and flame retardant effect of cotton fabrics which were performed through both conventional finishing and chemical foaming system were examined.

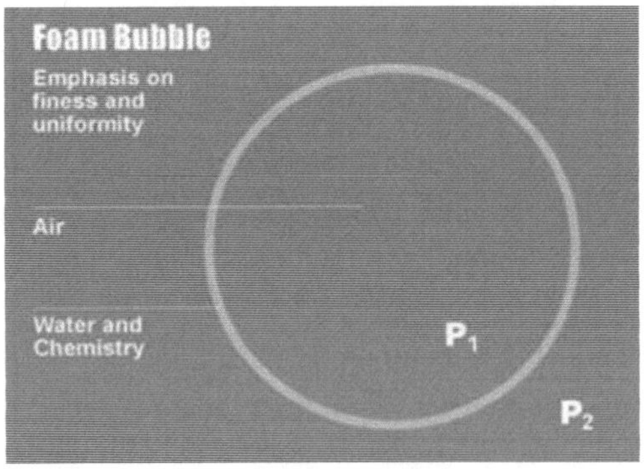

Fig.1. Structure of foam bubble used in textile industry [11]

II. EXPERIMENTAL

A. *Material-Method*

400 g/L Dialkylphosphonocarboxylic acid amide based chemical as FR agent, 12 g/L phosphoric acid, 25 g/L emulsion of polyalkylene and 55 g/L melamine formaldehyde as auxiliary chemicals (Huntsman, USA) applied on twill 100 % cotton fabrics (Cotton Incorp., Cary, USA) with both chemical foam system and padding application. Foam parameters as wet pick-up ratios were varied in order to determine and optimize the flame retardant effect of the fabrics. Wet pick-up ratio was 60 % in padding system whereas two different wet-pick up ratios were performed as 35% and 30% in foaming system, respectively. As a foaming device, Gaston Systems Chemical Foaming System (CFS, Gaston, USA) was used as showed in Fig. 2, and the test results were compared with conventional padding system. After the required applications, according to AATCC Test Method 193-2007, repeated washing cycles were performed and after each washing cycle, 45 min drying procedure were carried out up to 50 times. After every 5 or 10 washing and drying cycles, each performance tests such as vertical burning test (ASTM D6413/D6413M-12), color spectrums (Spectraflash SF 600X Datacolor Reflectance Spectrophotometer) and tearing strength tests (ASTM D1424) were also carried out. As characterization tests, SEM and SEM-EDX (Hitachi S- 3200N) analysis were performed. Codes of treated cotton fabrics were written as following:

PFR: Cotton fabric padded with flame retardant agent CPF1: Cotton fabric foamed with flame retardant agent with a liquor-ratio of 35%

CPF2: Cotton fabric foamed with flame retardant agent with a liquor-ratio of 30%

5W: 5 times washed and 5 times dried sample

10W: 10 times washed and 10 times dried sample

20W: 20 times washed and 20 times dried sample

30W: 30 times washed and 30 times dried sample

40W: 40 times washed and 40 times dried sample

50W: 50 times washed and 50 times dried sample

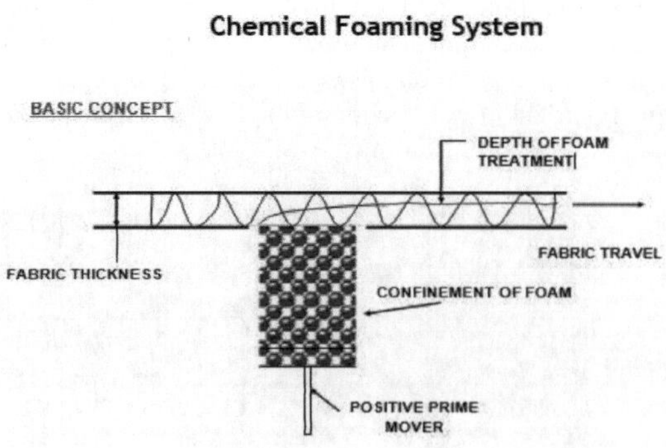

Fig.2. Basic concept of chemical foaming system, CFS [11]

III. DISCUSSION AND RESULTS

A. Vertical Burning Test Results

Vertical burning test results showed that untreated cotton fabric was burned totally and flame retardant effect of all treated cotton fabric was improved including padding system showed in Fig. 3, however; in foam application system there was a significant increase in flame retardancy effect even at low wet-pick up ratios as showed in Fig. 4 and Fig. 5.

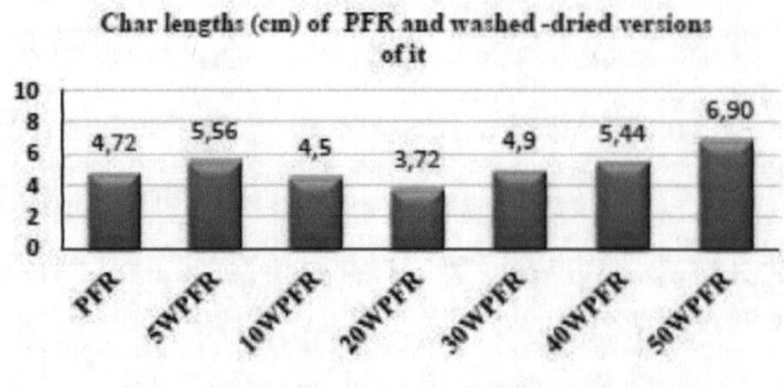

Fig.3. Char lengths (cm) of PFR and washed-dried versions of it up to 50 cycle after vertical burning test

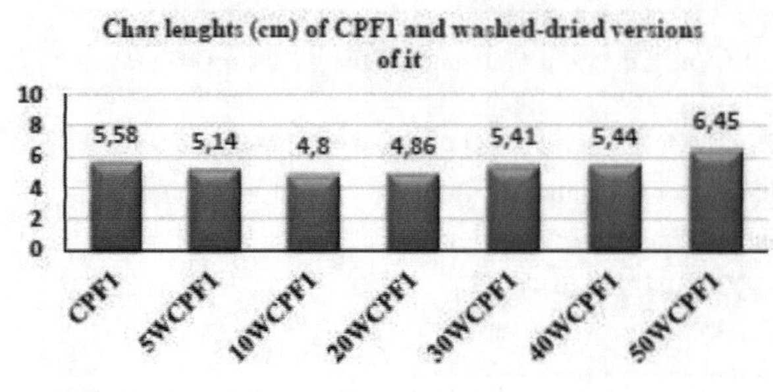

Fig.4. Char lengths (cm) of CPF1 and washed-dried versions of it up to 50 cycle after vertical burning test

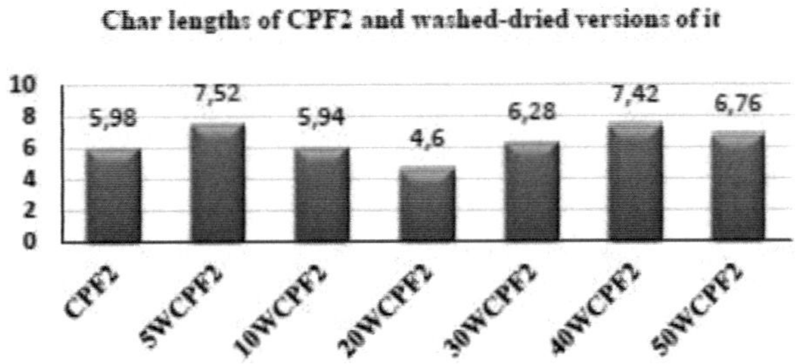

Fig.5. Char lengths (cm) of CPF2 and washed-dried versions of it up to 50 cycle after vertical burning test

B. Tearing Strength Test Results

Tearing test results showed that there was a decrease in all FR treated fabrics but this decrease was more significant in PFR sample as it was seen in Table 1. It was considered that this decrease was observed because of melamin-formaldehyde used in FR recipe of this study which prevented the movement of the fibers of the cotton fabric when tearing force was applied.

Table 1. Tearing strength test results of fabrics

	Decrease in Tearing Test Result (%)		
	Warp direction		**Weft Direction**
PFR	25	**PFR**	**26**
CPF1	34	**CPF1**	**28**
CPF2	**35**	**CPF2**	**29**

C. Color Spectrum Test Results

Color spectrum values, yellowness indexes and ΔE values of PFR, CPF1 and CPF2 were showed, respectively in Table 2, Table 3 and Table 4. There was a significant change in ΔE values of washed and dried versions of PFR samples whereas this change was less in the washed-dried versions of foamed cotton fabrics. When color spectrums of all treated unwashed fabrics (PFR, CPF1, CPF2) were examined, it could be clearly seen that the ΔE values were close to each other. In all samples, yellowness indexes decreased when the number of repeated washing-drying cycles was increased.

Table 2. Color spectrums, yellowness indexes (YI) and ΔE values of PFR

Sample	L*	a*	b*	YI-1925D	ΔE
PFR	59,55	3,04	14,38	40,00	1,35
5WPFR	60,68	3,02	13,76	38,06	2,49
10WPFR	61,09	2,97	13,54	37,34	2,94
20WPFR	61,62	2,94	13,20	36,29	3,56
30WPFR	61,88	2,73	13,27	36,11	3,78
40WPFR	61,99	2,80	13,11	35,67	3,94
50WPFR	62,80	2,83	12,87	34,95	4,72

Table 3. Color spectrums, yellowness indexes (YI) and ΔE values of CPF1

Sample	L*	a*	b*	YI-1925D	ΔE
CPF1	57,43	3,14	15,32	43,30	1,55
5WCPF1	58,01	3,02	14,60	41,24	1,29
10WCPF1	58,18	3,13	14,62	41,30	1,11
20WCPF1	59,19	3,00	14,10	39,51	1,43
30WCPF1	59,77	2,92	13,87	38,61	1,88
40WCPF1	59,85	2,92	13,66	38,09	2,07
50WCPF1	60,35	2,87	13,73	37,98	2,33

Table 4. Color spectrums, yellowness indexes (YI) and ΔE values of CPF2

Samples	L*	a*	b*	YI-1925D	ΔE
CPF2	57,18	3,27	15,52	44,04	1,75
5WCPF2	58,00	3,09	14,72	41,58	1,20
10WCPF2	57,97	3,16	14,61	41,43	1,23
20WCPF2	58,53	3,13	14,57	41,03	0,99
30WCPF2	59,34	3,00	13,99	39,18	1,56
40WCPF2	59,56	3,03	13,97	39,08	1,65
50WCPF2	60,11	2,96	13,77	38,29	2,12

D. SEM Results

SEM micrographs at x2000 magnification were seen in Fig.6 which indicates untreated, PFR and CPF1 samples, respectively. The surface of untreated sample was smooth and clean while there were residues of chemical on the padded cotton fabric. Chemicals applied via foaming system were seen as small black particles on foamed cotton fabric in Fig. 6c.

a) Untreated b) PFR

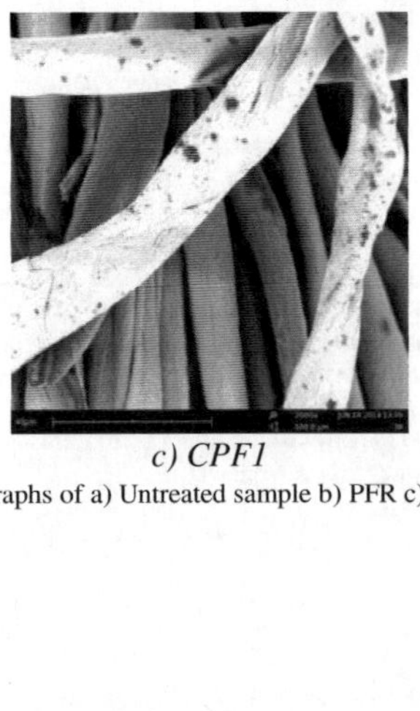

c) CPF1

Fig.6. SEM micrographs of a) Untreated sample b) PFR c) CPF1

E. SEM-EDX Test Results

SEM-EDX test result showed clearly in Fig. 7 that phosphorus based flame retardant chemical applied on cotton fabric via foaming system with the sharp peak of phosphor (P).

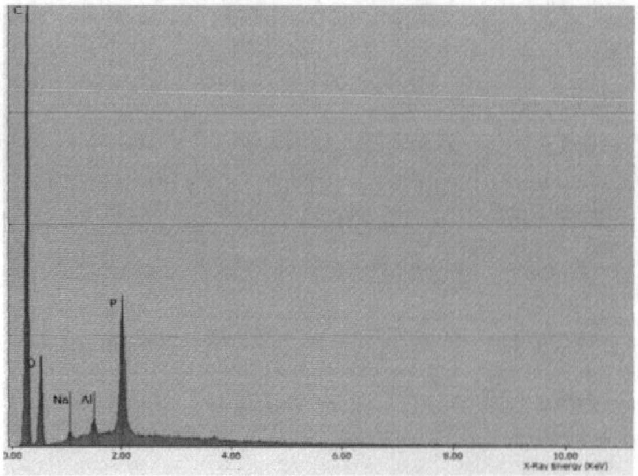

Fig.7. SEM-EDX analysis of CPF1

IV. CONCLUSION

As a conclusion, a significant improvement on flame retardant effect of cotton fabrics was achieved via chemical foam application. It was observed that improved flame retardant effects could be performed on 100 % cotton fabrics with low wet-pick up ratios. Even though there was a decrease in tearing strength test results after each FR finishing applications, in chemical foaming system these decreases were less than the conventional finishing treatment. According to test results, even at the low wet- pick-up ratios, a significant increase was observed in flame retardant effect of cotton fabrics that were treated with FR agents via foam application. Moreover, all treated fabrics were durable against 50 washing and 50 drying processes.

ACKNOWLEDGMENT

Authors are thankful for great support taken from Gaston Systems, Cotton Incorporated and NCSU in USA.

REFERENCES

[1] Schindler, Wolfgang D and Peter J Hauser 2004 *Chemical finishing of textiles* Elsevier.
[2] Ömeroğullari Z and Kut D 2012 Tekstilde güç tutuşurluk *Journal of Faculty of Engineering of Uludag University* 17 pp 27-41
[3] Chen, Qin, Charles Q Yang, and Tao Zhao 2016 The chemical bonding and fire performance of the nylon/cotton blend fabrics treated with a hydroxy-functional organophosphorus oligomer *Polymer Degradation and Stability* 128 pp 237-244
[4] Weil E D and Levchik S V 2008 Flame retardants in commercial use or development for textiles *Journal of Fire Sciences* 26 pp 243-281.
[5] Giuliana T C 1978 Chemical modification of polymers with flame retardant compounds *Journal of Polymer Science: Macromolecular Reviews* 13 pp 283-353
[6] Horrocks A R 2011 Flame retardant challenges for textiles and fibres: new chemistry versus innovatory solutions *Polymer Degradation and Stability* 96 pp 377-392
[7] Xinying C and Yang C Q 2009 Flame retardant finishing of cotton fleece fabric: Part V. Phosphorus- containing maleic acid oligomers *Fire and Materials* 33 pp 365-375
[8] Giuseppe R et al. 2017 Phosphorus-silica sol-gel hybrid coatings for flame retardant cotton fabrics t *Tekstilec* 60 pp 29-35
[9] Chapple S., and R. Anandjiwala 2010Flammability of natural fiber- reinforced composites and strategies for fire retardancy: a review *Journal of Thermoplastic Composite Materials* 23 pp 871-893
[10] Elbadawi A M and Pearson J S 2003 Foam technology in textile finishing *Textile Progress* 33 pp 1-31
[11] Gaston Systems, Chemical Foaming Systems Notes, 2012

IOP Conf. Series: Materials Science and Engineering **460** (2018) 012042 doi:10.1088/1757-899X/460/1/012042

Microwave Assisted Surface Treatment of Okra Fibers

A.A. İZGİ[1], E.D.KOÇAK[2], B.YILMAZ ŞAHİNBAŞKAN[2]

[1]*Marmara University, Institute of Pure and Applied Sciences, Department of Polimer Science and Technology, Goztepe/Istanbul, TURKEY*

[2]*Marmara University, Faculty of Technology, Department of Textile Enegineering, Goztepe/ Istanbul, TURKEY*

Corresponding author: aybenizgi@marun.edu.tr

Abstract-All over the world, the importance of the sustainable and biodegradable materials is increased very fast. The use of agricultural stem waste as a technical textile material can be made a huge contribution for save the environment. The aim of this research was to pre-treat the okra fibers with microwave energy. The fibers were obtained from Marmara Region agricultural waste by use a specially designed machine. They were microwave-assisted surface treated with 7 % sodium bisulphate and 10 % acetic acid. The results of tensile strength, fiber diameter, SEM and X-Ray analyses were compared to the conventionally treated fiber samples. Finally, microwave-assisted treatments were quite good. Bisulphate treatment was more effective than treatments with acetic acid. When the SEM images was investigated, a lot of porosity was observed on the surface of acetic acid treated fiber and the tensile strength values increased bisulphate treatment.

Key words- **agriculture waste, microwave energy, okra fibers, surface modification**

I. INTRODUCTION

In the strengthening of new generation composites and nonwoven surfaces, natural fibers are increasingly regarded as an environmentally friendly alternative to synthetic fibers. These fibers are considered in the aerospace industry, construction and automotive sectors instead of glass and carbon fibers which are harmful to the environment and expensive [1].The properties of the natural fibers are listed in Table 1.

Table 1. Natural fibers properties[2].

	Characteristic	Properties
NATURAL FIBERS PROPERTIES	Physical	Low density, Surface topology, Texture, Form and geometry, good thermal insulation performance, specific heat, electrical conductivity, high sound absorption coefficient.
	Chemical and Biological	Chemical composition (cellulose, lignin, etc.), Batch quality, availability, planting limitless, odder emission etc..
	Mechanical	Good Mechanical Properies (Elastic modulus, shear modulus, Poisson's ratio, Yield strenght, elongation to break, specific modulus of elasticity, tensile strenght, spesific shear modulus etc..)
	Technical	Processing knowledge, friendly processing, processing energy consumption, processing cost, transferring cost, raw fiber cost,cost of energy input.
	Environmental	Eco-firendly, goverment support, bio- degradability, social positive view.

Okra (Layd's finger) plant is a plant with lignocellulosic content, known as Hibiscus esculentus L. and Okra (*Abelmoschus esculentus* (L.) Moench), a member of the family Mallow (Malvaceae). Okra plant is considered as one of the abundant sources of natural fibers [3].

Generally, when the okra plant is grown as a vegetable, the plant body is left on the fields after the plants are collected. After collecting the fruits of the okra plant, the remaining amount of agricultural waste is 2.11 tons / da [4]. The okra remaining as agricultural wastes are left in the soil as organic matter. In parallel with the population increase in the world, the demands of the

human beings are increased. However, with the increasing demand, the amount of agricultural wastes also increases. For the environment, there is a need to effectively recycle and re-use these wastes and to reduce environmental pollution with new resources.

Chemical modifications of natural fibers are intended to improve adhesion to the polymer matrix and increase the fibrillization to hold the fibers together. Approaches such as alkaline treatment, acetylation, silane treatment, benzolization, peroxide treatment, permanganate treatment, enzyme treatment and isocyanate treatment are applied by researchers [1].

The main factor in the interfacial adhesion strength is the linkage between functional groups on the fiber surface and in the nonwoven or composites matrix. Fiber surfaces must be modified to increase the number of functional groups on the surface of inert textiles [5].

Microwave energy is an alternative heating method widely used for pre- treatment, dyeing, finishing, drying, fixing, sterilization, surface modification and grafting of textile materials. Compared to conventional heating methods, it provides fast, effective, uniform, and energy efficient heating due to absorption in textiles [6]. Using of microwave technology in textile industry is related with heating up, pretreatment, drying, condensation, dyeing and printing [7].

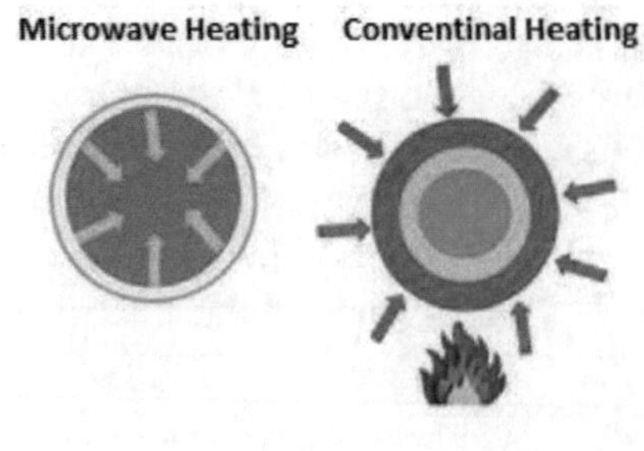

Figure 1.Microwave heating versus conventional heating.

In this study, acetic acid and sodium bisulphate was applied on okra fibers with two different kinds of processes which are conventional and microwave energy methods. The results of the okra fibers have been evaluated in terms of their mechanical and morphological properties.

II. EXPERIMENTAL

A. Fibers

Okra fibres were collected from Marmara region. Their overall lengths were 120 cm. These fibers were separated and cut according to their diameter and length. Okra body waste fibers were obtained by separating from plant extracts with the aid of a special machine designed to obtain fiber from lignocellulosic plants (Figure 2). It was waited in the water-filled container to remove residue and adhering dirt (20 days). The fibers were shaken under running water until cleaned and they were dried in an oven at 100 °C for 2 hours. After dried, fiber were conditioned for 24 hours prior to testing under 20 ±2 °C and 65 ±2 %RH conditions.

Figure 2. Removal of fibers from okra stem.

B. *Surface Modification*

Okra fibers have been treated into two methods in which are given as conventional method and microwave method in Table 2. Conventional process was applied in laboratory condition. Kerman' Laboratory sample heater (220 Volt and 3000 Watt) was used for conventional method. Microwave method was carried out with a Kenwood Mw440 at a frequency of 2.45 GHz. The microwave oven was set to Medium-Low (M-L) power of 350 W. The samples were placed in a sealed glass holder and treated by the microwave energy according to the experimental design.Treated okra waste fibers are shown in Figure 3.

Table 2. Process conditions.

<table>
<tr><th>Method</th><th>Time (min)</th><th>Temperature (°C)</th><th>Concentration (%)</th><th>Rinsing Process</th><th>Drying Condition</th><th>Chemicals</th></tr>
<tr><td rowspan="2">Conventional</td><td rowspan="2">10,20,30,40</td><td>60</td><td>7</td><td rowspan="4">25 °C, 10 min, pH 7 with distilled</td><td rowspan="4">At room temperature</td><td>Sodium bisulphate</td></tr>
<tr><td>60</td><td>10</td><td>Acetic acid</td></tr>
<tr><td rowspan="2">Microwave</td><td rowspan="2">3,5,7,10</td><td>60</td><td>7</td><td>Sodium bisulphate</td></tr>
<tr><td>60</td><td>10</td><td>Acetic acid</td></tr>
</table>

Figure 3. Treated okra waste fibers.

IOP Conf. Series: Materials Science and Engineering **460** (2018) 012042 doi:10.1088/1757-899X/460/1/012042

C. *Testing and Characterization of Okra Fibers After Chemical Treatments*

Tensile strength (N) values of the untreated/treated okra fibres were determined according to ASTM D 3822 test method by using Instron 4411 (50 N Load, 10 mm/min speed) tensile test instrument. Fiber diameters have been measured with Projectina CH-9495 microscope in Physical Test Laboratory of Marmara University Faculty of Technology, Department of Textile Engineering. Treated and untreated fibers morphological views have been taken with JEOL JSM-T330 electron microscope.Before and after surface treatment of okra body waste fibers, crystallinity was measured at 40 kV and 40 mA by D8 Bruker aXS Advance Germany brand X-ray diffraction (XRD). Diffraction density was measured at 0° - 50° at a scan rate of 2° / sec. The Crystallinity index in the results of the measurements was carried out according to equation 1.

$$Crystallinity\ index\ (\%) = ((I002 - Iam)\ /\ I002)\ x100 \quad (1)$$

III. RESULTS AND DISCUSSIONS

A. Results of Mechanical Testing

The values of the tensile strengths of the fibers treated with sodium bisulphate and acetic acid are shown in Figure 4 and 5, respectively.

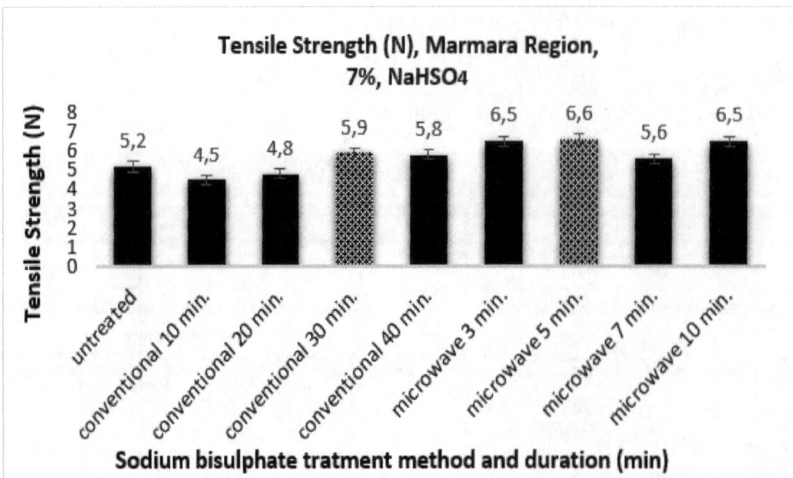

Figure 4. The tensile strength values of sodium bisulphate treated fibers with conventional and microwave energy methods.

The highest strength value of the okra fibers after surface modification compared to the untreated fiber was obtained at 7% concentration and 30 minutes of operation according to the conventional method with sodium bisulphate.

The strength was increased after surface modification with 7% sodium bisulphate for 5 minutes by the microwave method according to the raw fiber.

IOP Conf. Series: Materials Science and Engineering **460** (2018) 012042 doi:10.1088/1757-899X/460/1/012042

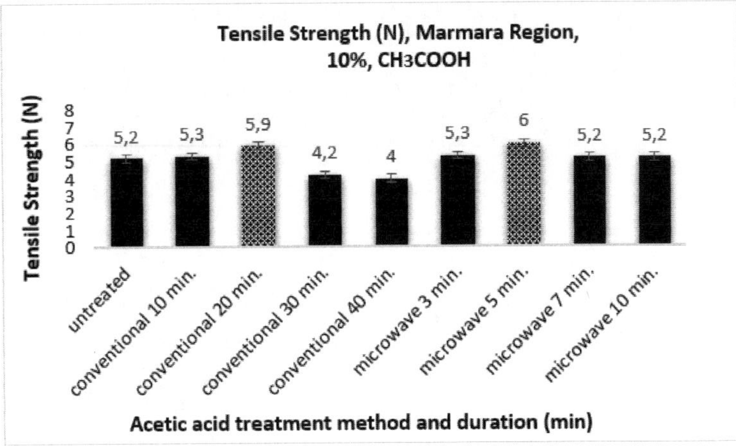

Figure 5. The tensile strength values of acetic acid treated fibers with conventional and microwave energy methods.

In the surface modification of the okra fibers with acetic acid by conventional method, generally the strength reduction is seen relative to the untreated fiber. Decrease in strength was determined depending on the increase in the processing time. The highest strength value was obtained at 10% concentration and 20 minutes of operation.

Improvement in strength is generally seen in surface modification of okra fibers with acetic acid by microwave method compared to untreated fiber. At 7 and 10 minutes of operation, the tensile strength values are at the same level as the untreated fiber. The highest strength was obtained at 10% concentration and 5 minutes of operation.

When the strength values were examined, it was determined that the highest strength values for both chemicals were obtained by the modification made by the microwave method.

The treated fibers which have highest strenght values were selected and their SEM, X-Ray analysis and diameter images were given in the following results.

B. Fibers Diameter

The physical diameters (μm) and longitudinal images of okra body waste fibers obtained from the Marmara Region were obtained by using light microscopes. The diameter changes images of untreated and treated fibers with different chemicals/process parameters are shown in Figure 6-8.

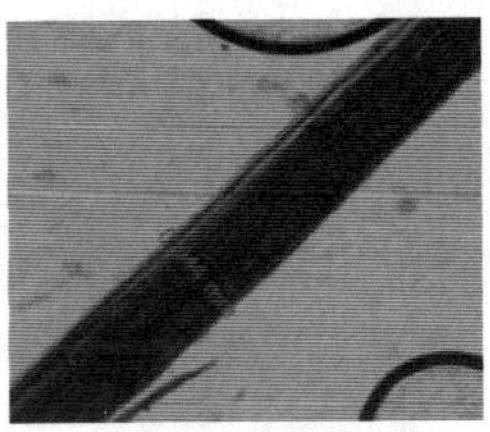

Figure 6. Untreated okra waste fiber.

IOP Conf. Series: Materials Science and Engineering **460** (2018) 012042 doi:10.1088/1757-899X/460/1/012042

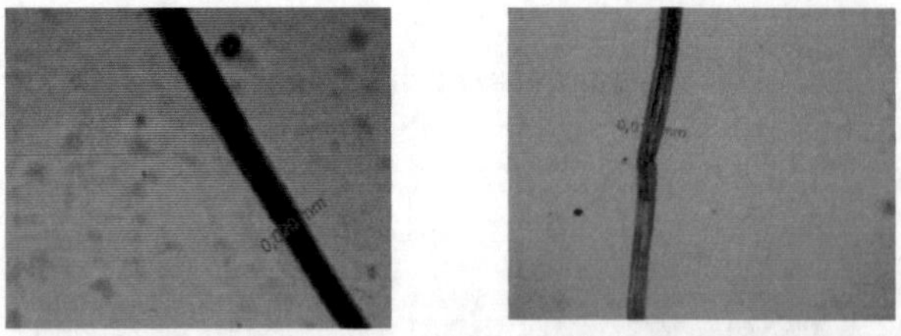

Conventional method 30 min **Microwave method 5 min**

Figure 7. Longitudinal images of Marmara region okra waste fibers surface modified with 7 % $NaHSO_4$ by different methods.

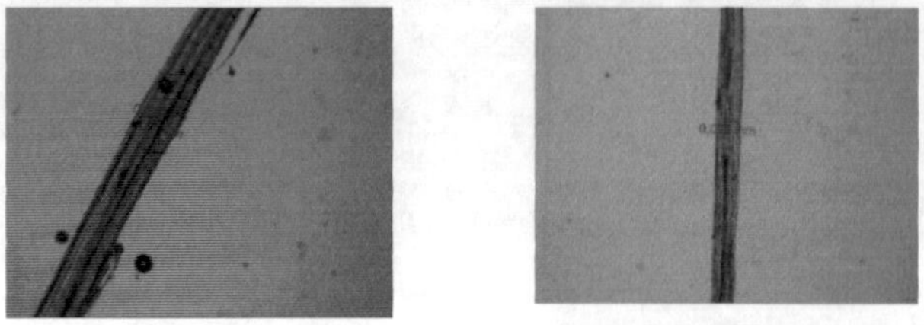

Conventional method 20 min **Microwave method 5 min**

Figure 8. Longitudinal images of Marmara region okra waste fibers surface modified with 10 % CH_3COOH by different methods.

When the fiber diameters investigated, it was found that the microwave method applied samples gave the best results for both chemicals. Microwave-treated fibers contain impurity removal and high amounts of fiber fibrillation. Thus, the fibers treated with the microwave method are more thin than the fibers treated with the conventional method.

C. Crystallinity of Okra Fibers (XRD)

Crystallinity index values obtained from surface modification of okra fibers with different chemical and both methods are given in Figures 9-13. The results were compared with the untreated sample.

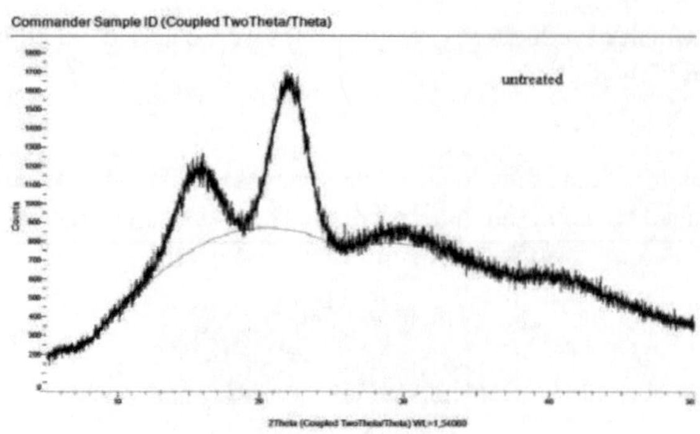

Figure 9. X-ray diffractometer results of untreated okra fibers.

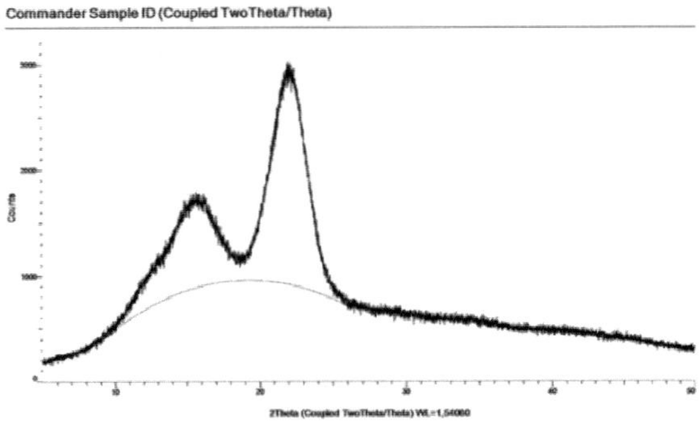

Figure 10. X-ray diffractometer results of conventionally treated fibers with 7% $NaHSO_4$ for 30 min.

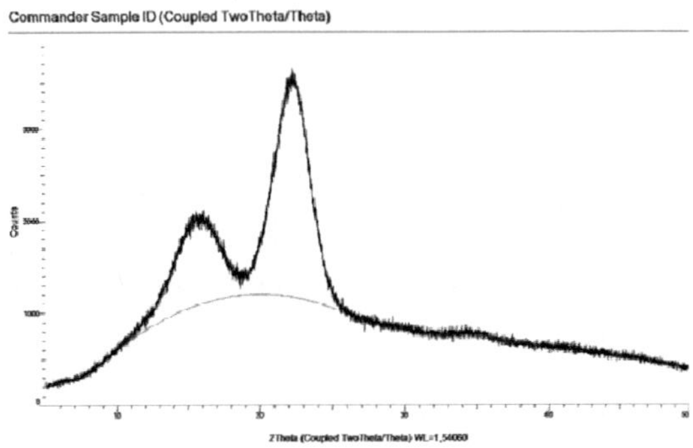

Figure 11. X-ray diffractometer results of treated fibers by microwave method with 7 % $NaHSO_4$ for 5 min.

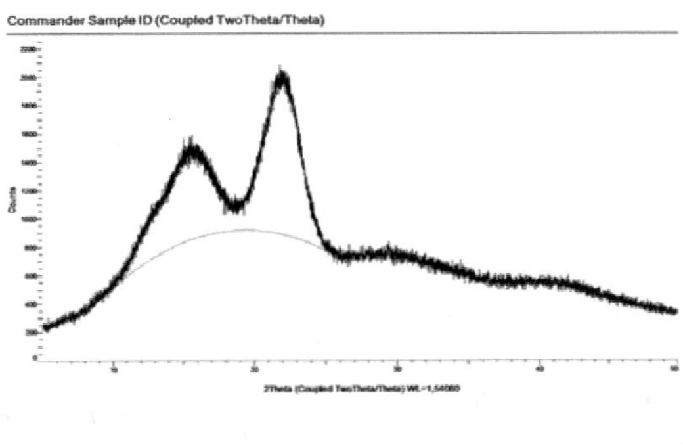

Figure 12. X-ray diffractometer results of conventionally treated fibers with 10% CH_3COOH method, 20 min.

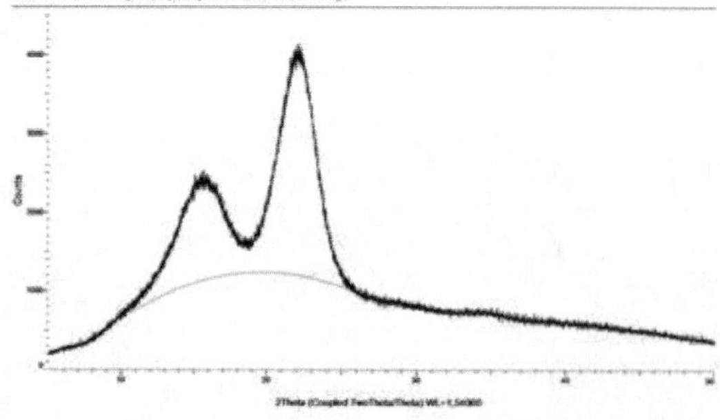

Figure 13. X-ray diffractometer results of treated fibers by microwave method with 10% CH_3COOH for 5 min.

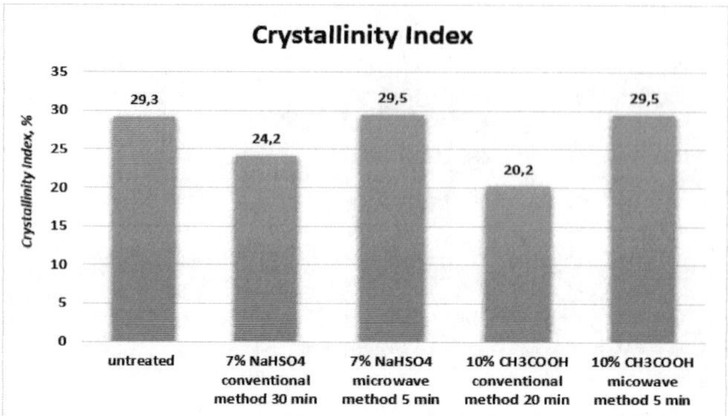

Figure 14. The crystallinity index (%) values of the fibers treated with sodium bisulphate and acetic acid.

The crystallinity index (%) values of okra stem fibers treated with 7% sodium bisulphate and 10% acetic acid at different times are seen in the Figure 14 . Accordingly, microwave treated fibers with sodium bisulphate and acetic acid have the highest crystallinity index value.It has been found that microwave treated okra stem fibers are shown higher crystallinity than the conventionally treated okra fibers. Innovative method have more effective results by the microwave method in a shorter time.

D. SEM Images

The SEM images of untreated / treated fibers are shown in Figure 15-19.

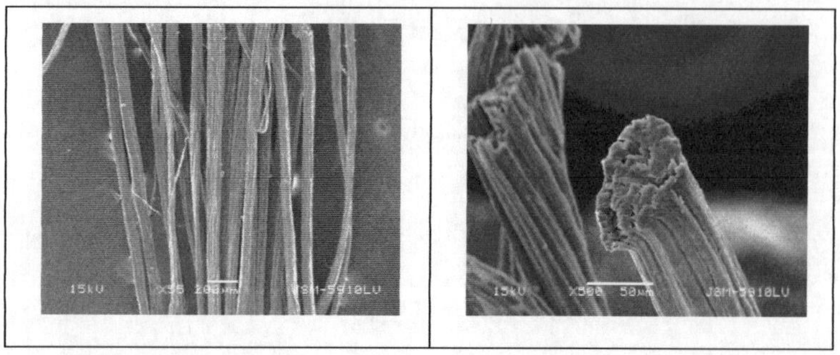

Figure 15. Longitudinal and cross-sectional SEM images of the untreated okra fiber.

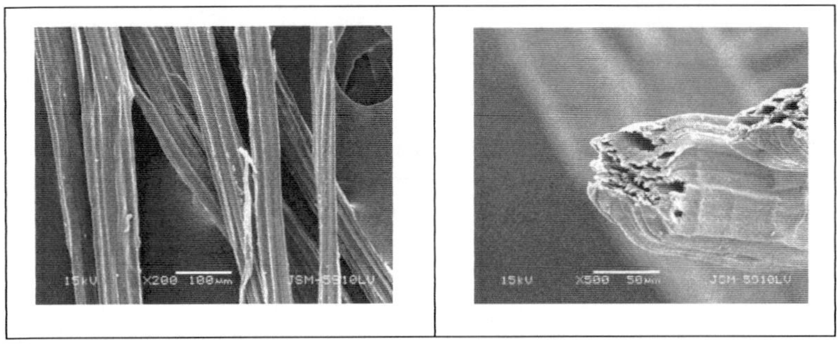

Figure 16. Longitudinal and cross-sectional SEM images of the okra body fibers treated with sodium bisulphate by conventional method.

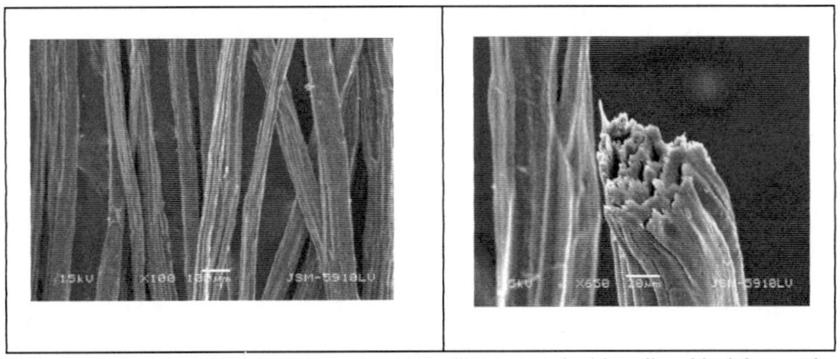

Figure 17. Longitudinal and cross-sectional SEM images of the okra fibers treated with sodium bisulphate and microwave energy by method.

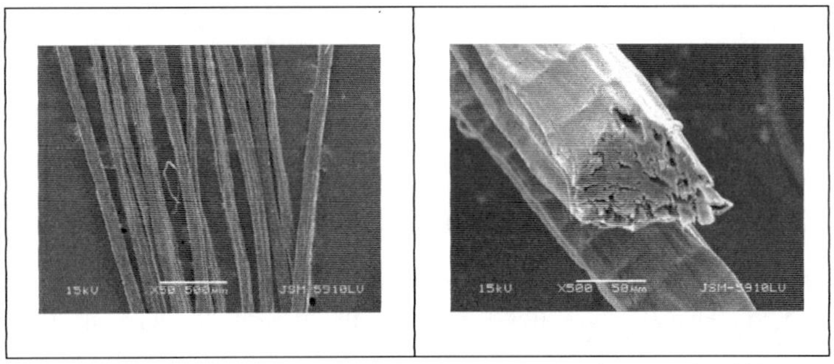

Figure 18. Longitudinal and cross-sectional SEM images of the okra fibers treated with acetic acid by conventional method.

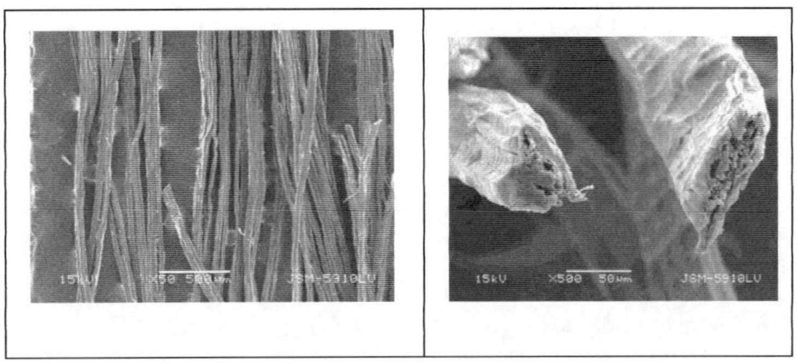

Figure 19. Longitudinal and cross-sectional SEM images of the okra fibers treated with acetic acid by microwave energy method.

SEM images of okra fibers after conventionally and microwave assisted treatments with both chemicals are shown in Figure. 15-19. High amount of porosity and fibrillation are observed in okra fibers treated with acetic acid by microwave method.

Following the treatment with sodium bisulphate and acetic acid by means of microwave, the outer layers of parenchyma cells have been removed to expose the inner fibres[8]. Impurities and waxes were effectively removed from the fiber surface with both metdos and chemicals compared to untreated fibers.

Conclusions

Two different processes carried out with acetic acid and sodium bisulphate chemicals were proved to be successful. The modification of fiber surface with microwave energy method had better results than conventionally surface modifications. The fibre strength were increased by microwave method.

According to the tensile strength, fiber diameter, SEM and X-Ray analysis results of untreated and treated okra fibers, higher tensile strength properties were obtained after treatments by microwave processes. Microwave process was more effective than the conventional process for surface modification of okra fibers and provided energy and time saving. Also microwave energy is more effective than conventional treatment in chemical reactions.

Considering all these results, surface treatments with sodium bisulphate and acetic acid by microwave energy have a positive effect on mechanical and morphological properties of okra fibers which were obtained from Marmara Region agricultural waste.

Acknowledgements

Authors are highly thankful to TÜBİTAK (The Scientific and Technological Research Council of Turkey) for aiding our project #215M984. This study is a part of this project.

References

[1] Asumani O.M.L, Reid R.G, Paskaramoorthy R, The effects of alkali–silane treatment on the tensile and flexural properties of short fibre non-woven kenaf reinforced polypropylene composites, Composites: Part A 43, 1431–1440,(2012).

[2] Al-oqla F.M, Sapuan, S.M, Natural fiber reinforced polymer composites in industrial applications: feasibility of date palm fibers for sustainable automotive industry, Journal of Cleaner Production, 66, 347-354, (2014).

[3] Koçak, E.D, Mıstık, S.İ, Merdan, N, Nayci Duman, M, Nonwoven Production from Agricultural Okra Wastes and İnvestigation of Their Thermal Conductivities, IOP Conference Series: Materials Science and Engineering, 254,(2017).

[4] "*Türkiye İstatistik Kurumu*" [Online]. Available: http://www.tuik.gov.tr/PreTabloArama.do. [Accessed: 09-May-2017].

[5] Li X, Tabil L.G, Panigrahi S, Chemical Treatments of Natural Fiber for Use in Natural Fiber-Reinforced Composites: A Review, Journal of Polymers and the Environment, 15,25–33, (2007).

[6] Yılmaz Şahinbaşkan B, Koçak E.D, Merdan N,Akalın Mehmet, Dyeing of Polypropylene Blends by Using Microwave Energy. Journal of Engineered Fibers and Fabrics, 12, 20-27,(2017)

[7] Kocak D, Merdan N, Yüksek M. and Sancak E, Effects of Chemical Modifications on Mechanical Properties of Luffa cylindrica, Asian Journal of Chemistry, Vol.25, No.2 637-641,(2013),

[8] Merdan N, Sancak E, Kocak D. and Yuksek M, Effect of Applied Different Surface Modification Processes with Cellulose Enzyme on Properties of Luffa Fibres, Asian Journal of Chemistry, Vol.24, No.3,975-980,(2012).

New Approach Towards m-PEG Grafting Onto Commercially Available Nylon 6 To Resist Bacterial Adhesion On Surface

S. Swar[1], V. Zajícová [2,3], J. Müllerová[2,3], P. Šubrtová[2] and I. Stibor[2,3]

[1] *Department of New Technologies and Applied Informatics, Faculty of Mechatronics, Informatics and Interdisciplinary Studies, (Technical University of Liberec), Studentská 1402/2, 46117 Liberec 1, Czech Republic*

[2] *Institute for Nanomaterials, Advanced Technologies and Innovation (Technical University of Liberec), Bendlova 1409/7, 46001 Liberec 1, Czech Republic*

[3] *Department of Chemistry, Technical University of Liberec, Faculty of Science, Humanities and Education (Technical University of Liberec), Studentská 1402/2, 46117 Liberec 1, Czech Republic*

Corresponding author: dearsumita@gmail.com

Abstract: Biofilm formation, caused particularly by pathogenic bacteria like *methicillin-resistant Staphylococcus aureus* (MRSA) on medical devices, is imposing threat to public health. There is thus an ever growing demand for designing materials that are both cytocompatible and resistant to biofilm formation as well as bacterial infections. Surface functionalized polyamides, such as Nylon 6, are widely used as biomaterial due to its strength, flexibility, toughness and cytocompatibility. The undertaken study is focused on the surface functionalization of Nylon 6 by reducing the surface with borane-tetrahydrofuran complex (BH_3-THF), followed by grafting with poly(ethylene glycol) methyl ether tosylate (mPEG-OTs) via a novel lithiation approach. The modified Nylon 6 surfaces were characterized by various techniques such as water contact angle (WCA) analysis, atomic force microscopy (AFM), scanning electron microscopy (SEM) and Fourier transform infrared spectroscopy (FTIR) to confirm the modification of Nylon 6 surface. Evaluation of bacterial adhesion on the pure and modified surface against biofilm active pathogenic bacterial strain: *Staphylococcus aureus* (S.A.) CCM 3953 was accomplished. The functionalized Nylon 6 surfaces showed significant resistance towards bacterial adhesion compared to pure Nylon 6.

Keywords— **grafting, lithiation, Nylon 6, pathogenic bacteria, surface functionalization.**

I. Introduction

The successes in modern biomedical industry is closely related to the increasing field of polymeric biomedical devices [1]. Despite the considerable success in biomedical field, one of the key challenges is the device associated infections due to microbial attachment to the polymeric surface leading to biofilm formation [2]. Amongst various pathogenic bacteria, *methicillin-resistant Staphylococcus aureus* (MRSA) has been a great concern in health sectors [3]. There are various approaches to deal with this problem by controlling hydrophobicity, surface roughness, surface functional groups and electrostatic interactions [4]. One of the key strategies is to prevent bacterial adhesion by immobilizing biocompatible hydrophilic polymer poly(ethylene glycol) (PEG) and its derivatives such as poly(ethylene glycol) methyl ether (mPEG) [5,6]. The research studies have reported that higher molecular weight PEG and longer brushes can resist bacterial adhesion more strongly [7].

The newly developed synthetic polymers like polyesters and polyamides were introduced in medical application during World War II [8]. Nylon 6 is widely used in catheters, wound dressing materials, suture material, dialysis membrane etc. but without surface modification, it may cause other complications like thrombogenesis [9]. Therefore, surface functionalization with specific functional groups is necessary to improve the performance as biomaterial. Both the physical (UV radiation or plasma) and chemical methods can be used for surface functionalization whereas, both the methods have their own advantages and disadvantages [10,11].

The present research is focused on amide group reduction on Nylon 6 surface with borane complex (BH_3-THF) [12], followed by a novel method of grafting poly(ethylene glycol) methyl ether (mPEG) onto reduced Nylon 6 surface using poly(ethylene glycol) methyl ether tosylate (mPEG-OTs) and *tert*-butyllithium (*t*-BuLi). The changes in WCAs of modified Nylon 6 surface indicated the surface morphology alteration that was further confirmed by atomic force microscopy (AFM), scanning electron microscopy (SEM) and Fourier transform infrared spectroscopy (FTIR). The bacterial adhesion assessment was performed using standard protocol CSN EN ISO 6222.

II. Procedures

A. *Materials*

Nylon 6 film, with the thickness of 0.015 mm, was supplied by Goodfellow Cambridge Ltd., (Huntingdon, England). Borane-Tetrahydrofuran complex (1 M, BH3-THF), poly(ethylene glycol) methyl ether tosylate (M_n = 2000 g/mol, mPEG-OTs), *tert*-butyllithium (1.7 M in hexane, *t*-BuLi) and THF (anhydrous, ≥99.9%, inhibitor-free) were obtained from Sigma-Aldrich Co. Solvents tetrahydrofuran (99.8% THF), dimethyl sulfoxide (99% DMSO), acetone (99.5%), ethanol (96%), 2-propanol (99.8%), hexane (99.9%) and hydrochloric acid (35% HCl) were purchased from Lach:Ner, s.r.o., CZ. Sodium hydroxide (NaOH) was supplied by PENTA Ltd. CZ. All washings were performed by deionized water.

Staphylococcus aureus (S.A.), CCM 3953 strain was used for bacterial adhesion tests. Soybean Casein Digest Medium: HIMEDIA®REF, Plate Count Agar without Dextrose (BIORAD a.s., CZ) and sterile physiological solution (6 % NaCl) were used for bacterial inoculum preparation. Globular protein Bovine Serum Albumin (BSA, Sigma-Aldrich, CZ) was used as a support for better S.A. proliferation.

B. *Methods*

Sample preparation

Nylon 6 samples (size 2 cm × 2 cm), were thoroughly rinsed with water, ethanol, 2-propanol, acetone, THF and hexane in sonicating bath for 3 mins per solvent. The washed samples were dried at 50°C/ 3 h in vacuum and stored for further use.

Nylon 6 surface reduction and grafting

The amide functional groups present in Nylon 6 were reduced to secondary amine by BH_3-THF according to procedure described by Herrera-Alonso *et al.* [12]. The reduced samples have been referred as Nylon 6-NH. The modified samples were washed with THF, 1 M HCl, deionized water, 1 M NaOH, deionized water, THF, ethanol, acetone and hexane, each by sonication for 3 minute. The reduced Nylon 6 films were dried at 50°C overnight in vacuum and used for grafting. Dry THF (50 ml) was added into the Schlenk flask containing Nylon 6-NH samples. Then *t*-BuLi (0.6 ml, 1 ml and 2 ml) was introduced into it for 2 h lithiation at r.t./ 250 rpm in inert atmosphere. After lithiation, the solution of remaining *t*-BuLi was removed carefully by syringe, maintaining the inert condition. 2g (1 mmol) mPEG-OTs, dissolved in dry THF (50 ml), was added into the same flask and allowed to react overnight. The grafted samples were washed with THF, ethanol, acetone and hexane; dried in the vacuum. Modified samples have been referred as Nylon 6-N-mPEG.

Surface Characterization

The WCAs was measured using a portable computer-based instrument with special purpose software following ISO 27448:2009 test method (See System E, Advex Instruments, CZ). The topography of the unmodified and modified Nylon 6 films was studied in the air at atmospheric pressure with NanoWizard® 3 NanoScience AFM (JPK Instruments, Germany). The surface morphology was determined as secondary electron images (1 kV) by SEM (ZEISS, Sigma Family, Germany). The Nicolet™ iS™10 FT-IR Spectrometer (Thermo Scientific™, USA) was used to examine the changes in surface composition after the modifications of Nylon 6.

Bacterial adhesion test

UV A sterilized pure Nylon 6, Nylon 6-NH and Nylon 6-N-mPEG samples were added into 50 ml of bacterial inoculum homogenous solution (3.0 x 10^7 CFU/ ml). Total 5 sets were prepared with *S. aureus* (S.A.). The bacterial solutions, as well as control, were incubated for 48 h at 37° C. Each sample was taken out from the inoculum and transferred to 100 ml Fisher bottle containing 20 ml sterile physiological solution (6% NaCl). The bottles were shook vigorously for an hour to remove surface adhered bacteria in solution. 1 ml each of the solution at different concentrations (10^0, 10^{-1}, 10^{-2}, 10^{-3} and 10^{-4}) was directly used for bacterial proliferation in agar medium and subsequently stored in the incubator 48 h/ 37° C. In the case of control, the concentration line was as follows (10^0

– 10^{-9}) where $10^0 = 3.0 \times 10^7$ CFU/ ml. The results are mentioned in TABLE III.

III. RESULTS

The reduction of amide functional groups on Nylon 6 surface to secondary amine (Nylon 6-NH), followed by grafting onto Nylon 6-NH forming Nylon 6-N-mPEG are shown in schematic representation (Fig. 1).

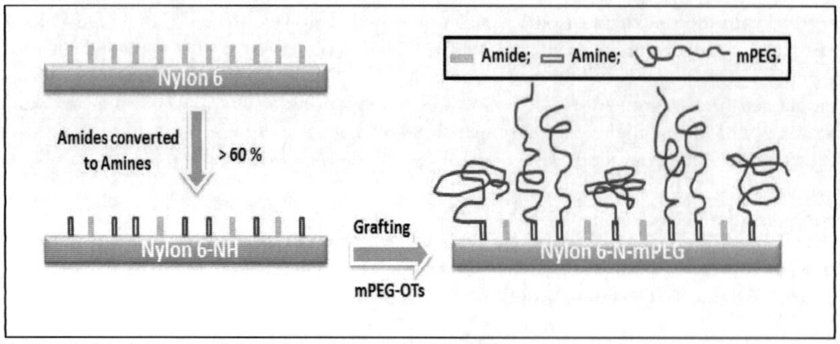

Fig. 1. Schematic presentation of Nylon 6 surface modification to Nylon 6-N-mPEG.

The WCAs measurements with standard deviations (± SD) related to the unmodified and modified Nylon 6 films are presented in TABLE I. The significant increase in WCAs was noticed after Nylon 6 film modification into Nylon 6-NH. On the contrary, the mean WCAs have been decreased significantly, as expected due to the mPEG chain immobilization, after grafting with mPEG-OTs via lithiation. The *t*-BuLi concentration played an important role in grafting intensity, thus differed the WCAs of three different grafted samples. Nylon 6-NH activated by 0.6 ml *t*-BuLi is more hydrophilic after grafting compared to the other two grafted samples activated with higher concentrations of *t*-BuLi.

TABLE I
Surface functionalization of Nylon 6: WCA data

Sample	Mean WCA (°) ± SD
Nylon 6	71.4 ± 4.3
Nylon 6-NH	82 ± 4.8
Nylon 6-N-mPEG (0.6 ml *t*-BuLi)	44.5 ± 3.8
Nylon 6-N-mPEG (1 ml *t*-BuLi)	46 ± 4.2
Nylon 6-N-mPEG (2 ml *t*-BuLi)	60.1 ± 4.8

The AFM analyses showed significant changes in surface topography which were reflected in surface roughness (Ra) values for measured areas (1 μm x 1 μm and 10 μm x 10 μm) in case of the pure Nylon 6, modified Nylon 6-NH and Nylon 6-N-mPEG. The obtained Ra values are presented in the TABLE II. The Nylon 6 surface roughness decreased noticeably after reduction, whereas increased after grafting. The lithiation concentration did not show any major effect on Ra values for all the grafted samples.

TABLE II
The changes in surface roughness (Ra) values for pure and functionalized Nylon 6

Surface area	Nylon 6	Nylon 6-NH	Nylon 6-N-mPEG
$(1 \times 1)\ \mu m^2$	5.1 nm	1.3 nm	1.3 nm
$(10 \times 10)\ \mu m^2$	34.8 nm	7.2 nm	12.8 nm

Surface morphologies were examined by SEM images before and after Nylon 6 surface modification (Fig. 2a – 2d). SEM images clearly showed a successful grafting on reduced Nylon 6 surface. The grafting intensity difference was visible between Nylon 6-N-mPEG (0.6 ml *t*-BuLi) and Nylon 6-N-mPEG (2 ml *t*-BuLi). This explained the difference in WCAs among the various grafted samples. The higher concentration of *t*-BuLi (2ml) during Nylon 6-NH lithiation caused denser grafting layer than lower *t*-BuLi concentration (0.6 ml). This was further supported by FT-IR spectra shown in Fig. 3.

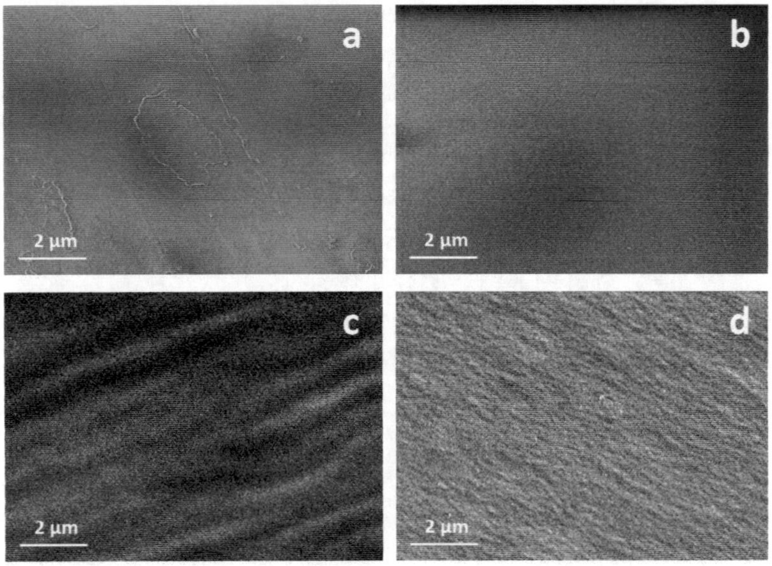

Fig. 2. SEM images of a) Nylon 6; b) Nylon 6-NH; c) Nylon 6-N-mPEG (0.6 ml *t*-BuLi); d) Nylon 6-N-mPEG (2 ml *t*-BuLi).

Fig. 3 shows the comparison between the FT-IR spectra of pure Nylon 6 and functionalized Nylon 6-NH and Nylon 6-N-mPEG films. In FT-IR spectra, the characteristic stretching vibrations of amide group (1660 cm^{-1} and 1541 cm^{-1}), the aliphatic groups (2858 cm^{-1} and 2931 cm^{-1}) and secondary amine groups (3293 cm^{-1}) for Nylon 6 were also visible in case of both reduced and grafted samples. This occurred due to 1 μm penetration depth of FT-IR. A new band appeared at the range of 2420 - 2250 cm^{-1} corresponded to the imine groups introduced on the surface due to chemical treatment with borane. This indicated the step towards amide reduction to amine. FT-IR spectra of all three Nylon 6-N-mPEG showed that the imine stretching band disappeared after grafting. The characteristic PEG stretching band (1087 cm^{-1}) for Nylon 6-N-mPEG was significant for the successful grafting. The stretching band intensity at 1087 cm^{-1} in spectra correlated the grafting intensity of Nylon 6-N-mPEG samples.

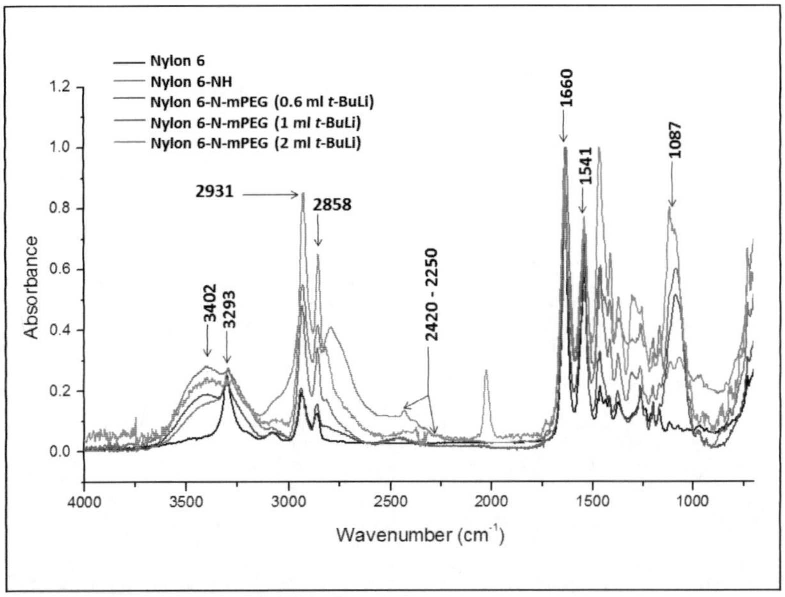

Fig. 3. FT-IR spectra showing pure, reduced and grafted Nylon 6.

The bacterial adhesion resistance to the surface of the pure Nylon 6, Nylon 6-NH and grafted Nylon 6-N-mPEG samples against *S. aureus* was evaluated by counting the number of bacterial colonies formed in agar media incubated for 48 h/ 37° C. The results are summarized in TABLE III. The grafted Nylon 6-N-mPEG (0.6 ml *t*-BuLi) sample showed significant resistance towards *S. aureus* adhesion compared to other two grafted ones. Nylon 6-NH also exhibited efficiency to resist bacterial adhesion, but pure Nylon 6 was very prone to bacterial adhesion as it showed high number of bacterial colony formation similar to the control.

TABLE III
The bacterial adhesion resistance against *S. aureus* by pure and modified Nylon 6 surfaces

	Number of bacterial colonies formed at various bacterial inoculum concentrations				
Sample	**10^0**	**10^{-1}**	**10^{-2}**	**10^{-3}**	**10^{-4}**
Control	*	*	*	*	975
Nylon 6	*	*	*	904	100
Nylon 6-NH	13	11	7	0	0
Nylon 6-N-mPEG (0.6 ml *t*-BuLi)	5	0	0	0	0
Nylon 6-N-mPEG (1 ml *t*-BuLi)	370	59	7	3	0
Nylon 6-N-mPEG (2 ml *t*-BuLi)	*	672	78	8	0

* implies uncountable bacterial colonies (> 1500)

IV. Conclusion

The Nylon 6 surface reduction and grafting were confirmed by the microscopic as well as spectroscopic analyses. The novel method of grafting mPEG on reduced Nylon 6 surface via lithiation has shown a new, simple yet efficient approach to functionalize Nylon 6 surface for providing effective resistance against *S. aureus* bacterial adhesion. The grafting intensity could be tuned by various concentrations of *tert*-butyllithium to obtain desirable efficacy. Even Nylon 6 surface reduction forming Nylon 6-NH proved to be efficient to resist the bacterial adhesion. Thus, our new approach has great potential for other biomedical polyamide surface functionalization to resist bacterial adhesion significantly.

Acknowledgment

This work was supported by the Ministry of Education, Youth and Sports of the Czech Republic and the European Union - European Structural and Investment Funds in the frames of Operational Programme Research, Development and Education - project Hybrid Materials for Hierarchical Structures (HyHi, Reg. No. CZ.02.1.01/0.0/0.0/16_019/ 0000843), the OPR&DI project "Extension of CxI facilities" (CZ.1.05/2.1.00/19.0386) and the project SGS 21207, Faculty of Science, Humanities and Education, Technical University of Liberec.

References

[1] S. Swar, V. Zajícová, M. Rysová, I. Lovětinská-Šlamborová, L. Voleský, I. Stibor, "Biocompatible surface modification of poly(ethylene terephthalate) focused on pathogenic bacteria: Promising prospects in biomedical applications," *J. Appl. Polym. Sci.*, vol. 134, pp. 1–11, Mar. 2017.
[2] R.M. Donlan, "Biofilm formation: a clinically relevant microbiological process," *Clin. Infect. Dis.*, Vol. 33, no. 8, pp. 1387–1392, Sep. 2001.
[3] J. Gallo, M. Holinka, C. S. Moucha, "Antibacterial Surface Treatment for Orthopaedic Implants," *Int. J. Mol. Sci.*, vol. 15, no. 8, pp. 13849–13880, Aug. 2014.
[4] I. Banerjee, R.C. Pangule, R.S. Kane, "Antifouling Coatings: Recent Developments in the Design of Surfaces That Prevent Fouling by Proteins, Bacteria, and Marine Organisms," *Adv. Mater.*, vol. 23, no. 6, pp. 690–718, Feb. 2011.
[5] A. Roosjen, H.C. Van Der Mei, H.J. Busscher, W. Norde, "Microbial adhesion to poly(ethylene oxide) brushes: Influence of polymer chain length and temperature," *Langmuir.*, vol. 20, no. 25, pp. 10949–10955, Sep. 2004.
[6] K. D. Park, Y. S. Kim, D. K. Han, Y. H. Kim, E. H. B. Lee, H. Suh, K. S. Choi, "Bacterial adhesion on PEG modified polyurethane surfaces," *Biomaterials.*, vol. 19, no. 7–9, pp. 851–859, Apr. 1998.
[7] B. Dong, S. Manolache, A. C. L. Wong, F. S. Denes, "Antifouling ability of polyethylene glycol of different molecular weights grafted onto polyester surfaces by cold plasma," *Polym. Bull.*, vol. 66, no. 4, pp. 517–528, Mar. 2011.
[8] M.F. Maitz, "Applications of synthetic polymers in clinical medicine," *Biosurface and Biotribology.*, vol. 1, no. 3, pp. 161–176, Aug. 2015.

[9] C. Mao, W. Zhao, C. Zhu, A. Zhu, J. Shen, S. Lin, "In vitro studies of platelet adhesion on UV radiation-treated nylon surface," *Carbohydr. Polym.*, vol. 59, no. 1, pp. 19–25, Jan. 2005.
[10] X. Jia, M. Herrera-Alonso, T. J. McCarthy, "Nylon surface modification. Part 1. Targeting the amide groups for selective introduction of reactive functionalities," *Polymer (Guildf).*, vol. 47, no. 14, pp. 4916–4924, June 2006.
[11] B. Dong, H. Jiang, S. Manolache, A. C. L. Wong, F. S. Denes, "Plasma-mediated grafting of poly(ethylene glycol) on polyamide and polyester surfaces and evaluation of antifouling ability of modified substrates," *Langmuir.*, vol. 23, no. 13, pp. 7306–7313, May 2007.
[12] M. Herrera-Alonso, T. J. McCarthy, X. Jia, "Nylon surface modification: 2. Nylon-supported composite films," *Langmuir.*, vol. 22, no. 4, pp. 1646–1651, Jan. 2006.

Preparation of bio-functional textiles by surface functionalization of cellulose fabrics with caffeine loaded nanoparticles.

D. Massella [1,2,3,4], A. Ancona[1], N. Garino[1], V. Cauda[1], J. Guan[4], F. Salaun[2,3], A. A. Barresi[1] and A. Ferri[1].

[1]*Department of Applied Science and Technology, Politecnico di Torino, Corso Duca degli Abruzzi 24, 10129, Turin, Italy.*

[2]*University of Lille Nord de France, F-5900 Lille, France.*

[3]*ENSAIT, GMTEX, F-59100, Roubaix, France.*

[4] *College of Textile and Clothing Engineering, Soochow University, Suzhou, Jiangsu, 215123, China*

Corresponding author: daniele.massella@polito.it

Abstract: In recent years transdermal drug delivery has aroused significant interest as a sustained and non-invasive way of administering active substances. The advancements of nanotechnology allowed the development of novel pharmaceutical formulations overcoming skin barrier. Furthermore, such nano-system can be combined with conventional fabrics to pave the way to a new generation of wearable drug delivery devices: bio-functional garments. First the NP were produced by flash nanoprecipitation technique (FNP), the production process was optimized to produce particles with suitable size for transdermal applications. The nanoparticles were characterized in terms of drug content by UV-visible spectroscopy and in term of antioxidant activity by Electron Paramagnetic Resonance spectroscopy (EPR) coupled with spin trapping technique. The NPs were used to functionalize cotton and viscose-micromodal fabrics and the transdermal release properties were tested in vitro by Franz's Cell experiment. FNP was proven to be an effective technique to produce tunable size particles. Moreover, the nanoencapsulated drug exhibited antioxidant activity. The Franz's Cell test evidenced a controlled release behavior, providing evidence that the bio-functional textile is suitable for applications where sustained release and antioxidant properties are required.

Keywords: **Antioxidant activity, Bio-functional textiles, Caffeine, Drug delivery, Encapsulation, Flash Nanoprecipitation, Hydrophilic, Nanoparticles.**

I. Introduction

Nowadays many efforts of pharmaceutical research have been focused on the development of innovative tools and approaches to administer drugs in "smart" way i.e. improving therapeutic efficacy while minimizing side effects [1]. Among these innovative approaches to drug delivery, the possibility of delivering active substances through skin aroused significant interest [2]. In facts skin is the largest tissue of the human body and has significant surface area available for drug administration; moreover, it plays the important role of protecting the body from external factors, which could significantly reduce side effects during drug administration [3]. Therefore, skin can be used as a route of administration for local and systemic drugs; according to Food and Drug Administration the first case is defined as topical administration and consists in the application of a drug to a particular area on the outer surface of the body, while the latter is defined as transdermal administration and involves the delivery through the dermal layer of the skin to the systemic circulation. Despite the numerous advantages in using skin for topical and transdermal drug delivery, such administration route is challenging for several compounds because of the structural complexity of the skin barrier [4]. Recent developments in nanotechnology and nanomedicine offer several approaches to overcome the above mentioned problem [5–8]. In addition, these advanced drug delivery systems can be combined with conventional textiles to develop wearable drug delivery devices; this new class of material defined as bio-functional textiles has shown the capability to improve the transdermal administration of active molecules [9]. Among bio-functional textiles much attention has been gained by transdermal patches that employ nano-carrier based drug delivery systems [10]. These patches aim to enhance drug permeation through the skin barrier in order to reach the blood circulatory system; such approach allows to obtain a sustained and constant drug release in a non-invasive way, while reducing the drug toxicity [11]. Cellulosic substrates and derivatives are the main potential candidates used in this technology, due to their natural biocompatibility, and the absence of side effects, such as allergies and skin irritations. Moreover, they are readily available and biodegradable, making them a suitable substrate for bio-functional textile preparation [12]. Due to skin barrier complexity only a limited number of active compounds with a lipophilic character can penetrate the outer layers of the skin. Thus, the preparation of a bio-functional textile usually requires encapsulation of the active substance before functionalizing the textile [13], the use of properly designed nanoparticles protects the drug from surrounding agents until it reaches the target tissue. Polymeric nanoparticles (NPs) are among the most studied nanocarriers given the large availability of polymeric materials that fulfill the requirements of biological applications such as non-toxicity, biodegradability and biocompatibility [14]. Furtherly, these carriers have been proved to be effective in entrapping and releasing active substances [15], [16]. Poly-ε-caprolactone (PCL) has peculiar features that makes it widely applied in tissue engineering, implantable devices, cell cultures and drug delivery [17], [18]. This polymer is commonly biodegraded in body fluids by cleavage of the ester bond: the degradation mechanism consists in a first step in which the molecular weight decreases due to chain scission and a second one where an actual weight loss is observable; the overall process can take several weeks to be completed. Thus this material is suitable for applications where long lasting releases are needed [19], and it has been also successfully employed in transdermal applications [20], [21]. Unfortunately, a wide application of such polymeric devices is still a challenging issue due to the scarce productivity and scalability of NPs production processes [22], and this issue is even more critical dealing with bio-functional textiles given the high amount of NPs that would be required in textile functionalization. The Flash Nanoprecipitation technique (FNP) [23] is a suitable process to overcome the above mentioned issue given its simplicity, fastness, productivity and good reproducibility of the obtained results [24]. FNP exploits the difference of solubility of a polymer in two miscible fluids and partition coefficient in polymer-solvent system. The polymer is dissolved in an organic solvent which collides against a water jet in a micro reactor to generate a highly turbulent mixing, the low affinity of the polymer for the aqueous phase leads to its precipitation in the form of nanoparticles [25]. The factor governing the particle formation phenomenon is mainly the mixing conditions. Reactors such as the confined impinging jet mixer (CIJM) have been appositively designed in order to control the process conditions [26]. On the other hand, the chemical nature of the active substance also influences the encapsulation yield. Thus, several studies have shown that the entrapment of hydrophobic substances (LogP >3.2) such as curcumin, paclitaxel, menthol and vitamin E can be realized with FNP, while to our knowledge only few works deal to encapsulate hydrophilic substances [27]. The aim of this study is to determine the process parameters allowing for encapsulating a hydrophilic substance, caffeine (CAF), which represents an opportunity for transdermal delivery, due to its potential application in dermatology [28]. CAF is able to protect skin from damages caused by UV light, and displays antioxidant activity. Furthermore, it was found that CAF can promote the lipolysis at cellular level and act as an anti-cellulite drug [29], [30]. In the present work a FNP methodology is proposed to produce CAF loaded PCL nanoparticles. The process parameters as well as the formulations were studied in order to achieve NPs of a size similar to the one skin annexes. The tested formulations were characterized by measuring their encapsulation efficiency (EE) and loading capacity (LC), while the antioxidant activity of the NPs was determined by Electron Paramagnetic Resonance Spectroscopy (EPR) coupled with spin trapping technique. The particles were then used to functionalize two different textiles substrates, a cotton and a viscose-micromodal blend. The release properties of the functionalized textile were tested *in vitro* in a vertical Franz diffusion cell.

II. Materials and Methods

Materials

Poly-ε-caprolactone (PCL) (molecular weight of 14000 g/mol), and caffeine (CAF), selected as matrix and active substance respectively, were purchased from Sigma Aldrich. Acetone with purity ≥ 99,5% meeting European Pharmacopedia standards (Sigma Aldrich) was used as solvent. Phosphate buffer solution used in the release test was prepared from sodium chloride anhydrous ≥ 99%, potassium chloride ACS grade ≥ 99,5%, sodium phosphate dibasic dehydrate ≥ 99% and potassium dihydrogen phosphate ACS reagent ≥ 99%, also purchased from Sigma Aldrich. For the spin trapping reaction 5,5-dimethyl-1-pyrroline N-oxide (DMPO), hydrogen peroxide (30%) and iron (II) sulphate were purchased from Sigma Aldrich. Ultrapure water was produced by mean of a Milli-Q RG system by Millipore R (Billerica, MA) and employed in all the experimental procedures. Textile materials tested were knitted cotton fabrics single jersey made of 100% cotton and a blend of viscose/micromodal 70/30%; they were kindly supplied by Eusebio S.p.a (Varese, Italy).

A. *Nanoparticle preparation*

The nanoparticles were prepared by FNP in a CIJM (1 mm inlet tube diameter, 5 mm chamber diameter, 11.2 mm chamber height). A stream of polymer organic solvent solution was mixed with an anti-solvent, water (Fig. 1). The collision of the two jets induced polymer precipitation in the form of nanoparticles.

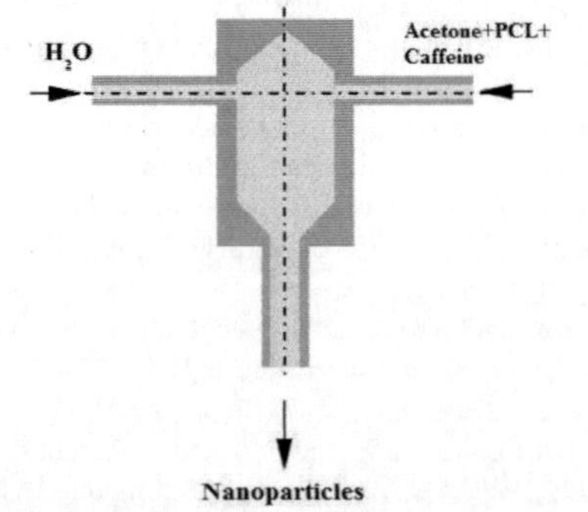

Fig. 1. Scheme of the FNP employed for the production CAF loaded PCL Nanoparticles.

Defined volumes of acetone solution with the proper concentration of PCL and CAF, and water were placed in syringes and fed to the CIJM micro-mixer by a syringe pump (KDS200, KD Scientific, Holliston, MA) at a flow rate varying from 20 to 80 ml/min; the concentration of CAF and PCL used in the inlet solutions are shown in Table I. Samples of 6 ml (3 ml for each stream) were taken, collected in a glass vial containing 3 ml of quenching water placed downstream of the mixer. The collected suspension and the quenching water were kept under magnetic stirring for two minutes. The collected nanoparticle suspension was used for further characterizations.

TABLE I
FORMULATIONS TESTED.

Formulation	PCL concentration (mg/ml)	CAF concentration (mg/ml)
F1	4.5	9.0
F2	6.0	9.0
F3	10.0	7.6

B. Fabrics functionalization

Fabrics were functionalized by imbibition. The NPs were firstly separated from liquid, in order to obtain about 2 ml of concentrated NP suspension. 0.5 ml of each suspension were then added dropwise over a disk of textile material of about 2 cm in diameter. Particular care was taken to spread the NPs uniformly over the textile surface and to let the fabric dry between the drop additions.

C. Characterizations and analytical methods

- Particle size analysis

Particles size distribution was measured by means of DLS Zetasizer Nanoseries ZS90, Malvern Instruments (Malvern, UK). Samples were prepared by diluting 0.1 ml of as produced nanoparticles suspensions in 1 ml of ultrapure water. All samples were measured in triplicate under controlled temperature at 25.0 ± 0.1 °C.

- Determination of Loading Capacity and Encapsulation Efficiency.

LC is defined as the mass of encapsulated (m_{en}, in g) drug divided by the mass of the whole polymeric nanoparticles system (m_{tot}, in g) as given by (1); it is an indicator of the amount of drug that can be incorporated in a given amount of nanoparticle formulation. EE is defined as the amount of encapsulated caffeine over its input quantity in the process (m_{in}, in g) as expressed by (2); it is an index of the efficiency of the nanoparticles production process.

$$LC(\%) = \frac{m_{en}}{m_{tot}} \times 100 \qquad (1)$$

$$EE(\%) = \frac{m_{en}}{m_{in}} \times 100 \qquad (2)$$

In order to calculate LC% and EE% a methodology based on the one described in [10] with slight modification was employed. Briefly the NPs suspension was placed in a rotary evaporator RE 300 at 50 °C, under vacuum for

15 min, to remove acetone; then, it was centrifuged for 30 min at 15800 g in a SL 16 Thermo scientific centrifuge in order to separate the liquid from the solid NPs. 0.1 mL supernatant sample was diluted in 25 ml of water and analyzed by UV-Vis spectroscopy in a 6850 UV/Vis spectrophotometer, Jenway (Stone, Staffordshire, UK). EE% and LC% were calculated indirectly after measuring the amount of unencapsulated caffeine in the supernatant.

- Antioxidant activity

The antioxidant activity of nanoparticles was studied by EPR coupled with the spin-trapping technique. For this analysis an appositively prepared sample with equal CAF and PCL concentration of F2 but produced by dissolving CAF in water was used; the NP was accurately separated from the liquid, taking also care of removing all traces of acetone. Hydroxyl radicals were generated in situ by Fenton reaction and trapped using the spin-trap DMPO. H_2O_2 (20 µL, 10mM), DMPO (100 µL, 50 mM) an aqueous solution of nanoparticles at different concentrations (78 µL) and $FeSO_4$ (2 µL, 10 mM) were mixed in an Eppendorf tube. The resulting solution was mixed, transferred to a quartz microcapillary tube and placed in the EPR cavity for measurement. After 5 minutes since the addition of $FeSO_4$, the spectra were recorded on a Bruker EMXnano X-Band spectrometer (Bruker, Billerica, MA, USA). The EPR measurements conditions were as follows: Frequency, 9.74 GHz; scan width, 100 G; receiver gain, 60 dB; time constant, 1.28 ms; sweep time, 80 s; scan, 1. After acquisition, the spectra were processed using the Bruker Xenon software (Bruker, Billerica, MA, USA) for baseline correction and the total number of hydroxyl radicals trapped was quantified using the SpinFit software (Bruker, Billerica, MA, USA).

- *In vitro* release Test

The release test from the functionalized fabrics was conducted on static vertical Franz diffusion cells (PermeGear, Hellertown, PA, USA). The Franz cells were constituted by an upper donor chamber and a lower receptor chamber (volume 11.4 mL) with a contact surface of 1.5 cm^2. The acceptor compartment was filled with pH 7.4 phosphate buffer solution kept at temperature of 33°C by a heating jacket. A cellulose acetate membrane with porosity of 0.45 µm was used to mimic skin. Prior release test the membrane was boiled in water for 1 h to enhance its wettability. 1 ml samples of the acceptor fluid were withdrawn at a fixed time intervals and analyzed for caffeine by spectrophotometry; thereafter, an equal volume of the fresh PBS solution was replaced into the cell. Experiments were performed in quadruplicate.

III. Results and discussion

A. *Particle size*

To be suitable for a transdermal administration, NPs size should less than the one of skin annexes (approximately 450 nm). The results of the production of the three formulations at varying inlet flow rate are plotted in Fig. 2.

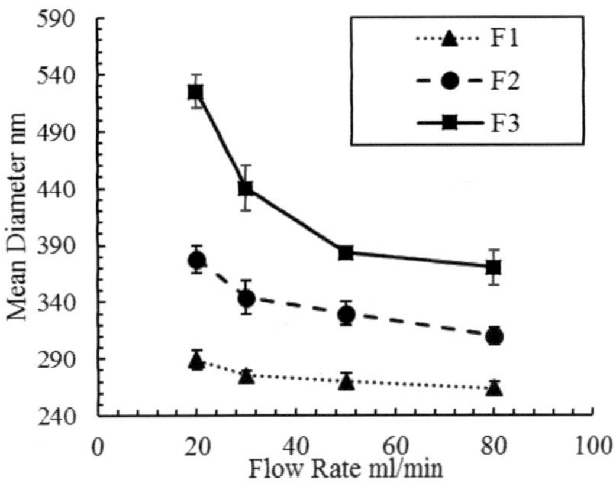

Fig. 2. Trends in particle diameter at varying flow rate for the different formulation tested.

The particle size tends to decrease as the flow rate is increased. Such result is in accordance with theoretical works in which particle formation occurs by two successive steps of nucleation and growth [31]. As a matter of facts, increasing solvent and antisolvent flow rate (FR) causes a higher turbulence inside the reactor chamber, which in turns increases the number of nuclei formed. This reduces the maximum size reachable by each particle. Comparing different formulations, it is noticeable an inverse correlation between the initial PCL concentration and the diameter, this mainly due to greater amount of polymer that can precipitate over the already formed nuclei causing the particles to enlarge during the growth stage.

B. Development of textile treatment

Given the suitability of the NPs systems for the transdermal release in terms of diameter, the choice of the optimal formulation was mainly based on the particle content. In Table II the values of EE and LC for the different formulations prepared at FR= 80 ml/min are reported.

TABLE II
EE AND LC OF THE DIFFERENT FORMULATIONS AT FR = 80 ML/MIN.

Formulation	EE%	LC%
F1	11.3	7.7
F2	17.4	10.4
F3	13.1	7.6

The values of the EE below 20% are an expected result given the high affinity of caffeine with water; as previously reported FNP is commonly used to encapsulated hydrophobic drug that precipitate together with the polymer due to their scarce water solubility. In this study, instead the CAF can be encapsulated only by a mechanical entrapment that occurs during the precipitation of the polymer and such phenomenon leads to lower EE values. The F2 displays the highest results both in term LC and EE, this mainly due to the good balance between initial caffeine and polymer concentration. The low PCL concentration in F1 does limit the available polymer chains for the entrapment while the high PCL content in F3 increases the overall mass of the system reducing the LC. Because of its capability of incorporating higher amounts of caffeine F2 was chosen as optimal formulation for the textile functionalization and release test. The effectiveness of the imbibition protocol was already proved in our previous work by SEM analysis [10].

C. Antioxidant activity

Hydroxyl radical is a highly reactive chemical species with very short half-life in the order of nanoseconds [32]. It is known to be the most biologically active free radical, damaging DNA and initiating lipid peroxidation [33]. In this study, the antioxidant activity of CAF loaded nanoparticles in scavenging hydroxyl radicals was studied. Hydroxyl radicals were generated using the Fenton reaction ($Fe^{2+} + H_2O_2 \rightarrow Fe^{3+} + OH\bullet + HO^-$) and the oxidation of DMPO spin trap by OH• was studied by EPR [34]. Results are shown in Table III. Addition of empty PCL nanoparticles did not significantly protect the spin-trap DMPO from oxidation by Fenton-generated OH• radicals: indeed, the acquired DMPO-OH signal intensity did not vary upon the addition of the nanoparticles. On the contrary, addition of CAF-loaded PCL nanoparticles protected the spin-trap DMPO from oxidation, markedly reducing the spin-adduct DMPO-OH signal. The protective effect is concentration dependent over the range 1 mg/mL - 4.5 mg/mL CAF-loaded nanoparticles, reaching 49% of DMPO-OH signal decrease for the highest NPs concentration. Notably, the corresponding higher amounts of empty PCL nanoparticles displayed negligible OH radicals scavenging activity, thus suggesting that the observed hydroxyl radical scavenging ability was only due to the CAF loaded in the nanoparticles. Overall, these preliminary data show that the encapsulation of CAF in PCL nanoparticles results in a marked hydroxyl radical scavenging ability, thus suggesting PCL-loaded NPs as innovative nanoparticle-based antioxidants.

TABLE III
OH RADICALS CONCENTRATION NORMALIZED TO THE CONTROL

Water (Control)	Empty NPs (4,6 mg/mL)	CAF-loaded NPs (5,75 mg/mL)
100%	98%	49%

D. In vitro release test

The Franz's Cell experiment aimed to assess the potential of the developed bio-functional textile as a transdermal delivery device. The normalized cumulative release curves are plotted in Figure 3, a sample obtained by functionalizing the fabric with a solution of free caffeine were used as reference (labeled as CAF COT and CAF

MIC). The NP used for the release test were formulation F2 produced at a FR of 80 ml/min.
It is noticeable that two factors play a role in slowing and controlling the release. One is ascribable to the textile substrate, as the CAF MIC sample present a slower and release compared to CAF COT, which reaches a plateau in less than 2 hours. This fact is mostly due to the smaller diameter of micromodal fibers, which means higher surface; this will make the fabric act as a reservoir that controls the diffusion from the fabric toward the membrane. Under the same substrate (micromodal), it can be observed that the encapsulation process furtherly slows down the release as the kinetics is slower for caffeine-PCL nanoparticles than free caffeine. This is due to a series of phenomena occurring: first, the particle must be released from the textile and interact with the membrane, then the NP may either pass through the membrane delivering CAF to the acceptor fluid or get stuck in the membrane pores; in this case CAF has to diffuse out the NP that may act as a reservoir inside the skin. During the first hour the CAF MIC and NP MIC curve present the same slope, and this is due to CAF diffusing out of the particles and passing directly through the membrane.

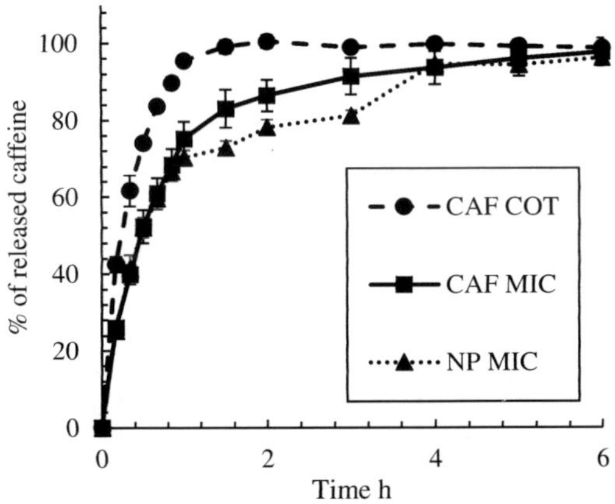

Fig. 3. Cumulative release curves for textiles functionalized with CAF and NP F2 formulation.

Between 1h and 4h the kinetic of the NP system becomes slower, here the particles get stuck in the membrane and CAF is released from the NP facing higher resistance to mass transfer. The overall contribution of those different phenomena make the micromodal-NP system the most effective in controlling the drug release.

IV. Conclusion

The present research inquired the preparation of a bio-functional textile. The employment of the polymeric nanoparticles in a transdermal delivery system combined with the proper textile material showed an effective control of the drug release kinetics. The use of FNP as a high productivity process was proven suitable even when dealing with a hydrophilic drug. The assessment on the antioxidant properties of the NP combined with the transdermal release of the textile-NP system provides interesting preliminary results about the possibility of administering caffeine through daily worn garments. This could represent an improvement with respect to conventional transdermal systems such as creams and ointments that must be locally applied several times a day. Once worn next to skin, the fabric can deliver CAF for several hours without any further action of the patient to be taken.

Acknowledgments

The European Union is kindly acknowledged for financing this research project under the frame of the program Erasmus Mundus Joint Doctorate in Sustainable Management and Design of Textiles (Project SMD-Tex 2016-41). The contribution to the experimental work of Giulia Delpiano is acknowledged. The staff of the THN Lab in Politecnico di Torino is kindly acknowledged for fruitful scientific discussion and contribution to the work. DM gratefully acknowledges Ermenegildo Zegna for supporting his mobility by means of the action "EZ founder scholarship".

References

[1] F. Leone and R. Cavalli, "Drug nanosuspensions: a ZIP tool between traditional and innovative pharmaceutical formulations," *Expert Opin. Drug Deliv.*, vol. 12, no. 10, pp. 1607–1625, Oct. 2015.

[2] L. F. Santos, I. J. Correia, A. S. Silva, and J. F. Mano, "Biomaterials for drug delivery patches," *Eur. J. Pharm. Sci.*, vol. 118, pp. 49–66, Jun. 2018.
[3] R. Goyal, L. K. Macri, H. M. Kaplan, and J. Kohn, "Nanoparticles and nanofibers for topical drug delivery," *J. Controlled Release*, vol. 240, pp. 77–92, Oct. 2016.
[4] T. W. Prow *et al.*, "Nanoparticles and microparticles for skin drug delivery," *Adv. Drug Deliv. Rev.*, vol. 63, no. 6, pp. 470–491, May 2011.
[5] L. A. DeLouise, "Applications of nanotechnology in dermatology," *J. Invest. Dermatol.*, vol. 132, no. 3, pp. 964–975, Mar. 2012.
[6] B. Dumontel *et al.*, "Enhanced biostability and cellular uptake of zinc oxide nanocrystals shielded with a phospholipid bilayer," *J. Mater. Chem. B*, vol. 5, no. 44, pp. 8799–8813, 2017.
[7] G. Canavese *et al.*, "Nanoparticle-assisted ultrasound: A special focus on sonodynamic therapy against cancer," *Chem. Eng. J.*, vol. 340, pp. 155–172, May 2018.
[8] M. Argenziano *et al.*, "Vancomycin-loaded nanobubbles: A new platform for controlled antibiotic delivery against methicillin-resistant Staphylococcus aureus infections," *Int. J. Pharm.*, vol. 523, no. 1, pp. 176–188, May 2017.
[9] M. Mihailiasa, F. Caldera, J. Li, R. Peila, A. Ferri, and F. Trotta, "Preparation of functionalized cotton fabrics by means of melatonin loaded β-cyclodextrin nanosponges," *Carbohydr. Polym.*, vol. 142, pp. 24–30, May 2016.
[10] D. Massella, F. Leone, R. Peila, A. Barresi, and A. Ferri, "Functionalization of cotton fabrics with polycaprolactone nanoparticles for transdermal release of melatonin," *J. Funct. Biomater.*, vol. 9, article no 1, 15 pp, Dec. 2017.
[11] S. A. Joshi, S. S. Jalalpure, A. A. Kempwade, and M. R. Peram, "Fabrication and in-vivo evaluation of lipid nanocarriers based transdermal patch of colchicine," *J. Drug Deliv. Sci. Technol.*, vol. 41, pp. 444–453, Oct. 2017.
[12] E. Pinho, M. Henriques, R. Oliveira, A. Dias, and G. Soares, "Development of biofunctional textiles by the application of resveratrol to cotton, bamboo, and silk," *Fibers Polym.*, vol. 11, no. 2, pp. 271–276, Apr. 2010.
[13] M. Martí, V. Martínez, L. Rubio, L. Coderch, and J. L. Parra, "Biofunctional textiles prepared with liposomes: *in vivo* and *in vitro* assessment," *J. Microencapsul.*, vol. 28, no. 8, pp. 799–806, Dec. 2011.
[14] A. Kumari, S. K. Yadav, and S. C. Yadav, "Biodegradable polymeric nanoparticles based drug delivery systems," *Colloids Surf. B Biointerfaces*, vol. 75, no. 1, pp. 1–18, Jan. 2010.
[15] M. Bazzano *et al.*, "Synthesis of polymeric nanocapsules by radical UV-activated interface-emulsion polymerization," *J. Polym. Sci. Part Polym. Chem.*, vol. 54, no. 20, pp. 3357–3369, Oct. 2016.
[16] F. Artusio *et al.*, "Polymeric nanocapsules via interfacial cationic photopolymerization in miniemulsion," *Polymer*, vol. 139, pp. 155–162, Mar. 2018.
[17] F. Cesca *et al.*, "Fabrication of biocompatible free-standing nanopatterned films for primary neuronal cultures," *RSC Adv*, vol. 4, no. 86, pp. 45696–45702, 2014.
[18] T. Limongi *et al.*, "Laboratory injection molder for the fabrication of polymeric porous poly-epsilon-caprolactone scaffolds for preliminary mesenchymal stem cells tissue engineering applications," *Microelectron. Eng.*, vol. 175, pp. 12–16, May 2017.
[19] V. R. Sinha, K. Bansal, R. Kaushik, R. Kumria, and A. Trehan, "Poly-ϵ-caprolactone microspheres and nanospheres: an overview," *Int. J. Pharm.*, vol. 278, no. 1, pp. 1–23, Jun. 2004.
[20] K. Madhaiyan, R. Sridhar, S. Sundarrajan, J. R. Venugopal, and S. Ramakrishna, "Vitamin B12 loaded polycaprolactone nanofibers: A novel transdermal route for the water soluble energy supplement delivery," *Int. J. Pharm.*, vol. 444, no. 1–2, pp. 70–76, Feb. 2013.
[21] C. Alonso, M. Martí, C. Barba, M. Lis, L. Rubio, and L. Coderch, "Skin penetration and antioxidant effect of cosmeto-textiles with gallic acid," *J. Photochem. Photobiol. B*, vol. 156, pp. 50–55, Mar. 2016.
[22] A. Lamprecht, "Nanomedicines in gastroenterology and hepatology," *Nat. Rev. Gastroenterol. Hepatol.*, vol. 12, no. 4, pp. 195–204, Mar. 2015.
[23] B. K. Johnson and R. K. Prud'homme, "Flash nanoprecipitation of organic actives and block copolymers using a confined impinging jets mixer," *Aust. J. Chem.*, vol. 56, no. 10, p. 1021, 2003.
[24] C. J. Martínez Rivas *et al.*, "Nanoprecipitation process: From encapsulation to drug delivery," *Int. J. Pharm.*, vol. 532, no. 1, pp. 66–81, Oct. 2017.
[25] A. A. Barresi, M. Vanni, D. Fissore, and T. Zelenková, "Synthesis and preservation of polymer nanoparticles for pharmaceutical applications," in *Handbook of Polymers for Pharmaceutical Technologies*, V. K. Thakur and M. K. Thakur, Eds. Hoboken, NJ, USA: John Wiley & Sons, Inc., 2015, pp. 229–280.
[26] F. Lince, D. L. Marchisio, and A. A. Barresi, "Strategies to control the particle size distribution of poly-ε-caprolactone nanoparticles for pharmaceutical applications," *J. Colloid Interface Sci.*, vol. 322, no. 2, pp. 505–515, Jun. 2008.
[27] D. Massella, E. Celasco, F. Salaün, A. Ferri, and A. Barresi, "Overcoming the limits of flash nanoprecipitation: effective loading of hydrophilic drug into polymeric nanoparticles with controlled structure," *Polymers*, vol. 10, Issue 10, article no 1092, 19 pp, Oct. 2018.
[28] M. Dias, "Topical delivery of caffeine from some commercial formulations," *Int. J. Pharm.*, vol. 182, no. 1, pp. 41–47, May 1999.
[29] G. Ginsberg, D. Hattis, A. Russ, and B. Sonawane, "Physiologically Based Pharmacokinetic (PBPK) modeling of caffeine and theophylline in neonates and adults: implications for assessing children's risks from environmental agents," *J. Toxicol. Environ. Health A*, vol. 67, no. 4, pp. 297–329, Feb. 2004.
[30] C. Puglia *et al.*, "Design of solid lipid nanoparticles for caffeine topical administration," *Drug Deliv.*, vol. 23, no. 1, pp. 36–40, Jan. 2016.
[31] A. D. Lavino, N. Di Pasquale, P. Carbone, and D. L. Marchisio, "A novel multiscale model for the simulation of polymer flash nano-precipitation," *Chem. Eng. Sci.*, vol. 171, pp. 485–494, Nov. 2017.
[32] L. A. Pham-Huy, H. He, and C. Pham-Huy, "Free radicals, antioxidants in disease and health," *Int. J. Biomed. Sci. IJBS*, vol. 4, no. 2, pp. 89–96, Jun. 2008.
[33] B. Lipinski, "Hydroxyl radical and its scavengers in health and disease," *Oxid. Med. Cell. Longev.*, vol. 2011, pp. 1–9, 2011.
[34] Y. Qiu *et al.*, "Antioxidant chemistry of graphene-based materials and its role in oxidation protection technology," *Nanoscale*, vol. 6, no. 20, pp. 11744–11755, 2014.

Carbon Containing Material in Non-Woven Fabrics for Thermoinsulation of Winter Apparel

V.T. Lipik[1], G. Yang[2]

[1]*Nanyang Technological University, School of Material Science and Engineering, 50 Nanyang Avenue, 639798, Singapore*
[2]*Nanyang Technological University, Institute for Sports Research, 50 Nanyang Avenue, 639798, Singapore*

Corresponding author: vitali@ntu.edu.sg

Abstract Nowadays different types of carbon containing fibers become more and more popular in textile. Pure carbon fibers, fibers with carbon particles, graphene and even carbon containing dyes began to apply in textile for different purposes. Fibers may contain from about 1% of black carbon for antistatic property up to 40% of graphene for special mechanical and moisture management properties. Carbon has pro and contra as thermoinsulating material. It possesses great ability to absorb heat but, at the same time, carbon is good thermoconductive material, and heat escapes quickly due to high thermal conductivity. Heat loss from human body is realized through several channels. For example, thermoconductive heat loss is only 10-15%, whereas irradiative heat loss can reach up to 80% from total heat loss value. Based on this information, we developed non-woven carbon containing compositions which is targeted for maximizing capture of outgoing irradiation heat and minimizing heat loss in thermoconductive channel. In this work, we presented result of development of different carbon containing non-woven compositions for insulation. As a result, some of our non-woven paddings with density 100 g/m^2 have clothing insulation value (CLO) equal 4, which makes them close to natural down insulation. We also provided wide characterizations of non-woven samples such as stiffness, IR absorption and reflection, washing stability, analysis of economic feasibility of industrial production.

***Keywords*—Thermoinsulation, Non-Woven, Carbon Fibers, Winter Apparel**

I. INTRODUCTION

Sales of winter apparel in countries with cold climate may take up to 35% of total sport apparel, and up to 60% of total casual apparel sales. And income level of buyers does not play a significant role in these numbers because winter apparel is a necessity [1, 2]. Shell fabric is the most expensive component in winter apparel. Thermoinsulation padding normally takes second place in price constituent of winter apparel at mass market. But thermoinsulation can be the most expensive part of special apparel, for example if winter jacket is designed for specific types of winter sport [3].

10-20 years ago, there were no specific requirements to synthetic thermoinsulations paddings. Nowadays there is sharp jump of interest to special textile insulation materials with series of functions and new properties. The most valuable and required properties of non-woven insulation is the decrease of stiffness and thickness, keeping at the same time high value of thermoinsulation. Special materials began to be used to achieve such requirements. The application of polyolefins and acrylic fibers is growing; carbon containing fibers start to be applied more and more widely; fibers with special additives like aerogel are manufactured already [4-5].

Heat loss of a person is realized by five main channels, three of which - radiation, convection and heat conductivity should be taken into considered at the development of non-woven paddings for thermoinsulation. The distribution of heat losses in the three above-mentioned channels depends on many factors, and primarily, the intensity of physical activity plays decisive role in this distribution. Approximately, in a state of rest or insignificant activity, the heat loss is distributed as follows: by irradiation channel - 50-60%; by convection - 20-30%; by thermoconductivity - 10-20%. But during high physical activity heat loss by irradiation can grow up to 80% [6]. The use of carbon-containing materials makes possible effective blocking of irradiation heat loss, because carbon perfectly absorbs waves in a wide range of spectrum. In this work, we presented the results of researches on development of non-woven thermoinsulation with the application of carbon-containing materials, directed for the improvement of heat-insulating capability.

II. METHODS AND MATERIALS

Carbon fibers made based on polyacrylonitrile (PAN) with 10 um in diameter and 50 mm in cut length were supplied by Haining Anjie Composite Material Co. LTD (China). Arselon fabric, carbon fabric and carbon fibers made based on viscose with diameter of 10 um in form of wisp were purchased from OJSC "SvetlogorskKhimvolokno" (Belarus). The rest of carbon containing fibers were supplied from Ziran Non-Vowen Co. Ltd (China). Polyester fibers with 0.8 denier and 38 mm length, bicomponent fibers PE/PP (PE- shell, PP –core) with 2 denier and 51 mm length, and polypropylene fibers with 6 denier and 51 mm length were purchased from

OJSC Mogilevkhimvolokno (Belarus). Textile fabrics were purchased from Gim Joo Textile Fabric Pte Ltd (Singapore).

Measurement of thermoinsulation property/clothing insulation or CLO has been done using Thermetrics Sweating Guarded Hotplate SGHP-8.2 according to ASTM D1518-14. Absorption and reflection measurement have been done using PerkinElmer UV/Vis spectrometer LAMBDA 950. The structure of carbon fibers was studied via Field Emission Scanning Electron Microscopy (FE-SEM, jsm-6340F).

Heat retention test has been done at room temperature by the following way. Sample of non-woven was placed under 30 W lamp with 10 cm distance. Lamp was on during first 20 minutes of observation, and off during second 20 minutes of observation. Temperature of the surface of non-woven was detected by IR camera Optris PI640. The stiffness of non-woven samples was measured by means of digital Pneumatic stiffness tester (GT-C70B/GESTER) according to ASTM D4032-94.

Samples of non-woven fabric were prepared using laboratory Carding Machine (Mesdan). The weight density for each nonwoven padding was maintained at 100±5 g/m^2. After that, 3 pieces of 30 × 30cm samples were cut from the large rectangular piece collected by the roller drum, then followed by glue spraying. If bicomponent fibers are applied, thermobonding of samples has been realized at 140°C during 10 min in fabric laboratory drier (Mesdan).

III. RESULTS

Carbon containing materials possess different abilities to absorb and reflect heat in different parts of spectrum. We have chosen a few types of carbon-containing fibers for the preparation of samples of non-woven fabrics. Our target was to identify materials which absorb the heat by the most effective way. The ability to absorb and reflect waves in visible and near IR parts of spectrum by different carbon containing fibers is presented on Fig. 1.

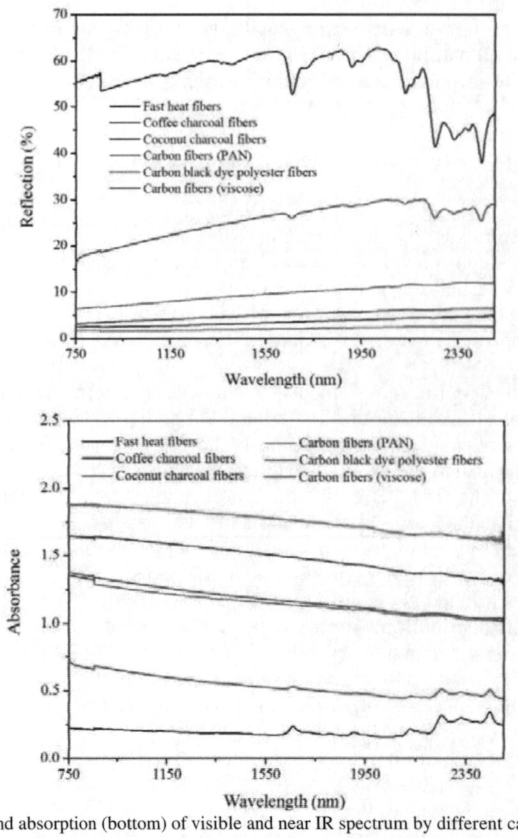

Fig. 1. Reflection (top) and absorption (bottom) of visible and near IR spectrum by different carbon-containing materials.

Considering the data presented in Fig. 1, we can see separation of carbon containing material into two groups: absorbing and reflecting heat. Obviously, general carbon fibers have good capability to absorb heat, which is very beneficial for usage in thermoinsulation. Based on our measurements, any traditional material used in textiles cannot compete with carbon fibers in heat absorption (Fig. 2).

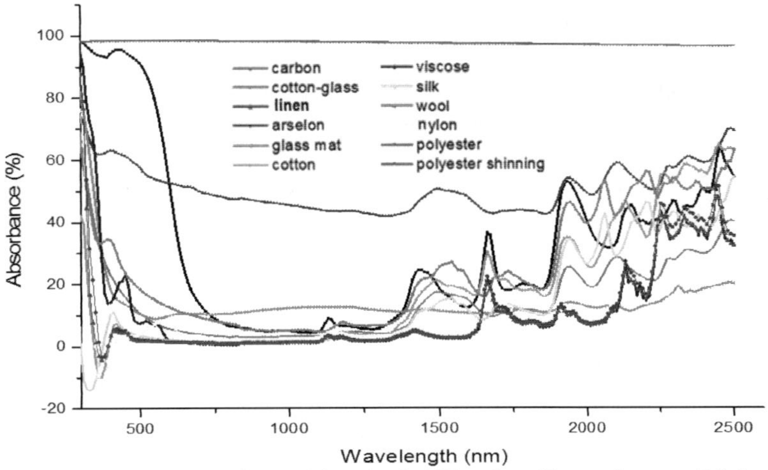

Fig. 2. Capability of different textile materials to absorb visible and near IR parts of spectrum (A below 0)

However, unlike widely used textile materials, carbon fibers have very high thermal conductivity [7-8]. Therefore, absorbed heat trapped by carbon fibers does not stay long time in the material and runs away, i.e. carbon layer quickly heats up but it cools down quickly as well. Fabric made of pure carbon fiber cedes in thermal insulation compared to traditional textile materials (Table I).

TABLE I
CLO values for different woven fabrics

Type of material	Density, g/m^2	CLO
Nylon	120	0.82
Polyester	130	0.81
Carbon fabric	250	0.88
Linen	190	0.92
Wool	180	0.88
Viscose	115	0.83
Silk	90	0.88

Regardless on different densities of fabrics presented in Table I, it is seen that carbon fabric should be at least twice heavier to be comparable in thermoinsulation with other traditional textile materials. Evidently, the application of carbon fibers and fabric in pure version is not rational. In our study we used carbon fibers manufactures on the base of polyacrylonitrile. Diameter of carbon fibers was about 10 μm (Fig. 3).

Fig. 3. Carbon fibers under SEM microscope

Carbon fibers should be combined with other materials which should be able to decrease the heat loss realized by thermal conductivity channel. To find out the optimum amount of carbon fibers in non-woven, we made first series of non-woven composites by thermobonding. Composition included two components: bicomponent fibers PE/PP with 2 denier 51 mm and carbon fibers. All non-woven samples had the same density – 100 g/m^2. Results of CLO measurement and stiffness of obtained samples are presented in Table II.

Table II
CLO values and stiffness of two components non-woven compositions: PE/PP bicomponent fibers with different amount of carbon fibers

Amount of carbon fibers, % wt	0	2.5	5	10	15	20	40	60
CLO	1.89	2.38	2.48	2.59	2.55	2.29	2.08	1.94
Stiffness, N	12.2	12.0	8.5	6.9	5.2	3.8	3.2	2.9

Based on the data presented in Table II, it is logically seen, that there is no reason to add more than 10% of carbon fibers into non-woven structure. If the quantity of carbon fibers is higher than 10%, the fabric loses the capability to keep the heat.

We prepared one non-woven composite with carbon fibers adding third component – glass fibers which possess good capability to reflect heat [9]. The ratio of bicomponent, glass and carbon fibers was 1:2:1. Measured CLO of this composite was equal to 3.0. It gave us an idea to develop composite non-woven material each component of which would capture the heat escaping by certain heat loss channel: irradiation, convection or thermal conductivity. As a result, we prepared non-woven sample using three following materials. Polyester microfibers with 0.8 denier and 38 mm length were used to decrease heat loss by convection. Polypropylene fibers with 6 denier and 51 mm length were taken to reduce heat loss by thermal conductivity, because polyolefines possess highest thermal capacity among all common polymers: thermal capacity of polypropylene exceeds one of polyester by more than 40% (Cp(PET) = 0.4393 kJ/kg*K, Cp(PP)=0.6238 kJ/kg*K) [10]. Carbon fibers were added to catch the heat going out by irradiation.

The ratio of components was defined based on our experience and existing rules of mixing thin and thick fibers in non-woven paddings as 75:25 to obtain high insulation property [11]. Weight ratio among microfibers, polypropylene and carbon fibers in our sample was taken as 70:20:10. CLO value for this composite with weight density 100 g/m^2 was equal to 4.0, which is very high for synthetic non-woven insulation.

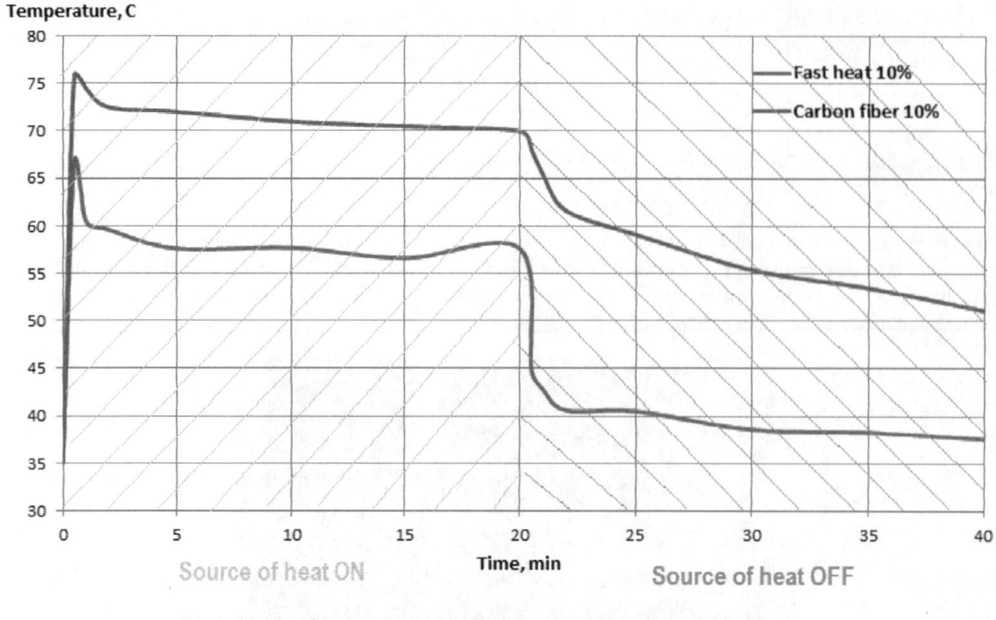

Fig. 4. Result of heat retention testing for non-woven with carbon and fast heat fibers.

Developed composite revealed another interesting property. It was able to absorb heat with fast increase of temperature. This property is valuable at the market of non-woven insulation. To compare our development with existing industrial analogue, we prepared another non-woven composite replacing carbon fibers (10% wt) into fast heat fibers obtained from Ziran Non-Woven Co. Ltd and leaving 90% of microfibers. This composition did not contain polypropylene fibers. Both non-woven composites were checked for the capability to retain heat at the irradiation by lamp with power of 30 W from the distance of 10 cm. First 20 minutes lamp was on and surface of non-woven was heated up; second 20 minutes lamp was off, and the surface was cooled down. Temperature of non-woven surface was detected by infrared camera. Results of experiments are presented in Fig. 4.

This test has shown the fact that carbon fibers provided the ability to heat up to higher temperatures compared with commercially available additives. Second valuable conclusion is about importance of addition of polypropylene fibers which played a role of heat retention component, providing lower cooling speed of non-woven composite.
This non-woven fabric was prototyped at the industrial level at the factory Ziran Non-Woven Co. Ltd (China). Samples obtained at industrial equipment have less regular than ones obtained at laboratory conditions by means of carding machine. Therefore, measured thermal insulation (CLO) of industrial samples is all time lower compared with one for non-woven samples prepared in laboratory conditions. But our non-woven composition prototyped at industrial level has shown CLO value equal to 3.03, which is extremely good for industrial paddings. Practically, based on our own experience, all industrial non-woven paddings with density 100 g/m^2 have CLO values below 2.3. Threshold value of stiffness for non-woven paddings with weight 100 g/m^2 after which material is losing the attractiveness is 3.5-3.8 N. Stiffness of our industrial prototype was 3.2 N. Thus, it fits to request of designers and technologist. Industrial sample was tested on washing stability according to the standard GB/T 8629-2001, receiving grade 4, which is acceptable. Economical evaluation of production of our developed non-woven padding was positive. Considering market price of carbon fibers made based on PAN as 15 USD/kg, the manufacturing of such non-woven padding for apparel is justified. The price of manufacturing, based on the factory evaluation, will be about 0.6-0.62 USD per running meter.

IV. CONCLUSION

In this paper, we presented the result of the development of nonwoven thermal insulation with the addition of carbon fibers. Carbon fibers, added to non-woven composite in the amount of 10% by weight allowed the effective capture of irradiation heat loss from the human body. Since this type of heat loss is dominant, its reduction is strongly beneficial in the development of winter apparel.

At the development of non-woven thermoinsulation, polypropylene fibers were added to level heat loss specified by high thermal conductivity of carbon fibers. Non-woven sample containing 10% by weight of carbon fibers showed CLO value of 3.03. The sample passed well washing stability test. The stiffness of obtained non-woven sample was 3.2 N, which allows for usage in winter clothing without restrictions. Economic evaluation showed the efficiency of production of this non-woven thermoinsulation.

REFERENCES

[1] *Textile in sport,* Editor R. Shishoo, 1st Edition, Woodhead Publishing Limited, 2010, 364 p.
[2] *Textile for sportswear, Editor R. Shishoo, 1st Edidtion. Woodhead Publishing, May 2015, 290 p.*
[3] J. H. Hammond , A. Raman, *Sport Obermeyer Ltd,* Review of Harvard Business School, 9-695-022, 15 Aug. 2006, 19 p.
[4] *The Global market for wearable and smart textile to 2017.* Part 13. Graphene and related two dimensional materials. Reference number ISO/TS 80004-13:2017(E).
[5] G. Song, *Thermal insulating properties of textile and clothing,* in: Editor J. T. Williams, Textiles for cold weather apparel, Woodhead Publishing Lim., New York, 2009, pp. 19-33.
[6] V. T. Bartel, *Physiological comfort of sportswear,* in: Editor R. Shishoo, Textiles in sports, Woodhead Publishing Lim., New York, 2005, pp. 177-202.
[7] T. M. Tritt, *Thermal conductivity: Theory, properties and Application.* Springler Science & Business Media, 2005.
[8] *Tables of physical constants: Directory.* Editor I. K. Kikoin, Moscow, Atomisdat, 1976.
[9] F. T. Wallenberger, J.C. Watson, H. Li, *Glass fibers.* ASM Handbook. Vol 21: Composites, ASM International, 2001.
[10] J. Wen. *Thermal capacity of polymers.* Chapter 9. P.145-154. In Physical properties of Polymers: Handbook. Second Edition, Springer. Editor J. E. Mark, 2006, 1078 p.
[11] *Handbook of non-wovens,* Editor S. Russel, Woodhead Publishing Lim., 1st Eddition, Elsevier, New York, Dec. 2006, 544 p.

Structural Health Monitoring of Composites with Newly Developed Textile Sensors *In Situ*

I. Jerkovic[1], A. M. Grancaric[1], and V. Koncar[2]

[1]*University of Zagreb, Faculty of Textile Technology, Department of Textile Chemistry and Ecology, Prilaz baruna Filipovica 28a, Zagreb, Croatia*
[2]*Ecole Nationale Supérieure des Arts et Industrie Textiles, Gemtex laboratory, 2 allée Louise et Victor Champier, Roubaix, France*

Corresponding author: amgranca@ttf.hr

Abstract Smart textile structures are promising solution nowadays for *in situ* structural health monitoring of composite parts that have important place in transportation industry. Composites made from a polymeric matrix and a fibrous reinforcement have been increasingly studied during the last decade. Conductive yarns as textile sensors can be integrated in textile structures by diverse technologies and crucial issue is that sensors integration does not modify their general behavior. In this work interest is focused on the structural health monitoring of textile reinforced thermoplastic composites with newly developed textile sensors *in situ* based on the conductive polymer complex poly(3,4-ethylenedioxythiophene)-poly(styrenesulfonate) (PEDOT:PSS).

***Keywords*—Smart Textile, Structural Health Monitoring, Textile Reinforced Composites, Textile Sensors**

I. Introduction

SMART textile structures are promising solution nowadays for *in situ* structural health monitoring of composite parts that have important place in transportation industry [1]. Composites made from a polymeric matrix and a fibrous reinforcement have been increasingly studied during the last decade [1-3]. In composite applications, the low material density is of environmental interest because fuel consumption and CO_2 emissions are directly related to vehicle weight [4], [5]. Conductive yarns as textile sensors can be integrated in textile structures by diverse technologies [6]. Textile sensors inside the textile reinforcements have to present all the characteristics of traditional textile materials: flexibility, lightweight and capability of adopting the geometry of the reinforcement. Therefore, it is important that sensors integration does not modify their general behavior [7], [8]. Textile materials are very flexible and easily deformable in all directions, and sensors used should be able to support, often all at the same time, tensile, shear, bending and even compression deformations [7], [9], [10]. Researches on electro-conductive textiles and related up graded composites prepared from conductive polymers such as Polypyrrole (PPy), Polythiophene (PT), Polyaniline (PANI) and Poly(3,4-ethylenedioxythiophene) (PEDOT) are developing quickly in the recent years due to their optimal compromise among electrical conductivity, stability, processability and high potential applications in different fields [11-13].

In this work interest is focused on the structural health monitoring of textile reinforced thermoplastic composites with newly developed textile sensors *in situ* based on the conductive polymer complex poly(3,4-ethylenedioxy thiophene)-poly(styrenesulfonate) (PEDOT:PSS).

II. Methods and Procedures

E-glass/polypropylene (GF/PP) and E-glass (GF) yarns by PD Fiberglass group (Glasseiden GmbH, Oschatz, Germany) were used for textile sensors production. Fineness of GF/PP commingled yarn is 842 tex (with GF/PP mass content of 71%:29%), while fineness of GF yarn is 830 tex. Conductive coating for yarn treatment was defined according to the study of percolation threshold of the polymer complex PEDOT:PSS. The percolation threshold determination was based on conductive dry films' electrical properties analysis to develop highly sensitive sensors [1]. A novel piece of laboratory equipment based on a conceptual design was taken to ensure effective and equally distributed coating of electroconductive polymer without destruction of textile properties [1]. Aluminum roll to roll device and plexiglass bath were used for textile sensors production needed for *in situ* structural health monitoring of composites during tensile loading [14]. GF/PP and GF textile sensors (Fig. 1) were integrated during weaving 2D fabric, 4-end satin (warp density, 4 ends/cm and weft density, 6 ends/cm), thickness ~2.660 x 10^{-3} m, in weft direction using computer controlled ARM hand weaving loom (Biglen, Switzerland).

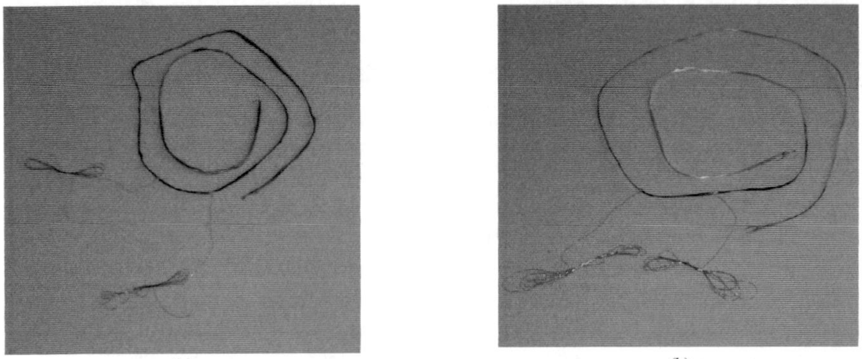

(a) (b)

Fig. 1. GF/PP textile sensor (a) and GF textile sensor (b).

Three-layered textile preforms with integrated textile sensors (Fig. 2) were consolidated at the Dolouets heating press (Soustons, France) under the pressure of 4-5 MPa and the temperature of 185 °C during 5 minutes followed by cooling at 100 °C in 2-3 minutes. The dimensions of developed composites were 21 cm x 8 cm x 0.15 cm.

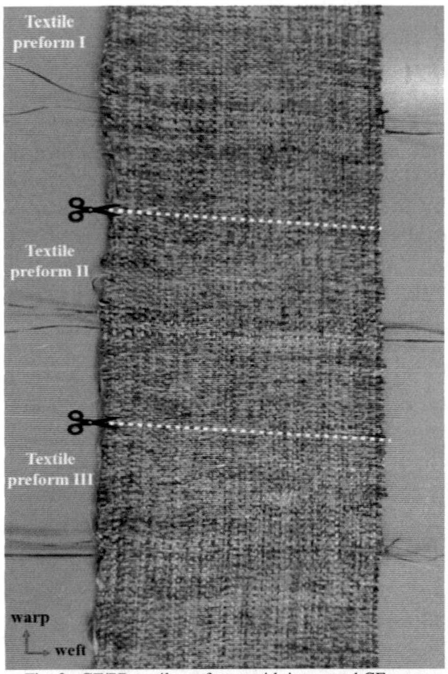

Fig. 2. GF/PP textile preforms with integrated GF sensors.

The tensile measurements of composites were performed on the Instron 5900 (Norwood, USA) testing machine with a speed of 2 mm/min and 115 ±1 mm difference between clamps. Electrical resistance measurements were carried out simultaneously. Keithley® KUSB-3100 data acquisition digital I/O counter/timer and two resistance boxes were connected to a computer (QuickDAQ software) to record the electrical resistances of two textile sensors integrated in each composite specimen during tensile loading *in situ*.

III. Results

A. Electrical Resistance of Textile Sensors After Production and Insertion in 2D Fabrics

Electrical resistance of textile sensors (Table 1 and 2) after production is in range between 350 Ω - 500 Ω for GF/PP textile sensors while 330 Ω - 880 Ω for GF textile sensors. Electrical resistance of textile sensors after insertion in 2D fabrics is slightly higher compared to results after their production. It is in range between 390 Ω - 550 Ω for GF/PP textile sensors while 380 Ω - 1000 Ω for GF textile sensors.

TABLE I
ELECTRICAL RESISTANCE OF GF/PP TEXTILE SENSORS AFTER THEIR PRODUCTION AND INSERTION IN 2D GF/PP FABRIC

Textile sensor	Electrical resistance of textile sensor after production (Ω)	Electrical resistance of textile sensor after insertion in 2D fabric (Ω)
GF/PP-Sy08sp03-1	490	550
GF/PP-Sy08sp03-2	500	550
GF/PP-Sy08sp03-3	370	410
GF/PP-Sy08sp03-4	350	390
GF/PP-Sy08sp03-5	440	510
GF/PP-Sy08sp03-6	390	450
Mean	423	477
Standard deviation	63	70

TABLE II
ELECTRICAL RESISTANCE OF GF TEXTILE SENSORS AFTER THEIR PRODUCTION AND INSERTION IN 2D GF/PP FABRIC

Textile sensor	Electrical resistance of textile sensor After production (Ω)	Electrical resistance of textile sensor after insertion in 2D fabric (Ω)
GF-Sy08sp03-1	790	920
GF-Sy08sp03-2	730	870
GF-Sy08sp03-3	550	700
GF-Sy08sp03-4	580	740
GF-Sy08sp03-5	330	380
GF-Sy08sp03-6	880	1000
Mean	643	768
Standard deviation	198	221

B. Electrical Resistance of Textile Sensors in Consolidated GF/PP Textile Preforms – Textile Reinforced Composites

Electrical resistance of GF/PP textile sensors after consolidation of GF/PP textile preforms prepared from related 2D fabric and right before electromechanical tests is in range between 16 kΩ - 240 kΩ (Table 3). Higher difference in electrical resistance values between two GF/PP sensors was observed in the first GF/PP composites. Low electrical resistances of GF/PP textile sensors were detected for the second GF/PP composite.

TABLE III
ELECTRICAL RESISTANCE OF GF/PP TEXTILE SENSORS AFTER CONSOLIDATION OF 2D TEXTILE PREFORMS AND BEFORE ELECTROMECHANICAL TESTS OF COMPOSITES

Textile sensor	Electrical resistance of textile sensor after consolidation of 2D textile preform in composite (kΩ)	Electrical resistance of textile sensor before electromechanical test of composite (kΩ)	Mean (kΩ)	Standard deviation (kΩ)
GF/PP-Sy08sp03-1	30	26	28	2
GF/PP-Sy08sp03-2	240	190	215	35
GF/PP-Sy08sp03-3	19	16	17	2
GF/PP-Sy08sp03-4	20	18	19	1
GF/PP-Sy08sp03-5	130	117	124	9
GF/PP-Sy08sp03-6	87	92	90	4

Electrical resistance of GF textile sensors after consolidation of GF/PP textile preforms prepared from related 2D fabric and right before electromechanical tests is in range between 3 kΩ - 67 kΩ (Table 4). Higher difference in electrical resistance values between two GF sensors integrated was observed in the third GF/PP composite.

TABLE IV
ELECTRICAL RESISTANCE OF GF TEXTILE SENSORS AFTER CONSOLIDATION OF 2D TEXTILE PREFORMS AND BEFORE ELECTROMECHANICAL TESTS OF COMPOSITES

Textile sensor	Electrical resistance of textile sensor after consolidation of 2D textile preform in composite (kΩ)	Electrical resistance of textile sensor before electromechanical test of composite (kΩ)	Mean (kΩ)	Standard deviation (kΩ)
GF-Sy08sp03-1	19	14	17	3
GF-Sy08sp03-2	59	18	38	29
GF-Sy08sp03-3	48	38	43	7
GF-Sy08sp03-4	27	22	24	4
GF-Sy08sp03-5	3	4	3	0.2
GF-Sy08sp03-6	67	46	57	15

C. Electromechanical Results of Textile Reinforced Composites with Integrated Textile Sensors

According to electromechanical results, gauge factor for all composites was determined from a polynomial equation of degree 2. The first GF/PP composite (Fig. 3) showed similar electrical resistance-elongation curves of GF/PP textile sensors. For the second GF/PP composite, curves of electrical resistance variation of GF/PP textile sensors versus elongation showed several interruptions in the second part due to beginning of the delamination and cracks inside the composite. Breakage of one GF/PP sensor was occurred during *in situ* structural health monitoring of the third GF/PP composite.

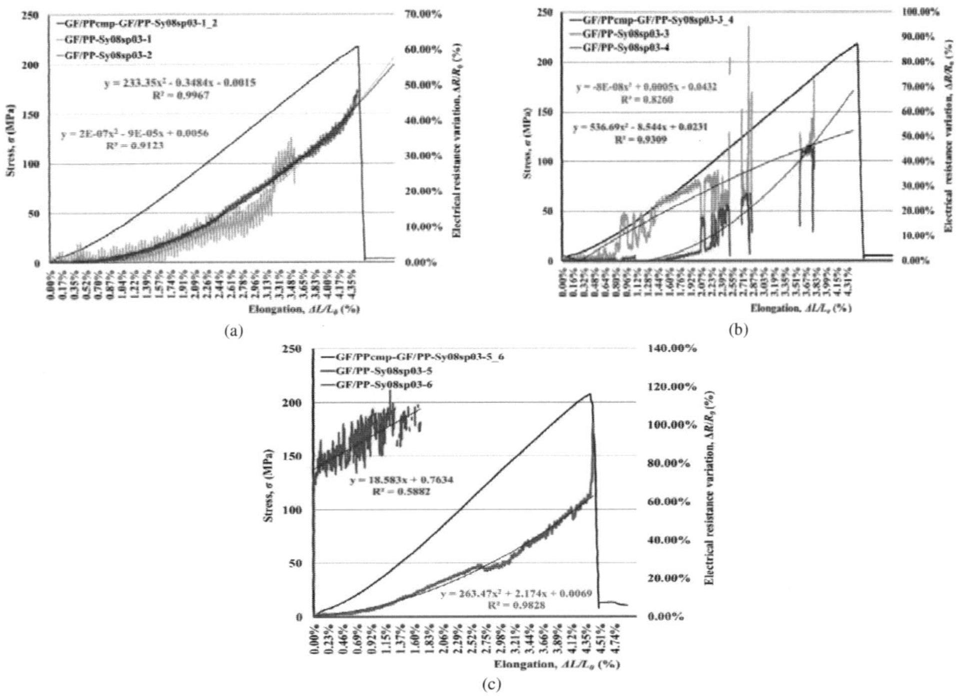

Fig. 3. Results of electromechanical measurements of GF/PP composites with integrated GF/PP textile sensors - stress of composite (σ) and electrical resistance variation of textile sensors ($\Delta R/R_0$) versus elongation ($\Delta L/L_0$): (a) GF/PPcmp-GF/PP-Sy08sp03-1_2, (b) GF/PPcmp-GF/PP-Sy08sp03-3_4, (c) GF/PPcmp-GF/PP-Sy08sp03-5_6

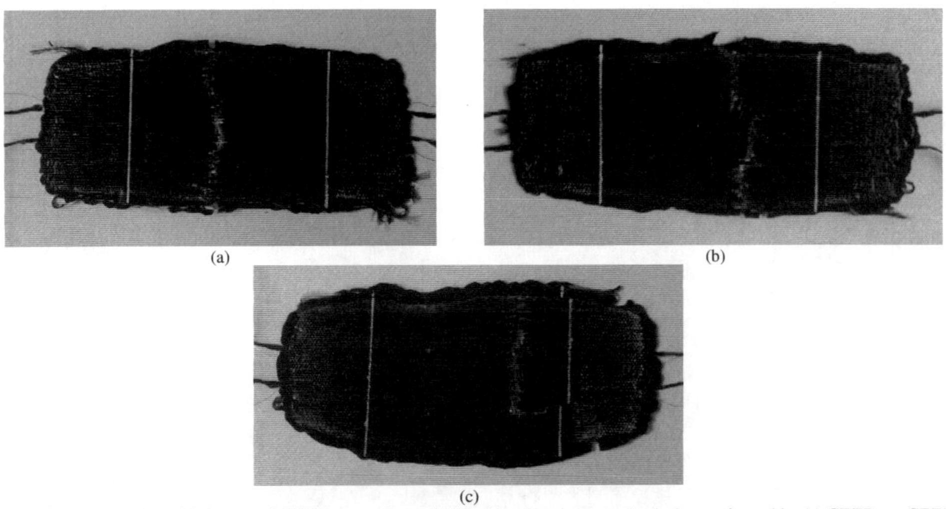

Fig. 4. GF/PP composites with integrated GF/PP sensors in weft direction after electromechanical test - front side: (a) GF/PPcmp-GF/PP-Sy08sp03-1_2, (b) GF/PPcmp-GF/PP-Sy08sp03-3_4, (c) GF/PPcmp-GF/PP-Sy08sp03-5_6

Maximum peaks of GF/PP textile sensors curves appeared before the maximum stresses achieved for the composite specimens. Stress at break of GF/PP composites with integrated GF/PP textile sensors was in range between 207 MPa and 217 MPa while elongation at break between 4.40 % and 4.45 %. Composite breakage "in a line" (Fig. 4) can be seen after electromechanical test between clamps for the first GF/PP composite with integrated GF/PP textile sensors what is not in case for the other two composite specimens.

The first GF/PP composite with integrated GF textile sensors (Fig. 5) showed also similar electrical resistance-elongation curves, what is not in case for the other two composites.

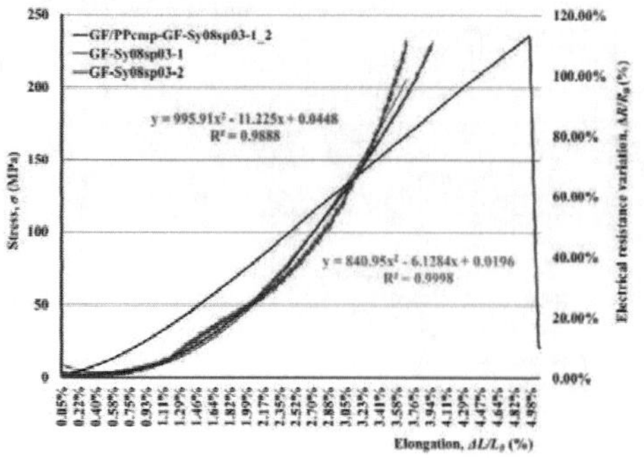

(a)

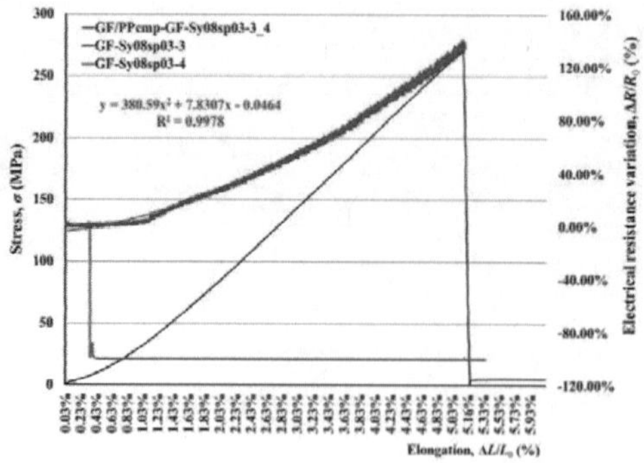

(b)

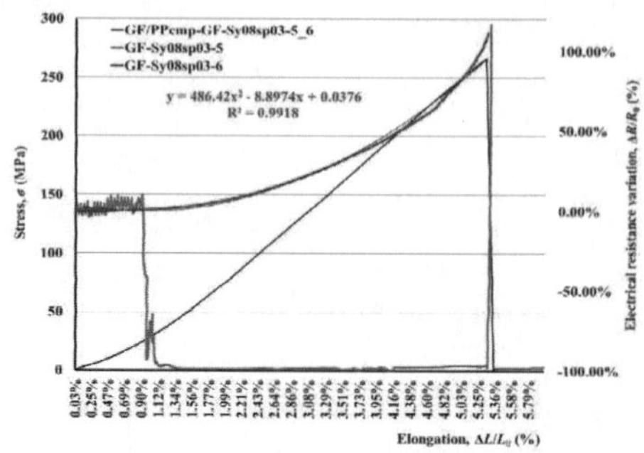

(c)

Fig. 5. Results of electromechanical measurement of GF/PP composites with integrated GF textile sensors - stress of composite (σ) and electrical resistance variation of textile sensors ($\Delta R/R_0$) versus elongation ($\Delta L/L_0$): (a) GF/PPcmp-GF-Sy08sp03-1_2, (b) GF/PPcmp-GF-Sy08sp03-3_4, (c) GF/PPcmp-GF-Sy08sp03-5_6

(a)

(b)

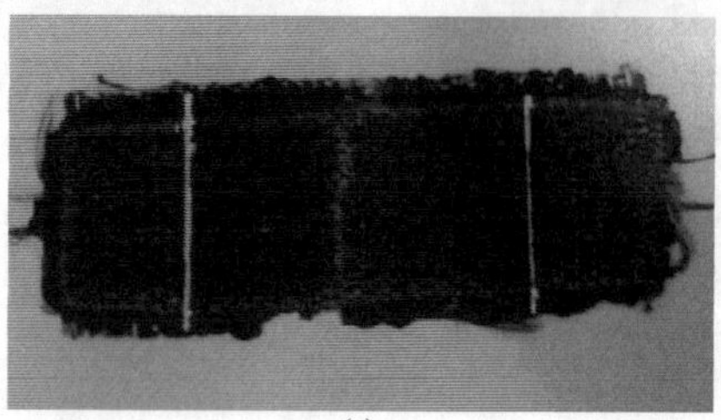

(c)

Fig. 6. GF/PP composites with integrated GF sensors in weft direction after electromechanical test - front side: (a) GF/PPcmp-GF-Sy08sp03-1_2, (b) GF/PPcmp-GF-Sy08sp03-3_4, (c) GF/PPcmp-GF-Sy08sp03-5_6

For the second GF/PP composite, breakage of one GF sensor was occurred during *in situ* structural health monitoring. In case of the third GF/PP composite, due to interruptions occurred inside the composite at the beginning of electromechanical test, breakage of one GF textile sensor was detected as well.

Maximum peaks of GF textile sensors curves appeared before or at the maximum stresses achieved for the GF/PP composites. Stress at break of GF/PP composites with integrated GF textile sensors was in range between 235 MPa and 269 MPa while elongation at break between 4.97 % and 5.34 %. GF/PP composites with integrated GF sensors were not displayed breakage "in a line" between clamps after electromechanical tests (only partially the last specimen) (Fig. 6).

IV. Conclusion

Textile sensors showed resistance to high temperature and pressure as given electrical resistance values after consolidation of textile preforms as a part of these structures what is important in order to obtain *in situ* structural health monitoring of composites during tensile loading. Textile sensors showed also possibility to follow the tensile loading and to detect damages in the composites *in situ*. Earlier fractures inside the structures were noticed for some composites. Several interruptions due to beginning of the delamination and cracks inside were detected as well.

Acknowledgment

The paper is a part of the EU project "MAPICC 3D" results within the call NMP-FP7- 2010-3.4-1, numbered with 263159 entitled: *One-shot Manufacturing on large scale of 3D up graded panels and stiffeners for lightweight thermoplastic textile composite structures*. The authors would like to thank the European Commission for funding of the project.

References

[1] I. Jerkovic, V. Koncar, A. M. Grancaric, "New textile sensors for in situ structural health monitoring of textile reinforced thermoplastic composites based on the conductive poly (3, 4ethylenedioxythiophe ne)-poly(styrene sulfonate) polymer complex," *Sensors*, vol. 17, no. 10, pp. 1-2297-21, 2017.

[2] N. Encinas, M. Lavat-Gil, R. G. Dillingham, J. Abenojara, M. A. Martíneza, "Cold plasma effect on short glass fibre reinforced composites adhesion properties," *Int. J. Adhes. Adhes.*, vol. 48, pp. 85–91, 2014.

[3] B. S. Kandola, L. A. T. H. A. Krishnan, "Fire performance evaluation of different resins for potential application in fire resistant structural marine composites," in *Proc. 11th International Symposium Fire Safety Science* 2014, Christchurch, New Zealand, 2014, pp. 769–780.

[4] G. Marosi, K. Bocz, B. Szolnoki, B. Bodzay, K. Madi, G. Harakály, H. Erdélyi, E. Zimonyi, "Fire retardancy of biofibre-glassfibre and self-reinforced polymer composites," in *Proc. 15th European Conference on Composite Materials*, Venice, Italy, 2012, pp. 1–5.

[5] D. Teixeira, M. Giovanela, L. B. Gonella, J. S. Crespo, "Influence of flow restriction on the microstructure and mechanical properties of long glass fiber-reinforced polyamide 6.6 composites for automotive applications," *Mater. Des.*, vol. 47, pp. 287–294, 2013.

[6] A. M. Grancaric, I. Jerkovic, V. Koncar, C. Cochrane, F. M. Kelly, D. Soulat, X. Legrand, "Conductive polymers for smart textile applications," *First published online J Ind Text*, pp. 1–31, 2017.

[7] A. M. Grancaric, I. Jerkovic, C. Dufour, X. Legrand, V. Koncar, "Glass/polypropylene sensor yarns integration during weaving of 2D structure," in *Proc. 8th Central European Conf. on Fiber-grade Polymers*, Chemical Fibers and Special Textiles, Zagreb, Croatia, 2015, pp. 73–78.

[8] I. Cristian, S. Nauman, C. Cochrane, V. Koncar, "Electro-conductive sensors and heating elements based on conductive polymer composites in woven structures," in *Advances in Modern Woven Fabrics Technology*, Chapter 1, S. Vassiliadis, Ed. Rijeka: InTech, 2011, pp. 1–22.

[9] S. Vassiliadis, M. Rangoussi, D. Meimaris, K. Prekas, C. Provatidis, "Electrically conductive textile yarns and their contact behavior," in *Proc. 2nd ITMC International Conference* 2009, Casablanca, Morocco, 2009, pp. 129–133.

[10] C. Cochrane, V. Koncar, M. Lewandowski, C. Dufour, "Design and development of a flexible strain sensor for textile structures based on a conductive polymer composite," *Sensors*, vol. 7, no. 4, pp. 473-492, 2007.

[11] I. Jerkovic, C. Dufour, X. Legrand, X. Tao, F. Boussu, A. M. Grancaric, V. Koncar, "E-glass/polypropylene sensor yarns developed by roll to roll coating procedure," in *Proc. 5th ITMC International Conference* 2015, Casablanca & Marrakesh, Morocco, 2015, pp. 68–74.

[12] N. Trifigny, F. M. Kelly, C. Cohrane, F. Boussu, V. Koncar, D. Soulat, "New PEDOT: PSS – based piezo – resistive sensors applied to reinforcement glass fibres for in situ measurement during the composite material weaving process," *Sensors*, vol. 13, no. 8, pp. 10749 – 10764, 2013.

[13] M. Subhankar, A. Chatterjee, "Polypyrrole based electro – conductive cotton yarn," *J. Textile Sci. Eng.*, vol. 4, no. 6, pp. 1– 1000171–4, 2014.

[14] A. M. Grancaric, I. Jerkovic, M. Leskovac, C. Dufour, F. Boussu, V. Koncar, "Surface free energy of sensor yarns and textile reinforced thermoplastic composites," in *Proc. 16th World Textile Conference AUTEX* 2016, Ljubljana, Slovenia, 2016, pp. 6–11–1-6–11–6

IOP Conf. Series: Materials Science and Engineering **460** (2018) 012047 doi:10.1088/1757-899X/460/1/012047

Smart clothing to increase safety of people with dementia

S. Uran[1] and J. Geršak[2]

[1]*University of Maribor, Faculty of electrical engineering and computer science, Laboratory for cognitive systems in mechatronics, Koroška cesta 46, 2000 Maribor, Slovenia*
[2]*University of Maribor, Faculty of mechanical engineering, Research and Innovation Centre for Design and Clothing Science, Smetanova ulica 17, 2000 Maribor, Slovenia*

Jelka.Gersak@um.si

Abstract: Human population is aging. Therefore, considerable amount of attention has to be given to the problems specific to aging. Dementia is one of such problems. There is a strong relation between dementia and gait disorders. With aging the amount of people falling during standing, walking or climbing stairs dramatically increases. Because of this, the elderly people have a greater need for care and assistance and are more likely to be admitted to a nursing home after a fall, although they would like to stay at their own home and be independent. Although a number of researchers are dealing with this problematics so from the point of the medical diagnostics, as from view of possibility of integrating fall detection device into clothing, there are still unresolved problems on the area of providing safety of people with dementia.
This contribution deals with the integration of a fall detection system being supplied by printed batteries into clothing, designed with medical function in mind for safety of elderly people and patients with dementia. The first part presents a fall detection and long-term motion monitoring system and considers its power consumption. Energy consumption will be considered based on an amount of current consumed by particular parts of fall detection and long-term monitoring system. The second part gives an overview of various batteries suitable to supply the fall detection and long-term motion monitoring system integrated into clothing, while the third part presents suggestions for fall detection and long-term motion monitoring system integration into clothing.

***Keywords*— Dementia, fall detection system, printed batteries, smart clothing.**

I. INTRODUCTION

Dementia is becoming an increasing problem of modern society, as the number of patients, due to the aging of population, is steadily increasing. According to the World Health Organization (WHO), worldwide around 50 million people have dementia, and there are nearly 10 million new cases every year. The most common form of dementia is Alzheimer's disease and may contribute to 60–70% of cases [1]. Dementia is one of the major causes of disability and dependency among older people worldwide. There is a strong relation between dementia and gait disorders. With aging the amount of people falling during standing, walking, climbing stairs, etc. dramatically increases. For example, at the age of 60 only 15 % of people has gait disorders and 85 % of people has no gait disorders. The situation is reversed at the age of 85 when 82 % of elderly people have gait disorders and only 18 % has no gait disorders [2], [3]. Elderly people have a greater need for care and assistance and are more likely to be admitted to a nursing home after a fall, although they would like to stay at their own home and be independent [4]. Falls are a problem for elderly people because of fall related injuries as well as a fear of falling. The latter is related to the fear from 'long lie' which means involuntary laying in the floor for an hour or more after the fall because 'long lie' may result in medical complications [3].

Many wearable (electronic) devices with the ability to detect falls and capturing of gait were developed **recently** [3], [5-10]. Some of them were developed for medical diagnostics while some were developed with functionality to start intervention after a fall and thereby support elderly people to be independent and to live in their own homes.

Research experiments as well as medical diagnostics are time limited and usually short in time. In addition, both are usually performed within specialized environment. In such cases mounting of specialized sensor equipment on the body by band is acceptable. To guarantee continuous safety of elderly people and especially people with dementia a much better solution is integration of fall detection system into human clothing. In this area may be exposed to research of an integration of fall detection device into clothing, i.e. custom vest or smart skirt [11], [12].

This contribution deals with the integration of a fall detection system being supplied by printed batteries into clothing, designed with medical function in mind for safety of elderly people and patients with dementia. The first part presents a fall detection and long-term motion monitoring system and considers its power consumption. Energy consumption will be considered based on an amount of current consumed by particular parts of fall detection and long-term monitoring system. The second part gives an overview of various batteries suitable to supply the fall detection and long-term motion monitoring system integrated into clothing. Battery capacity given in mAh and battery life will be considered and batteries selected. The third part presents suggestions for fall detection and long-term motion monitoring system integration into clothing.

II. Fall detection and Long-term Motion Monitoring Wearable System

There have been many fall detection and motion monitoring wearable systems hardware and software configurations designed, built and evaluated. The construction of the majority of fall detection and motion monitoring wearable systems resembles the construction of sports activity monitoring device. They are designed as a self-contained one unit (box) rigid device attached to the human body by a band or a patch. And they possess the ability to store measured data and to send measured data by wireless communication to a stationary computer or a mobile phone. Such a design makes the fall detection and motion monitoring wearable systems suitable also for use as a short-term gait diagnostic device in medicine. Unfortunately, such a design of fall detection and long-term motion monitoring wearable device is not suitable for elderly people and people with dementia. It is also not suitable for long-term unobtrusive motion monitoring of people who are under long-term monitoring due to a risk of dementia development. In the case of fall detection and motion monitoring system design for people with dementia the short-term memory impairment of the people of dementia appearing already at an early stage of the disease should be considered. People with dementia and elderly people might be able to remember events that took place years ago, but they cannot remember where they left an item, why they entered a particular room or what they were supposed to do on a given day. Due to this they frequently forget to take fall detection and motion monitoring device with them. Therefore, the best solution is the integration of the fall detection and motion monitoring device into their clothing.

Today electronic devices are developing very fast. Main development directions of electronic devices are increasing integration, reduction of power consumption, flexibility and stretchability. Increasing integration and reduction of power consumption lead to reduction in size and weight of electronic devices and/or to increase of lifetime of battery powered electronic devices. New technologies of organic and printed electronic contribute to introduction of flexibility and stretchability of electronic devices. All mentioned trends contribute to easier integration of wearable electronic devices into the clothing.

To provide functionality of human fall detection and motion monitoring the wearable device consists of motion sensor(s), a microcontroller, an additional memory, a wireless radio frequency communication and a power supply. The motion sensor, microcontroller and additional memory are connected by a wired communication. The fall detection and motion monitoring wearable device is connected by wireless communication to the stationary computer or to a mobile phone being a gateway to the Internet (Internet of Things), Fig. 1. Particular functions of the fall and motion monitoring device may be integrated – depending on an implementation of electronics.

Motion sensors – Inertial measurement unit (IMU) used in fall detection and long-term monitoring device might consist solely from 3-axis accelerometer or from 3-axis accelerometer and 3-axis gyroscope or from 3-axis accelerometer, gyroscope and compass. To detect human body posture 3-axis accelerometer or 3-axis compass may be used. The body posture using 3-axis accelerometer is determined by the use of earth's gravity acceleration.

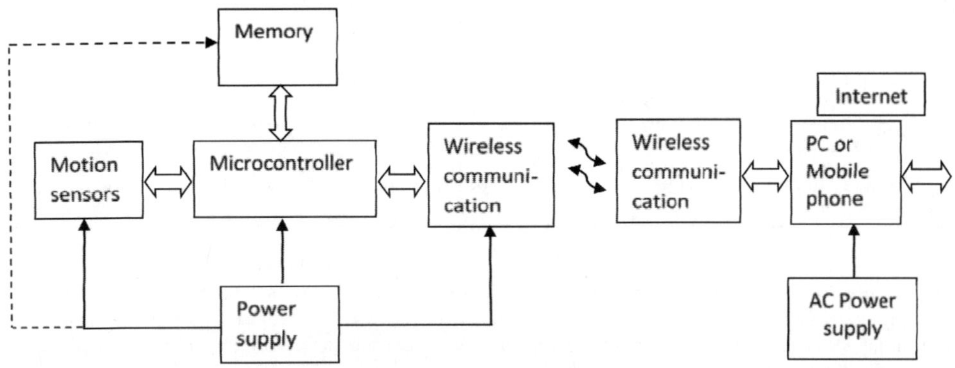

Fig. 1. Fall detection and long-term motion monitor functionality schematic.

Therefore, fall detection devices are sensitive to the position and orientation on the body it is to where they are worn or attached to the body or clothing [11]. Fix attachment of the fall detection device is necessary. The most suitable position of a fall detection device motion sensor was found to be at the waist [6], [9]. Other considered body or clothing attachments up to day include attachment on the shoes [7], on an ankle [5], on a tight, to the chest [10], upper arm, on a wrist. For instance, ActivePal which has become a reference device for activity of daily life (ADL) measurements is attached to a tight by a patch. A solution of walking pattern detection by an accelerometer inside a smart (mobile) phone worn in a small pocket was proposed in [8]. The smart mobile phone used in this latter case was placed and oriented in the pocket by a responsible person. Measured motion data are sent from motion sensors to the microcontroller. Motion sensors are frequently used in battery powered electronic devices therefore they are designed to operate with minimal current consumption while fulfilling application demands. They posses low power modes, programmable data output rates, programmable measurement ranges, interrupts, auto-wake/sleep function, etc. all in order to enable minimal current consumption. Table 1 shows data for current consumption of two motion sensors: the 3-axis accelerometer MMA8451Q and MPU6050 6-axis motion (3-axis gyroscope + 3-axis accelerometer) MEMS motion tracking device with regard to acceleration low power mode operating currents.

The MPU6050 consumes 3.8 mA at full power, it is with gyroscope measurements at all rates and acceleration at 1000 Hz. Both motion sensors could be supplied with 3V.

The most frequently used options for sending data to the microcontroller are SPI, I2C and UART communication. Energy consumption increases from SPI is (the lowest), over I2C to the highest consumption of UART.

TABLE I
MOTION SENSOR CURRENT CONSUMPTION

MMA8451Q		MPU6050	
Acceleration data rate[Hz]	Current min/max. [μA]	Acceleration data rate[Hz]	Current [μA]
1.56	6 /165	1.25	10
6.25	6/165	5	20
12.5	6/165	20	70
50	14/165	40	140
200	44/165	-	-

Power consumption of the microcontroller (MCU) depends on processor type (8-bit, 16-bit, 32-bit), the clock/speed of calculations, the amount of calculations (processing power) needed in an algorithm used for fall detection if it is implemented on the microcontroller and many other power-saving features, like: unneeded peripheral module disable, low power watch dog, idle and sleep mode, programmable clock frequencies, etc. MCU sleep mode halts all clock operation, halts all code execution and almost all peripheral operation. In sleep mode MCU current consumption is reduced to a minimum. Idle mode halts the code execution but allows peripheral modules, for instance: communication, to continue operation. Table 2 shows current consumption data for power down mode of 8-bit PIC18F25K22 microcontroller and 16-bit dsPIC33FJ64MC802 micro-controller with digital signal processing engine (40-bit data registers).

TABLE II
8-BIT MCU PIC18F25K22 AND 16-BIT MCU DSPIC33FJ64MC802 CURRENT CONSUMPTION

8-bit PIC 18F25K22 at 3V supply			16-bit dsPIC33FJ64MC802 at 3.3 V supply		
Module mode	Current min/max [μA]	Clock	Module mode	Current min/max [μA]	Clock
MCU sleep	0.01/13	all stopped	MCU sleep	28/87	all stopped
MCU idle	14/35	32 kHz, secondary	MCU idle	-	-
MCU idle	80/150	1 MHz, primary	MCU idle	-	-
MCU idle	750/1000	20 MHz, primary	MCU idle	8000/10000	20 MHz, primary
MCU run	19/50	32 kHz, secondary	MCU run	-	-
MCU run	200/300	1 MHz, primary	MCU run	-	-
MCU run	2650/3500	20 MHz, primary	MCU run	18000/22000	20 MHz, primary
watch dog	0.5/0.5	-	watch dog	10/15	-

The data in Table 2 show that greater processing power means also considerably higher current consumption. According to the available battery power we might decide whether to implement fall detection algorithms on the battery powered fall detection and motion monitoring device or on the stationary computer which is also a gateway to internet and has power is available through the wall outlet or on mobile phone. Depending on this decision we also select the appropriate microcontroller (with high or low computing power) for fall detection and motion monitoring device.

Microcontroller sends processed or unprocessed data to the wireless communication module using SPI, I2C or UART communication. If the motion data are processed on the microcontroller of the fall detection and motion monitoring device, then alarms to start intervention via wireless communication must be send in the case of a detected fall. Otherwise quite a lot of motion data (accelerations, speed, orientation) should be periodically sent to the stationary computer and internet gateway where a fall detection algorithm is run. Therefore, current consumption of wireless communication is another important data for the battery supplied fall detection and motion monitoring device design. Table 3 shows current consumption of frequently used wireless communications all using 2.4 GHz ISM band Data in table 3 show that current consumption of wireless communications is considerably higher than the consumption of microcontroller and motion sensors. WiFi has the highest current consumption, Standard Bluetooth and ZigBee have middle current consumption with peak currents between 30 mA and 60mA. The lowest current consumption have NRF24L01 and Bluetooth low energy with peak current consumption around 10 mA. Due to the nature of wireless protocols (many short time peaks) the average current consumption of wireless communication is much lower than show values of maximum current consumption.

TABLE III
CURRENT CONSUMPTION OF WIRELESS COMMUNICATIONS WIFI, BLUETOOTH, NRF24L01 AND ZIGBEE

Wireless communication	Range	Tx Current consumption	Rx current consumption	Current in sleep	Data rate
WiFi (WGM 110)	450 m	261 mA	81 mA	22 μA	72 Mbps
low power WiFi (RN171)	150 m	120 mA	40 mA	4 μA	24 Mbps
Bluetooth (WT12)	30 m	60 mA	7 mA	< 1 mA	2 Mbps
Bluetooth low energy(RN4871)	150 m	10 mA	10 mA	2.9 μA	10 kbps
nRF24L01	100 m	11.3 mA	12.3 mA	0.9 μA	2 Mbps
ZigBee [CC2530]	30 m	30 mA	24 mA	1 μA	250 kbps

Therefore, for estimation of average current consumption the current consumption values in sleep (longer time periods between peaks) are important. If Bluetooth low energy RN4871 (based on IS1870) is used then according to data during advertising peak currents of 12 mA appear, but the average current during advertising is around 230 μA. Such a current consumption is comparable with maximal current consumption of motion sensor.

An overall power consumption of the fall detection and long-term monitoring device is obtained as a sum of average current consumption values of motion sensor, microcontroller including SPI or I2C communication and wireless communication. Based on the obtained data and an assumption of 8-bit microcontroller based fall detection device that 3V supply is needed which is capable of providing short-time 10 mA peak currents and average current around 1 mA.

III. POWER SUPPLY FOR FALL DETECTION AND LONG-TERM MONITORING WEARABLE DEVICE

For the supply of fall detection and long-term monitoring wearable device batteries are needed. Standard batteries are cylindrical, for instance AA or smaller AAA type of batteries. AA and AAA batteries are rather big, heavy and thick to be used in wearable devices integrated into clothing. More convenient batteries for integration into clothing are coin battery cells and flat Li/ion batteries. Both are flat, rather small and thin, but still rigid. The last type of batteries we are going to consider are printed batteries. The shape of printed batteries seems to be the best for integration into clothing. Printed batteries are thin, it is less than 1 mm thick, and flexible. They could be attached to a curved surface with a minimum radius of 25 mm.

Table 4 shows comparison of considered battery data, like dimensions, weight, voltage, capacitance and internal resistance. All batteries in Table 4 are primary except VARTA V 150H which is rechargeable battery. Enfucell batteries are printed batteries thin as a paper sheet.

For good operation of fall detection and long term monitoring device Tabel 3 shows that fairly high current is needed for wireless communication. The lowest maximal values of current supply needs Bluetooth low energy (RN4871) 10mA, but the lowest speed of data transfer. Next lowest maximal current supply needs nRF24L01, it is up to 13mA and highest speed of data transfer 2 Mbps. And after nRF24L01 the maximal current supply needs ZigBee (30mA, 250 kbps data transfer speed).

An overview of batteries in Table 4 shows that new Enfucell batteries 1.5V PLUS could provide sufficient current even for maximal current demands. But we have to connect two batteries in series to get the 3V voltage supply level for the fall detection device. Internal battery resistance doubles and the maximal available current decreases if two batteries are connected in series. To be able to run fall detection and long-term monitoring device with some margin in maximal current we connect additional two batteries in parallel to the previous two batteries. Altogether we now use four batteries to supply our fall detection and long-term monitoring device. The following power consumption of the fall detection and long-term monitoring device can be estimated, Table 5.

Estimated battery voltage drop due to maximal current peaks is expressed as:

$$U = Un - R{*}I_{max} = 3 - 50{*}0.015 = 3 - 0.75 = 2.25 \text{ V} \quad (1)$$

According to data of motion sensor, microcontroller, Bluetooth and memory all four work satisfactory at voltage supply 2.25 V. Taking into account the average power consumption (current) I_{ave}, equation (2):

$$I_{ave}\,[\text{mA}] = \text{sum}\,(I_{motion\_sensor} + I_{MCU} + I_{Bluetooth} + I_{memory}) = 0.2 \text{ mA} + 3 \text{ mA} + 0.3 \text{ mA} + 0.25 \text{ mA} = 3.75 \text{ mA} \quad (2)$$

TABLE IV
AVAILABLE ELECTRICAL POWER BY BATTERY TYPE, SIZE AND CAPACITY

Name	Shape	Weight m [g]	Dimensions [mm]	Thickness h [mm]	Un [V]	R [Ω]	I [mA]	Im [mA]	C [mAh]
VARTA HE AA	cyl	23.8	50.0	14.0	1.5	0.2	200.0	1000	2960
VARTA HE AAA	cyl	11.0	44.0	10.0	1.5	0.2	20.0	1000	1250
VARTA V 150H	oval	6.0	25.5	5.725	1.2	0.8	28.0	280	150
VARTA CR2032	coin	3.0	19.8	3.0	3.0	16.0	0.2	10	230
ENFUCELL PLUS	sheet	2.9	72 x 60	0.7	1.5	25.0	1.2	18-20	90
ENFUCELL	sheet	2.9	72 x 60	0.7	1.5	50.0	0.6	8-10	90
ENFUCELL PLUS	sheet	1.4	60 x 42	0.7	3.0	150.0	0.4	6	10

TABLE V
ESTIMATED POWER CONSUMPTION OF THE FALL DETECTION AND LONG-TERM MONITORING DEVICE

Un [V]	R [Ω]	I_{ave} [μA]	I_{max} [mA]	C [mAh]	Dimensions [mm]	Weight [g]
3	50	3750	15 mA	180	288 x 120	23.2

The battery lifetime $t_{b\text{-}lif}$ can be expressed according to equation (3):

$$t_{b\text{-}lif} = C[mAh] / I_{ave}[mA] = 180/3.75 = 48 \text{ h} \quad (3).$$

At the assumed power (current) consumption, the four printed batteries sewn into the clothing would have a lifetime of 48 hours working round the clock. Then they would be replaced. The 48 hours seems a reasonable long time to start implementation of the device.

IV. INTEGRATION OF THE FALL DETECTION AND LONG-TERM MONITORING DEVICE INTO CLOTHING

In previous works was found that the most suitable position of the fall detection device is around the waist. Therefore, we decided to integrate the fall detection and long term monitoring device into women's trousers. The specific intended integration of a fall detection system, supplied by printed batteries into clothing, required development a functional pattern, which will provide easy to use. For this purpose was ergonomically designed women's trousers, with minimum inhibitory effect on movement and provide suitable comfort and performance to the user.

The sketch of women's trousers, designed with medical function in mind for safety of elderly people and patients with dementia, enable the integration of a fall detection system and printed battery into lumbar part of the belt. The sketch of women's trousers with a designed position for the integration of a fall detection system and printed battery is shown in Fig. 2.

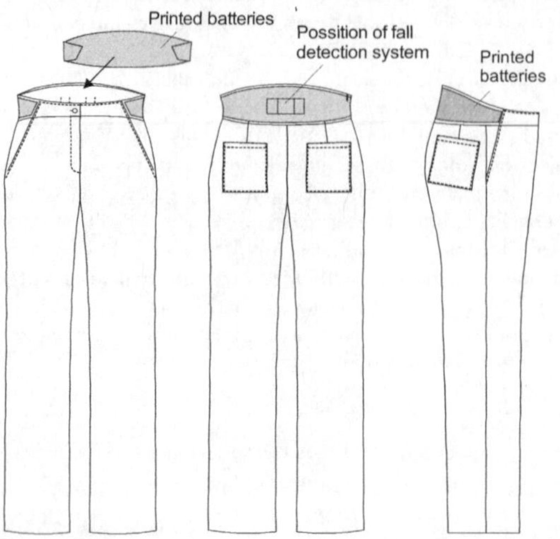

Fig. 2. Sketch of women's trousers with a designed position for the integration of a fall detection system and printed battery

From analysis of the results of pattern engineering of the women's trousers with a designed position for the integration of printed battery is clear that the designed construction solution provides the surface of the battery up to 660 cm^2, depending on the trousers size. The dimensions of the battery surface according to the women's trousers size is shown in the Table. 6.

TABLE VI
DIMENSIONS OF THE BATTERY SURFACE ACCORDING TO THE WOMEN'S TROUSERS SIZE

Size designation	Surface area for printed batteries [cm^{-2}]
36	416.73
38	441.86
40	467.48
42	493.56
44	520.14
46	547.21
48	574.77
50	602.83
52	631.39
54	660.45

V. CONCLUSION

The integration of a fall detection system supplied with a printed battery embedded into clothing, designed with medical function in mind for safety of elderly people and patients with dementia, has been investigated. The special attention was devoted to studying the power consumption for the fall detection from the long-term motion monitoring standpoint, which will be integrated into the women's trousers, as well as to pattern engineering of the women's trousers. The model of women's trousers was specially engineered to the requirements for the integration of a fall detection system and printed batteries. Energy consumption was investigated from point of view of an amount of current consumed by particular parts of fall detection system.

The results of pattern engineering of the women's trousers with a designed position for the integration of printed batteries indicates that the designed construction solution provides the surface of the battery up to 660 cm^2, depending on the women's trousers size.

According to obtained surface area for printed batteries, and related assumed power (current) of printed batteries, integrated into the women's trousers, which corresponds a lifetime to about 90 hours working round the clock, the next step of our research will be focused on the implementation of the fall detection system supplied with a printed batteries embedded into clothing.

REFERENCES

[1] WHO (2017, December), Dementia. [Online]. Available: http://www.who.int/mediacentre/factsheets/fs362/en
[2] H. Axer, M. Axer, H. Sauer, O.W. Witte, G. Hagemann, "Falls and gait disorders in geriatric neurology," *Clinical Neurology Neurosurgy,* no. 112, pp. 265-274, 2010.
[3] S. Chaudhuri, H. Thomson, G. Demiris, "Fall Detection Devices and their Use with Older Adults: A Systematic Review," *J Geriatr Phys Ther*. vol. 37, no. 4, pp. 178-196, Oct-Dec.2014, doi:10.1519/JPT.0b013e3182abe779.
[4] M. Chan, D. Esteve, J-I. Fourniols, C. Escriba, E. Campo, "Smart wearable systems: Current status and future challenges", *Artificial Intelligence in Medicine*, vol. 56, no. 3, pp. 137-156, November 2012.
[5] T. Kirste, A. Hoffmeyer, P. Koldrack, A. Bauer, S. Schubert, S. Schröder, S. Teipel, "Detecting the Effect of Alzheimer's Disease on Everyday Motion Behavior" *J. Alzheimers Dis,* vol. 38, no. 1, pp. 121-132, 2014, doi: 10.3233/JAD-130272.
[6] F. Wu, H. Zhao, Y. Zhao, H. Zhong, "Development of a Wearable-Sensor-Based Fall Detection System," *International Journal of Telemedicine and Applications*, pp. 1-11, February 2015, doi: 10.1155/2015/576364.
[7] Y. Hsu, P. Chung, W. Wang, C. Pai, C. Lin, H. Wu, J. Wang, "Gait and Balance Analysis for Patients With Alzheimer's Disease Using an Inertial-Sensor-Based Wearable Instrument," *IEEE J Biomed Health Inform,* vol. 18, no. 6, pp. 1822-1830.
[8] A. Nieto-Reyes, R. Duque, J. L. Montala, C. Lage, "Classification of Alzheimer's Patients through Ubiquitous Computing *Sensors*", vol. 17, no. 7, 1679 doi: 10.3390/s17071679.
[9] A. Sucerquia, J. D. Lopez, F. Vargas-Bonolla, "SiSFall: A Fall and Movement Dataset," *Sensors,* vol. 17, no. 1, 198, 2017, doi: 10.3390/s17010198.
[10] C. Crispim-Junior, V. Joumier, L. Hsu, M. Pai, A. Dechamps, P. Robert, F. Bremond, "Alzheimer's patient activity assessment using different sensors", *Gerontechnology,* vol. 11, no. 2, pp. 266-267, 2012.
[11] A. K. Bourke, P. W. Van de Ven, A. E. Chaya, G. M. Olaighin, J. Nelon, "Testing of a long-term fall detection system incorporated into a custom vest for elderly", in *Proc. IEEE EMBS Conf.* 2008, Vancouver, Canada, 2008, pp. 2844-2847.
[12] N. Mezghani, Y. Ouakrim, Md. R. Islam, R. Yared, B. Abdulrazak, "Context aware adaptable approach for fall detection based on Smart textile", in *Proc. Biomedical & Health Informatics* (BHI), 2017 IEEE EMBS International Conference on Proc. IEEE EMBS BHI, Orlando, 2017, pp. 473-476. DOI: 10.1109/BHI.2017.7897308.

Thermal Analysis of PVA Nanofibrous Membranes

Merve Kucukali Ozturk[1], Banu Nergis[1], Cevza Candan[1]

[1]*Istanbul Technical University, Faculty of Textile Technologies and Design, Textile Engineering Department, Inonu Cad. No: 65 34437 Gumussuyu Beyoglu Istanbul, Turkey*

kucukalim@itu.edu.tr

Abstract Thermal behavior of electrospun PVA nanofibrous membranes was studied by thermal analysis techniques i.e., differential scanning calorimetry (DSC) and thermogravimetric analysis (TGA). For the production of nanofibrous membranes needleless (roller) electrospinning method was used. DSC graphs showed that there is a decrease in melting temperature of the PVA polymer when crosslinking agent is used. Moreover, the change in the polymer concentration doesn't influence the T_m. Considering the TGA results, it is realized that while crosslinked PVA nanofibers have weight losses at about 240°C, pure PVA nanofibers have a weight loss at about 290°C.

***Keywords*—Crosslinking agent, differential scanning calorimetry, PVA nanofibrous membrane, thermogravimetric analysis.**

I. Introduction

NANOFIBERS form a new field for researchers due to their outstanding nanostructure characteristics i.e., high surface area, small pore size, and the possibility of their producing three dimensional structures [1]. Their properties provide them to be used as advanced materials in the application of technical textiles. Investigation on the thermal stability of nanofibers is required in order to have safety and stability information for handling, storage, and usage of nanofiber based combined structures. Thermal analysis techniques i.e., TGA, and DSC are used for experimentally determining thermal properties of the polymers [2].

In this study, PVA nanofibrous membranes were fabricated by electrospinning method. The effects of the polymer concentration and the usage of crosslinking agent on morphology and diameter of the fibers were investigated by SEM technique. Thermal behavior of electrospun PVA nanofibrous membranes was studied by differential scanning calorimetry (DSC) and thermogravimetric analysis (TGA).

II. Experimental Study

A. *Materials*

Nanofibrous membranes from polyvinylalcohol (PVA) polymer were produced. The water solution of polyvinyl alcohol PVA (Mw = 80,000 -100,000 g/mol) having both concentration of 12% and 14% was used for electrospinning. Glyoxal and phosphoric acid were added to the solution as crosslinking agents. The solution containing PVA, distilled water, glyoxal, and phosphoric acid was vigorously stirred at room temperature. A water solution of polyvinyl alcohol PVA (Mw=80,000-100,000 g/mol) having a concentration 12% was also prepared without crosslinking agent addition. Non-ionic surfactant (SLOVASOL 247/9) was added to the solution to decrease the surface tension to simplify the electrospinning process and increase the productivity.

B. *Nanofiber Production*

For the production of nanofibrous membranes needleless (roller) electrospinning method, which has been known as Nanospider trade name, was used [3]. In roller electrospinning (Fig. 1), a slowly rotating cylinder is partially immersed in a polymer solution. Polymer solution is connected to a high voltage source and collector is usually grounded. In electrospinning process, polymer solution is taken to the surface of the roller because of its rotation. With suitable high voltage, many Taylor cones are simultaneously created on the roller surface and produce nanofibers. The nanofibers are then transported towards the collector [4].

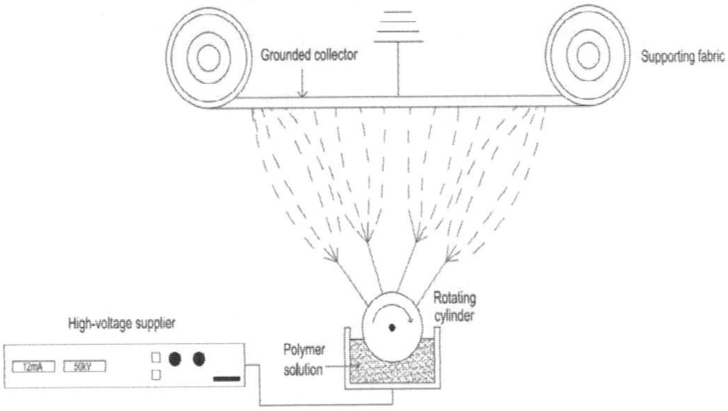

Fig. 1. Schematic diagram of roller electrospinning method used for nanofibrous membrane production.

The process parameters were set as shown in Table I below.

TABLE I
PROCESS PARAMETERS FOR PVA MEMBRANE PRODUCTION

Process parameters	Value
Roller length (mm)	145
Roller diameter (mm)	20
Roller angular velocity (rpm)	2
Distance between electrodes (mm)	100
Source voltage (kV)	50
Relative humidity (%)	34
Temperature (^{0}C)	19

C. Scanning Electron Microscopy (SEM)

The fiber morphology and fiber diameter of the electrospun nanofibers were determined using scanning electron microscopy (SEM). A small section of the fiber mat was placed on the SEM sample holder and sputter-coated with gold using Quorum Q150R Rotary-Pumped Sputter Coater. Carl Zeiss Ultra Plus Field Emission SEM was used to take the SEM photographs. The nanofiber diameter was determined by taking 100 measurements using Image J software.

D. Differential Scanning Calorimetry (DSC)

DSC analysis provide information about the amount of energy absorbed or released by a material as it is heated, cooled or held at constant temperature. The thermal properties of nanofibrous membranes were studied by DSC. Differential scanning calorimetry analysis was performed by using a DSC-6 Perkin–Elmer differential scanning calorimeter and using 8 mg of various electrospun materials. Materials were heated from 25°C to 300°C at a heating rate of 10°C /min and then, the samples were cooled to 25°C at a rate of 10°C /min. Samples wait 1 minute at 25°C and then heated to 300°C at a heating rate of 10°C /min and the samples were cooled to 25°C.

E. Thermogravimetric Analysis (TGA)

Thermogravimetric Analysis (TGA) measures weight/mass change (loss or gain) and the rate of weight change as a function of temperature, time and atmosphere. Thermal stability of the materials can be determined by measurements. The thermal property was analyzed with Mettler Toledo TGA/SDTA851e Analyzer. Nanofibers having weight between 6-7 mg were prepared and put into aluminum oxide crucibles. Then, closed crucibles put into heating tunnel and experiment was started. TGA analysis was performed at 25°C –500°C with 10°C/min in air.

III. Results And Discussion

A. *Morphology*

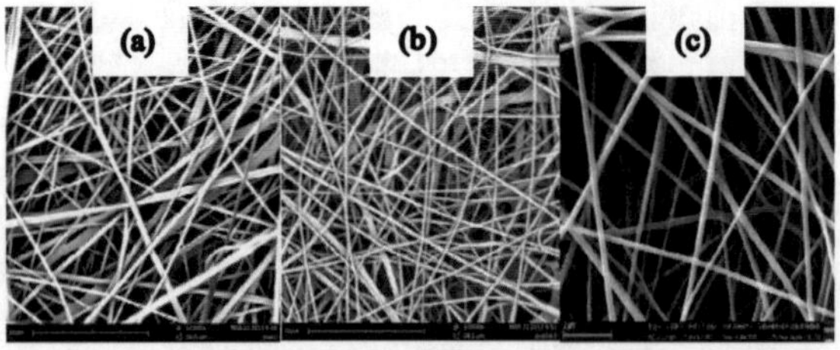

Fig. 2. SEM images of a) crosslinked PVA (12%) b) crosslinked PVA (14%) c) pure PVA (12%), magnification is 10k.

Fig. 2 shows the morphology of PVA nanofibers. As can be seen in the figure, some fibers were observed to stick each other forming an interconnected fibrous structure. The formation of bonded fibrous structure was because of the insufficient solvent (water) evaporation from the polymer jets. This is a common observation for PVA membrane production when using the needleless electrospinning process that involves many jets operating simultaneously in a very limited space [5]. However, it is evident that crosslinking agent affects the morphology of PVA nanofibers as SEM images show that the beaded structure diminished when crosslinking agent was not used.

Fiber diameters of the crosslinked PVA (12%), crosslinked PVA (14%), and pure PVA (12%) are 240±30, 300±20 and 290±20, respectively.

B. *Differential Scanning Calorimetry (DSC)*

DSC thermograms of the PVA electrospinning fibers are shown in Figure 3. The pure PVA fibers showed a relatively large and sharp endothermic curve with a peak at 196°C. However, for crosslinked PVA nanofibers, the peak shifted toward the low temperature.

Table II lists thermal properties of the fibers. It was observed that the peaks of the endothermic curves shifted toward the high temperature from 185 to 196°C and the ΔH_m values also increased from 29 to 52 J/g without using crosslinking agent in the polymer solution. This demonstrated that presence of the crosslinking agent in the polymer solution leaded to a worse condition for crystallization of the fiber. With increasing polymer concentration, the peak of the endothermic curve remained constant at 185°C and the ΔH_m values also increased from 29 to 50 J/g.

TABLE II
DSC Data Obtained From The Electrospun Fibers Of PVA

	T_m(°C)	ΔH_m (J/g^{-1})
Crosslinked PVA (12%)	185.5	29.82
Crosslinked PVA (14%)	185.1	50.77
Pure PVA (12%)	196.3	52.75

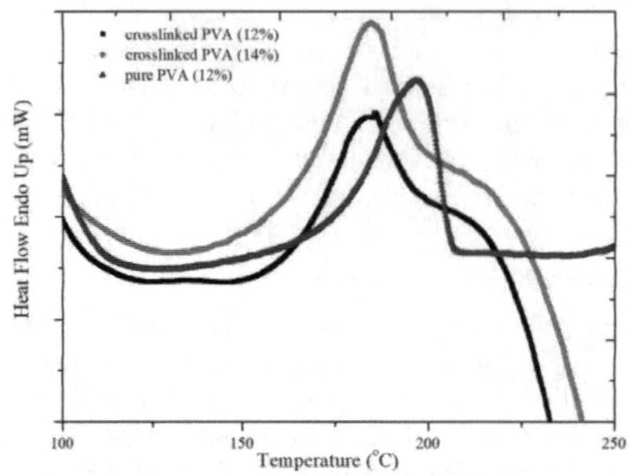

Fig. 3. DSC graphs of crosslinked PVA (12%), crosslinked PVA (14%), and pure PVA (12%).

C. *Thermogravimetric Analysis (TGA)*

Fig. 4 to 6 show the TGA graphs of PVA nanofibers that are drawn by Mettler Toledo Software Program. Graphs show the weight change of the samples with increasing temperature. Onset temperature that is marked in the figure is the temperature where a sharp decrease of sample weight starts.

As can be seen from Fig. 4 to 6, the degradation temperature of the crosslinked PVA nanofibers is between 230°C and 245°C. On the other hand, the degradation temperature shifted to 290°C when pure PVA solution was used for membrane production. This fact shows that crosslinking agents begin to degrade at lower temperature.

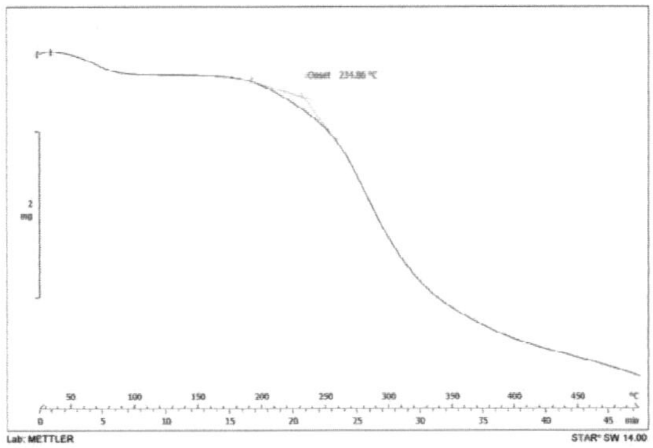

Fig. 4. TGA graph of crosslinked PVA (12%).

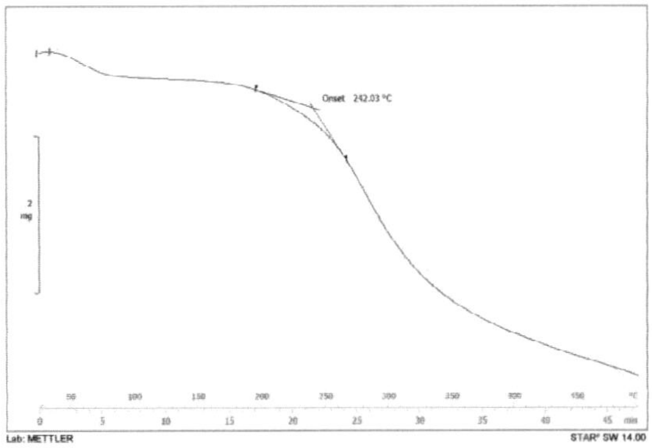

Fig. 5. TGA graph of crosslinked PVA (14%).

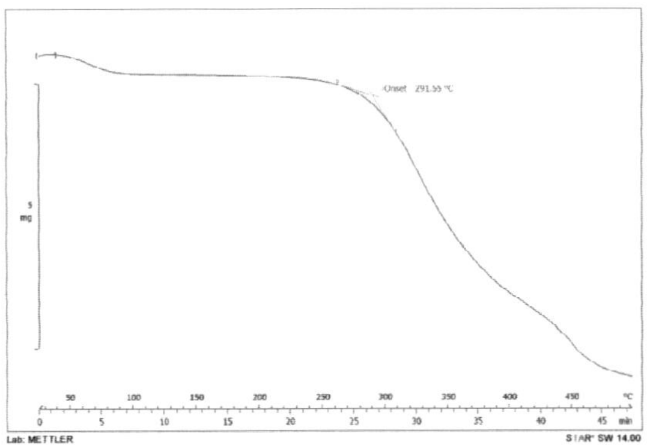

Fig. 6. TGA graph of pure PVA (12%).

IV. Conclusion

The effects of the use of crosslinking agent and the solution concentration on structure and morphology of the electrospun nanofibrous membrane were investigated. The result indicated that the average diameter of the fiber gradually increased with increasing polymer concentration. Moreover, pure PVA (without crosslinking agent) showed more uniform morphology and less bead formation. DSC analysis demonstrated that the peaks of the endothermic curves shifted toward the high temperature and the ΔH_m values also increased without using crosslinking agent in the polymer solution. TGA analysis showed that with using crosslinking agents, the degradation temperature shifted to lower temperature.

REFERENCES

[1] C. T. Lim, "Nanofiber technology: current status and emerging developments", *Prog. Polym. Sci.*, vol. 70, pp. 1–17, Jul. 2017.
[2] S. M. Pourmortazavi et al., "Thermal decomposition kinetics of electrospun azidodeoxy cellulose nitrate and polyurethane nanofibers", *J. Therm. Anal. Calorim.*, vol. 119, no. 1, pp. 281–290, 2015.
[3] O. Jirsak, K. Kalinova, and D. Stranska, 'Nanofibre Technologies and Nanospider Applications', in *5th Int. Nanotechnology Symposium*, Karlsruhe, Germany, 2006.
[4] M. Kucukali-Ozturk, B. Nergis, and C. Candan, 'Design of layered structure with nanofibrous resonant membrane for acoustic applications', *J. Ind. Text.*, p. 1528083717708483, May 2017.
[5] X. Wang, H. Niu, X. Wang, and T. Lin, "Needleless Electrospinning of Uniform Nanofibers Using Spiral Coil Spinnerets", *J. Nanomater.*, vol. 2012, p. e785920, Mar. 2012.

Improvement of Mechanical Properties of Jute/E-glass Fabric Reinforced Hybrid Composites

H. Sezgin[1] and O. B. Berkalp[1]

[1]*Istanbul Technical University, Faculty of Textile Technologies and Design, Textile Engineering Department, Inonu Cad. No:65 Gumussuyu, Beyoglu Istanbul, Turkey*

Corresponding author: sezginh@itu.edu.tr

Abstract : Composite structures, that are one of the most critical materials of today, are being improved day by day with high performances besides their lightness. Textile reinforced composite structures constitute a key share among all composite forms. In these structures, textile components provide strength and dimensional stability to the composite material. In this study, the effects of different types of Multi-walled carbon nanotubes (MWCNT, MWCNT-OH,MWCNT-COOH) on mechanical properties (tensile strength and impact strength) of jute/E-glass fabric reinforced hybrid composites were examined. Unsaturated orthophthalic polyester resin was used as the matrix material. Vacuum assisted resin transfer molding technique was utilized for the preparation of four-plied composite specimens. In sample codes, J and G refer to jute and E-glass fabrics, respectively.

***Keywords*—Composites, E-glass, jute, mechanical properties, multi-walled carbon nanotube.**

I. Introduction

During the last four decades, polymer matrix composites are gaining importance in both conventional and high technology application areas due to their high specific stiffness and strength, low cost, corrosive resistance and high dimensional stability properties [1]. Polymer matrix composites are mostly reinforced with textile structures. The utilization of textile materials in composite structures is increasing day by day under favour of excellent mergence of their relative low costs and high performance [2–5].

Composite materials are subjected to very different loading conditions and this situation sometimes forces the reseachers to utilize various types of reinforcement materials in the composite structure. Such composites, which are reinforced with two or more types of materials, are known as hybrid composites. By using two or more types of reinforcement materials, better mechanical properties and reduced material costs compared to the conventional composites can be achieved [1, 6-9]. These hybrid composite structures combine the beneficial properties and eliminate the undesirable properties of each component [4, 10].

Nanofillers are priviledged filler materials to develop the properties of composite structures. They have very high specific surface areas (more than 1000 m^2/g) and aspect ratios compared to micro-scaled fillers [11]. Nanoclays, nanofibers and nanotubes are the forms of nanofillers that are used in composite industry [12].

II. Experimental Study

A. Materials

Jute and E-glass plain weave fabrics were used as the reinforcement materials. Fabric parameters are given in Table 1.

TABLE I
FABRIC PARAMETERS

	Jute	E-glass
Areal density (± SD) (g/m^2)	200 (±15.2)	200 (±5.3)
Thickness (± SD) (mm)	0.99 (±0.10)	0.41 (±0.01)
Fabric count (e.p.c. x p.p.c)	4 x 4	4 x 3
Warp yarn count (± SD) (Tex)	375 (±36.3)	288 (±10.3)
Weft yarn count (± SD) (Tex)	234 (±25.8)	409 (±11.5)

The matrix system consists of unsaturated orthophthalic polyester resin. Also, cobalt was used as accelerator, methyl ethyl ketone peroxide (MEKP) was utilized as hardener in the resin system.

Three different types of MWCNTs (pristine MWCNT, MWCNT-OH and MWCNT-COOH) were added to the matrix material to enhance the mechanical properties.

B. Methods

Vacuum assisted resin transfer molding technique (Fig. 1) was used for the preparation of four-plied composite specimens. Composite fabrication was realized at room temperature (20°C±2°C).

Fig. 1. Vacuum assisted resin transfer method.

Mixing of the MWCNT and polyester was performed by ultrasonic mixer. Table 2 shows the specimen codes and layer sequences of composite samples.

Tensile strength tester and impact strength tester were used to evaluate the mechanical properties of fabric reinforced composites. Tensile testing was realized by Shimadzu AG-IS test machine according to ASTM D638-10 standard. Devotrans Charpy Impact Test Machine was utilized to evaluate the impact resistance of the composite samples according to BS EN ISO 179:1997 standard.

TABLE II
COMPOSITE SAMPLES

Specimen Code	Layer Sequence	Filler Type
JGGJ	Jute/E-glass/E-glass/Jute	-
JGGJ-MWCNT	Jute/E-glass/E-glass/Jute	MWCNT
JGGJ-MWCNT-OH	Jute/E-glass/E-glass/Jute	MWCNT-OH
JGGJ-MWCNT-COOH	Jute/E-glass/E-glass/Jute	MWCNT-COOH
GJJG	E-glass/Jute/Jute/E-glass	-
GJJG-MWCNT	E-glass/Jute/Jute/E-glass	MWCNT
GJJG-MWCNT-OH	E-glass/Jute/Jute/E-glass	MWCNT-OH
GJJG-MWCNT-COOH	E-glass/Jute/Jute/E-glass	MWCNT-COOH
JGJG	Jute/E-glass/Jute/E-glass	-
JGJG-MWCNT	Jute/E-glass/Jute/E-glass	MWCNT
JGJG-MWCNT-OH	Jute/E-glass/Jute/E-glass	MWCNT-OH
JGJG-MWCNT-COOH	Jute/E-glass/Jute/E-glass	MWCNT-COOH

III. RESULTS

A. Tensile Strength Results

The tensile strength results of jute/E-glass fabric reinforced composite samples are given in Figure 2. When the effect of fabric stacking sequence on the tensile strength of composite structures were investigated, it was seen that samples which have jute fabrics at the inner layers (GJJG) had the lowest strength.

When the effect of addition of different types of MWCNTs to the tensile strength was considered, it was revealed that all types of MWCNTs had increasing effect on the tensile strength values of samples. In these samples with different fabric sequences, the highest values were obtained with pristine MWCNT treated samples (JGGJ-MWCNT, GJJG-MWCNT and JGJG-MWCNT) at all three fabric stacking sequences.

Moreover, it come in sight that samples that taken from warp direction reached higher tensile strength values than the weft direction samples.

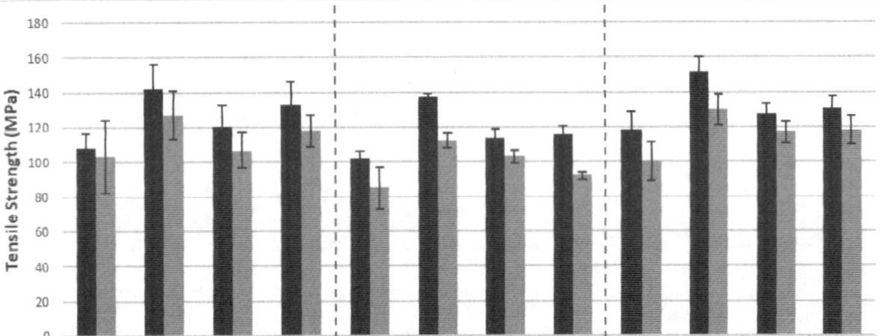

	JGGJ	JGGJ-MWCNT	JGGJ-MWCNT-OH	JGGJ-MWCNT-COOH	GJJG	GJJG-MWCNT	GJJG-MWCNT-OH	GJJG-MWCNT-COOH	JGJG	JGJG-MWCNT	JGJG-MWCNT-OH	JGJG-MWCNT-COOH
■ Warp	108.06	141.86	120.56	132.65	101.65	137.29	113.47	115.84	117.96	151.40	127.29	130.29
■ Weft	103.21	127.08	106.84	117.85	84.84	112.46	102.87	91.85	100.12	129.82	116.82	118.15

Fig. 2. Tensile strength results of composite samples.

B. Impact Strength Results

Figure 3 indicates the impact strength results of jute/Eglass fabric reinforced hybrid composite samples. The effect of the various fabric stacking sequences on the impact strength of hybrid composite structures is obviously seen from the Figure. While samples which have E-glass fabrics at the outer layers of the structures reached the highest impact strengths, samples which have E-glass fabrics at the inner layers had the lowest impact strengths among all samples. Besides, like the tensile strength results, warp direction samples reached higher impact strengths than the weft direction samples.

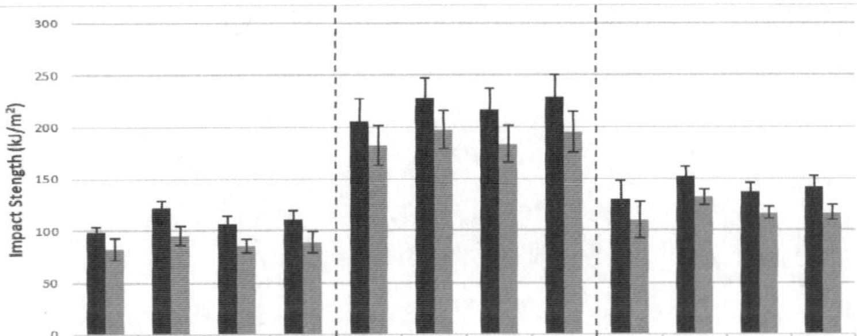

	JGGJ	JGGJ-MWCNT	JGGJ-MWCNT-OH	JGGJ-MWCNT-COOH	GJJG	GJJG-MWCNT	GJJG-MWCNT-OH	GJJG-MWCNT-COOH	JGJG	JGJG-MWCNT	JGJG-MWCNT-OH	JGJG-MWCNT-COOH
■ Warp	98.94	122.11	106.70	110.98	205.42	227.86	216.70	229.17	130.00	152.89	136.78	141.99
■ Weft	82.74	95.57	85.84	89.17	182.53	197.35	183.57	194.91	110.43	132.03	116.97	117.14

Fig. 3. Impact strength results of composite samples.

When three groups with different fabric sequences were examined, it was found that the highest impact strengths were obtained with pristine MWCNT treated specimens in the groups with JGGJ and JGJG fabric sequences, whereas the impact strength values of the MWCNT and MWCNTCOOH reinforced specimens in the GJJG specimens were very close to each other and found higher than those of MWCNT-OH sample. As a result, it can be said that all MWCNT types show increasing effects on impact strengths of hybrid composite samples.

IV. CONCLUSION

It can be concluded that;

- The specimens taken from the warp direction showed comparatively higher mechanical properties than the samples taken from the weft direction.
- Addition of MWCNTs (MWCNT, MWCNT-OH and MWCNT-COOH) to the composite structure enhanced the mechanical properties of composite materials.
- Among MWCNT types, the best results were achieved with pristine MWCNT treated composite samples.
- High impact strength values were obtained by placing the high strength fabrics (E-glass) on the outer layers of the composites, while the lowest tensile strengths were obtained at the same sequences.

REFERENCES

[1] K. S. Pandya, C. Veerraju and N. K. Naik, "Hybrid composites made of carbon and glass woven fabrics under quasi-static loading," Mater. Des., vol. 32, no. 7, pp. 4094–4099, 2011.
[2] M. Bodaghi, C. Cristóvão, R. Gomes and N. C. Correia, "Experimental characterization of voids in high fibre volume fraction composites processed by high injection pressure RTM," *Compos. Part A Appl. Sci. Manuf.*, vol. 82, pp. 88–99, 2016.
[3] Y. Q. Ding, Y. Yan, R. McIlhagger and D. Brown, "Comparison of the fatigue behaviour of 2-D and 3-D woven fabric reinforced composites," *J. Mater. Process. Technol.*, vol. 55, no. 3, pp. 171–177, 1995.
[4] K. F. Hasan, M. Islam, M. N. Morshed, M. I. I. Iqbal and J. H. W. Wu, "Dynamic mechanical behavior & analysis of the jute-glass fiber reinforced polyester hybrid composites," *Am. J. Appl. Phys.*, vol. 1, no. 1, pp. 1–12, 2016.
[5] V. Mishra and S. Biswas, "Physical and Mechanical Properties of Bidirectional Jute Fiber Epoxy Composites," *Procedia Eng.*, vol. 51, pp. 561–566, 2013.
[6] G. Agarwal, A. Patnaik, R. Sharma and J. Agarwal, "Effect of stacking sequence on physical, mechanical and tribological properties of glasscarbon hybrid composites," *Friction*, vol. 2, no. 4, pp. 354–364, 2014.
[7] M. Jawaid and H. P. S. Abdul Khalil, "Cellulosic/synthetic fibre reinforced polymer hybrid composites: A review," Carbohydrate Polymers, vol. 86, no. 1, pp. 1–18, 2011.
[8] D. Romanzini, A. Lavoratti, H. L. Ornaghi Jr., S. C. Amico and A. J. Zattera, "Influence of fiber content on the mechanical and dynamic mechanical properties of glass/ramie polymer composites," *Mater. Des.*, vol. 47, pp. 9–15, 2013.
[9] H. Sezgin and O. B. Berkalp, "The effect of hybridization on significant characteristics of jute/glass and jute/carbon-reinforced composites," *J. Indus. Text.*. https://doi.org/10.1177/152808371 6644290, 2016.
[10] T. Hamouda, A. H. Hassanin, A. Kilic, Z. Candan and M. S. Bodur, "Hybrid composites from coir fibers reinforced with woven glass fabrics: Physical and mechanical evaluation," *Polym. Compos.*, 2015.
[11] F. H. Gojny, M. H. G. Wichmann, B. Fiedler, W. Bauhofer and K. Schulte, "Influence of nano-modification on the mechanical and electrical properties of conventional fibre-reinforced composites," *Compos. Part A Appl. Sci. Manuf.*, vol. 36, no. 11, pp. 1525–1535, 2005.
[12] V. C. S. Chandrasekaran, S. G. Advani and M. H. Santare, "Role of processing on interlaminar shear strength enhancement of epoxy/glass fiber/multi-walled carbon nanotube hybrid composites," *Carbon*, vol. 48, no. 13, pp. 3692–3699, 2010.

Effect of PVA Addition on SO_2 Adsorption Properties of GO Fibers

İlkay Özsev Yüksek[1], Özge Alptoğa[2], Pelin Altay[1], Nilgün Karatepe[4], Ayşen Önen[3], Nuray Uçar[1]

[1]*Istanbul Technical University, Textile Engineering Department, Istanbul, Turkey*
[2]*Istanbul Technical University, Polymer Science and Technology Programme, Istanbul, Turkey*
[3]*Istanbul Technical University, Faculty of Sciences and Letters, Chemistry Department, Istanbul, Turkey*
[4]*Istanbul Technical University, Institute of Energy, Istanbul, Turkey*

Corresponding author: İlkay Özsev Yüksek , email: yukseki@itu.edu.tr

Abstract Since late 20th century, the drawbacks of pollution became obvious which increases the demand for environmental friendly and ecological approaches in all disciplines. Not only using environment friendly materials and procedures, but also detecting any pollution, and potential solutions are also common area of current research. SO_2, one of the significant topics in waste management, is reported to cause acidic rain. To abolish SO_2 waste, excess temperature is required. The current efforts are to develop materials that can remove SO_2 at comparatively low temperatures. The aim of this study is to examine a potential use of graphene oxide (GO) fibers and composite polyvinyl alcohol (PVA) incorporated GO fibers in SO_2 filters. So as to observe the effect of PVA incorporation on composite fiber properties, PVA incorporated GO fibers and neat GO fibers were prepared. GO fibers were prepared with modified Hummers method followed by wet spinning through three coagulation baths. For composite fiber formation, 10% PVA was added to coagulation bath three. So formed fiber morphologies were analyzed with SEM, AFM and BET. Electrical conductivities were calculated from electrical resistances measured with conductivity tester. The SO_2 adsorption and desorption properties were measured in a quartz reactor. PVA addition did not have a significant effect on electrical conductivity whereas increases the SO2 adsorption capacity and reduces the surface roughness, pore size, pore volume and surface area.

Keywords—graphene oxide, SO_2 adsorption, electrical conductivity, PVA.

I. INTRODUCTION

SO_2 , an industrial and exhaust waste, is one of the major pollutants causing environmental damage. Increase in SO_2 level rises the greenhause effect and results with acidic rain. Current efforts are to develop gas sensors, air filters or seperators to remove SO_2 ; however, a major drawback is high temperature needed [1-3]. Graphene, with its high surface area, is a potential material for use in SO_2 adsorption [3-5].

Some of the main application areas of graphene based fibers are;

- Filtration and seperation
- Sensors
- Catalysts
- Functional textiles
- Energy storage
- Nanotechnology
- EMI shielding

In this study, adsorption and desorption properties of graphene oxide (GO) fiber and GO fiber incorporated with PVA, were examined for their potential use in gas filters.

IOP Conf. Series: Materials Science and Engineering **460** (2018) 012050 doi:10.1088/1757-899X/460/1/012050

II. MATERIALS, METHODS AND PROCEDURES

A. Materials

Graphene-GIC (expandable, 30 μm thickness, 300 μm diameter, %99 purity) was obtained from Grafen Co. , $KMnO_4$ (158.03 g/mol), H_2SO_4 (98.08 g/mole) and HCl (37%) were supplied from Merck. $NaNO_3$ (84.99 g/mole) and PVA were provided by ZAG.

B. Methods and Procedures

Graphene oxide was produced from grapheme oxide dispersion obtained by Modified Hummers method described in the studies by Ucar et al [6-8].

Natural graphite flakes), $NaNO_3$ and H_2SO_4 were mixed and stirred uniformly in an ice bath with a magnetic stirrer. $KMnO_4$ was added to the mixture slowly and gradually. After the completion of $KMnO_4$ addition, the dispersion was stirred in the ice bath. Distilled water was added to the solution gradually to prevent any chemical risks. The reaction temperature was raised, distilled water and H_2O_2 were added to the solution while the temperature is being lowered at the same time. H_2O_2 addition was performed at once instead of gradual addition, as the solution was considerably diluted. Addition of H_2O_2 turned the solution into a yellow – blackish color. The next day, the diluted top portion of the solution was decanted and splitted into centrifuge tubes evenly. Graphene oxide solution was washed with HCl solution in centrifuge (Nüve, NF800R) to remove impurities. For pH adjustment, the solution went through several washings with distilled water until the solution pH reached pH 5-6. Then, the solution is dispersed by mechanical homogenizer (WiseTis Homogenizer, HG-15D) to prepare the GO dispersion for production of GO fibers by wet spinning method.

The graphene oxide fibers were produced with wet spinning in three coagulation baths containing various amounts of ethanol and water. In order to obtain PVA incorporated GO fiber, 10% PVA was included to the last coagulation bath.

The coagulation bath conditions are shown in Figure 1 and the production process of the nonwoven surface discs is shown in Figure 2.

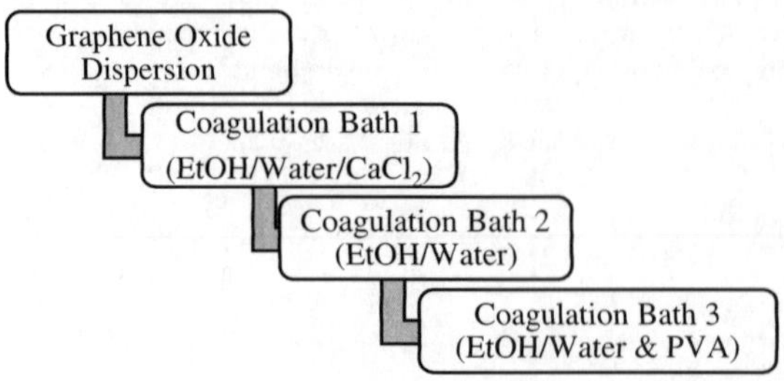

Fig 1. The coagulation conditions

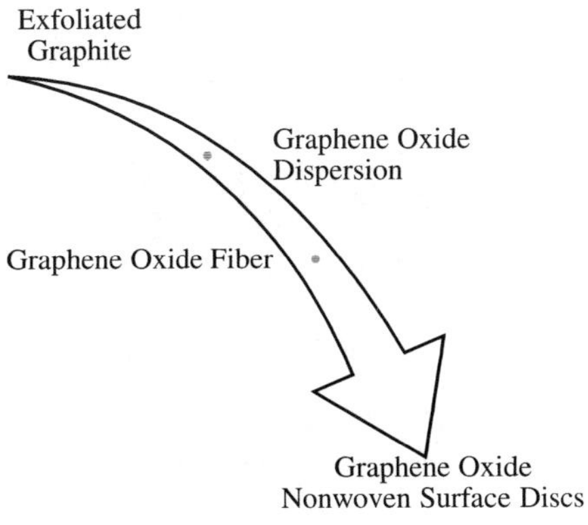

Fig 2. The production process of the nonwoven surface discs

Fiber surface morphologies were characterized with QUANTA FEG 250 Scanning Electron Microscopy (SEM). Before SEM measurements, the samples were coated with Au/Pd. AFM analysis was performed with Ambios Technology, Santa Cruz, CA. A silicone nitrate tip was used with AFM-tapping mode. SO_2 adsorption capacities were analyzed by obtaining fiber web disc of GO fiber / PAN incorporated GO fiber in the last coagulation bath. Absorption and desorption measurements were performed by using a quartz reactor. The flow rate of SO_2 gas was set to 1.47 ml/min and for N_2 gas 150 ml/min for 60 minutes. Also, titration method was applied after desorption measurements to determine the maximum adsorption capacities. The surface area calculations were performed according to Brunauer-Emmett-Teller (BET) method. The pore size and surface area measurements were conducted with Quantachrome Quadrosorb SI instrument.

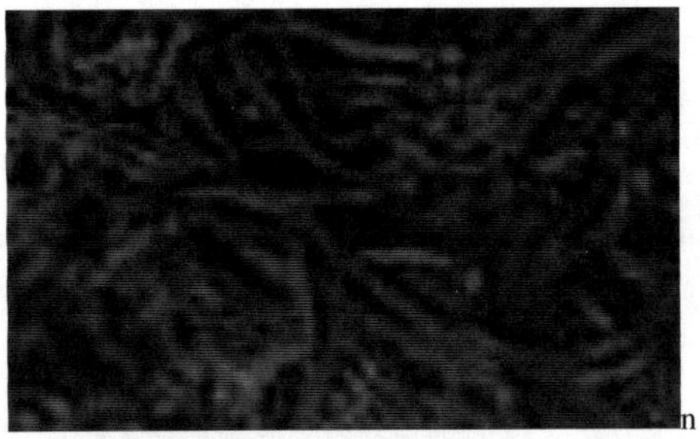

Fig 3. GO nonwoven surface

III. RESULTS

SEM images were used to characterize the surface morphologies and SEM images of the fibers are shown in Figure 4 and average electrical conductivity and AFM test results are shown in Figure 5 and Table 1 respectively.

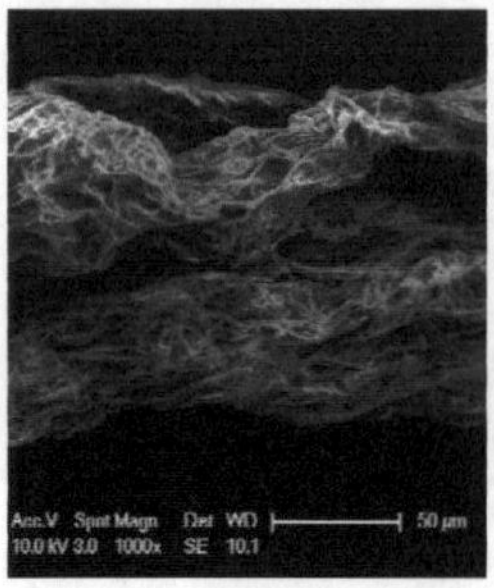

a) GO Fiber

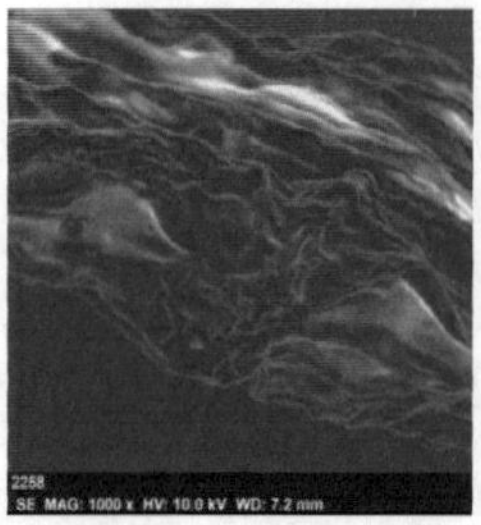

b) PVA incorporated GO Fiber

Fig. 4. SEM images of GO fiber and PVA incorporated GO fiber

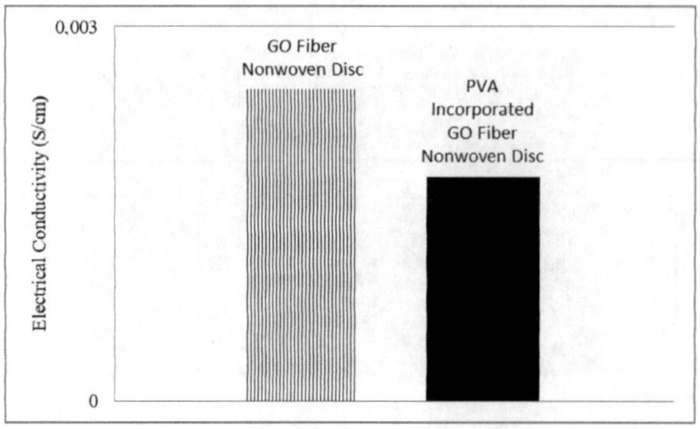

Fig 5. Electrical conductivity test results

TABLE 1. AFM test results

Sample	Rpv (μm)
GO Fiber	4,963
PVA incorporated GO Fiber	3.238

It is evident from both SEM images and Rpv values that the surface roughness decreases with the incorporation of PVA. As PVA itself is not as conductive as GO, when PVA is incorporated to GO fibers, the electrical conductivity also is not affected clearly by the addition PVA since both of sample has 10^{-3} S/cm.

TABLE 2.

BET test results

Sample	Surface area (m2/g)	Average pore diameter (nm)	Total pore volume (cc/g)
GO Fiber	24,971	5,399	$3{,}37*10^{-2}$
PVA incorporated GO Fiber	7,056	1,128	$1{,}99*10^{-2}$

BET results are shown in Table 2. As can be seen, incorporation of PVA reduces the surface area, total pore volume and average pore diameter.

In order to measure SO_2 adsorption capacity, fibers are spun in a continue fiber set-up and nonwoven surface discs were obtained.

The SO_2 adsorption test results are shown in Figure 6

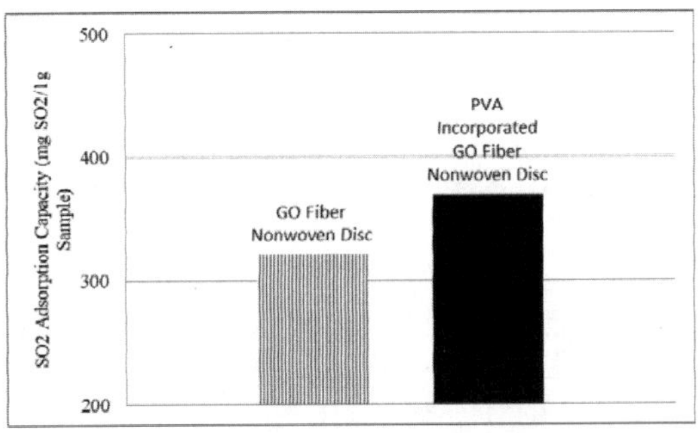

Fig 6. The SO_2 adsorption test results

As can be seen in Figure 6, addition of PVA increased SO_2 adsorption properties of nonwoven discs that might be because of PVA related functional groups PVA. In the literature, the adsorption capacity of nanosized GO was found to increase with reduced temperature, and measured 325 mg/g at highest at 15 ^{0}C [9].

IV. CONCLUSION

In this study, effect of PVA incorporation on SO_2 adsorption properties of GO fibers was investigated. It is found that addition of PVA reduces the surface roughness, surface area, average pore size and pore volume. There is no significant effect of PVA on electrical conductivity, however the effect of PVA on SO_2 adsorption is significant, leading to higher SO_2 adsorption.

V. ACKNOWLEDGMENT

This study has been supported by TUBITAK (Project number:114M524)

REFERENCES

[1] Varghese SS, Varghese SH, Swaminathan S, Singh KK and Mittal V, Two-Dimensional Materials for Sensing: Graphene and Beyond, Electronics, 2015, 4, 651-687.

[2] Jimenez-Cadena G, Riu J and Rius FX, Gas Sensors Based on Nanostructured Materials, The Royal Society of Chemistry, Analyst, 2007, 132, 1083-1099.

[3] Shao L, Chen G , Ye H, Wu Y, Qiao Z, Zhu Y and Niu H, Sulphur Dioxide Adsorbed on Graphene and Heteroatom-Doped Graphene: A First-Principles Study, Eur.Phys.J.B, 2013, 86:54.

[4] Xu C, Cui A, Xu Y, Fu X, Graphene Oxide – TiO2 Composite Filtration Membranes and Their Potential Application for Water Purification, Carbon, 62, 2013, 465-471.

[5] Stengl V , Bakardjieva S, Grygar TM, Bludska J and Kormunda M, TiO_2-Graphene Oxide Nanocomposite as Advanced Photocatalytic Materials, Chemistry Central Journal, 2013, 7:41.

[6] Ucar N, Gokceli G, Yuksek IO, Onen A, Karatepe Yavuz N. The Effect of Coagulation Time, Number of Coagulation Bath and Ingredients on Properties of Continuous Graphene Oxide Fiber, accepted for publication in IJFTR.

[7] Ucar N, Gokceli G, Yuksek IO, Onen A, Karatepe Yavuz N. Graphene Oxide and Graphene Fiber Produced by Different Nozzle Size, Feed Rate and Reduction Time with Vitamin C, J Ind Tex, December 2016.

[8] Ucar N, Can E, Yuksek IO, Olmez M, Onen A and Karatepe Yavuz N. The Effect of Exfoliation and Plasma Application on the Properties of Continuous Graphene Oxide Fiber, Fullerene Sci. Technol. , June 2017.

[9] Babu D J, Kuhl F G, Yadav S, Markert D, Bruns M, Hampeb M J and Schneider J J. Adsorption of Pure SO_2 on Nanoscaled Graphene Oxide, RSC Adv., 2016, 6, 36834–36839.

Investigation the surface modification of Polyamide fabrics using Low-Pressure Plasma Technique

Zuhaib Hassan[1], Dr. Ozgur Atalay[2] , Prof. Dr. Fatma Kalaoglu[2]

[1,2]*Faculty of Textile Technologies and Design, Istanbul Technical University, 34437, Istanbul, Turkey*

Corresponding author: kalaoglu@itu.edu.tr, atalayoz@itu.edu.tr

Abstract Plasma technique can be used to improve the surface characteristics of polyamide fabric with elastane through surface modification. In this study, the Polyamide fabrics with elastane treated with low-temperature, low-pressure, radio frequency argon plasma were investigated. After treated with the Plasma, in order to observe the effect of plasma treatment, fabrics were characterized and examined by the Scanning Electron Microscope (SEM). The changes in surface roughness and etching effect was observed. SEM analyses showed that the surface roughness increased due to the etching effect of the plasma treatment, leading to the improvement of the making the fabric conductive via different coating approaches for E-Textile applications.
From the experimental results, it was observed that higher surface roughness can be achieved by the increasing the plasma exposure time. The polyamide fabric with examined at different plasma exposure time 2, 4 and 20 minutes with the argon gas.
The obtained samples are promising for making the fabric conductive via different coating approaches for E-Textile applications.

Keywords— **plasma treatment, polyamide, Scanning electron microscope (SEM).**

I. Introduction

THE major end use application of textile fibres is its use in apparel fabrics. Now a days, wearing comfort and the functional requirements of clothing have increased for daily wear. Knitted fabrics are widely used in apparel products for their extraordinary properties like elasticity, extension, higher strength, better solvent, flame and heat resistance and hand and wearing comfort for whole the year and seasons.

Polyamide with elastane fabric adjust easily with the body's natural movements, but the original shape is recovered when it is allowed to relax. A knitted fabric has always elastic properties even without elastane.

Knitted fabrics with elastane can easily recover its original shape due to elastane in the fabrics. Fabrics containing spandex yarn have a wide application value, especially because of their increased extensibility, elasticity,the high degree of recovery, good dimensional stability and simple care [1-3].

Plasma treatments were an extensively used in textile processing technology is a clean and dry process that requires no water or organic solvent during processing. Plasma may be used for the surface modification and it changes the surface properties without affecting the bulk properties of materials, and less energy is consumed compared to conventional treatments [4]. Low-temperature plasma or gas discharge plasma is the subject of this study because most of the textile materials are heat sensitive polymers [5].

Plasma-induced surface modification is an effective way to produce functional textiles. Numerous studies indicated that plasma treatment improves the adhesion characteristics of surfaces. These studies stated that the etching effect of plasma gases caused nano- or micro-roughness, which contributes to the enhancement of adhesion on the surfaces[6,7].

The adhesion improvement of polyamide laminated fabric through argon and CO2 plasma treatment was also studied. It was observed that the adhesion properties of polyamide fabrics improved with plasma induced surface roughness compared to the untreated samples [8].
Plasma technology can be used to improve the surface characteristics of polypropylene through surface modification [9].

II. METHODS AND PROCEDURES

The polyamide fabric was cut into small pieces of about 8 cm × 10 cm by the scissors in the dust-free zone in order to get accurate results and the following samples were treated.

1. Untreated fabric
2. Treated fabric with plasma for 2 min by Argon gas
3. Treated fabric with plasma for 4 min by Argon gas
4. Treated fabric with plasma for 20 min by Argon gas

Argon (purity higher than 99.99 %) was used as processing gas with a gas flow rate of 0.3 l/m. Plasma chamber pressure was fixed at 0.3 mbar and 13.56 MHz radiofrequency was utilized during the process. Fabric samples were placed over the tray in the plasma chamber. Then, argon gas was fed in plasma chamber using various plasma treatment conditions (discharge power: 40 W, 60 W and 80 W, and exposure time: 2, 4 and 20 min). The one side of fabric was exposed to plasma.

Before plasma treatment, samples were washed with detergent for about 20 min, rinsed with distilled water, dried at 30°C in an oven for half an hour and were conditioned in desiccators.

Table 1:

Plasma Discharge Pressure (mbar)	Plasma Gas	Frequency (Radio, Mhz)	Plasma Discharge Power (Watt)	Plasma Exposure Time (min)
0.3	Argon	13.56	40	2 4 20
0.3	Argon	13.56	60	2 4 20
0.3	Argon	13.56	80	2 4 20

III. RESULTS

The SEM has been extensively used in analysis of the surface characteristics of plasma treated textiles. Qualitative information providing an image of the surface at high magnification is obtained from SEM. Accelerated beam of high-energy electrons generate a variety of signals at the surface. These signals are collected and transferred to the screen in order to obtain high-resolution images. The SEM uses a magnification ranging from 20X to approximately 30,000X. If the material is not electrically conductive, like most of the textiles, material surface has to be coated with conductive elements such as carbon, gold, platinium etc. before the tests. Here for the characterization of the samples, we coated the fabric samples with the gold and platinium coating in order to make the sample conductive before the test.

Plasma induced surface modification of fabrics were analyzed by SEM measurements. The increase in surface roughness after plasma treatment may facilitate the interaction between the adhesive and the coated layers, and thus contribute the mechanical adhesion of the coating.

As can be seen from Figures below that increasing the plasma power and exposure time enhanced roughness of the fabrics. As the plasma treatment time was increased from 4 min to 20 min, the surface roughness increased at a plasma power of 80 W. The highest surface roughness were obtained at a plasma power of 80 W and plasma time of 20 min. The results showed that the plasma treatment time was more effective in improving the adhesion than the plasma power. The results are in agreement with recent studies showing that the degree of plasma modification depended on the plasma exposure time and discharge power [10].

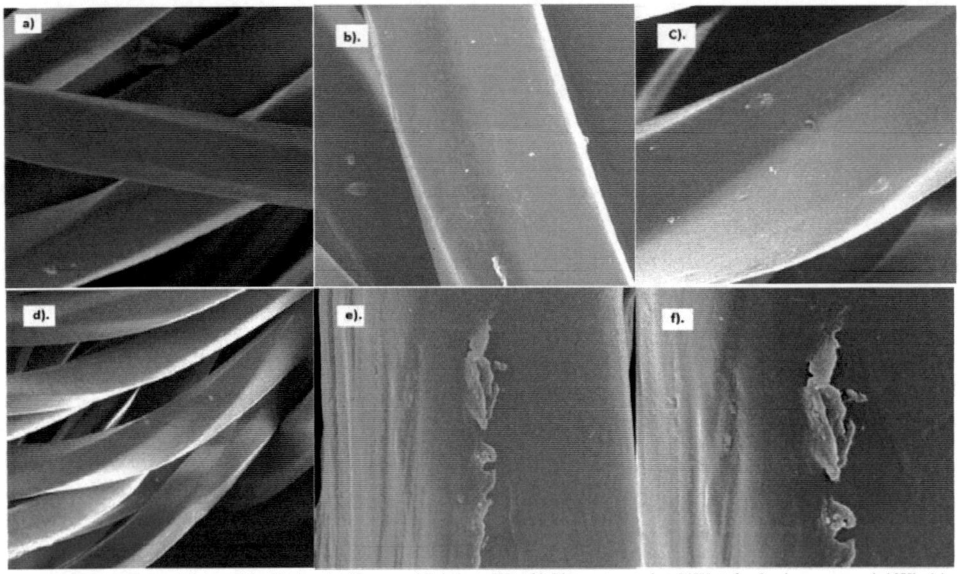

Figure 1. SEM images of Polyamide fabrics. (a) Untreated Fabric (x5K), (b) Plasma treated at 80Watt for 2 min exposure (x10K), (c) Plasma treated at 80Watt for 4 min exposure (x10K), (d) Plasma treated at 80Watt for 4 min exposure (x3K), (e) Plasma treated at 80Watt for 20 min exposure (x20K), (f) Plasma treated at 80Watt for 4 min exposure (x30K).

IV. CONCLUSION

This research work mainly focuses to analyze and investigate the surface modification of the polyamide fabric with elastane. Plasma technique are very useful to improve the surface characteristics of polyamide fabric with elastane through surface modification. SEM analyses showed that the surface roughness increased due to the etching effect of the plasma treatment, leading to the improvement of the making the fabric conductive via different coating approaches for E-Textile applications.

Additionally, this study will highlight the reasons of variation in innovation, and suggest solutions to firms to increase their innovation performance, which not only will increase their performance, but national economic performance too.

V. ACKNOWLEDGMENT

The Authors wish to acknowledge the Prof. Dr. Dilek Toprakkaya Kut, who gave permission for using her Lab at Uludag University for some sort of experiments. Without her permission, it was not possible.

VI. FUTURE WORKS

Polyamide fabrics with elastane can be further utilize and by different coating methods make the fabric stretchable. The further study will help to findout the new era of stretchable conductive fabrics in E-Textile industry, because E-Textile has fastest growing industry now a days.

Conductive materials with high flexibility and stretchability have gained much attention in both academia and industry now a days.

REFERENCES

[1] Bayazit, M.A. "Dimensional and physical properties of cotton/spandex single jersey fabrics", Textile Research Journal, Vol. 73 No. 1, 2003, pp. 11-14.

[2] Mcintyre j e, Synthetic Fibres: Polyamide, Polyester, Acrylic, Polyolefi n, Woodhead Publishing Limited, The Textile Institute, England, 2005, p 21, 42.

[3] Šajn, D., Geršak, J. and Bukošek, V, "A study of the relaxation phenomena of fabrics containing elastane yarns", Proceedings of the 2nd International Textile, Clothing and Design Conference – Magic World of Textiles, Dubrovnik, 3-6 October,2004 pp. 605-610.

[4] Wei Q. Surface modification of textiles: Surface modification of textiles by plasma treatments, Cambridge: Woodhead Publishing, 2009.

[5] Morent, R., De Geyter, N., Verschuren, J., De Clerck, K., Kiekens, P., Leys, C. Non-thermal plasma treatment of textiles, Surface & Coatings Technology, 202, 2008, 3427–3449.

[6] Krump H, Simor M, Hudec I, Jasso M, Luyt AS. Adhesion strength study between plasma treated polyester fibres and a rubber matrix. Applied Surface Science 2005; 240 (1-4): 268-274.

[7] Krump H, Hudec I, Jasso M, Dayss E, Luyt AS. Physical-morphological and chemical changes leading to an increase in adhesion between plasma treated polyester fibres and a rubber matrix. Applied Surface Science 2006; 252 (12): 4264-4278.

[8] Özdoğan E. The effect of plasma treatment over the adhesion property of polyamide 6 fabrics. Tekstil ve Konfeksiyon 2006; 2: 128-133.
[9] Wei Q, Xu Q, Cai Y, Wang Y. Evaluation of the interfacial bonding between fibrous substrate and sputter coated copper. Surface & Coatings Technology 2008; 202: 4673–4680.
[10] Cireli A, Kutlu B, Mutlu M. Surface modification of polyester and polyamide fabrics by low frequency plasma polymerization of acrylic acid. Journal of Applied Polymer Science 2007; 104 (4): 2318–2322.

Mechanical Assessment of Fire-off on CO/PET Fabrics

Raziye Atakan, Gulay Ozcan

Istanbul Technical University, Faculty of Textile Technologies and Design, Department of Textile Engineering, Gumussuyu, 34437, Istanbul, Turkey.

ratakan@itu.edu.tr

Abstract. The present study investigates the mechanical properties of newly developed FR chemical Fire-off in order to assess textile service-life features. 350 g/L Fire-off finished bath was prepared and CO/PET fabric samples were treated via impregnation method. In order to investigate FR performance of treated fabrics LOI test was performed. Subsequently, fabric strength, abrasion properties and outward appearance; tensile properties, abrasion, pilling performance and whiteness of Fire-off treated fabrics were tested and evaluated. Mechanical test results showed that there was a slight increase in strength, elongation, abrasion and pilling. On the other hand, whiteness and air permeability of fabrics were decreased after Fire-off treatment.

1. Introduction

Imparting flame retardancy to cotton/polyester (CO/PET) fabrics is difficult because of well-known scaffolding effect between PET and cotton when burned. PET (thermoplastic) shrinks at its melting point and run away from the fire, whereas cotton forms char. The molten PET generally tends to wick on cotton char, and prevent its melt dripping [1-10]. It is known that phosphorous based flame retardancy (FR) products can enhance polyester and cellulose FR properties through such FR mechanisms of dehydration, phosphorylation, and phosphate-ester decomposition. They form a network crosslinking the cotton and PET, supress the release of volatile combustible fragments and thus enhance the formation of char [11].

Nowadays, scientific community have mainly focused on the development of less or nontoxic phosphorus-based FRs, which represent an adequate alternative to their highly toxic halogen-based counterparts [12-15]. A remarkable research papers has attempted to develop formaldehyde-free FR replacements due to the formaldehyde concern. Despite the efficiency of already existing FR treatments such as Proban, Pyrovatex, Trevira CS, researchers from various disciplines from both academia and industry are still looking for alternative choices, while taking into consideration the requirements such as ease of application, durability, comfort, cost effectiveness, lower or non-toxicity and comparable textile service-life features. Development of any new flame retardant compound, in general, should comprise both material performance and toxicity levels for a complete assessment [11, 16].

In our previous studies, a novel halogen-free FR chemical called Fire-off was synthetized and applied to cotton, polyester and CO/PET blend fabrics. According to the findings, the thermal stability of the fabrics and their flammability behaviours strongly affected by Fire-off finishing and all treated samples lead to favour formation of char [17]. Fire-off is also an eco-friendly product not containing formaldehyde, either. Besides its thermal and toxicity assessments, durability, comfort, tensile properties and appearance of FR treated fabrics are also in significance for practical end uses and that should be further stage for commercialization of a developed FR product.

In the present of the study, mechanical properties of Fire-off has been investigated in order to assess textile service-life properties. Concerning to fabric strength, abrasion properties and outward appearance; tensile properties, abrasion, pilling performance and whiteness of untreated and Fire-off treated fabrics were tested and evaluated. Thermal test results showed that there was a slight increase in strength, elongation, abrasion and pilling. On the other hand, whiteness and air permeability of fabrics were decreased after Fire-off treatment.

2. Materials and Method

2.1. Materials
The experiments were conducted using scoured 50/50% CO/PET fabrics (227 and 292 g/m^2, twill) supplied from Ata Textile in Turkey. 50/50 % was chosen, because flammability of this blend is much more critic with flammability and burning behaviour than other blend ratio of CO/PET fabrics such as 70/30, 65/35 etc. As chemical, Fire-off (PVP (PR)-P-DCDA) supplied from Eksoy Chemical was used for FR treatments of fabrics.

2.2. Fabric Finishing Procedure
FR bath with 350 g/L Fire-off was prepared in water. The fabric samples were impregnated by 2 dip-2 nip padding using a laboratory padder at room temperature. The wet pick-up was approximately 100±2%. After padding process, they were dried at 100ºC for 3 min, and subsequently cured in an oven at 180ºC for 3 min.

2.3. Evaluation of flame retardancy
Limiting Oxygen Index (LOI) test was performed for cotton and CO/PET blends according to BS EN ISO 4589-2 standard using Concept Equipment Oxygen Index Module to assess their FR properties. Before to test, specimens were conditioned for 24 h under the temperature of 20 ±2°C and relative humidity of 65±5%.

2.4. Determination of tensile property
Tensile Properties of treated and untreated CO/PET fabrics were tested according to ISO 13934-1 (2013) using an Instron Tension tester to assess fabric strength and elongation alteration due to the FR treatment with Fire-off.

2.5. Determination of abrasion and pilling resistance
ISO 12947-2 (2017) and ISO 12945-2 (2002) standard procedures were carried out to assess abrasion and pilling properties of untreated and treated CO/PET fabrics, respectively.

To perform abrasion test, the fabric samples were subjected to 50.000 or 100.000 abrasive cycles in a Martindale apparatus with a pressure of 9 kPa. In piling test, 125, 500, 1000, 2000, 5000 and 7000 cycles were performed, respectively. Visual evaluation values were determined by 3 observer and average results of 3 specimens were calculated.

2.6. Determination of air permeability
The air permeability of untreated and treated fabrics were measured by Prowhite Air Test II with a pressure applied of 100 Pa, according to ISO 9237 (1995). Average results of 5 specimens were calculated.

2.7. Determination of whiteness
The whiteness index values of untreated and treated fabrics were measured on a spectrophotometer with a commercial name of Datacolor. The whiteness index values evaluated according to the CIE lab.

3. Results and Discussion

3.1. Evaluation of the flame retardancy
To evaluate FR properties of the treated fabrics, LOI of untreated and treated samples was also tested and the data obtained were given in Table 1.

Table 1. LOI of CO/PET fabrics.

Fabric Type	Untreated fabrics	Fire-off treated fabrics
CO/PET (227 g/m^2)	18.7	27.1
CO/PET (292 g/m^2)	18.8	26.7

The LOI values of fabrics were increased from around 18.5 to 27 after Fire-off treatment that indicate FR properties.

3.2. Determination of tensile property
Tensile Properties of CO/PET fabrics before and after treatment were given in Table 2.

Table 2. Tensile properties of CO/PET fabrics.

Fabric Type	Untreated fabrics		Fire-off treated fabrics	
	(N)	(%)	(N)	(%)
CO/PET (227 g/m^2)	1376.51	21.07	1404.20	27.20
CO/PET (292 g/m^2)	1365.52	19.59	1274.13	22.00

Even though, it is expected that FR treatments deteriorate tensile properties of the fabrics, it is clearly seen from the data in Table 3.4, tensile strength of CO/PET samples do not change considerably. On the other hand, elongation values of CO/PET samples increase slightly after the Fire-off treatment.

3.3. Determination of abrasion and pilling performance
Abrasion resistance results can be seen in Table 3 and Figure 1.

Table 3. Abrasion resistance results of CO/PET fabrics.

Fabric type	FR (g/L)	25,000-30,000 cycles	30,000-35,000 cycles	35,000-40,000 cycles	40,000-45,000 cycles	45,000-50,000 cycles
CO/PET (227 g/m^2)	-	+	-	-	-	-
CO/PET (292 g/m^2)	-	+	+	+	+	+
CO/PET (227 g/m^2)	350	+	+	+	+	+
CO/PET (292 g/m^2)	350	+	+	+	+	+

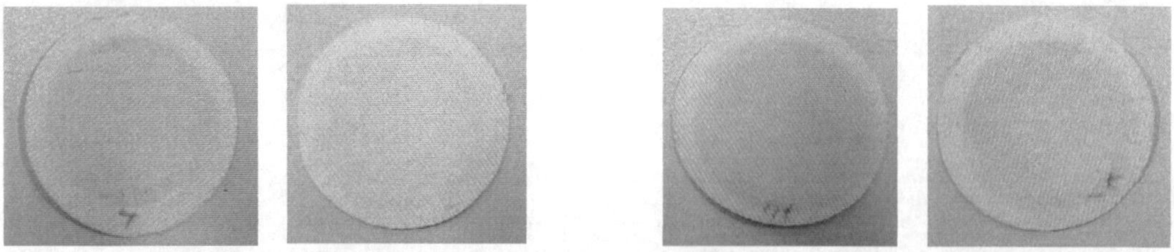

CO/PET (227 g/m²) CO/PET (292 g/m²)
A: Untreated samples after 50.000 cycles.

CO/PET (227 g/m²) CO/PET (292 g/m²)
B: Treated samples after 100.000 cycles.

Figure 1. Untreated and treated sample pictures after abrasion test.

Untreated lightweight CO/PET (227 g/m^2) fabric failed in 30,000 and 35,000 cycles. However, 292 g/m^2 CO/PET fabric did not have any breakage after 100,000 abrasion cycles. On the other hand, treated fabrics passed to abrasion test with no breakage and all demonstrated a super resistance to abrasion.

Visual assessment results of pilling test are shown in Table 4.

Table 4. Piling results of CO/PET fabrics.

Fabric type	FR treatment	Visual assessment values (After 7000 cycles)
CO/PET (227 g/m²)	-	4
CO/PET (292 g/m²)	-	3-4
CO/PET (227 g/m²)	350 g/ L	4-5
CO/PET (292 g/m²)	350 g/ L	4

Visual assessments show that there was no pilling occurrence on CO/PET fabric surfaces after Fire-off treatment while lightweight untreated CO/PET fabrics (227 g/m^2) have hairiness, heavier weight of untreated CO/PET fabric (292 g/m^2) have hairiness and also pilling on some part of fabric surface. After treatment of Fire-off, little or no hairiness was observed on fabric samples.

3.4. Determination of air permeability

Air permeability results of fabrics were given in Table 5. Air permeability should be maintained in any ideal chemical treatment. However, results indicate that fabrics after FR treatment are less permeable to air (> 12-23%) due to the chemical deposition onto surface, as it is expected.

Table 5. Air permeability results of CO/PET fabrics.

Fabric type	Untreated	Fire-off treated
CO/PET (227 g/m²)	61	23
CO/PET (292 g/m²)	30	12

3.5. Determination of whiteness

Table 6 exhibits the effects of Fire-off treatment on fabric whiteness index. Fire-off tends yellowing effect on the fabric (35-41 %), it may be due to acidic condition (pH =3-4) that in the agreement of Li et al (2011) [9].

Table 6. Whiteness index data of CO/PET fabrics.

Fabric Type	Untreated	Fire-off treated	Change %
CO/PET (227 g/m²)	73.41	42.70	41
CO/PET (292 g/m²)	68.05	44.02	35

4. Conclusion

In order to detect any change on strength, elongation, pilling and abrasion values, whiteness index of fabrics, Fire-off treated fabrics were tested mechanically for product quality. Moreover, relative handles of the Fire-off treatments was observed for their practical application, particularly for apparel end uses. Tensile test results demonstrated that treated fabric strength did not drastically changed and elongation of CO/PET increased after Fire-off treatment. It was found that Fire-off treatment significantly increase abrasion resistance of CO/PET fabrics. As a result of visual assessment of pilling test, treated samples had none pilling and only little hairiness on surface of fabrics as opposed to untreated ones. Concerning to air permeability, treated fabrics showed lower permeability to air as expected. Although, Fire-off induces yellowing effect on fabrics, the change on whiteness of treated fabrics were not dramatic. Furthermore, FR treatment in 350 g/L concentration is quite high chemical add-on the fabrics; it was surprisingly observed that fabric handles were not affected drastically from the treatment.

To sum up, changes in mechanical properties of fabrics after the Fire-off treatment demonstrate a definitive trend of increasing elongation, abrasion, pilling properties and decreasing air permeability and whiteness.

Acknowledgment

The authors gratefully acknowledge the funding by I.T.U BAP, under grants of 39775 project. We wish to thank Ata Textile for fabric supply.

References

[1] Hendrix JE, Drake GL, Reeves WA (1971) Effects of Moisture on Oxygen Index (OI) Values for Textiles. *Textile Research Journal* (41): 854.

[2] Jarvis CW and Barker RH (1978) Flammability of Cotton Polyester Blends. In: Atlas SM, Pearce EM and Lewin M (eds) *Flame Retardant Polymeric Materials*. New York and London: Plenum Press, pp.133-135.

[3] Day M, Suprunchuk T, Wiles DM (1983) A Combustibility Study of the Gaseous Pyrolysates Produced from Some High-Performance Fabrics. *Journal of Applied Polymer Science* (28):3681-3693.

[4] Pan N, Postle R (1995) Strengths of Twisted Blend Fibrous Structures: Theoretical Prediction of the Hybrid Effect. *Journal of the Textile Institute* (86):559-580.

[5] Schindler WD and Hauser PJ (2004) *Chemical Finishing of Textiles.* Cambridge-England: Woodhead Publishing Ltd.

[6] Horrocks AR, Kandola BK, Davies PJ, Zhang S and Padbury SA (2005) Development in Flame Retardant Textiles. *Polymer Degradation and Stability* (88): 3-12.

[7] Bourbigot S (2008) Flame retardancy of textiles: new approaches. In: Horrocks AD and Price D (eds) *Advances in Fire Retardant Materials.* Cambridge-England: Woodhead Publishing Ltd, pp.10.

[8] Li Q-L, Wang X-L, Wang D-Y, Xiong W-C, Zhong G-H, Wang Y-Z (2010) A Novel Organophosphorus Flame Retardant: Synthesis and Durable Finishing of Poly(Ethylene Terephthalate)/Cotton Blends. *Journal of Applied Polymer Science* (117):3066-3074.

[9] Li Q-L, Wang X-L, Wang D-Y, Xiong W-C, Zhong G-H, Wang Y-Z (2011) Durable Flame Retardant Finishing of PET/Cotton Blends Using a Novel PVA-Based Phosphorus-Nitrogen Polymer. *Journal of Applied Polymer Science* (112)342–353.

[10] Reeves W A and Marquett Y B (1979) Lightweight, Durable-Press Cotton and Polyester/Cotton with Ignition Resistance, *Textile Research Journal* 49, 163.

[11] Salmeia KA, Gaan S, and Malucelli G, (2016) Recent Advances for Flame Retardancy of Textiles Based on Phosphorus Chemistry *Polymers* 8, 319; doi:10.3390/polym8090319.

[12] Gaan S, Mauclaire L, Rupper P, Salimova V, Tran T.-T and Heuberger M. Thermal degradation of cellulose acetate in presence of bis-phosphoramidates. *J. Anal. Appl. Pyrolysis* 2011, 90, 33–41.

[13] Hirsch C, Striegl B, Mathes S, Adlhart C, Edelmann M, Bono E, Gaan S, Salmeia, K.A, Hoelting L, Krebs A et al. Multiparameter toxicity assessment of novel DOPO-derived organophosphorus flame retardants. *Arch. Toxicol.* 2016, 1–19.

[14] Sun S, He Y, Wang X, Wu D. Flammability Characteristics and Performance of Halogen-Free Flame-Retarded Polyoxymethylene Based on Phosphorus-Nitrogen Synergistic Effects. *Journal of Applied Polymer Science.* 2010, Cilt 118:611-622.

[15] Lu S. Y. and Hamerton I. Recent Developments in the Chemistry of Halogen-free Flame Retardant Polymers. *Progress in Polymer Science.* 2002, Vol. 27:1661-1712.

[16] Horrocks AR (2011) Flame retardant challenges for textiles and fibres: New chemistry versus innovatory solutions. *Polymer Degradation and Stability* (96):377-392.

[17] Atakan R, Bical A, Celebi E, Ozcan G, Soydan N and Sarac A (2018) Development of a Flame Retardant Chemical for Finishing of Cotton, Polyester and CO/PET Blends. *Journal of Industrial Textiles*, DOI: 10.1177/1528083718772303.

Seam Strength and Washability of Silver Coated Polyamide Yarns

E. Ismar[1], S. Zaman[2], S.K. Bahadir[3], F. Kalaoglu[4], and V. Koncar[5]

[1] *Department of Nano Science & Nano Engineering, Istanbul Technical University, Maslak, Turkey*

[2,5]*GEMTEX Laboratory, École Nationale Supérieure des Arts et Industries Textiles, Roubaix, France, University of Lille, Lille, France*

[3,4]*Department of Textile Engineering, Istanbul Technical University, Gumussuyu, Turkey*

Corresponding author: ismar@itu.edu.tr

Abstract Seam strength has a huge impact on the durability of the garment. Conductive yarn usage in a stitching while development the e-textile structures is often used to connect electronic modules to sensors and actuators. Most common conductive yarn, used for the joining of e-textile products, is silver coated polyamide yarn. In this standpoint, chosen woven fabric stitched with silver coated polyamide yarn and seam strength was measured in both directions, weft, and warp and seam efficiency was calculated. Moreover, washing procedure was applied to the stitched samples to investigate the changing behavior of mechanical strength and electrical conductivity through the washing. Durability of the e-textiles is one of the key factors of maintaining the product through the washing. In this perspective, mechanical performances of the seam were tested together with the washing and electrical resistance values of the seam were recorded. After the washing procedure, seam elongation was decreased for both warp and weft directions. As expected, washing cycle has a negative effect on the silver coated yarns thus, some signal damage was recorded. However, electrical conductivity of the silver coated yarns is not totally destroyed after the gentle washing cycle.

Keywords— **Conductive thread, Conductive yarn, E-textiles, Silver coated yarn, Washing of e-textiles**

I. Introduction

Textile industry gains different forms due to the emerging fields such as e-textiles. Conventional textile products are combined with the smart solutions to obtain an electronic feature on the wearable textile products. E-textiles mainly consist of two parts; one is textile and other is an electronic part. To interconnect the electronic parts among them and with power supplies, conductive threads are used. These conductive threads are the backbone of the smart textile system. Efficient smart textile products should be strong enough to bear the strength and also electrically conductive enough to guaranty a good signal quality and proper power supply.

To transmit data from textile sensors or towards flexible actuators, in the field of medical, sports or leisure applications, conductive threads should be able to transmit all information [1]. E-textile prototypes are often developed in research laboratory worldwide, but they exhibit poor washability and reliability limiting their access to the market. Consequently, conductive threads together with sensors actuators and electronic modules should withstand numerous laundering cycles. They are used for instance to supply LED arrays, ECG monitoring systems embedded to underwear etc. Interconnection of conductive threads with sensors actuators and electronic modulus is also a major problem. They are used to draw conductive patterns, in order to make textile flexible circuit boards. For pattern making, usually embroidery and stitching techniques are used [2], [3].

Different types of conductive threads are available on the market. Mainly, conductive yarns are produced by spinning and coating processes. For spinning processes, raw material can be in melt or wet form, made of conductive polymer solution, used to obtain conductive fibers [4], [5]. For coating processes; electrically conductive materials or metal powders are used as coating compounds to cover the yarns and provide the electrical conductivity [6], [7]. For instance; silver coated polyamide yarns or PEDOT:PSS coated yarns. However, the most common and the easiest threads to stitch are silver coated polyamide yarns. They are available on the market with different diameters and different number of plies. These threads are stitched or embroidered on the fabric. In both cases, stitch efficiency is one of the major factors responsible for durability and reliability of these structures.

Stitching has a key role on the e-textile designs as a joining element [8] thus, seams should be robust enough to withstand everyday use and laundering.

Stitching processes should also be conducted gently to prevent the thread damages during it. If conductive threads are damaged through the stitching process, their electrical resistance may become high or even infinitive provoking the loss of electrical conductivity. Another possible damaging factor is a washing of e-textile samples. Gentle washing cycles should be applied [9], in washing standard ISO 6330, different washing times are discussed depending on type of washing materials. For instance, maximum washing time is for cotton and minimum for silk.

A lot of research works have been done on smart textile products. Only few of them focus on the seam importance of conductive threads. Seam is one of the major key factors in joining of e-textiles. If conductive yarn used as a seam is damaged, all prototypes fail. Signals generated by sensors or sent to actuators within smart textile structures are conducted through these conductive threads. In our study, linear electrical conductivity and mechanical strength of conductive threads are discussed and effects of washing on these parameters are examined.

Mechanical performance of seam was measured by taking into account the effect of washing performance of the basic joining unit. Also, electrical conductivity is recorded for both directions (weft and warp) before and after washing samples in order to examine and establish the relationship between stitching direction and electrical conductivity of silver coated polyamide yarns.

II. Procedures

Cotton/Polyester woven fabric, 127.4 g/cm^2 is selected for the experimental studies. Shildtex f117/17 silver coated polyamide yarn was used as a conductive yarn. This is two ply yarn with 17 fibers in each ply. Its linear resistance before stitching is 300 ohm/m. Linear resistance was measure by Keithley 8009 device. Basic sewing machine from SINGER was used to produce the stitched samples. Plain stitching type was followed and stitch steps are about 4.7 stitches per cm as mentioned in the standard. For measuring the seam strength, ASTMD 1683 standard was followed for preparing the samples. Fabric strength and seam strength are separately measured to calculate the seam efficiency. MTS Criterion Model 43 device was used to measure the tensile strength of fabric and seams.

$$E = Ss/Fs \times 100 \qquad (1)$$

All samples (warp and weft direction stitched) are washed in the domestic type of laundering machine and parameters were set to 40 rpm washing speed, 600 rpm spin speed and 30°C water temperature. Tumble dryer was not used for drying of the samples. After the washing process resistance values of stitched yarns and seam strength were measured and analyzed.

III. Results and discussion

Mechanical performance of unstitched fabrics was examined to calculate the seam strength. Accordingly, mechanical test was applied in both warp and weft direction fabrics. Elongation percentage of fabric without seam and with seam in warp and weft direction is shown in Figure 1 and error bars represents the fluctuating value range. Elongation in warp direction was relatively higher comparing to weft direction.

Afterwards, seam stress was measured for the unwashed samples. Linear resistance of these samples was determined in kohm per 5 cm. These samples are then washed and dried at room temperature. For seeking the washing effect on the stitched samples, seam strength and conductivity measurements were realized. Resistance values of stitched samples before and after washing are given in Table 1 with distance between the probes set to 5cm for all the samples. Five samples were tested for warp and five for weft direction and calculated average values are presented together with their standard deviations. Figure 2 gives the comparison of resistance values for warp and weft directions. After washing procedure, a warp direction seams show more important variations in resistance values.

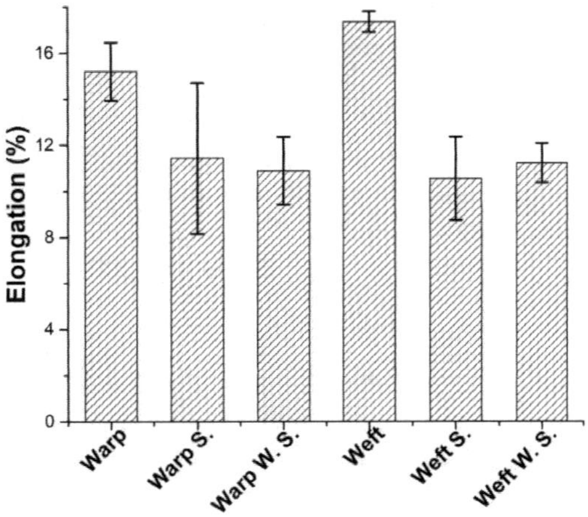

Fig. 1. Elongation at break values for both directions of fabric, S. refers to seam and W. refers to washed sample.

Seam has a key role on the serviceability of the apparel thus, it should be robust enough [10]. Silver coated polyamide yarns are one of the most common conductive yarns. In this study the seam strength of the yarns before and after washing for the aim of its usage in e-textiles as a joining element is analyzed.

Stretchable electronic textile modules have an importance to maintain the comfort and wearability of textile based electronic materials [9]. Due to that reason, conductive threads should exhibit flexibility. Silver coated polyamide yarns are stitched on the both directions and elongation at break values of weft direction seam represents more extension than for the warp direction. However, elongation values decreased after the washing cycle for both directions and values approached.

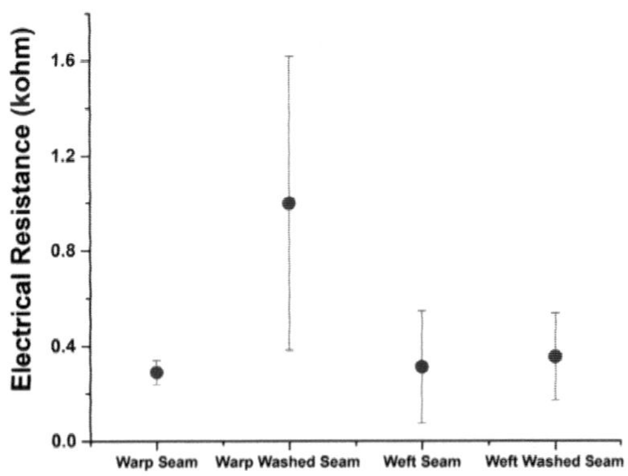

Fig. 2. Compression of resistance values before and after washing for both directions.

Average electrical resistance value for warp direction seam increased dramatically after the first washing process, when compared with the weft direction seam. Moreover, the standard deviation of the washed seam on warp direction became higher due to the effects of the washing process. Increased standard deviation on the electrical resistance points the poor reliability of the joining element. Mechanical performance values should be evaluated together with the electrical resistance values before choosing the stitching direction of conductive yarn.

TABLE I
ELECTRICAL RESISTANCE VALUES OF STITCHED SILVER COATED YARNS, BOTH WEFT AND WARP DIRECTIONS BEFORE AND AFTER WASHING.

	Electrical Resistance (KΩ)			
	Before Washing		After 1st Washing	
	Warp	Weft	Warp	Weft
	0.21	1.04	1.10	0.52
	0.35	0.22	0.34	0.11
	0.28	0.34	1.56	0.32
	0.29	0.71	-	0.47
	0.32	0.73	-	-
Average	0.29	0.608	1.00	0.35
STDEV	0.05	0.33	0.62	0.18
CV (%)	18	54	62	52

Table 2 and table 3 show the seam strength, seam efficiency and breaking force for fabric and conductive seams before and after washing. Seam efficiency for warp direction washed samples was highest in all cases. Seam efficiency increased after washing in both warp and weft direction. Figure 3 show the tensile strengths of seam in warp and weft directions after washing. These values are almost in the same range.

TABLE II
Tensile test results for warp and weft directions.

Direction	Fabric Breaking Force (N)	Seam Strength
Warp	405.61±61.12	
Weft	430.96±57.70	
Warp Seam		281.47±147
Weft Seam		246.63±61
Warp Washed Seam		305.88±71
Weft Washed Seam		295.46±60

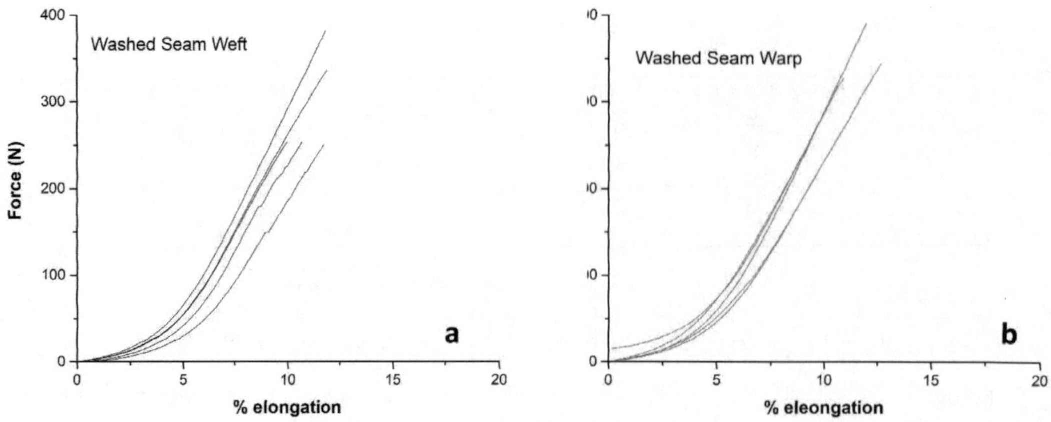

Fig. 3. Tensile test results of the stitched fabrics in (a) weft direction, after washing, (b) warp direction, after washing.

Strength of the seams slightly increased after the washing cycle, while average elongation at break values decreased for the washed samples, on the other side breaking force increased for both warp and weft directions.

TABLE III
SEAM EFFICIENCY CALCULATION RESULTS.

Direction	Seam Efficiency (%)
Warp Seam	69.39
Weft Seam	57.23
Warp Washed Seam	75.41
Weft Washed Seam	68.56

IV. CONCLUSION

Mechanical and electrical resistance have been measured and it was seen that: first cycle of gentle washing cycle has not a huge impact on the seam strength of silver coated polyamide yarns. Results point out that; silver coated polyamide yarn can resist the gentle washing cycle. However, the electrical resistance is affected inversely due to the damage of coating layer. It means that coated silver layer has not any effect on the mechanical performance of the polyamide yarn, and silver coating only supports the electrical conductivity of the polyamide yarn. Weft direction stitching is more flexible for unwashed samples, as expected. After, the washing cycle, elongation values decreased for both weft and warp directions. Silver coated polyamide yarns are strong enough to stitch and wash for both directions of the fabric.

Silver coated yarns can be a good alternative as a joining element for flexible electronics, they exhibit a certain elongation, but they preserve their electrical conductivity feature.

ACKNOWLEDGMENT

Authors would like to acknowledge the E-TEXWELD project which is the "European Union's Horizon 2020 research and innovation programme under the Marie Skłodowska-Curie Grant agreement no. 644268".

REFERENCES

[1] K. A. Asanovic, T. A. Mihajlidi, S. V. Milosavljevic, D. D. Cerovic, and J. R. Dojcilovic, "Investigation of the electrical behavior of some textile materials," J. Electrost., vol. 65, no. 3, pp. 162–167, Mar. 2007.
[2] M. Stoppa and A. Chiolerio, "Wearable Electronics and Smart Textiles: A Critical Review," Sensors, vol. 14, no. 7, pp. 11957–11992, Jul. 2014.
[3] I. Kazani et al., "Washable screen printed textile antennas," in Advances in Science and Technology, 2013, vol. 80, pp. 118–122.
[4] S. J. Pomfret, P. N. Adams, N. P. Comfort, and A. P. Monkman, "Advances in processing routes for conductive polyaniline fibres," Synth. Met., vol. 101, no. 1, pp. 724–725, May 1999.
[5] C. Cochrane, V. Koncar, M. Lewandowski, and C. Dufour, "Design and Development of a Flexible Strain Sensor for Textile Structures Based on a Conductive Polymer Composite," Sensors, vol. 7, no. 4, pp. 473–492, Apr. 2007.
[6] B. Kim, V. Koncar, E. Devaux, C. Dufour, and P. Viallier, "Electrical and morphological properties of PP and PET conductive polymer fibers," Synth. Met., vol. 146, no. 2, pp. 167–174, Oct. 2004.
[7] Z. Hua, Y. Liu, G. Yao, L. Wang, J. Ma, and L. Liang, "Preparation and Characterization of Nickel-Coated Carbon Fibers by Electroplating," J. Mater. Eng. Perform., vol. 21, no. 3, pp. 324–330, Mar. 2012.
[8] F. K. Ko, "Seaming and joining methods," Geotext. Geomembr., vol. 6, no. 1, pp. 93–107, Jan. 1987.
[9] T. Vervust, G. Buyle, F. Bossuyt, and J. Vanfleteren, "Integration of stretchable and washable electronic modules for smart textile applications," J. Text. Inst., vol. 103, no. 10, pp. 1127–1138, Oct. 2012.
[10] A. K. Choudhary and A. Goel, "Effect of Some Fabric and Sewing Conditions on Apparel Seam Characteristics," Journal of Textiles, 2013. [Online]. Available: https://www.hindawi.com/journals/jtex/2013/157034/. [Accessed: 16-Apr-2018].
[11] U. Chowdhary and D. Poynor, "Impact of stitch density on seam strength, seam elongation, and seam efficiency," Int. J. Consum. Stud., vol. 30, no. 6, pp. 561–568, Nov. 2006.
[12] M. Akter and M. R. Khan, "The effect of stitch types and sewing thread types on seam strength for cotton apparel," Int. J. Sci. Eng. Res., vol. 6, no. 7, pp. 198–205, 2015.

High Frequency Attenuation Characterization of Knitted E-Textile Structures

S. Kursun Bahadır[1], Ö. Atalay[2], F. Kalaoglu[3], S. A Mitilineos[4], and S. Vassiliadis[5]

[1,2,3] *Istanbul Technical University, Faculty of Textile Technologies and Design, Textile Engineering Department, Istanbul Turkey*

[4,5] *University of West Attica, Department of Electronics Engineering, Athens, Greece*

Corresponding author: kursuns@itu.edu.tr

Abstract In this research study, knitted fabrics were produced on an E=12 gauge electronic flat bed knitting machine for the High frequency attenuation characterization in the frequency range of 1 GHz -1,5 GhZ. Conductive yarns with different linear resistances were knitted into non-conductive base fabrics made from double covered PA 6.6 core spun lycra yarns. Two different design approaches have been applied for manufacturing of samples. In the first approach, conductive yarn has been knitted in plain arrangement into elastomeric interlock base fabric. In latter case, the base fabric was produced with elastomeric yarns in an interlock arrangement and a conductive yarn was embedded in this substrate to create a series of single loop structures. Effect of design approaches and conductive yarn linear resistance on high frequency attenuation properties of conductive knitted samples was investigated. It was observed while design differences have more effect on attenuation characteristics of samples, linear resistance values of conductive yarns have also slightly affected the properties of samples.

Keywords— **Attenuation, Conductive Knits, E-Textiles, High Frequency, Conductive Yarns**

I. Introduction

INTEGRATION of conductive materials into textiles is promoting a new range of functional textile materials over traditional textile applications, which are woven [1], knitted [2], and embroidered [3] to produce fabric for both structural and aesthetic purposes. Conductive materials, i.e., conductive yarns [4], conductive coating [5] or printing materials [6], can be used for creation of textile-based sensors [7], transmission lines [8] or textile-based antennas [9] within the electronic textile structures. Since electromagnetic waves radiates from electronic devices threats to human health, one specific application area of conductive textiles is creation of textile structures with electromagnetic shielding properties and there are many studies shows that harmful effect of electromagnetic field [10-12]. These harmful effects includes but not limited to increase the chance of developing a brain tumor, cancers and problems associated with nervous system. Thus, conductive textile structures are suitable candidates due their lightweight, flexibility, as well as ease of manufacturing of desired products for wide range applications thanks to the developments in textile machinery and technology.

There are a few studies that investigate electromagnetic shielding properties of conductive textile structures. Coating of textile structures with conductive materials is one way to produce these structures and metal deposition on textile surfaces can be used to generate electromagnetic shielding. Kardarian et al. created silver coated textile structures by in situ synthesis of silver particles on textile surface combination with sintering process and they concluded that sintering process increased the shielding effectiveness of the structures [13]. Electroless plating is also widely preferred method to produce metal-coated electromagnetic shielding fabrics [14]. Inherently conducting polymers, i.e., polypyrrole, polyaniline can also be applied on textile surfaces to develop shielding fabrics [15]. Another main approach for the creation of textile-based electromagnetic shielding structures is the embedding of conductive yarns into knitted or woven structures [16]. As a distinction from the coated structures, they offer an integration of the conductive part during the manufacturing stage of the fabric. Thus, this approach reduces the production stage to one step. In addition to this, since conductive materials relatively more expensive, processing of conductive yarns along with nonconductive yarns to form textile structures is desirable in terms of cost issues.

The main purpose of this research was to investigate high frequency characterization of various knitted textile structures. One of the structures has been successfully employed as a strain sensor in our previous study [7]. Thus, this study will also reveal possible usage of same design for different application areas. The following section describes production of knitted strain samples followed by the measurement method for the high frequency

IOP Conf. Series: Materials Science and Engineering **460** (2018) 012054 doi:10.1088/1757-899X/460/1/012054

attenuation characterization in the range of 1 GHz and 1,5 GHz .The third part reports the results obtained from the experimental procedure and discussion of the electromagnetic shielding properties of the knitted structures.

II. Materials and Methods

A. Materials

In this study, in order to do the high frequency attenuation characterization of e-fabrics, two silver conductive yarns with linear resistances of 50 ohm/m and 235 ohm/m were used to form electrical circuits in knitted fabric samples. The linear resistance of the conductive yarns was measured in ohm per meter (ohm/m) using a TTi 1906 computing multimeter. Polyamide covered lycra yarn with a yarn density of 800dtex was used to form an insulating area in the structure. The samples were produced by Shima Seiki SES124-S 12GG flat knitting machine. Two different knitting design were used to test the fabrics. The knitted designs of the samples with their original images are shown in Fig.1. The grey colored area shows the conductive zones which are designed using silver plated PA yarns.

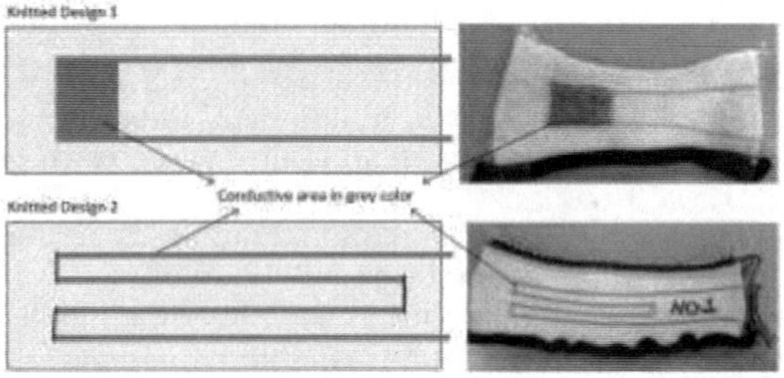

Fig. 1. Knitted sample designs

B. Measurement method

In order to measure the high frequency attenuation of the samples, a straight wave guide with its two coaxial to waiveguide converter,WR650 was used. Signals generated from port of the vector network analyser (VNA- Rohde & Schwarz ZVA24) were sent to waveguide and then, the signals passing through the waveguide are transmitted via knitted samples to the input port of VNA . The measurement set up is shown in Fig.2. As seen in the figure, during the measurements all the fabric samples were placed in vertical position inside the waveguide.

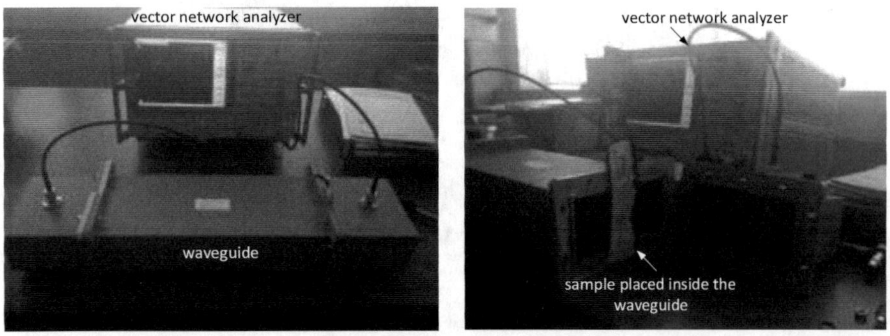

Fig. 2. Measurement set-up and placement of the fabric inside the waveguide

III. Results and discussion

Using the measurement setup of Figure 2 the attenuation through the waveguide is measured by the vector network analyzer (VNA). The spectrum of the guided wave that is excited by the VNA's generator and which propagates through the waveguide at the TE10 polarization mode, is analyzed by the instrument. The attenuation due to the inserted yarn is derived as the difference between the reference response and the sample's response at each frequency point of the two curves. Using this procedure the spectral attenuation of each sample of design 2 in the range of 1.1 to 1.7GHz is depicted in Figure 3.

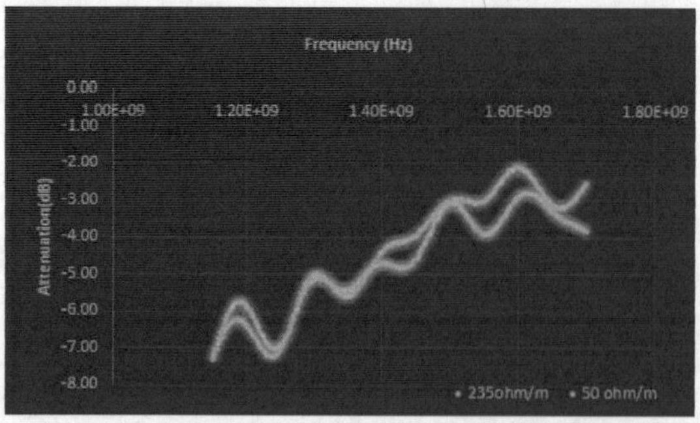

Fig. 3. Attenuation of knitted e-fabric design 2 including silver plated PA conductive yarn with a linear resistance of <235ohm/m and 50ohm/m

In particular, the higher conductive sample exhibits a higher attenuation in the upper part of the examined frequency band. Although this is justified partially by the increased reflectivity of the lower resistive yarn, the dimensions and the shape of the specific design are also critical. In particular, the attenuation increases in the lower band due to the small separation between the conductive lines in comparison with the wavelength at low frequencies. As a result, both samples react as a solid reflector with the same attenuation. On the other hand, at the upper frequency band, the wavelength decreases and the pattern of the conductive yarn design is more recognizable by the electromagnetic wave. As a result, the conductivity of the yarn of each sample is crucial and justifies both, the higher attenuation of the lower resistive yarn and the overall lower attenuation at the upper band.

Furthermore, a ripple is also evident for both samples throughout the spectrum. This ripple is caused by the expected interference phenomena caused by the interaction of the waveguided field with the textile obstacle. It should be also noted that the interference phenomena are weaker for the lower conductive sample due to the decreased reflectivity through the sample, which is mainly evident at the upper frequency band.

In Figure 4 the sample with the conductive area is examined (design1). The position and the area of the conductive sample, explains the reduced attenuation in comparison with the samples of the design 2.

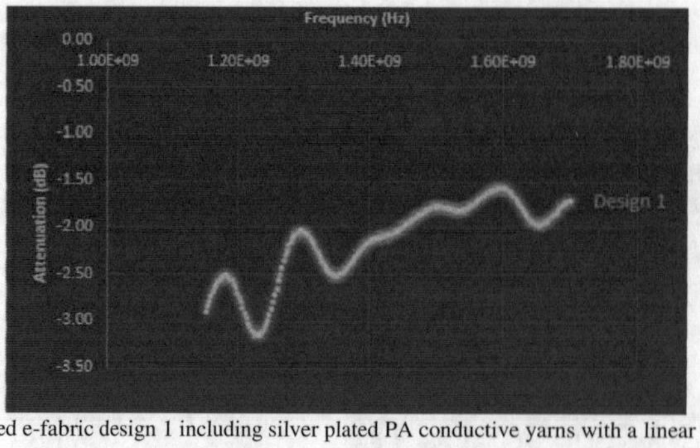

Fig. 4. Attenuation of knitted e-fabric design 1 including silver plated PA conductive yarns with a linear resistances of <235ohm/m

The spectral response of the attenuation is rather similar to the one discussed before with the difference of a lower attenuation throughout the spectrum. This is an anticipated result for two reasons: first, due to smaller effective reflection area in comparison with the previous samples, and second because the reflection area is not distributed deeply inside the waveguide, therefore, it has a small impact on the field profile of the wave guided mode.

Acknowledgment

This project has received funding from the European Union's Horizon 2020 research and innovation programme under the Marie Skłodowska-Curie Grant agreement no. 644268.

References

[1] A. Dhawan, "Woven fabric-based electrical circuits: Part I: Evaluating interconnect methods", *Textile Research Journal, vol.* 74, no.4, pp. 913-919, 2004.

[2] R. Ma, J. Lee, D. Choi, H. Moon, and S. Baik,"Knitted fabrics made from highly conductive stretchable fibers", *Nano*

letters, vol. 14, no.4, pp. 1944-1951, 2014.
[3] A. Tsolis, W. G. Whittow, A. Alexandridis, and J.C. Vardaxoglou, "Embroidery and related manufacturing techniques for wearable antennas: challenges and opportunities," *Electronics*, vol.3, no.2, pp. 314-338, 2014
[4] O. Atalay, W.Kennon,and E. Demirok, "Weft-knitted strain sensor for monitoring respiratory rate and its electro-mechanical modeling". *IEEE Sensors Journal*, vol. 15, no. 1, pp. 110-122, 2015.
[5] H.W. Cui, K. Suganuma, and H. Uchida "Highly stretchable, electrically conductive textiles fabricated from silver nanowires and cupro fabrics using a simple dipping-drying method". *Nano Research*, vol. 8, no.5, pp. 1604-1614, 2015.
[6] H. Jinno, K. Kuribara, M. Kaltenbrunner, N. Matsuhisa, T. Someya, T. Yokota, and T. Sekitani, "Printable elastic conductors with a high conductivity for electronic textile applications", *Nature communications*, vol. 6, pp. 7461, 2015.
[7] O. Atalay, A. Tuncay, M.D. Husain, and W.R. Kennon"Comparative study of the weft-knitted strain sensors", *Journal of Industrial Textiles*, vol.46, no.5, pp. 1212-1240, 2017.
[8] S.K. Bahadir, S. Jevsnik,, "Optimization of hot air welding process parameters for manufacturing textile transmission lines for e-textiles applications: Part I: Electro-conductive properties. *Textile Research Journal*, vol. 87, no.2, pp. 232, 2017.
[9] S. Shao, A. Kiourti, R.J. Burkholder, and J.L. Volakis "Broadband textile-based passive UHF RFID tag antenna for elastic material". *IEEE Antennas and Wireless Propagation Letters,* vol. 14, pp. 1385-1388., 2015.
[10] M. Ulubay, A. Yahyazadeh, Ö.G. Deniz, E.G. Kıvrak,B.Z. Altunkaynak, G. Erdem, and S. Kaplan "Effects of prenatal 900 MHz electromagnetic field exposures on the histology of rat kidney". *International journal of Radiation Bbiology*, vol. 91, no.1, pp.35-41, 2015.
[11] M.L. Pall, "Microwave frequency electromagnetic fields (EMFs) produce widespread neuropsychiatric effects including depression". *Journal of Chemical Neuroanatomy*, vol.75, pp. 43-51, 2016.
[12] S. Dasdag, M.Z. Akdag,, Erdal, M. E., Erdal, N., Ay, O. I., Ay, M. E.,and K. Yegin. "Effects of 2.4 GHz radiofrequency radiation emitted from Wi-Fi equipment on microRNA expression in brain tissue", *International Journal of Radiation Biology*, vol. 91, no.7, pp. 555-561, 2015.
[13] K. Kardarian T. Busani, I., Osório, I., Domingos, H., R. Igreja, R.Franco, and J. Cortez "Sintering of nanoscale silver coated textiles, a new approach to attain conductive fabrics for electromagnetic shielding". *Materials Chemistry and Physics*, vol. 147, no.3, pp. 815-822, 2014.
[14] S.M. Kim, I.Y. Kim, H.R. Kim, "Production of electromagnetic shielding fabrics by optimization of electroless silver plating conditions for PET fabrics." *The Journal of The Textile Institute*, vol. 108, no.6,pp. 1065-1073, 2017.
[15] S. Maity, S., and A. Chatterjee, "Polypyrrole functionalized polyester needlepunched nonwoven fabrics for electromagnetic interference shielding". Polymer Composites., 2017.
[16] D. Duran, and H. Kadoğlu, "Electromagnetic shielding characterization of conductive woven fabrics produced with silver-containing yarns", *Textile Research Journal*, vol. 85, no. 10, pp. 1009-1021, 2015.

IOP Conf. Series: Materials Science and Engineering **460** (2018) 012055 doi:10.1088/1757-899X/460/1/012055

Structure Properties Change of Ready to Use Nonwoven Wiping Materials over Storage Time

X. Song[1], L. Vossebein[2] and A. Zille[1*]

[1]2C2T – Centro de Ciência e Tecnologia Têxtil, University of Minho, Campus de Azurém, 4800-058 Guimarães, Portugal.

[2]Faculty of Textile and Clothing Technology, Niederrhein University of Applied Sciences, 41065 Mönchengladbach, Germany.

Corresponding author: azille@2c2t.uminho.pt

Abstract. An efficient cleaning and disinfection practice plays a crucial role in preventing cross-contamination in food industry, domestic situations as well as nosocomial environment. Pre-impregnated disinfecting wipes (ready-to-use disinfectant wipes) are broadly applied in above-mentioned conditions. Regarding the effectiveness of the disinfectant pre-impregnated wipes, research still stays in case study phase and no comprehensive investigation has been carried out to evaluate the structure and function change of the ready-to-use disinfecting wipes over storage time. This work studied the ageing of the disinfecting wipe over storage time. Chloramine as surface disinfectant and 3 commercial wiping materials of polyester (PET), 55% cellulose/45%PET and pure cellulose have been selected. The FTIR (Fourier-transform infrared spectroscopy) and DMA (Dynamic mechanical analysis) result revealed that oxidation occurred on the textile substrates during the storage of the pre-impregnated disinfecting wipes with special emphasis to the cellulose polymer. Moreover, the occurred oxidation changed the mechanical properties of the cellulose-containing wipes increasing their viscous properties over the elastic ones.

1. Introduction

Healthcare-associated infections (HAIs) caused by the transfer of nosocomial pathogens from high-touch environmental surfaces and medical devices are responsible for significant patient morbidity, mortality and economic cost [1–3]. Nosocomial pathogens shed by patients can contaminate hospital surfaces at concentrations sufficient for transmission, surviving for extended periods and persisting despite attempts to remove them [4]. An effective cleaning and disinfection practice plays a key role in preventing cross-contamination and spread of HAIs [5–7]. Traditionally, healthcare staff has used the "bucket method", which consists of towels saturated with diluted disinfectant contained in a bucket. This method exhibits several limitations such as improper disinfectant dilution, inadequate saturation, uneven moisture distribution, unknown material compatibility and possible contamination from reusing [8,9]. Among the most effective surface disinfection methods, the nonwoven ready-to-use disinfectant wipes are increasingly accepted for decontamination of high-touch surfaces because of its convenience and reliable performance [10,11].

Though some research has been investigated on the effectiveness of commercial available disinfecting wipes in practical use [1,12–14]. Despite, several absorption issues due to active ingredients onto textile materials have been previously reported in the literature, the knowledge regarding the ageing of

wipe materials in presence of disinfectant under storage conditions remains vague [15-17]. The selection of an inappropriate wipe material could interact with the adsorbed active ingredient resulting in lower or even abolished disinfectant efficacy.
This project studied the ageing of the disinfecting wipe over storage time. Chloramine, as surface disinfectant and 3 commercial wiping materials, composed of polyester, cellulose, and their combinations have been selected for the investigation. The bulk disinfectant solution and the wipes before and after disinfectant contact were analysed by the means of FTIR (Fourier-transform infrared spectroscopy) and DMA (Dynamic mechanical analysis).

2. Material and methods

2.1. Sample preparation
The tests were carried out at standard condition of 65% relative humidity (RH) and 20ºC. The surface disinfectant chloramine-t-trihydrate (C7H8ClNO2S.3H2O.Na) solution from Acros Organics® was prepared at the concentration of 10% (w/w) in distilled water. Each textile wipe sample (Table 1) was prepared with the weight of 1 g ± 0.5%. Every experiment includes a control raw sample, a control treated in water and a sample iterated with the disinfectant solution. The wipes were tested for immersion times at 1, 3, 7, 15, and 31 days. All the experiments were replicated 2 times.

Table 1. Composition and Structure of the Used Wipes

Substrate	Composition	Structure
W1	100% PET	Nonwoven hydroentangled
W2	55% cellulose/45%PET	Nonwoven hydroentangled
W3	100% cotton	1/1 plain weave

2.2. Fourier transform infrared spectroscopy (FTIR)
A Shimadzu FTIR spectrophotometer (IR-Affinity 1) with an attenuated total reflectance accessory (ATR) was used to record the FTIR spectra of the fabric samples. Spectra were collected in the region of 4000-700 cm^{-1} and at the resolution of 4 cm^{-1} with 45 scans at room temperature. All the treated wipe samples were dried in an oven at 40°C for 24h before testing.

2.3. Dynamic Mechanical Analysis (DMA)
DMA was performed using a DMA 7100 from Hitachi® (Japan) in programmed tension mode with a heating rate of 3 °C min^{-1} scanned from 30 to 200 °C at 4 Hz of frequencies. The geometry of the testing sample was 20 mm length, 10 mm width and 0.002 mm thickness. Specimens were prepared in duplicate to conduct mechanical analyses. These analyses were carried out under nitrogen purge of 200 ml min^{-1}. The wipe samples were dried in an oven at 40°C for 24h before testing.

3. Result and discussion
When chloramine is brought into contact with water, it slowly breaks down to generate hypochlorous acid and hypochlorite, which in turn releases chlorine and oxygen that are responsible for the bactericidal and bacteriostatic action. However, they are also strong oxidizing agents that can damage the textile fibres. It is reported that the hypochlorous acid and hypochlorite that supposed to work on the microbicidal effect diminishes when they interact with the textile substrate [15-17].

The ATR-FTIR spectrum of untreated W1 and W2 fabrics exhibits peaks of the polyester component (W1) and of the polyester blend (W2) at 1710 cm^{-1} assigned to stretching vibration of C=O group in ester, 1250 cm^{-1} assigned to asymmetric stretching of aromatic ester, 710 cm^{-1} attributed to aromatic C-H bending vibrations and 871 cm^{-1} attributed to C–C out of plane bending vibrations of the benzene rings (Fig. 1) [18]. The spectra of W2 and W3 displays the strong bands at 1160, 1100 and 1020 cm^{-1} assigned to the vibrations of the C-O-C bond of the glycoside bridges of the cellulose structure. The broad and strong bands at 3340 and 3270 cm^{-1} observed in W3 and with minor intensity in W2 are attributed to the stretching vibration of the hydroxyl (OH) group of the cellulose structure [20]. The strong peaks in W3 at 1150, 1100 and 1020 cm^{-1} are from the vibrations of the C-O-C bond of the glycoside bridges of the cellulose structure [19].

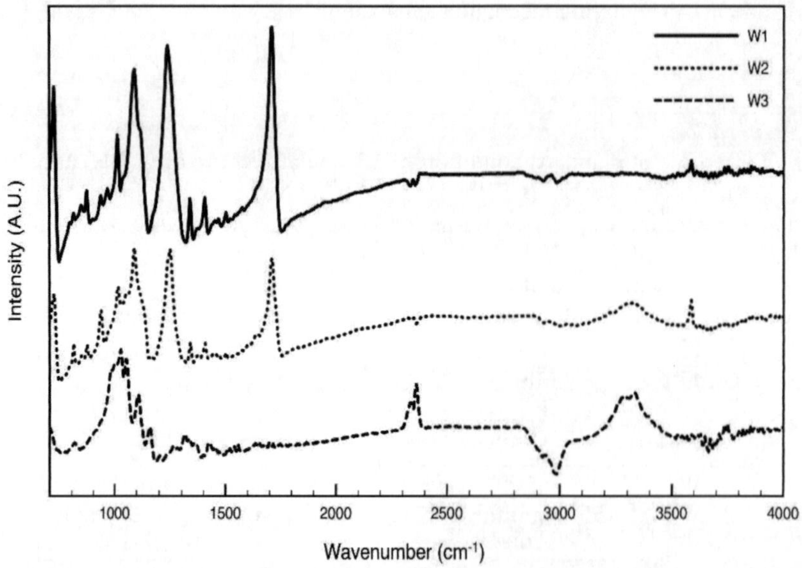

Figure 1. ATR-FTIR spectrum of W1, W2 and W3 samples in the range between 700 and 4000 cm^{-1}.

FTIR analysis was performed to understand the change of the wipe samples after different treatment time with disinfectant solution. The measures were taken at different immersion times at 1, 3, 7, 15 and 31 days. All the samples treated with disinfectant solution showed the formation of a new band at 3587 cm^{-1} that was attributed to the oxidation action of chloramine on the textile material [21,22]. No change in the FTIR spectra was noted in control wipes treated with distilled water in the same conditions. The new band can be assigned to the hydroxyl band -OH due to the oxidation of the polyester and cellulose surface and gradually increase with the immersion time [23]. Fig. 2-a and 2-b below exhibit the FTIR result from W1 and W2 immersed in the chloramine solution at different times. The observed oxidation peak displays an increasing behaviour up to 7 days. After this period the band intensity starts to decrease. This phenomenon can be explained by a removal of polyester molecules on the surface. When an entire layer of the oxidized polymer chain is removed from the surface the bulk unreacted PET is exposed to the surface reducing the oxidation peak. However, since W1 is composed of pure polyester and the W2 by a mixture of cellulose and polyester, the higher oxidation of W2 can be explained by the higher local chloramine concentration due to cellulose adsorption.

Cellulose is also oxidized by chloramine and W3 sample showed more changes in the FTIR spectra (Fig. 2-c) compare to the others. Different from PET, the oxidation increases over time showing the higher peak at 31 days (data not shown). However, the cellulose sample shows other changes in its chemical structure especially between 700 and 1500 cm^{-1} (Fig. 2-c). New peaks at 1250, 1140, 1090, 940 and 810 cm^{-1} appear with the simultaneous disappearing of the intense peaks of the glycoside bonds of the cellulose structure. These changes indicate a radical change in the cellulose structure with the formation of carbon moieties and the chemical substitution of the hydroxyl groups due to the formation of methylated derivative [24].

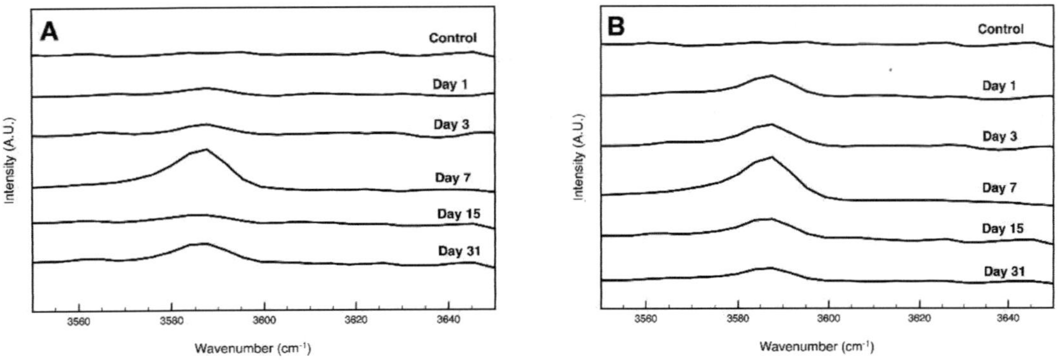

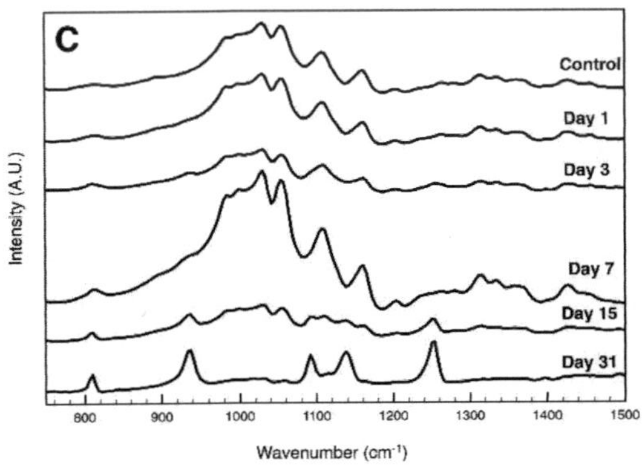

Figure 2. FTIR spectrum of W1 (A), W2 (B) and W3 (C) ageing in chloramine over storage time in the range between 3550 and 3650 cm^{-1} for W1 and W2 and between 700 and 1500 cm^{-1} for W3.

The dynamic mechanical analysis (DMA) was applied to study the effect of temperature on the mechanical properties of the wiping materials. The DMA parameters including tan delta, loss and storage moduli provide important information about the stiffness of the polymer, molecular motion, relaxation process, structural hetero groups, and morphology of the polymer blend systems [25]. Fig. 3 shows that in the cellulose-containing samples (W2 and W3) after chloramine treatment showed significant differences. In W1 (Fig. 3-a) the water and chloramine treated samples do not show significant differences. W1 sample is made of pure PET and despite the disinfectant is able to chemically interact with the PET structure (see FTIR discussion) this do not change the mechanical properties of the wipe. These samples showed a decrease in their moduli from 100 ºC up to 160 ºC. In the W2 sample (Fig. 3-b), the storage modulus at 30 °C compared with the 65 GPa of the samples treated in pure water have decreased to 1 GPa, which is a decrease of 98%. Similar behaviour was observed for the loss modulus decreasing from the 6.5 GPa in water to 0.1 GPa in chloramine. These measures maintained stable up to 100 ºC. After this temperature, the storage and loss moduli of the water treated sample start to decrease due to the increased mobility of the polymer chains. However, the chloramine treated samples after 100 °C, showed a slight increase of the storage and loss moduli attributed to the chloramine-altered intermolecular bonding that limits the mobility of the polymer chains in the wipes [26]. The pure cellulose sample (Fig. 3-c) shows similar storage and loss moduli at 30 ºC. However, this sample is clearly temperature dependent showing divergent behaviour in the water and chloramine samples. The storage and loss moduli of the water treated sample increased with the temperature reaching values of one order of magnitude higher. Contrarily the storage and loss moduli of chloramine treated samples showed a decrease in moduli from 120ºC to 160ºC. The storage modulus decreased from 20 GPa to 4 GPa and the loss modulus from 1.5 GPa to 0.45 GPa.

The damping factor or tan delta is the ratio between the loss and storage modulus in a viscoelastic material. A high tan delta value is indicative of a material that has a high, non-elastic strain component, while a low value indicates one that is more elastic. As expected the W1 sample did not show a significant difference in tan delta between water and disinfectant treated samples (Fig. 4). An increase in the viscous properties is noted until 120 °C then the system became more elastic reaching a plateau at 180°C. The tan delta of the sample W2 also showed similar results however the presence of disinfectant seems to reduce the temperature of the viscous peak increase probably due to the presence of disinfectant-oxidized surface modifications that increase the stiffness of the material at lower temperatures.

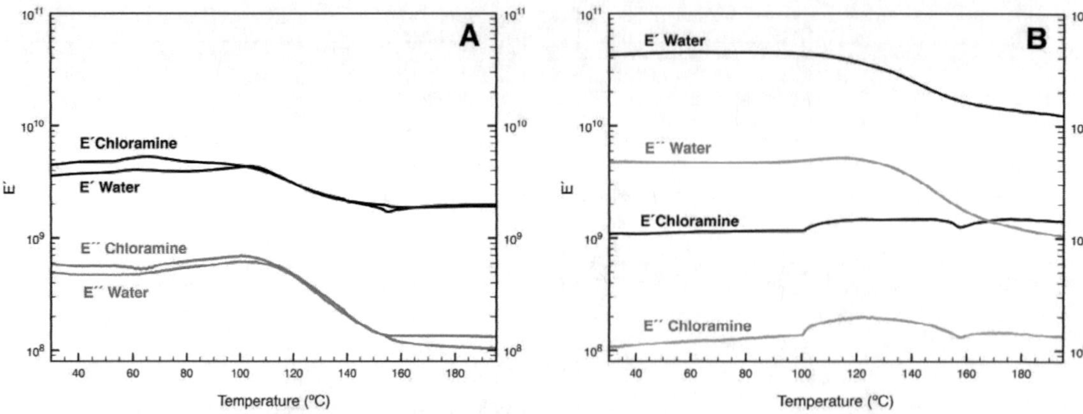

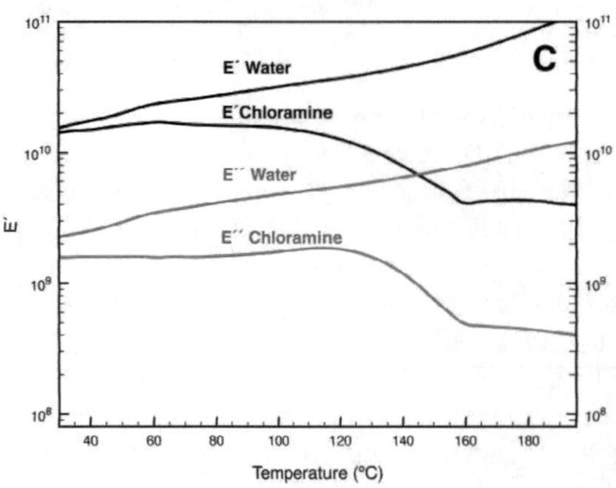

Figure 3. Temperature dependence at 4 Hz of storage (E') and loss (E") modulus of W1 (A), W2 (B) and W3 (C) after 7 days of immersion in water and chloramine

Despite the shift in temperature, the values of the Tan δ remained very similar suggesting that in the blend the thermoplastic PET network is not perturbed by the non-thermoplastic cellulose polymer network in terms of its viscoelastic properties [27]. Sample W3 is clearly the most affected in its mechanical properties by the chloramine action. The water control did not show significant events in tan delta analysis while the disinfectant-treated sample significantly increased its viscous component between 100 and 160 °C. This seems to occur in response to molecular interactions, mainly hydrogen bonding, between the PET oxidized molecules and the cellulose or between the cellulose structure itself promoting conformational changes that alter the mechanical behaviour of the wipes [28].

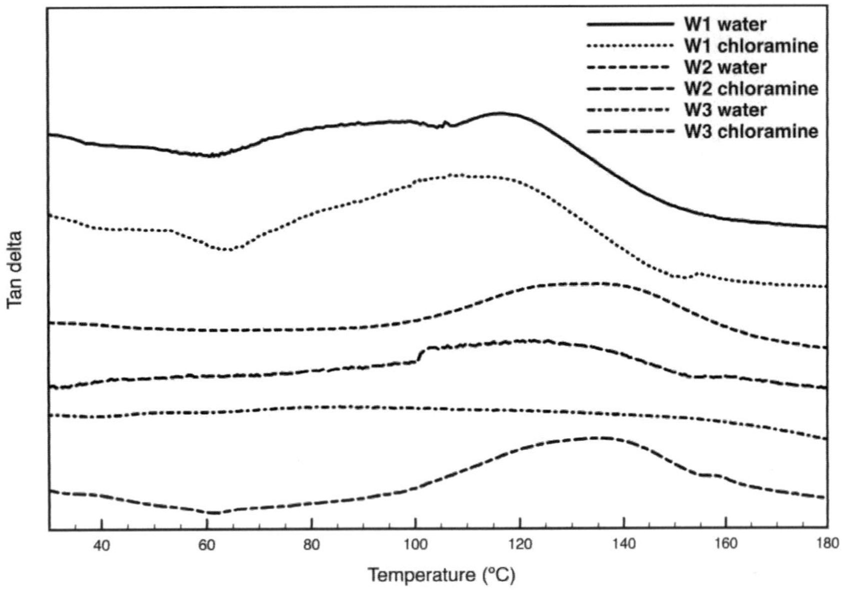

Figure 4. Temperature dependence of tan delta of W1, W2 and W3 after 7 days of immersion in water and chloramine.

4. Conclusion

The results of this study fill in the gap of the absorption issue between active ingredients and wiping materials over storage time. The chloramine is able to oxidize both the wipe materials with a higher action on cellulose structure than on PET. A significant change in mechanical properties was observed for cellulose-containing wipes while the PET viscoelastic properties did not show significant changes. The next step of the work will be to measure the effectiveness of the disinfectant antimicrobial action after adsorption on the wipes at different times of exposure. In addition, the generated outcome knowledge will provide a reference of the guideline for the hospital cleaning and disinfection in practice as well as ensure hospitals daily workflow from unnecessary risk of infection outbreak and complement the products' user manual of disinfectant and wipes in the market.

Acknowledgement

Xinyu Song (SFRH/BD/130028/2017) acknowledges Fundação para a Ciência e Tecnologia (FCT), Portugal, for its doctoral grant financial support. Andrea Zille acknowledges FCT through the Investigator FCT Research contract (IF/00071/2015) and the project POCI-01-0145-FEDER-007136 and UID/CTM/00264 under the COMPETE and FCT/MCTES (PIDDAC) co-financed by European funds (FEDER) through the PT2020 program.

References

[1] Gold KM, Hitchins VM. Cleaning assessment of disinfectant cleaning wipes on an external surface of a medical device contaminated with artificial blood or Streptococcus pneumoniae. Am J Infect Control. 2013, 41(10):901–7.

[2] Kramer A, Schwebke I, Kampf G. How long do nosocomial pathogens persist on inanimate surfaces? A systematic review. BMC Infect Dis. 2006, 6(1):130.

[3] Meyer B, Göhring N, Wishart E. The contribution of surface disinfection to prevent health care acquired infections TS

[4] Otter JA, Yezli S, French GL. The Role Played by Contaminated Surfaces in the Transmission of Nosocomial Pathogens. Infect Control Hosp Epidemiol. 2011, 32(7):687–99.

[5] Donskey CJ. Does improving surface cleaning and disinfection reduce health care-associated infections? Am J Infect Control. 2013, 41(5):S12–9.

[6] Disinfectants used for environmental disinfection and new room decontamination technology. Am J Infect Control. 2013, 41(5):S36–41.

[7] Dancer SJ. Controlling hospital-acquired infection: focus on the role of the environment and new technologies for decontamination. Clin Microbiol Rev. 2014 Oct, 27(4):665–90.

[8] Wiemken TL, Curran DR, Pacholski EB, Kelley RR, Abdelfattah RR, Carrico RM, et al. The value of ready-to-use disinfectant wipes: compliance, employee time, and costs. Am J Infect Control. 2014, 42(3):329–30.

[9] Boyce JM, Sullivan L, Booker A, Baker J. Quaternary Ammonium Disinfectant Issues Encountered in an Environmental Services Department. Infect Control Hosp Epidemiol. 2016, 37(3):340–2.

[10] Boyce JM. Modern technologies for improving cleaning and disinfection of environmental surfaces in hospitals. Antimicrob Resist Infect Control. 2016, 5:10.

[11] Kundrapu S, Sunkesula V, Jury LA, Sitzlar BM, Donskey CJ. Daily disinfection of high-touch surfaces in isolation rooms to reduce contamination of healthcare workers' hands. Infect Control Hosp Epidemiol. 2012, 33(10):1039–42.

[12] Kenters N, Huijskens EGW, de Wit SCJ, van Rosmalen J, Voss A. Effectiveness of cleaning-disinfection wipes and sprays against multidrug-resistant outbreak strains. Am J Infect Control. 2017 Aug 1, 45(8):e69–73.

[13] Siani H, Cooper C, Maillard J-Y. Efficacy of "sporicidal" wipes against Clostridium difficile. Am J Infect Control. 2011 Apr, 39(3):212–8.

[14] Sattar SA, Bradley C, Kibbee R, Wesgate R, Wilkinson MAC, Sharpe T, et al. Disinfectant wipes are appropriate to control microbial bioburden from surfaces: Use of a new ASTM standard test protocol to demonstrate efficacy. J Hosp Infect. 2015, 91(4):319–25.

[15] Engelbrecht K, Ambrose D, Sifuentes L, Gerba C, Weart I, Koenig D. Decreased activity of commercially available disinfectants containing quaternary ammonium compounds when exposed to cotton towels. Am J Infect Control. 2013 Oct, 41(10):908–11.

[16] Hinchliffe DJ, Condon BD, Slopek RP, Reynolds M. The adsorption of alkyl-dimethyl-benzyl-ammonium chloride onto cotton nonwoven hydroentangled substrates at the solid–liquid interface is minimized by additive chemistries. Text Res J. 2017 Jan, 87(1):70–80.

[17] Bloß R, Meyer S, Kampf G. Adsorption of active ingredients of surface disinfectants depends on the type of fabric used for surface treatment. J Hosp Infect. 2010, 75(1):56–61.

[18] Palaskar, S., Kale, K. H., Nadiger, G. S., & Desai, A. N. Dielectric barrier discharge plasma induced surface modification of polyester/cotton blended fabrics to impart water repellency using HMDSO. J App Polym Science. 2011, 122(2):1092-1100.

[19] Algar I, Fernandes S, Mondragon G, Castro C, Garcia C, Astrain C, Gabilondo N, Eceiza A. Pineapple agroindustrial residues for the production of high value bacterial cellulose with different morphologies. J Appl Polym Science. 2015, 132(1): 41237

[20] Li SM, Fu LH, Ma MG, Zhu JF, Sun RC, Xu F. Simultaneous microwave-assisted synthesis, characterization, thermal stability, and antimicrobial activity of cellulose/AgCl nanocomposites. Biomass and Bioenergy. 2012, 47:516-521.

[21] Dinsdale RM, Almemark M, Hawkes FR, Hawkes, DL Composition and Biodegradability of Products of Wet Air Oxidation of Polyester. Environmental Science & Technology. 1999, 33(22):4092-4095.

[22] Murthy ARV, Rao BS. "Oxidation by chloramine-T," Proceedings of the Indian Academy of Sciences - Section A. 1952, 35(1):7.

[23] Duarte N, Lage H, Ferreira MJ. Three New Jatrophane Polyesters and Antiproliferative Constituents from Euphorbia tuckeyana, Planta Medica, 2008. 74(1):61-68.

[24] Kumar A, Negi YS, Choudhary V, Bhardwaj, NK. Characterization of cellulose nanocrystals produced by acid-hydrolysis from sugarcane bagasse as agro-waste. J Mats Phys Chem,

2014. 2(1):1-8.

[25] Rahimi, M. H., Parvinzadeh, M., Navid, M. Y., & Ahmadi, S. Thermal characterization and flammability of polyester fiber coated with nonionic and cationic softeners. J Surfac Deterg, 2011. 14(4):595-603.

[26] Babu KF, Narayanan, TN, Kulandainathan MA. Molecular motions aided thermally responsive biocompatible textile pads. Advan Materis Interf. 2014. 1(3): 1300139.

[27] Coelho D, Sampaio A, Silva CJ., Felgueiras HP, Amorim MTP, Zille A. Antibacterial Electrospun Poly (vinyl alcohol)/Enzymatic Synthesized Poly (catechol) Nanofibrous Midlayer Membrane for Ultrafiltration. ACS App Mat Interf, 2017. 9(38):33107-33118.

[28] Sudhamani SR., Prasad M S, Sankar KU. DSC and FTIR studies on gellan and polyvinyl alcohol (PVA) blend films. Food Hydrocol, 2003. 17(3):245-250.

AUTHOR INDEX

AUTHOR INDEX